THE SELLOUT

ALSO BY CHARLES GASPARINO

King of the Club

Blood on the Street

THE SELLOUT

How Three Decades of

Wall Street Greed and Government

Mismanagement Destroyed

the Global Financial System

CHARLES GASPARINO

HARPER
BUSINESS

An Imprint of HarperCollins*Publishers*
www.harpercollins.com

HarperCollins books may be purchased for educational, business, or sales promotional use. For information, please write: Special Markets Department, HarperCollins Publishers, 10 East 53rd Street, New York, NY 10022.

FIRST EDITION

Designed by Kate Nichols

Library of Congress Cataloging-in-Publication Data
Gasparino, Charles.
    The sellout: how three decades of Wall Street greed and government
mismanagement destroyed the global financial system / Charles Gasparino.—1st ed.
        p. cm.
    Includes bibliographical references and index.
    ISBN 978-0-06-169716-6
    1. Financial crises—United States—History—21st century. 2. Wall Street
(New York, N.Y.)—History—21st century. 3. Avarice—United States—History—
21st century. I. Title.
    HB3722.G37   2009
    332'.042—dc22   2009028097

09 10 11 12 13   OV/RRD   10 9 8 7 6 5

To Gin, for putting up with me . . .

CONTENTS

KEY PEOPLE

Lloyd Blankfein: Goldman Sachs' chairman and chief executive officer since June 2006 and a director since April 2003. Before taking the helm at Goldman, Blankfein had been the company's president and chief operating officer since January 2004. While other firms melted down following their forays into risky subprime loans, under Blankfein's leadership Goldman largely dodged the crisis.

James "Jimmy" Cayne: Former Bear Stearns chairman and CEO. Cayne joined Bear in 1969 as a retail broker, rising through the ranks to become one of the wealthiest CEOs on Wall Street. Though some question his financial knowledge, few question his smarts. Under Cayne's leadership, Bear's stock price jumped from $16 a share to a high of $172. JPMorgan Chase ultimately purchased the company for $10 a share in 2008; Cayne lost most of his vast fortune.

Ralph Cioffi: Former Bear Stearns hedge fund manager. The collapse of his two funds, the Bear Stearns High-Grade Structured Credit Strategies Fund and the Bear Stearns High-Grade Structured Credit Strategies Enhanced Leverage Fund, signaled the start of the credit crisis and the end of Bear.

James "Jamie" Dimon: JPMorgan Chase chairman and CEO since December 2005. Dimon joined JPMorgan when it purchased Bank One, where he had been chairman and chief executive officer since March 2000. Before going to Bank One, Dimon held several senior positions at Citigroup, learning at the side of Sandy Weill. Dimon was one of the few CEOs to largely avoid the subprime crisis.

Patrick Dunlavy: Former head of fixed-income sales at Salomon Brothers. Dunlavy's long career at the firm witnessed both the rise (and fall) of Lew Ranieri and, many years later, Tom Maheras.

Ahmass Fakahany: Former co-president and chief operating officer of Merrill Lynch. Fakahany resigned from Merrill in January 2008, following billions of dollars in losses related to risky subprime loans. Fakahany's chief responsibility was risk management.

Laurence Fink: Chairman and chief executive officer and founder of BlackRock. Before starting BlackRock, Fink was co-head of Taxable Fixed Income at First Boston, where he was one of the first to trade mortgage-backed securities. Together, Fink and Lew Ranieri pioneered the market for mortgage-backed securities. Fink's vast Rolodex straddles the world of Washington and Wall Street.

Gregory Fleming: Former president and chief operating officer of Merrill Lynch. From 2003 to 2007, Fleming was the head of Merrill's investment banking unit. A cool operator, Fleming bridged the cultural gap between Merrill's old guard and O'Neal's new regime. In the company's final moments, Fleming played an integral role in the Merrill Lynch–Bank of America merger.

Richard Fuld: Former chairman and CEO of Lehman Brothers. Before the firm's bankruptcy, Fuld's fourteen-year reign at the helm of Lehman was the longest of any Wall Street CEO. A brash trader who loved to take risks; his disastrous push into real estate ultimately led to the firm's collapse.

Timothy Geithner: Current Treasury secretary and former president of the New York Federal Reserve Bank. Though he is now charged with reviving the financial system, Geithner's professional career has been largely devoid of any private experience. Described by colleagues as intense, Geithner is no stranger to financial crises, having served in the Clinton Treasury Department during the financial problems of the 1990s.

Alan "Ace" Greenberg: Former chairman and CEO of Bear Stearns. Greenberg began working at Bear in 1949, eventually becoming CEO in 1978. Greenberg would lose the CEO title and later the chairmanship to Jimmy Cayne in 1993. Known for his focus on risk management and the bottom line, Greenberg ran Bear during a period of massive expansion. Now eighty-two years old (in 2009), he continues to work for JPMorgan Chase, which acquired Bear in 2008.

Maurice "Hank" Greenberg: Former chairman and CEO of American International Group. Greenberg relinquished his post amid an accounting scandal in 2005. Now eighty-four, he currently is the chairman and CEO of C. V. Starr and Company, a financial services firm.

Alan Greenspan: Chairman of the Federal Reserve from 1987 to 2006. Some economists have pointed to his aggressive lowering of interest rates following the attacks of September 11, 2001, as the fuel that fed the housing crisis.

John Gutfreund: CEO of Salomon Brothers during the 1980s, known for his autocratic ways. His career came to a crashing halt when he was forced out as head of Salomon after he failed to notify regulators of a Treasury bond–bidding scandal.

Dow Kim: Former co-head of Merrill Lynch's Global Markets and Investment Banking units. Kim oversaw trading and investment banking at Merrill during its expansion into mortgage-related securities. Former Merrill CEO Stan O'Neal blames Kim for Merrill's massive mortgage losses.

Kenneth Lewis: CEO and president of Bank of America. Lewis is an empire builder who began his career as a credit analyst but eventually became one of the most important men in finance. Since taking the helm of Bank of America in 2001, Lewis has grown the company through a series of massive mergers, including the purchase of Fleet–Boston Financial, MBNA, Countrywide Financial, and ultimately Merrill Lynch.

John Mack: CEO of Morgan Stanley. Mack's career at Morgan Stanley spans nearly thirty years. He left the company in 2001 amid a power struggle with then-CEO Philip Purcell, eventually taking the helm at Credit Suisse First Boston. He returned to Morgan Stanley in 2005. Mack's sharp focus on cost cutting earned him the nickname "Mack the Knife."

Thomas Maheras: Former co-president of Citigroup's investment banking unit. Maheras began his career at Salomon Brothers and quickly rose to become one of the firm's top bond traders. As head of Citigroup's capital markets area, Maheras was at the helm as the bank delved into subprime securities.

John Meriwether: After a spectacular career as a bond trader at Salomon Brothers, Meriwether left the firm and in 1994 founded Long-Term Capital Management, a hedge fund that later collapsed and had to be bailed out by the rest of Wall Street (except Bear Stearns).

Michael Milken: A pioneer in the world of high-yield debt. Many credit Milken with inventing the junk bond market while at Drexel Burnham Lambert. Milken's career was derailed after he pled guilty to securities fraud, and he subsequently served two years of a ten-year prison sentence.

Michael Mortara: Head of mortgage trading at Salomon Brothers during the Lew Ranieri reign. Shortly after Mortara's departure, Salomon posted a sharp drop in profits due to losses in the fixed-income area. Mortara died in November 2000.

Angelo Mozilo: Former CEO of Countrywide Financial. The son of a Bronx butcher, Mozilo rose to leading one of the world's biggest mortgage lenders. Mozilo's meteoric rise and fall became a symbol of the excesses of the real estate market. He is best known for two things: his enthusiasm for subprime mortgages and his year-round tan.

E. Stanley O'Neal: Former CEO of Merrill Lynch, O'Neal led Merrill's fatal push into subprime mortgages while simultaneously working to destroy the firm's clubby insider culture, known as "Mother Merrill." The grandson of a slave, O'Neal is the highest-serving African American to have worked on Wall Street. O'Neal resigned amid massive losses due to the firm's foray into the mortgage-backed market.

Vikram Pandit: CEO of Citigroup. Pandit joined Citigroup after the bank purchased his hedge fund, Old Lane Partners, for $800 million in 2007. The deal was orchestrated by Citigroup CEO Chuck Prince, who had long sought Pandit's services in the hope of upgrading Citigroup talent. Prior to running his hedge fund, Pandit ran Morgan Stanley's investment banking business, where he oversaw sales and trading. He left Morgan after being passed over for promotion.

Henry "Hank" Paulson: Treasury secretary during the heart of the financial crisis and CEO and chairman of Goldman Sachs from 1999 to 2006. Few straddle the intersection of Wall Street and Washington as Paulson does. He was CEO of Goldman at the same time as Jimmy Cayne, Stan O'Neal, and Dick Fuld ran their respective firms, and in 2008 he engineered the federal bailout of America's financial institutions.

Charles "Chuck" Prince: Former CEO and chairman of Citigroup, Prince inherited Sandy Weill's far-flung financial kingdom in 2003. A former attorney with no background in trading mortgages, Prince stepped down from both positions due to the company's heavy losses in the mortgage market. Weill tapped Prince for the top job after Prince successfully shielded him from then Attorney General Eliot Spitzer's investigation of Citigroup's research practices.

Lewis Ranieri: Frequently called the "Godfather" of the mortgage bond, Ranieri, the former vice chairman at Salomon Brothers, reportedly coined the term "securitization," which refers to the packaging of mortgage loans into bonds, while running Salomon's mortgage desk. A college dropout, Ranieri joined Salomon as a $70-a-week mail room clerk. By his thirties, he was running Salomon's most important division: mortgages. He left the firm amid massive losses for the department.

Howard Rubin: Former Salomon Brothers trader who lost a reported $250 million while trading mortgage bonds at Merrill Lynch. Rubin was fired from Merrill but soon found himself on his feet at Bear Stearns.

Robert Rubin: Treasury secretary under Bill Clinton and co-chairman and co–senior partner at Goldman Sachs in the 1990s. Rubin later went on to become a director and senior adviser at Citigroup before stepping down in January 2009. Many credit Rubin with expanding Citigroup's push into the mortgage arena.

Osman Semerci: Merrill Lynch's global head of fixed income. Semerci was handpicked by O'Neal to ramp up Merrill's push into risky subprime mortgages. Semerci's rise coincided with the departure of two senior fixed-income executives, Jeff Kronthal and Harry Lengsfield, who were reportedly dismissed for not taking enough risk. O'Neal considered Semerci, a onetime bond salesman of Turkish descent, to be the image of the modern Merrill executive: young, international, and aggressive.

Warren Spector: Former co-president of Bear Stearns. Spector's twenty-four-year career at Bear came to a crashing halt after the implosion of the company's two subprime hedge funds. Spector headed Bear's mighty mortgage business, rising through the ranks to the top of the firm at a young age. Although some colleagues considered him aloof, few doubted his financial prowess.

John Thain: Former CEO of Merrill Lynch who sold the firm to Bank of America during the height of the financial crisis. He would later be ousted after an internal battle with BofA CEO Ken Lewis over the massive losses Merrill would experience—ones that would lead to a government bailout of the combined firm.

Michael Vranos: Founder of Ellington Capital Management. Vranos is considered one of the premiere mortgage-backed traders. He made a name for himself at Kidder Peabody, where he headed that company's mortgage desk. Vranos's Ellington Fund had a near brush with death after the implosion of Long-Term Capital Management.

Sanford "Sandy" Weill: Former CEO and chairman of Citigroup. Weill created the idea of the one-stop-shopping mega–financial conglomerate, engineering a series of mergers that eventually brought Citicorp and Travelers Group under the same roof, and in the process created the world's largest financial company. Famously obsessed with his firm's stock price, Weill announced his resignation in 2003 after investigators discovered he'd pressured a stock analyst, Jack Grubman, to raise his rating on AT&T, where Weill was on the board, in return for ensuring that Grubman's children got into an exclusive preschool.

KEY FIRMS

American International Group (AIG): The insurance giant (the largest insurer in the world) whose collapse threatened to take down the entire financial industry. The company was founded in Shanghai in 1919 and built into a colossus by Hank Greenberg; its troubles began when it moved away from offering standard insurance products and started insuring complex mortgage securities that left it vulnerable to the vagaries of the housing market. Eventually, the company had to be bailed out by the U.S. government.

Bank of America: One of the largest U.S. banks by revenue and deposits, Bank of America was founded as the Bank of Italy by an immigrant, Amadeo Giannini, in 1904. Bank of America (BofA for short) has been subject to a series of government bailouts following its purchases of Countrywide Financial and Merrill Lynch.

Bear Stearns: Founded by Joseph Bear, Robert Stearns, and Harold Mayer in 1923, Bear was traditionally a bond house. It was the first major victim of the credit crisis; its heavy focus on mortgage-backed securities and use of leverage eventually led to its collapse in March 2008. Prior to the subprime crisis, Bear had never posted an unprofitable quarter.

BlackRock: Founded in 1988 by Larry Fink as Blackstone Financial Management, changing to its current name in 1992, BlackRock has emerged as one of the few winners in the financial crisis, garnering key contracts from both private and government players to manage beaten mortgage portfolios. With $1.3 trillion under management, BlackRock is one of the world's largest asset management companies. Made in the image of its founder, BlackRock is renowned for its expertise in risk management.

Blackstone: A private-equity powerhouse founded in 1985 by Peter Peterson and Stephen Schwarzman. Larry Fink's BlackRock started as a subsidiary of Blackstone before being spun out. Blackstone was one of the first private-equity companies to list its shares publicly.

Citigroup: At one point the largest financial services company in the world, Citigroup was formed by Sandy Weill from the merger of Citicorp and Travelers Group in 1998. The combined company endured massive losses following the credit crisis and had to be bailed out by the government.

Countrywide Financial: The nation's largest mortgage lender, Countrywide was a leader in offering subprime loans to home owners. Led by its charismatic CEO, Angelo Mozilo, Countrywide became the corporate face of the housing bubble.

Drexel Burnham Lambert: At one point one of America's largest investment banks, Drexel Burnham Lambert pioneered the market for high-yield, or junk, debt in the late 1980s, led by Michael Milken. But as the leveraged buyout boom busted, investors lost faith in Drexel itself, and the firm filed for bankruptcy in 1990.

Goldman Sachs: The gold standard of investment banks, founded by Marcus Goldman in 1869, Goldman Sachs largely sidestepped the subprime crisis. Still, at the height of the financial crisis, it converted to a bank holding company, signaling the end of the modern investment bank. The last of the major investment banks to go public, in 1999, Goldman remains unrivaled for its risk management. Many former senior executives from Goldman have held a number of high-level positions in government.

JPMorgan Chase: Formed in 2000 when Chase Manhattan Bank purchased J.P. Morgan, JPMorgan Chase is one of the largest commercial banks in the world. With the Federal Reserve's backing, it purchased Bear Stearns for the paltry sum of $2 a share following the broker's collapse in March 2008. The final price was later raised to $10.

Kidder Peabody: At one point owned by General Electric, Kidder Peabody was a mortgage powerhouse in the 1990s, home to such superstar traders as Mike Vranos. But its heavy use of leverage and a bad bet on the bond market caused massive losses at the firm, leading GE to sell it to PaineWebber in late 1994.

Lehman Brothers: One of the oldest of Wall Street's investment banks, dating back to 1850, Lehman was spun off from American Express in

1994. Despite its heavy presence in fixed income, after 2000 Lehman made significant progress in diversifying its business model, buying the asset management firm Neuberger Berman in 2003 for $2.6 billion. Still, the company's fateful push into risky mortgages would spell its doom, and in September 2008, it filed for bankruptcy, setting off a global financial meltdown.

Long-Term Capital Management: Founded by a former Salomon head of fixed-income trading, John Meriwether, LTCM used heavy leverage to employ a number of complex fixed-income strategies. Though it was initially successful, its bad bets in the credit markets ultimately spelled huge losses, and because of the systemic risk that its failure posed, the hedge fund was ultimately bailed out by Wall Street.

Merrill Lynch: Known for its "thundering herd" of brokers, Merrill was a relative latecomer to the subprime arena. But under CEO Stan O'Neal, the company made an aggressive push into the sector, a move that ultimately led to its being acquired by Bank of America in September 2008.

Morgan Stanley: One of the most storied and celebrated investment banks in the United States, Morgan Stanley was able to weather the financial storm as an independent company, but not before changing its charter and becoming a bank holding company.

Salomon Brothers: One of Wall Street's leading investment banks in the 1970s and '80s. Salomon Brothers' bond traders were considered innovators in their field, pioneering the mortgage-backed bond market that ushered in the era of structured finance. Salomon would later be acquired by Travelers Group and then ultimately become part of Citigroup.

Smith Barney: The brokerage unit of Citigroup, later sold to Morgan Stanley, Smith Barney is one of the oldest retail brokers. In 1993, before it was a part of the Citigroup financial empire, Smith Barney was a stand-alone retail broker that was purchased by Travelers Group. In 1997, Smith Barney merged with Salomon Brothers to form Salomon Smith Barney.

Travelers Group: The insurance giant run by Sandy Weill that merged with Citicorp to form Citigroup in 1998. The combined entity became the world's largest financial services company.

THE SELLOUT

PROLOGUE

Charles, I need to speak to you," Jimmy Cayne, the CEO of Bear Stearns, said when I picked up my cell phone.

I was surprised because Cayne hadn't returned my calls in weeks. I had been covering Bear Stearns and been speaking to him for more than a decade, mostly when Bear Stearns was one of Wall Street's most successful companies. But more recently, as the firm fell on hard times, I had been one of the reporters to speculate early and often, primarily on CNBC, that Bear might not survive—that it would be forced to sell out to a larger player because of its uneven business model, which focused on the depressed market for mortgage-backed securities, which had recently imploded and showed few signs of coming back. And for that, Cayne had begun to despise me.

During our conversation, Cayne was pleasant but perfunctory. He said he wanted to tell me how he was going to move the company forward, bring it back from its troubles. But in order to do that he needed to see me in person.

We met at Shun Lee Palace in midtown Manhattan. Cayne was accompanied by Vincent Tese, Bear Stearns' lead board member and one of his closest friends and advisers. Cayne was obviously a regular; he was seated at a large table near the back of the restaurant, where we could speak in private as the waiters regularly refreshed his scotch on the rocks without his asking.

It was the evening of August 5, 2007. Cayne and the Bear Stearns board of directors had just fired the man long considered to be his heir apparent, Bear Stearns' president, Warren Spector, following some of the most tumultuous months in the firm's long history.

The upheaval in the market for bonds backed by mortgages had clearly taken its toll on the firm. Bear Stearns' share price had fallen by more than 40 percent since the beginning of the year as two of Bear's hedge funds that invested in mortgage-backed securities, Spector's specialty, had imploded and been forced to shut down. The demise of the hedge funds was initially considered by Bear Stearns' management, Cayne and Spector in particular, as an isolated event without any broader implications for the market or the entire firm.

It wasn't.

As the U.S. housing market began to unravel in late 2006 and early 2007, so did the value of nearly every mortgage bond Wall Street had created. When the hedge funds began to unravel, the prices of mortgage bonds fell to distressed levels, a crushing blow to most of Wall Street but particularly to Bear Stearns, which had aligned its business model so closely with the growth, and profitability, of this market. Warren Spector, of course, was the executive directly in charge of the mortgage business at Bear, and now he was paying the price for its failure.

When we sat down, I could see that this was a different Jimmy Cayne than I had known for so long. I had always considered Cayne, now seventy-three, a youthful man. He loved to smoke cigars, and as he stated on more than one occasion he had also smoked pot (Cayne once attempted to hand me what looked like a joint in the Bear Stearns elevator). Still, he worked out regularly and hung out late with clients and friends at his favorite New York restaurant, San Pietro, where he spun tales about his Wall Street exploits—how he'd kicked the crap out of the competition, the hated "Stanley Morgan," as he used to refer to the investment bank Morgan Stanley, and the even more hated "Goldman Sucks."

I never thought the pot smoking was particularly newsworthy because Cayne never seemed stoned at work, and it made him so much more interesting than all those square-jawed, robotic executives at most of the big firms (in just a few months it would become so after a news account prompted Bear's board to raise the issue). But now Cayne was thin, almost gaunt, with quite a few more wrinkles in his face than the last time we'd met. Gone were the usual swagger and cigar. He attributed the weight loss, implausibly, to a new diet and workout regimen.

And now, instead of mocking the competition, he spoke of the gravity of the situation: Bear Stearns was in deep trouble. It was a startling concession. Most CEOs spend their time spinning the truth: the future is always rosy, and profits will always soar.

Because of the hedge funds' collapse, Bear Stearns, not just the funds but the entire *firm*, had lost credibility, he said. Wall Street firms are businesses run on billions of dollars of borrowed money, and those billions can evaporate quickly when lenders stop trusting a firm and its management. And that's what was now happening to Bear.

"You don't understand what's going on here," Cayne said with desperation in his voice. "My future, my kids' future, my grandchildren's future is at stake!"

Cayne had never asked me for good coverage. I'm sure he thought it was beneath him, given the years of record profits and a stock price that had hit $170 a share. But now he wasn't just asking for good coverage, he was pleading with me to be nice.

I told Cayne I was neither on his side nor opposed to him. What was happening to Bear was part of a much bigger story. It was about Wall Street's ties to the dying business of underwriting mortgage debt, although at the time neither I nor indeed anyone aside from a few savvy short sellers (who make money by betting stocks will fall), had any idea just how big a story it would become because of the arcane accounting rules that made figuring out the scope of the toxic mortgage debt secretly held on the books of Bear and every firm on the Street impossible to fully understand.

What I did know was this: with the mortgage market dead—for who knew how long—it was becoming clear that Bear, the smallest firm among the major investment banks and the one most heavily focused on mortgages, wouldn't last as an independent entity. It would have to sell out to a larger bank or sell pieces of itself to sovereign wealth funds, the Chinese and Korean banks, or rich Saudis. Bear might be the first to go, but others would follow. The Street needed to consolidate; there were too many investment banks chasing the same deals.

What was odd was that Cayne didn't necessarily disagree with my analysis about either his firm or the rest of the financial business. In fact, he told me he was planning a trip to China in a few weeks to meet with potential investors to raise fresh capital.

And with that came our Chinese food, which he ate quickly and pensively, as if he had something more important on his mind.

The credit crunch would get worse, and so would Cayne, and so would Bear Stearns. My suspicions about his health problems were right. A little later, Cayne revealed that he suffered from a prostate infection that had spread through his body, and he nearly died. His health eventually improved, but not his job performance: losses were building, and Bear's stock continued to reel until his pensiveness at our lunch proved prophetic with his firing in early 2008.

I t was then that I thought back to that dinner and decided to write this book. It was during one of the many lulls in the credit crisis when it appeared that Bear and the rest of Wall Street might yet survive, although perhaps not without selling out to foreign investors and not without selling out the country, as the banks' steep losses began to reverberate through the economy.

What I set out to do in this book is pretty straightforward: to explore how the combination of government policies that encouraged home ownership even for people who couldn't afford to pay their mortgages and Wall Street's greed had led the country into economic despair and ended American dominance of the world's financial system.

I wanted to analyze the root causes of the problem, from the Wall Street rating agencies that had assigned their highest, triple-A ratings to all those risky bonds that were now blowing up to the duplicitous mortgage brokers who'd made pots of money selling mortgages to people without the means to repay them, and even to the home owners who had defaulted on their mortgages because they'd somehow failed to understand that they couldn't afford a $500,000 home on a $50,000 salary.

I wanted to shine a light on the government bureaucrats who'd crafted the housing policies that had led to the crisis, such as former Housing and Urban Development secretaries Henry Cisneros and Andrew Cuomo, who, with the active support of some members of Congress, had been the first to prod the government-backed agencies Fannie Mae and Freddie Mac to guarantee increasingly riskier types of loans that would eventually cost U.S. taxpayers tens of billions of dollars in losses.

I wanted to look at the people and companies that had feasted off this massive government intervention, such as Angelo Mozilo, the flashy, ever-tanned CEO of the mortgage lender Countrywide Financial, who through a combination of massive campaign contributions, heavy lobbying, and great salesmanship had convinced policy makers to adopt his more "flexible ways" of measuring the risk in a potential mortgage. The players would be the top executives at the Wall Street firms—people such as Cayne, but also, and just as significantly, former Merrill Lynch CEO Stan O'Neal and former Citigroup CEO Chuck Prince, who'd both lost their jobs around the same time as Cayne had.

There were also the lower-level executives, the top traders who'd raked in titanic bonuses piling up billions of dollars of mortgage debt, almost without any close supervision from above. There were the all too few who had refused to sell out, whose warnings and calls for change were often ignored or derided. And there was Wall Street's transformation of its business model from a stodgy advice-driven enterprise to one that took tremendous risks, rolling the dice frequently with ever-greater sums of other people's money on the table.

Most of all, I wanted to focus on Wall Street, which had made billions upon billions of dollars in profits from mortgage bonds and were now bleeding billions in losses.

But in early 2008, when I set out to write this book, I had little idea, as did few but the most prescient investors, economists, and analysts, of just how big the story would eventually become as the financial system almost crumbled to pieces in the fall of 2008, leading to lots of drinking and many sleepless nights trying to figure out just how and why the system had come to the brink of collapse.

So my goal in writing this book became to tie together—both to my satisfaction and, I hope, to yours as well—the many disparate causes of the crisis, to create as complete a story as I can of the Great Sellout of 2008.

The roots of the story lie farther back than that year, much farther back. One of them, in fact, can be traced all the way back, I discovered, to the 1970s and to a city hardly known as a major financial center, Cleveland, Ohio.

It's there we'll begin.

PART I
LET'S MAKE MONEY

1. FUN AND GAMES

Ask Pat Dunlavy to give you the defining moment of his long career at Salomon Brothers—the point in time when he started to *really* understand how the firm and the rest of Wall Street really works— and he'll tell you the story about "The Great Race of 1978."

Dunlavy was thirty years old. He was making a good living as a bond salesman in Salomon Brothers' Cleveland office. His customers were predominantly large pension funds and other institutional investors in the Midwest that bought and traded bonds. Because of his position, he had contact with some of the firm's power players in New York, including the firm's legendary CEO, John Gutfreund, and some of the most savvy bond traders he'd ever met, people such as Lew Ranieri and a brilliant and charismatic trader named John Meriwether, known throughout the firm simply as "J.M."

The Cleveland office occupied one of the largest buildings in Cleveland, fourteen stories overlooking a decaying downtown of abandoned buildings and steel mills. Like most securities firms, Salomon Brothers had its share of loudmouthed former jocks, particularly at its sales and trading desks. Daniel Benton, a salesman and former high school football player, was one of those (though certainly not the worst bloviator of the bunch). Benton was growing tired of being ribbed about his expanding waistline. At one point he made an officewide announcement. He challenged anyone in the office to a race up the building's fourteen floors. He said he would wipe the floor with any one of them.

Dunlavy, a former college football player, had been working out regularly. One afternoon he approached Benton. "You really want to race up all fourteen stories?" Benton said he did, and if Dunlavy was man enough, he should meet him downstairs in two weeks, just before the firm's Saint Patrick's Day celebration, and race him to the top.

"And get ready to lose," he added with a smirk.

But what started out as a prank between two over-the-hill football players in Salomon's Cleveland office grew into something much bigger. The office manager invited some of the office's biggest clients to watch. Employees brought their wives. News of the "great race" even spread to Salomon's New York office. Traders were now laying odds on which one of the guys would win or which of them would drop dead of a heart attack before making it to the top.

It was then that Dunlavy received a call from Meriwether. In 1978, John Meriwether hadn't yet achieved his notoriety through his depiction in Michael Lewis's book *Liar's Poker*, but his legend was growing. Most firms on the Street specialized in providing advice, to either companies or individual investors. Not Salomon Brothers. Its specialty was trading, and not just stocks but bonds as well. Because of high inflation, which eats away at the value of these so-called fixed-income investments, the bond market was a backwater during most of the 1970s. They were sold primarily to people who held them until they "matured" so they could collect regular interest payments.

But to traders at Salomon Brothers, the bond market was a big casino where they were the only real players. In the New York office, Lew Ranieri was perfecting a relatively new type of bond known as the mortgage-backed security—essentially a pool of mortgages in which the investor is paid an interest rate that flows from the mortgage payments of home owners (more on this in a bit). John Meriwether was just in his early thirties when he made his first major score on what was then considered a big gamble: buying New York City municipal bonds at the height of the city's fiscal crisis.

Those were the years of high crime and white flight; the city nearly defaulted on its short-term debt. Real estate prices were falling. People were afraid to ride the subways. President Gerald Ford told the city to drop dead. But Meriwether saw the city's promise and began snapping up its debt on the cheap. He took a calculated risk on New York, and his calculations were right on the money.

By the late 1970s Meriwether was taking even bigger risks. He was now part of the firm's newly created bond arbitrage desk, a unit of the firm that he headed and grew into a money machine. Bond arbitrage is a more complex form of trading than simply buying New York City municipal bonds at depressed levels; it involves trading various types of bonds and taking advantage of price differences when market conditions change. A spike in interest rates can cause government bonds to fall because of inflation fears but corporate bonds (particularly junk bonds) to rise because inflation may spur economic growth, which in the eyes of investors may make corporate debt more appealing. The trick is to guess which markets will fall and which will rise and place your bets accordingly. It's a high-reward but high-risk business; tens of millions of dollars can be made or destroyed in the blink of an eye.

Meriwether was one of the best arbs at Salomon, maybe on all of Wall Street. He was a math genius with a penchant for gambling, something he had been doing since he was a kid growing up on the South Side of Chicago.

When J.M. called, there wasn't much small talk, and today's call was no different. Meriwether said he wanted to know everything he could about the great race. He wanted precise details about Dunlavy's weight and height and Benton's as well. He asked questions about the building and if Dunlavy had been training for the event. Dunlavy said he had. He asked if Benton had been training. Dunlavy said he doubted it. "He's so confident he's going to win, he thinks he doesn't have to train," Dunlavy said.

"So will you beat this guy?" Meriwether asked.

"Absolutely," Dunlavy shot back. "If I can't drag my sorry ass up those stairs faster than Dan, I'll shoot myself!" Dunlavy said he also had some money on the line; he'd bet Benton $50 he would win.

After a pause, Dunlavy asked, "Why do you care about any of this?"

"I might want to get involved," Meriwether answered cryptically. "Let me think about it."

Dunlavy wasn't sure what the hell "get involved" meant, but a few days later Meriwether called again. He told Dunlavy he was now the official bookmaker for the great race. He was posting a "line," laying odds on who might be the favorite to win and taking bets from anyone in the organization, from traders to back-office personnel. Meriwether said he wanted to make big money off the great race.

Dunlavy wasn't totally surprised. He'd always viewed the typical trader as a gambler with a college degree. The best traders, however, are more like bookmakers—they use information to lay odds on favorites and underdogs. They arrange odds in their favor based on information that you can't get anywhere else. Meriwether was the best bookmaker on Wall Street and now in the most literal sense: he had been fielding bets for days. The firm as a whole believed the race was Benton's to lose. But based on what he'd heard from Dunlavy, Meriwether believed Benton couldn't possibly win.

Meriwether smelled money; he knew he could make a killing by getting as many people as possible to bet on Benton. To do that, he made Dunlavy the favorite and gave Benton the "spread," a bookmaker's term for extra points. In other words, those who bet on Dunlavy would have to cover the spread—Dunlavy would have to win by several seconds instead of just one. The trick was finding the right spread, one that would attract enough dumb money to bet on Benton.

And that's what Meriwether was focused on now, just as focused as he would be on figuring the odds of one of his most complex trades. "How many seconds do you think you can beat him by?" Meriwether asked.

"Well, it's hard to say," Dunlavy said. "I would guess ten seconds or so."

"Really?" Meriwether fired back. "How confident are you of that number?" Dunlavy said he had run up the stairs once already and finished in 73 seconds. Benton was an overweight ex-jock with bad knees. He'd barely finish, Dunlavy assured Meriwether.

"Okay, then," Meriwether snapped. "I'm going to let the word out that I'm taking you and giving Dan six seconds—I'll take all comers." After a long pause he told Dunlavy, "Don't let me down."

Dunlavy wasn't sure what he'd gotten himself into. He had been working at Salomon's Cleveland office for three years and had dreams of making it to the big time and working out of New York. Meriwether was a rising star in the company, a favorite of Gutfreund and the firm's powerful executive committee. In other words, the last thing he wanted to do was let Meriwether down.

In the coming days, Dunlavy continued to train. He did a couple more trial runs up the fourteen flights and heard through the company grapevine that Meriwether was taking in some fairly big bucks—not the multimillion-dollar trades he made on the bond desk but enough cash to make people realize he could win or lose serious money on the outcome of the race.

Dunlavy knew Meriwether loved to win money, but he also knew that for Meriwether, winning wasn't just about the money. Like any good trader, Meriwether was looking to make a point, and no doubt he wanted to prove to his bosses, his colleagues, and himself that he understood risk and odds better than them all.

Dunlavy remembers the day of the great race well. He wore green gym shorts and a yellow shirt with "THE GREAT RACE" lettered on the front. Benton wore white gym shorts and a blue-and-gray Georgetown University sweatshirt. After many photos were taken, a coin was tossed to see who would run first. Dunlavy won first dibs. At the command, he raced up the first seven floors pretty easily, and then, despite all the training, fatigue started to set in. When Dunlavy finally reached the top floor, he was gasping for air. His wife, Daryl, attended the event and stood over him as he collapsed across the finish line, worried he might have a heart attack.

Benton didn't fare much better, collapsing as he made it across the finish line as well.

A minute later, the results were announced over the "squawk box," the firmwide communications system where research calls and other important messages were announced. "Ladies and gentlemen, the Great Race has just concluded. The times are—Pat Dunlavy, sixty-six seconds . . . Dan Benton . . . seventy-four seconds . . . Happy Saint Paddy's to you all!"

There were many groans but a few cheers all over Salomon country.

Meriwether was one of those cheering. Dunlavy had covered the spread by two seconds. Before the day was over, J.M. called Dunlavy to congratulate him. Dunlavy had now recovered, but barely; it completely slipped his mind to ask Meriwether how much money he had won.

Pat Dunlavy is now retired from Salomon Brothers. During his twenty-three years at the firm, he saw it all: scandals, massive trading losses, huge paychecks, boardroom battles, and ultimately the firm's demise, when, in 1997, Salomon Brothers was purchased by the financier Sandy Weill on his way toward creating another fabled and ultimately troubled financial powerhouse, Citigroup. But through it all, Dunlavy looks back on the Great Race and concedes it taught him something about the culture of Salomon and Wall Street.

In a few years' time, he would be working in New York alongside Meriwether, Ranieri, and Gutfreund. They were all legends of Wall Street—and enormously rich, earning millions of dollars a year in salary and bonuses. They were successful because they were smart, of course. But they had something else going for them: the will—or, to be more precise, the desire—to take risks and gamble.

Most of it was with other people's money, which made the losses more palatable, although all of them had some skin in the game. But risk taking was an obsession at Salomon Brothers, particularly among the new breed of traders and managers during the late 1970s and early '80s, people like Meriwether and Ranieri and others who would fill the trading floor of every big Wall Street firm for the next three decades.

A good case could be made that the success of Salomon Brothers and its bond traders led to a broader revolution that swept Wall Street and sowed the seeds of the financial meltdown of 2008. Historically, and continuing through the 1960s and most of the 1970s, the big firms and partnerships that dominated Wall Street employed a business model that was decidedly low risk, focused mainly on selling advice to investors and large companies.

Things started to change as several forces began to converge. One was money—to be more exact, other people's money. As the big Wall Street firms converted from private partnerships to public companies in the 1980s, they were gambling no longer with their own money but with that of public shareholders, and literally overnight the bets got bigger and the use of borrowed funds, known as leverage, grew and grew. Another was the cost of maintaining the old business model of acting as an "agent" for customers, underwriting corporate stock and bond deals, and selling stocks to small investors. It simply didn't pay as competition caused fees to shrink dramatically beginning in the early 1980s and continuing for the next three decades.

Yet another was technology. As soon as the first computers made their way to the Street, traders began using technology to spit out information about markets and securities, to find historical patterns in the way markets behaved and crunch data in ways that would have been impossible just a few years earlier. Armed with this information, traders for the first time seemingly had a real edge; they could process enormous amounts of information to predict trends and prices and how they repeated over time.

This is the story of how those forces, combined with the very human elements of greed and lust for power, transformed Wall Street over a quarter century and set the stage for the meltdown of 2008 that led to what many economists believe is the worst economic crisis since the Great Depression. There were other factors that contributed to the financial mayhem that reached its peak in the fall of 2008: risk takers such as Meriwether, Ranieri, and the traders they spawned would take over the management of

the big financial firms; the government would entice Wall Street to innovate and create new types of debt on one hand and provide aid and comfort
to the risk takers to trade these newly created bonds on the other; regulators such as Fed Chairman Alan Greenspan, as well as various chairmen
of the Securities and Exchange Commission, appointed by Republicans
and Democrats alike, bowed to pressure from the banking industry, made
it easier to take enormous risk, and, despite periods of market unrest, never
forced Wall Street to adopt more restraint.

The Ronald Reagan–era tax cuts spurred the economy and the stock
market to new heights. But it was Fed Chairman Paul Volcker's policy
of squeezing inflation—one of the great economic achievements of
our time—that spurred the bond market and made the taking of risk
through trading various forms of debt and derivatives of debt the Wall
Street business model for the next three decades.

With inflation tamed, lower interest rates in the early 1980s meant
cheaper borrowing rates; risk wasn't just made easier by technology, it was
now cost effective. Lower borrowing rates meant that speculators could
borrow more—a concept known as "leverage"—and put less of their own
money down when trading in the open market. Because bonds had become
cheaper to sell, companies would rather sell debt if they needed to raise
cash than dilute current stockholders' holdings by issuing additional shares
to the public.

Wall Street began inventing new bonds—one was the junk bond, another was known as the mortgage-backed security—and the various iterations of each would reach massive proportions. Mortgage bonds would
prove particularly enticing for Wall Street because of the massive fees
generated in the creation of these securities and because of the alleged
social good they created. In its simplest form, the mortgage-backed security is nothing more than a bond packed with mortgages; payments are
funneled from the mortgage holder to the bondholder. The objective: to
get banks to take the loans off their books and sell them to Wall Street,
which would then sell them to investors so the banks could keep making
more loans.

All the bonds could be traded. Liquidity is the lifeblood of any market,
and as interest rates fell, bond trading exploded. As bonds were traded
and commoditized, profit margins naturally shrank, as they had in the

old-line businesses of underwriting and providing advice to small investors. That forced Wall Street to innovate further. New types of bonds were created—"derivatives" of the old bonds. The bonds became more complex and packed with riskier mortgages, for which home buyers paid higher rates of interest that were funneled through to investors, who demanded higher yields. The trades became more complex and larger, based on computer models that allegedly reduced risk to the bare minimum. Where would it all end? No one seemed to care. Money was being made, and no one seemed to think that someday it all might end. And just like that, Wall Street's business model had shifted from giving advice to taking on risk.

Young MBAs working on Wall Street no longer wanted to advise CEOs how to run their businesses; they wanted to use leverage to take over the companies, restructure their operations, and sell them at a profit.

Evaluating and taking risk were now necessary ingredients of working on Wall Street. Brokers didn't make big bucks by recommending to their clients some supersafe long-term investment; they made their fortunes by churning the accounts of their best customers, essentially trading shares that didn't need trading in order to generate commissions. Bankers didn't buy their second homes in the Hamptons simply by telling a company how to manage its cash flow; the trick was to get the typical CEO in the 1980s to grow a company by acquisitions, often by using debt to finance the deal.

Traders didn't receive $10 million bonuses because they were completely hedged against endless losses; they made money by weighing the odds and then making decisive, and massive, bets.

And they made those bets because they were now in their comfort zone of risk taking. The randomness of events meant little to this new breed of executive on Wall Street; even as the markets grew more complex and the stakes got higher with wild market swings, a concept known as volatility, risk continued to dominate the Wall Street mind-set. In the new world of Wall Street, randomness was a friend because it increased the odds of making the ultimate payday even bigger, even as it increased the odds of losing.

P at Dunlavy was now discovering this firsthand. Dunlavy had come a long way from the "Great Race" in Cleveland; by 1983, he was working for Salomon's number two executive, Tom Strauss. Given the speculative nature of their business, Dunlavy was most proud that the business didn't lose any money.

"Tom, we had a pretty good quarter," Dunlavy said. "We cut back on our risk position dramatically. We're in great shape, and most of all we haven't lost any money. It's amazing!"

Dunlavy was expecting promises of a big bonus, drinks at the end of the day, or at the very least a gold star on the lapel of his suit. What he got, instead, was a bucket of cold water thrown right in his face. "Well, if you're telling me you haven't had a loss yet, that doesn't make me happy. All that means is you're not taking enough risk."

Dunlavy was floored. It was a speech he would have expected from one of the firm's risk takers, someone such as the mortgage bond trading chief, Mike Mortara, or John Meriwether. But not Strauss, who was better known as the company stiff. Strauss himself made no secret of his concern about the risk taking at the firm, particularly in the mortgage department, which made his speech to Dunlavy seem so odd. But what Dunlavy was discovering was that Strauss, like just about everyone else at Salomon, from the CEO to the lowliest bond trader, had drunk the Salomon Kool-Aid. That Kool-Aid made taking risk second nature. It was why you worked at Salomon.

Dunlavy didn't get intimidated easily. So he just shook his head and smiled after Strauss's statement and said his guys would crank it up if that's what Strauss wanted.

When the conversation was over, Dunlavy couldn't help but think back to the Great Race. Meriwether had been willing to risk losing thousands of dollars while Dunlavy himself had wagered a measly $50 even though he'd known for a fact that he could beat an overweight, out-of-shape former high school lineman with bad knees. Maybe he wasn't cut out to work at a place like Salomon Brothers.

Dunlavy might not have been in love with risk, but he knew how to work with clients and sell them bonds, which, for the moment at least, was enough. By 1984 he was a managing director, one of the top two dozen or so senior executives. He held a series of more senior jobs in Salomon's sales and trading operation and by 1985 was working directly for the great Lew Ranieri as the sales manager for one of the hottest areas of the Wall Street money machine, one that would change the business in years to come: the Salomon Brothers mortgage bond department.

ew Ranieri, Salomon's mortgage bond chief, wasn't so much big (he was barely five feet eight in height) as he was wide—for most of the 1980s, he weighed close to 250 pounds and reminded Dunlavy of a bowling ball. He liked to eat junk food, curse, and worry about the competition—most of all a skinny Jewish kid from southern California named Larry Fink, First Boston's star mortgage trader. Salomon had Wall Street's largest mortgage department, First Boston the second largest, and not a day went by that Ranieri didn't scream or curse out Fink—and Fink was equally paranoid about and disparaging of his rival.

On the surface, there are probably no two people more different than Ranieri and Fink. Ranieri was a college dropout from Brooklyn. Fink was a self-described "Jew from Los Angeles," the son of a shoe store owner and a college professor. Ranieri's suits were rumpled, and his tie was never tight, while Fink was almost never seen out of his suit. Ranieri gorged himself on pizza, and to this day, Fink is a serious foodie with a taste for refined Italian cuisine. But one thing they had in common was their status as outsiders; Jews and Italians were still rarities on Wall Street government bond desks, where most of the big money had been made, and that fueled the other thing they had in common: the driving ambition that would lead these two opposites to develop and grow into a market that would eventually transform Wall Street.

In the early 1980s, the concept of "securitizing" mortgages into bonds was relatively new. It involved taking relatively illiquid loans (e.g., mortgages)—stuff that couldn't be sold to the general public—pooling them into a single bond, and selling them, thus removing them from the banks' books. Think of the typical mortgage-backed security as a stew. Taken separately, each ingredient might be pretty unappetizing, but cooked together, they make something worth eating. Likewise, no investor would want to own an individual mortgage: the risk of possible default would be too great, and, if interest rates fell, the home owner would try to refinance the mortgage, thus changing the terms of the investment.

But because in a mortgage bond so many loans are packaged together, the bond represents something of value. In theory, default risk can be minimized since not all the loans will go belly-up at once; in fact, chances are most won't, so the mortgage payments of the good loans will more than cover the losses of the bad ones and generate a safe return for the investors who purchased the bond. Furthermore, because the banks could sell off their loans as bonds, they could then make more loans, thus expanding home ownership and the credit available to consumers.

The first mortgage bond was sold in 1970 by a federal government agency called the Government National Mortgage Association, or Ginnie Mae, which was chartered by Congress to buy home loans from banks in an effort to spur more lending. The bond was known as a pass-through— the mortgages were pooled, and the loan payments were simply "passed through" to bondholders. The underwriting was a moderate success, and soon a few banks holding mortgages copied the Ginnie Mae deal.

But the market remained a backwater. The bond deals in their current form didn't really do what was ultimately needed to allow the banks to ramp up mortgage lending: remove the mortgages from the banks' balance sheets. The pass-through deals were simply a mechanism for the banks to borrow money, nothing more. The mortgages remained their obligations, and the banks still needed to hold capital against the loans.

That was all about to change.

In 1977, Lew Ranieri was called into the office of Salomon CEO John Gutfreund. Gutfreund was accompanied by a man named Robert Dall, one of the firm's senior partners and an early student of the mortgage market, and by the firm's chief economist, Henry Kaufman.

The topic of the conversation: a potentially hot new area of the bond market.

Gutfreund explained that Dall and Kaufman had done some detailed work on demographic changes and what they believed was coming: a massive expansion in housing driven by demographic shifts. The baby-boom generation needed housing and needed it soon, and Wall Street—or, to be more precise, one Wall Street firm—would be there to make the most of what was yet to come.

That's because the banks and the savings and loan industry (known as "thrifts") simply couldn't meet this demand, given their capital constraints, Gutfreund explained. But the mortgage bond market, which was still in its infancy, could. And Gutfreund wanted Salomon in on the ground floor.

With that, Gutfreund said he was reassigning Ranieri immediately to the firm's small but growing mortgage bond department.

What little Ranieri knew about the mortgage bond market, he didn't like; he traded corporate bonds, particularly those issued by utilities. There was barely any issuance of mortgage bonds and almost no trading in the simple pass-through securities that had been issued. In other words, how was he going to make money? He was now in his prime, a thirty-three-year-old bond trader in what was then one of the hottest sectors of the market. Now he was being reassigned to the Siberia of Wall Street.

Ranieri told Gutfreund politely (that was the only way you spoke to the autocratic Salomon CEO) that he didn't know what he could add to the effort. Translation: he really wanted no part of it. "Listen, I'm not a research guy," he said, "I'm a trader, and there's nothing here I can trade. How am I going to get paid?"

But Gutfreund was adamant. "Don't worry about getting paid," he told Ranieri. "We're a partnership, and we'll find a way to make it worth your while."

There have been many explanations as to why Ranieri was chosen to become the head trader in a market that had almost no trading. One, offered in Michael Lewis's *Liar's Poker*, was that Gutfreund and Dall believed no other trader at the firm had the will to make something out of nothing and gave him carte blanche to make it happen. This was, after all, a guy who had no college degree, who'd started as a part-timer in the mail room and fought his way to trading success.

Another explanation was provided to me by Larry Fink, a mortgage trader at First Boston, who would soon become Ranieri's chief competitor and nemesis. In the 1970s and early '80s, Wall Street still had its ethnic enclaves. Fink used to laugh at all the WASPs who held key positions in Wall Street's high-paying investment banking departments. Italian Americans and Jews were outsiders in the Street's hierarchy, and the nascent market of underwriting and trading mortgage bonds was one way to break in.

"We all were smart," Fink explained, "and we understood how to use computers that helped us crunch data. But make no mistake, we were the first guys who got into the mortgage business because we weren't really wanted anywhere else."

Ranieri had considered Gutfreund his mentor, but given Salomon's brutal meritocracy he wasn't taking any chances. He rushed to find his first deal, and a few months later, Salomon packaged a number of loans into a mortgage bond for Bank of America.

This deal was different because, for the first time, Dall, with Ranieri at his side, was able to sell the rating agencies on a concept that would be known on Wall Street as "securitization." The mortgages were now moved off the balance sheet of the bank into a pool that was "securitized," backed by so many mortgages or other collateral that the risk of default was minimal. The rating agencies agreed, and the bond won the coveted triple-A rating from the two largest rating agencies, Moody's and Standard & Poor's.

In theory, moving the loans off the balance sheets of the banks was supposed to be the beginning of the gold rush for Salomon. But the deal was a bust, as Ranieri would later say. Salomon had yet to work out all the kinks. There were tax implications that made selling mortgages to a separate entity costly for the banks. And there was still very little trading. Investors didn't really know what they were buying. Was it a bond issued by a bank or a bond issued by some amorphous entity created by underwriters to benefit their banking clients?

Other factors slowed the mortgage market's growth as well. The bane of the bond market is inflation and high interest rates, and the 1970s and early 1980s had plenty of both. (Inflation eats away at the value of the fixed income that bonds pay investors.) The Salomon mortgage department was suffering massive losses. Despite Gutfreund's promises of a grand commitment to the effort, he considered closing the department on several occasions. Ranieri, however, wasn't having it. He might have been reluctant at first, but now he had been converted into a true believer in the market's potential, and he fought with the firm's executive committee for another chance, arguing that securitization was the future of the bond markets and reminding them of the reasons they had started the department in the first place: housing was destined to explode, and the banks needed a mechanism to meet that demand. Securitization was the only solution, and Salomon had the biggest department on Wall Street.

Once Ranieri convinced his bosses to give the mortgage department another chance, he turned his attention to Washington, D.C., where he hired teams of lobbyists to change laws in order to create tax advantages for S&Ls when they sold mortgages. By the early 1980s, Ranieri's efforts began producing results as inflation disappeared, the economy began growing, and lower interest rates caused a massive rally in the bond market. All those baby boomers finally began taking out mortgages thanks to lower borrowing costs and the economic recovery, and the market that had remained dormant for so long exploded

As the market grew, the mortgage department, which had originally been Bob Dall's idea, became Ranieri's baby. It wasn't long before traders across Wall Street were giving Ranieri credit for its creation, and that made another brash young mortgage trader seethe.

Goddamn it, I'm getting whip chained!" Larry Fink screamed, his head rocking back as he watched his position evaporate with a sudden and unexpected change in interest rates. In 1977 Fink was just twenty-five years old. He had been trading mortgage bonds for two years and, like any young trader, was now experiencing one of many brushes with losing money. Being young and inexperienced, he had yet to develop what good gamblers and good traders call a "poker face." John Meriwether had a poker face when he lost money. So did Ranieri.

And Fink? He complained that he was being "whip chained."

The mortgage-trading desk was the smallest group in the First Boston bond department. It comprised the head of the department, a veteran trader named Thomas Kirch, and a couple of traders and bankers. The group didn't even have its own sales force; it borrowed salesmen from elsewhere in the company. But it did have Fink, whose outsized presence made up for the mortgage department's modest size. This latest Finkism occurred during a particularly rough day in the markets. Fink had lost money. How much wasn't immediately clear; it wasn't enough to get anyone fired, but it was enough to get Fink to do some serious soul-searching, and Kirch, who was his mentor, was a good place to start.

"Tom, what the fuck went wrong?" Fink asked his boss. "How could this happen?"

Kirch looked Fink in the eye and gave it to him straight: "Larry, shit like this happens all the time, and I can't tell you why. But one thing I do know is that the word is 'whip*sawed*,' not 'whip chained.' Get it straight!"

Fink was momentarily stunned and embarrassed; his boss had just humiliated him in front of his colleagues. That is, until he saw a faint smile on Kirch's face and Kirch erupted in laughter, as did the rest of the mortgage trading desk, Fink included.

Despite the laughter, there was a message that Kirch was trying to convey to the young upstart in the early years of his career: you can be really smart (which Fink was), go to the best schools (Fink had an MBA from the prestigious Anderson School of Business at UCLA), be able to add differential equations in your head (Fink was gifted with numbers), and *still* be a lousy trader if you can't accept risk and losing as a fact of life. Every young trader learns the lesson sooner or later or washes out. One has to be prepared to lose money and not cry about it.

It wasn't long before Larry Fink developed the maturity that Kirch was calling for, and when he did, his career took off—and with perfect timing, as the mortgage bond market developed into a profit center that neither Fink nor Kirch nor any of their supervisors at First Boston would have predicted.

Part of the reason the market took off, of course, was Fink himself. Sure he was great with numbers and very good at dealing with clients. But he was even better at conceptualizing how mortgage bonds could be sliced and diced into components that could be sold to investors with varying needs and risk tolerances, which is what he began doing until he came up with an idea that took the innovation that started at Salomon Brothers in the late 1970s to a new level.

In just a few years, the mortgage market had grown amazingly complex. It hadn't been so long before that Fink and his team had come up with the idea of combining various mortgages, slicing up the cash flow into different levels of risk, and selling those levels of risk—known on Wall Street as "tranches"—to investors. The investors with the greatest risk tolerance would buy the tranche with the highest risk and the biggest yield; those with low risk tolerance would buy the "triple A" tranche, made up solely of the highest-quality mortgages, those that were least likely to default. These bonds, at least according to the rating agencies, were as safe as those issued by the U.S. Treasury. (The reality would turn out to be much different, as investors would soon find out.) To add some additional assurance, the bankers would pack the high-level tranche with so many mortgages that it would be "overcollateralized," meaning that even if a few mortgages went into default, the investor would still receive interest payments in full because the other mortgage payments would make up the difference.

If Ranieri and Salomon could take credit for selling the first mortgage-backed bond for Bank of America in 1977, Fink and his crew now could take credit for selling the latest innovation in the market. Fink was spending lots of time in the office with his analysts trying to come up with just the right structure for his new creation, while reaching out to lobbyists in Washington who had been following recent legislation that made issuing bonds of this kind tax efficient. The new bond—known as a collateralized mortgage obligation, or CMO—was radical and revolutionary. Risk was reduced to a computer model that allowed Fink, and then the growing list of dealers who would soon copy his invention and further mold this new brand of mortgage security, to fit an investor's risk tolerance. Fink's first CMO client was Freddie Mac, the Federal Home Loan Mortgage Corporation. Freddie and its sister agency, Fannie Mae, the Federal National Mortgage Association, were "government-sponsored enterprises," or GSEs, created by Congress to encourage home ownership by buying up mortgages from banks, securitizing them, and selling the new mortgage-backed securities to investors, and they would soon become large, powerful players in the mortgage market.

The team on the receiving end of Fink's pitch included a woman named Marcia Myerberg, a finance officer at Freddie Mac, who had also been working with Lew Ranieri for years on various new structures involving mortgage debt. Ranieri had often said that Myerberg, as much as himself, was responsible for the creation of the mortgage-backed security and the burgeoning market that began to take hold in the mid-1980s. Myerberg for her part was equally effusive in crediting Ranieri for the market's growth.

But she hadn't seen anything quite like what Fink had come up with. The first CMO issue hit the market in 1983; initially they were planning a modest-sized issue of $400 million. By now Ranieri and Salomon Brothers were in the game, adding their input and variations to the final product. Freddie Mac polled potential investors, was astounded at the positive response, and decided to more than double the issue to $1 billion, tying the largest private bond issue ever (IBM had issued $1 billion in bonds a few years earlier).

Salomon and First Boston shared most of the underwriting duties on the deal, which included more bells and whistles than any other that had hit the market. More than that, the deal had set the stage for greater variations in the years to come, as well as a battle royal on Wall Street between Salomon and First Boston, and Ranieri and Fink in particular.

With the creation of the CMO, Fink became a Wall Street rock star, and he used his status to let the world know that it was he, not the great Lew Ranieri, who had created what was soon to be one of the hottest-selling products Wall Street had ever seen. Ranieri, of course, thought Fink was "full of shit," but one thing he couldn't deny was that his rival had arrived. By mid-1984 Fink was named head of the First Boston mortgage department, and was rivaling Ranieri's team in size, scope, and most of all, profitability.

As Fink would later say, "We stepped in shit," which is trading desk slang for being in the right place at the right time. Though the mortgage bond departments at most firms were backwaters, the landscape was about to change. By the late 1970s, the federal government passed a law, the Employment Retirement Income Security Act, or ERISA, that imposed regulations on private pension plans, mandating that they become fully funded and thus leading them to expand the scope of their investments beyond stocks and government and corporate bonds to include the nascent market of mortgage-backed securities. At the same time, insurance companies and annuities began looking for new types of securities as well, ones that offered higher returns, or "yields," than conventional U.S. Treasury bonds.

By 1983 Federal Reserve chairman Paul Volcker had done his job; he had squeezed the economy with high interest rates that had led to a steep recession but reduced inflation to tolerable levels. Now he let interest rates fall, sparking a massive rally in the bond market. Investors "reached for yield," looking for higher interest rates than could be found on conventional Treasury bonds, and they started snapping up mortgage-backed securities, junk bonds, and anything else that offered a high return.

Most of all, "fee compression" was everywhere on Wall Street. Investment banking fees declined sharply, as did stock-trading commissions. The so-called agency business model of providing advice to corporations or individuals made less and less money thanks to a combination of greater competition and less regulation of trading fees.

All this boded well for the nascent business of packaging and trading mortgage bonds; because it was not as well understood as other markets, the fees were greater. It also pushed Wall Street in the direction of using mortgage bonds to change its business model and take on more risk. Firms began to leverage their own balance sheets, cheaply borrowing multiples of the amount of capital they had to invest in these newfangled securities, carrying them on their own books, and pocketing the interest and price accumulation.

The practice became known in trading circles as the "carry trade," and, over the next twenty-five years, the carry trade would become one of the most efficient ways Wall Street made and then lost tremendous amounts of money.

For the moment, however, Wall Street wasn't worrying too much about losing money. Interest rates were falling, and the housing market was booming. Investors who hadn't understood what a mortgage-backed security was in 1977 couldn't get enough of them just a few years later. In the weighing of risk versus return, mortgage-backed securities had an edge over other types of higher-yielding debt such as junk bonds because the big investment-rating firms such as Moody's and Standard & Poor's rated them highly. Wall Street was having it both ways: it could earn fatter fees constructing the bonds because of their complexity and then could easily trade them because of their high ratings. Indeed, many earned the coveted triple-A rating because they were securitized with a diversity of mortgages and backed up by other forms of collateral, meaning the risk of default was thought to be minimal. Indeed, the biggest risk with mortgage bonds in the early 1980s was something known as "prepayment risk." As interest rates began to fall, mortgage holders began to replace their outstanding loans with new ones at lower interest rates, thus eating into the returns of the mortgage bond holders.

The mortgage market got another boost thanks to help from the federal government. In the early 1980s, the government gave a huge tax break to savings and loans if they could sell assets such as mortgages. The move took the mortgage business to new heights. Low interest rates drove mortgage lending on a massive scale; the favorable tax treatment gave every incentive for S&Ls to get them off their balance sheets and sell them to the big Wall Street firms that securitized them. It wouldn't be the last time government intervention would cause a boom in the housing market. It's a lesson in basic economics, the sort studied by every Economics 101 student worldwide: give people incentives to issue and take on debt, and they will respond—in a big way.

By the mid-1980s the mortgage desk was no longer the place where the white-shoe firms kept their loud, fat Italian Americans or brash Jews away from the clientele. It was the place to be, and the place to be on Wall Street is where the most money is made. Young traders from top schools bypassed the chance to do investment banking deals and trade government bonds for the chance to work with mortgage-backed securities and start earning high-six-figure salaries compared to the relative chump change handed out to newly minted banking rookies. The profit margins on mortgage desks rivaled those of some of the highest-margin businesses on Wall Street, primarily because no one understood the business but people like Ranieri, Fink, and the teams of PhDs and MBAs they hired to cobble together deals with complex formulas that weighed risk and return. Only junk bond traders such as the legendary Michael Milken, who financed leveraged buyouts using massive amounts of junk-rated debt, rivaled the mortgage desks in profit margins.

Despite the competitive tension—some of it petty (who played a bigger role in the development of the CMO), some of it not (who was more attentive to client needs)—life was good for Fink and Ranieri. First Boston and Salomon dominated the market for mortgage-backed bonds, and by the mid-1980s, both men were becoming incredibly rich. Maybe not as rich as Milken, who cut a deal that gave him a share of his firm's profits and take-home pay of $500 million for one year, but still enormously rich by 1980s Wall Street standards. Both Fink, then barely in his thirties, and Ranieri, a few years older, were earning salaries of about $3 million a year.

The more money they made, the greater their rivalry grew. With each passing day came new "innovation," or a claim of one, by either Ranieri's or Fink's troops. They bad-mouthed each other to investors and to colleagues at other firms every chance they got, becoming increasingly obsessed with the league tables, the charts that measured how many deals each underwrote. And as the competition grew, so did the amount of risk each used to squeeze out even more profits. Soon both were rolling the dice in ways that Wall Street had never before seen. Fink was just ten years out of college in 1985 and running a balance sheet that was leveraged as much as 15 to 1 and growing, meaning that for every dollar of the firm's capital he invested in the mortgage-backed market, he and his team were borrowing—and risking—at least $15 more. Ranieri's leverage was about the same, and their bets were becoming increasingly esoteric and growing swiftly in size, well into the billions of dollars.

Math geeks in their research departments pored over housing statistics, weighing the risks of defaults and prepayments, the twin evils of the mortgage bonds. The balance sheets of Salomon and First Boston were now carrying large blocks of mortgage-backed bonds on their books, borrowing what was then considered huge amounts of cash to finance holdings of the various iterations, or "derivatives," of mortgage debt that had been created in recent years, securities known as interest-only strips, or IOs, created from the interest payments of the bonds; or principal-only strips, or POs, which were just the principal portion of the bonds.

As long as the firms guessed right on interest rates, the carry trade was the most profitable bet on Wall Street, earning enormous amounts of money on the difference between the high yields of the mortgage bonds and the low interest rates paid for the borrowed money used. But a sudden change in interest rates could result in hundreds of millions of dollars in losses almost instantaneously.

At least for now, Fink and Ranieri were guessing mostly right, and the more money they made, the bigger their stature grew. Gossip around the halls of Salomon had Ranieri angling for the CEO's spot, while Fink, the skinny Jewish kid from Van Nuys who just a few years before had been screaming about being "whip chained" was now the odds-on favorite to run First Boston someday soon. In fact, Fink's ascendancy, both inside First Boston and in the mortgage market, was so swift and sudden that Ranieri demanded that his guys maintain round-the-clock surveillance on his chief nemesis's every move.

T hat fucking Fink!" screamed Ranieri. "If we don't keep doing this, he's going to be eating our lunch!"

Ranieri was attending a meeting along with other members of Salomon's executive committee, led by CEO John Gutfreund. At issue: whether the firm should continue risking its own capital in a relatively new, but growing and lucrative, part of the bond market known as "fixed-income derivatives," new financial products that were created or "derived" from the interest payments of another bond. The markets for derivatives were still a backwater, dominated for the most part by so-called hedge funds—pools of capital, open only to the wealthy, that risked huge sums of money by placing large bets on various markets using borrowed funds, or leverage, in order to magnify gains.

In the mid-1980s, hedge funds hadn't achieved the status they would in the next decade, so Salomon, being Salomon, saw an opening. The firm's top brass was weighing whether to expand its reach into esoteric trading and risk even farther in the nascent area of fixed-income derivatives known as the swap market. Swaps are a pretty simple concept. Start with the fact that some bonds are issued with a fixed rate of interest and some "float," meaning they rise or fall based on some index, like the London Interbank Offered Rate or the interest rates of Treasury bills. The fear of any bond investor or issuer is a sudden and remarkable change in interest rates that scrambles investment outlooks. The CFOs of major companies that had sold billions of dollars in bonds with floating interest rates were worried that a sudden spike in rates would ramp up their interest payments. Likewise, an investor with exposure to fixed-rate debt might worry that higher rates would lead to losses since the price of a bond in the open market declines when interest rates go up—so he might want interest rates that float lower. The swaps are essentially a way to bring those parties together to reduce their risk—the issuer holding floating-rate bonds who wants fixed-rate exposure and the investor with bonds issued at fixed interest rates who wants some of his debt to float.

For several years Salomon's involvement in the swap market was pretty routine; it simply served as a middleman in these transactions, earning fees by bringing those parties together. The plan on the table—the one that made Ranieri's mouth water—was to take its role a step further. Instead of merely acting as a traffic cop by arranging the swaps, collecting a fee but not risking any of its own money, Salomon would hold some of the positions on its books—it would be betting on the directions of interest rates with house money.

There are many reasons for Wall Street's growing addiction to risk that began in the 1980s. With low interest rates, it was cheap to take risk with borrowed money and ramp up the size of trades. Financial innovations and computer technology were supposed to give traders ways to hedge their bets and weigh odds as never before (and, as the Street would discover, with varying degrees of success). There was, of course, the greed factor, which couldn't be hedged away. The bigger the risk, the more money can be made, and when the market's rolling, most traders don't take into account the inevitable bigger losses that can also result.

Then there's the issue of money—or, to be more precise, other people's money. By the mid-1980s Wall Street's private partnerships were a dying breed; one by one, the big Wall Street private partnerships became public companies that grew in size by issuing stock to public shareholders, and with that they began gambling not with the house money but with public shareholders' money.

The once-staid Wall Street business model with its modest leverage was now an anachronism, and with Salomon, now a public company since 1981, the risk taking took off. Pat Dunlavy would later recall that Salomon was in essence running a huge hedge fund under the guise of an investment bank, given the growing clout of the Meriwether arbitrage group and the heightened risk taking of the mortgage-trading desk. The full-scale push into the swaps markets was another step in that progression.

Thus, it didn't surprise Dunlavy that Ranieri was one of the biggest proponents of the new move. At bottom, Ranieri was interested in the swap market because his biggest customers, the savings and loans that sold Salomon all the mortgages he was packaging together, were now major players in the bond market and were increasingly interested in using swaps to hedge their growing exposure. It was one thing for Salomon to act, as it had, as a mere middleman and put the swapping parties together. But if the firm were to act as a principal, offering its clients swaps as a service whenever they needed them, it would bring a mammoth new source of profit to the firm. There would be risk, of course, but why be on Wall Street if you're afraid of risk?

The meeting was held in the corporate boardroom, an ornate space with chandeliers and paintings of Salomon's partners going back to its founding. Helping explain how the swap market worked was Dunlavy, the former staircase racer, now a thirty-four-year-old senior executive who had regular contact with the firm's leaders and frequently chatted with Gutfreund about the bond market. But he had never appeared before the executive committee before. And now he was scared as he watched Gutfreund puffing away on his customary large Cuban cigar and all the firm's top executives staring at him with blank looks as if they didn't really understand—or care—what he was about to say other than the part about how the firm could make a lot of money.

Gutfreund asked no questions as Dunlavy drew on a whiteboard by hand a series of boxes and arrows demonstrating how the interest rate swap business worked and how the firm could make money from it by risking a modest amount of house money. At one point, someone asked about the amount of firm capital that would be needed; after all, more money spent on swaps would mean less capital for the other groups at Salomon Brothers, which would certainly see the new business as an intrusion on their own.

The only other question came from Henry Kaufman, a man known on Wall Street as "Dr. Doom" because during the late 1970s and early 1980s he had correctly predicted that the economy would stutter because of high inflation and low economic growth. Those days were long gone, but the name still followed him around because, unlike most of his colleagues, he was someone who saw danger just around the corner. Kaufman asked just how much risk Salomon would be taking; Dunlavy had used the term "notional amount" of $100 million in one of his examples of an interest rate swap. The notional amount of the swap is the actual size of the bond holdings the interest rates are based on. Interest rate swaps are described in the market based on the amount of bonds or the "notional" size of the deal, but in reality this concept means very little, because counterparties merely swap their rates—they don't exchange the amount of bonds. Kaufman thought the firm would be risking the notional amount of the bond deal in addition to the interest rates, thus holding billions of dollars' worth of the bonds on its balance sheet.

Dunlavy calmly explained that at any given point in time, the risk to Salomon would be the differential between the fixed and floating rates—not the "notional amount" of the bond in question. Still, Dunlavy said, the firm would face other risks—for instance, if interest rates made a sudden shift. That could hurt its position, and the firm would be on the hook for the losses.

But Salomon would never have those "notional" dollars at risk, Dunlavy said, and Kaufman seemed appeased. Others soon chimed in that this was a unique opportunity they couldn't let slip by. Ranieri agreed. "We will need this new tool in my business if we plan to keep our dominant share in the mortgage market. You can bet Larry Fink over at First Boston is thinking about this as we speak!"

Everyone in the room knew Ranieri had a vested interest in making sure the business worked even if it didn't appear on his profit-and-loss statement—this new business would only enhance his power inside the firm. But his comments about Fink carried weight. Ranieri wasn't the only one paranoid about First Boston's growing presence in the bond market. Ranieri couldn't stand the idea of Fink's beating Salomon to the punch on anything (Fink's claim to have developed the first CMO still had him steamed), and the case he made to the executive committee was that Fink wouldn't think twice about offering swaps to the S&Ls and stealing away all of Salomon's clients.

Right before the meeting concluded, Kaufman cautioned, "We will need to monitor our credit and counterparty risk very carefully. Let's go slow with this. We should set an initial limit in terms of notional risk."

The others just listened and nodded. Dunlavy had known the committee would approve the deal once Ranieri mentioned Fink and the threat he posed to Salomon in one of the most lucrative businesses on the Street. Once a trading limit was agreed to, they unanimously approved the proposal. Gutfreund thanked the group and said gruffly, "Now go make some money!" Salomon went on to make millions of dollars trading the swaps in the next several years, and Dunlavy was promoted as his reward for helping the firm break new ground by being among the first to embrace this new type of risk.

Just like that, the foundation had been laid for the creation of a completely unregulated, opaque market that would in a few years' time grow to immense proportions, first as a tool designed to control interest rate risk and then as something totally different, a way for a firm to bet on interest rate swings and move risk taking on Wall Street to a whole new level. As the use of swaps and other derivatives grew, controlling risk became an afterthought for those using the products. The derivatives market became a huge casino, with trillions of dollars being bet by the big banks on a daily basis and each Wall Street firm looking for a trading edge over the competition.

Few people outside even knew it was happening. Junk bonds had become increasingly controversial as the 1980s rolled on, as takeover artists used junk debt to finance buyouts of some of the nation's largest corporations with high-cost debt that would have to be repaid at some point. But at least the debt was disclosed on the balance sheets of the newly acquired companies and even traded on some exchanges; the New York Stock Exchange had traded junk debt for years, with the prices clearly disclosed in newspapers. The derivatives market, however, was hidden and opaque. No central clearinghouse existed for the increasingly complex bonds derived from other bonds, or the complex maneuvers to hedge risk.

Regulators like those at the Securities and Exchange Commission and congressional committees with oversight responsibility over Wall Street had little or no idea that this market even existed and that it was growing swiftly as Wall Street embraced risk as its business model and slowly pushed traditional advisory services into the background.

No one knew, and, at least for now, no one cared about the long-term costs of embracing risk and leverage because Wall Street lives for the moment. And for the moment, risk and leverage were making Wall Street a lot of money.

At Salomon Brothers, the size of the mortgage department matched the growing size of Ranieri's waistline. He now had under his control more than 150 traders, bankers, and salesmen. Salomon's size was something Fink would ridicule—"We do more with less," he loved to point out, arguing that Ranieri ran the Spanish Armada of mortgage departments: big, slow, and easily outmaneuvered.

And Fink was developing his own counterattack. His group at First Boston was smaller than Salomon's but, over time, not by much. When Fink had started, the mortgage department had had just a handful of traders. It had borrowed salesmen and bankers from other departments. By the mid-1980s, Fink had grown the department to more than a hundred, and he continued to hire the smartest people he could find, not just to devise risk programs but also to come up with different types of mortgage bonds.

In just a few years, the mortgage market had grown amazingly complex. It hadn't been so long before that Fink and his team had come up with new ways to cobble together cash flows from various mortgages to create tranches that reflected the ever-growing needs and desires of investors, including something called the "Z" tranche, which receives interest and principal only when the other tranches are paid off (which means it's sold at a large discount) and by stripping out the interest and principal payments into so-called interest-only and principal-only bonds.

Before long, the first CMO sold by Freddie Mac seemed almost quaint and anachronistic. Computers had made the creation of new mortgage bonds from the basic model seem almost like child's play, and it didn't take Fink long to figure out that if you can securitize home loans, you can securitize car loans and credit card receivables and just about any other claim on consumers.

In 1983 Fink's team was the lead underwriter on what was then the largest bond issue ever, for the General Motors Acceptance Corporation (GMAC). The structure was similar to that of a CMO, only instead of home mortgages the GMAC bonds were created from car loans.

The deal was a blowout. Fink ended up increasing the size of the GMAC deal from $3.2 billion to $4 billion because investors couldn't wait to get their hands on the bonds, which, according to the computer models, were relatively risk free.

GMAC was a watershed deal for several reasons. Fink had proved something to Wall Street that would become both a blessing and a curse for years to come. Banks for years had had to hold capital against these loans because in times of economic hardship, people default on their credit cards or miss car payments. But through the magic of securitization, where risk could be squeezed to infinitesimal levels, these loans could be removed from the banks' balance sheets and transferred into moneymaking instruments held by sophisticated investors. Everybody was happy: the banks could make more loans and increase their fee income, the Wall Street firms and investors could make money from packaging and holding the securities, and consumers could borrow more easily (and borrow and borrow), because banks could now keep on lending.

At least that's how it all looked on paper, which for the moment appeared to be the reality of the markets and the economy. Fink was in his glory as First Boston did the honors and set the pricing of the deal, allotted how many bonds each firm would get, and walked away with a $20 million fee, while back at Salomon Brothers, Lew Ranieri was apoplectic. His archrival not only held bragging rights over the creation of the CMO but had now sold the largest bond deal in Wall Street history.

Fink didn't think it could get sweeter than that. But it did. The mortgage securities market was making Larry Fink a rich man at a pretty young age, and the money kept rolling in as the market began to expand beyond anyone's wildest dreams. The number of new mortgage deals more than doubled between 1985 and 1988, reaching nearly $1 trillion. There were many reasons for the surge—a booming economy, low interest rates that spurred a housing boom, and, of course, the Wall Street alchemy of structured finance that allowed everyone involved—most notably Fink and Ranieri—to believe that the good times would never end.

But they always do.

The battle between Ranieri and Fink was growing exponentially. It didn't generate the headlines that some others did, such as the epic RJR Nabisco takeover battle or the war waged by junk bond king Michael Milken and his firm Drexel Burnham Lambert against the entire Wall Street establishment. But inside the mortgage bond business, the Ranieri-Fink wars of the 1980s were now escalating to an almost bizarre degree.

It was the last business day of June 1985. Fink was at home with his wife and children when he received a call that Ranieri and Salomon had allegedly priced a $250 million mortgage deal. If it had happened on any other day, Fink wouldn't care. But it hadn't. Salomon was gaming the league tables. Fink could smell it. The firm was pricing the deal on the last day of the second quarter—nearly the last minute of the quarter—which technically ended the following night, on Saturday, just to stick it to First Boston in the league tables. Wall Street executives love to brag about being at the top of the league tables—it's like a baseball team bragging about being in first place. But being at the top of the tables is more than a macho thing; clients flock to the firms that do the most business. It's like the Good Housekeeping Seal of Approval of Wall Street; if you do the most deals, you must be the best. And now that the stakes in the mortgage market were growing, being at the top of the league tables would mean much more business down the road.

In June 1985 Fink thought he had it won, at least for the first half of the year, with a seemingly comfortable lead over Salomon. It would take a near miracle—a last-minute deal—to change the ranking.

That near miracle, Fink was told, had just occurred. A trader on the mortgage desk gave him the full report: He had received a call from Ranieri's lead trader, Mike Mortara, that Salomon had just priced another $250 million offering, putting the firm comfortably in front.

Fink was in a panic when he heard the news. He had told his supervisors he had the contest locked up—the department was getting ready to celebrate with drinking and parties.

"You got to be fucking kidding me!" Fink screamed when he realized that Salomon might have edged them out. His orders were simple: call up every S&L, every bank, every client, cobble together another deal, and do it fast. The firm had just twenty-four hours to take back the lead before the quarter ended Saturday at midnight.

Well into the night, First Boston worked to put together a deal until it had its mandate: a $200 million mortgage bond deal ready to be packaged and sold upon Fink's orders. It had some buyers, and those it didn't have, it could find later. The firm would just hold the bonds in inventory and pocket the interest on its balance sheet. The deal would be messy—the firm might lose money if rates suddenly moved and investors headed for cover. But the deal would give it a comfortable lead in the league standings. It could worry about losses later.

Late Friday night, Fink was informed that the deal had been assembled. Then something struck him. The notion that Salomon had pulled such a ridiculous stunt made him sick. It also made him sick that he'd had to go out and scrape together some poorly conceived deal that might ultimately blow up on him, just to try to save face. His gut was telling him to do the deal and take first place away from Salomon; his head was telling him not to stoop to its level and do something stupid.

In the end he listened to his head. Fink called it off and decided to wait and do the deal the right way. At the time, he said the whole episode was "too insane . . . I didn't want to carry it further."

As it turned out, Fink had made the right move, the smart move. The call to the First Boston trading desk was a joke—a hoax; Salomon had never priced a deal Friday night. Mike Mortara at Salomon had made it up because he knew it would drive Larry Fink crazy. But even in second place, he had his victory, as he told Ranieri, his boss: he had driven Larry Fink crazy, and that was almost as much fun as winning.

Even as his team fought its bitter battle with Fink, Ranieri was fighting an internal battle at Salomon Brothers against people like John Meriwether. While all of Wall Street was on a roll—Drexel Burnham Lambert, the great junk bond powerhouse, earned a remarkable $1 billion in profit in 1985—there was still squabbling over limited resources. At Salomon, Meriwether's arbitrage group was making so much money that he demanded additional turf and began taking over areas of the bond market that had been run by other departments.

One thing stood out amid all this fighting: Ranieri, Fink, and Meriwether were all remarkably similar people. They were hypercompetitive, loved to trade, and, most of all, *loved* leverage. Fink may have decided against that one last-minute deal to save the league tables because of its balance sheet risk, but he was using the balance sheet in many other ways. Official statistics aren't available, but Fink confirms that his use of leverage, his borrowing to place trades and hold increasingly risky securities on his firm's balance sheet, increased just about every year during the 1980s. Pat Dunlavy says that leverage also increased at Salomon, systematically and dramatically, as Ranieri's and Meriwether's trading profits grew.

How much? The conventional wisdom among salesmen at Salomon was that the mortgage department had leveraged itself at times more than 15 to 1, though some former executives note that the only way Meriwether's arbitrage group could have made the money it did and paid the salaries it had (one trader eventually earned $23 million) was by using leverage that was at least twice as high. Dunlavy was astounded when he heard how much money and how much risk the firm was taking on. "I remember thinking, these guys care more about rolling the dice than they do servicing the customers, but it was obvious why they were doing it: the more they borrowed and traded, the more money they made."

Historically, Wall Street has had a love/hate relationship with leverage. Investment banks are not like traditional commercial banks, which have deposits and then borrow from those deposits to make investments such as loans to home owners. The typical investment bank is thinly capitalized, even by the changing standards of the 1980s, when firms such as Salomon, Morgan Stanley, and Bear Stearns started the trend of expanding their capital base by becoming public companies.

During good times, leverage on Wall Street soars; when the markets crash, firms typically leverage less. The 1980s were a good time to be on Wall Street, and now that most of the big investment houses (Salomon, Bear Stearns, Morgan Stanley, and Merrill Lynch) had become public companies, meaning they were risking shareholder money and not their own, as you might expect, leverage on the Street rose to levels never before seen.

With so much money being made in the bond market, some of the most risk-adverse firms on the Street saw leverage as the path to glory and started copying Salomon's and First Boston's business model of risk and leverage. In the 1980s, Bear Stearns was run by the odd couple of Wall Street: CEO Alan "Ace" Greenberg, legendary for his risk aversion, and President James Cayne, who had made his fortune betting that New York City bonds would recover after the fiscal crisis of the 1970s. They were now building a bond department to take advantage of the boom in the fixed-income markets through very high leverage, which soared to as high as 50 to 1 at some points during the mid-1980s. Even Merrill Lynch, the firm that had made its bones advising small investors on the market, had begun expanding its bond operations and growing through borrowing.

Making matters even more complicated was the fact that the people who allegedly best understood the dynamics of leverage—the risk of its turning against the borrower violently and unexpectedly to produce outsized losses equal to or exceeding the outsized gains—were the very same people who were making so much money using leverage in the first place. Why would they stop and consider the consequences when what they were doing was bringing them such outsized paychecks?

2. POWER AND PERKS

Bottom line, we want hungry, street-smart people, and that's why we want you." Bear Stearns President Jimmy Cayne was giving his standard pitch to a potential employee sometime in the mid-1980s, in his spacious office in lower Manhattan at the firm's longtime headquarters. Larry Friedlander was a top broker at Smith Barney with a specialty business: trading bonds on behalf of individual investors. For years, investors had wanted bonds as a safety net; they bought them and held them until maturity. But the great bond market rally that began around 1982 and grew larger and larger as the decade rolled on had begun to change the face of Wall Street and investing, spawning brash risk takers, such as Lew Ranieri, Larry Fink, and John Meriwether, who specialized in trading newly formed varieties of debt, and brokers such as Larry Friedlander, who traded bonds on behalf of clients as some brokers traded stocks—frequently and repeatedly, looking for increments in pricing when he could unload his inventory and price declines that were opportunities to buy.

Like Meriwether, Jimmy Cayne made his first big payday rolling the dice in the bond market. He had been at Bear just a few years when he made a similar bet on the resurgence of New York City debt at the height of the 1970s fiscal crisis, when city bonds were trading at just pennies on the dollar, and then made a small fortune as the city eventually recovered and its bond prices soared.

As a result, he fancied himself if not a bond expert, as someone who knew enough to appreciate a business model that exploited the current market trends, namely lower interest rates and higher bond prices, and that's what Friedlander was selling.

"We want you—name your title," Cayne said, according to people with knowledge of the meeting.

Friedlander shot back, "Executive vice president."

"Done," Cayne answered.

Friedlander also wanted a guarantee: $40,000 a month plus commissions. Cayne said he could handle that as well. Cayne, however, wasn't sure he had iced the deal. So he really began to turn on the charm. "Listen, you're going to love working here," he said. Cayne spoke about Bear as a big family, where Friedlander's ethnicity as a Jew wasn't a drawback as at other firms, such as Morgan and even the blue-blood Jewish firms such as Goldman "Sucks," as Cayne liked to call Goldman Sachs.

Cayne then gave Friedlander one last reason to work at Bear: "Come over here," he said as he opened the bottom drawer of his desk and pointed. "You should also know I get the best pot in New York City!"

Friedlander laughed nervously; as a former lawyer who had once defended criminals in the South Bronx for a famous Mob attorney, Murray Richman, Friedlander later told friends that he thought he had seen and heard everything, but Cayne had nearly blown him away. "I can't believe the president of Bear Stearns just did that," he thought as he shook Cayne's hand and said he would start as soon as possible.

Jimmy Cayne loved many things about being the number two executive at Bear Stearns, which by the mid-1980s was one of the fastest-growing firms on Wall Street. He loved the power and the perks that went with it, such as access to the executive dining room and the ability to schmooze with top clients and politicians looking for campaign contributions. He loved recruiting; he used to brag that it was the best part of his job, because he got to know the rising stars of the firm that way and helped him maintain his hold on power.

And he loved the fact that no one—except possibly his boss, Ace Greenberg—could stop him from enjoying his recreation, which consisted of smoking a big joint rolled with the best pot he could find. Most of all, he loved the fact that, at least in his mind, he was going to replace Greenberg as CEO in the not-so-distant future. Nothing was going to stop him.

Bear Stearns was founded by three men, Joseph Bear, Robert Stearns, and Harold Mayer. They were outsiders—entrepreneurs who weren't part of the Wall Street trading elites at Morgan Stanley, the lace-curtain Irish who had started Merrill Lynch, or the upper-crust German

Jews who ran Goldman Sachs. They were the "tough Jews," as they were known at the time on Wall Street, men who couldn't get a job working for the elites, so they set out to start a firm on their own and in their image.

Bear, Stearns, and Mayer were long gone by the time Jimmy Cayne joined the firm in 1969, but the firm's name and outsider image remained. Once the original three partners had moved on, the power shifted to a man named V. Theodore Low, who carried business cards without his name and phone number, and Salim "Cy" Lewis, a gun-slinging, six-foot-six stock trader who specialized in trading large blocks of shares. Block trading became one of the most profitable businesses at Bear, and by the mid-1960s Lewis became the firm's driving force.

Lewis was a prototypical trader: he was loud—obnoxiously so—but also cunning. He craved power, but he clearly understood talent. In the 1960s he hired and consistently promoted another brilliant stock trader named Alan "Ace" Greenberg, and by the early 1970s Greenberg had been anointed Lewis's successor.

By now Ace Greenberg had the reputation of being among the best stock traders on Wall Street, but he and Lewis sparred over everything. At bottom, they were completely different people. Lewis was a screamer with a bad temper. His job was his life. Greenberg had outside interests, including a long list of women he dated, including Barbara Walters. He had a personality that was decidedly un–Wall Street, which is known for its share of blowhards. Greenberg could be taciturn, and he often spoke in short sentences. He wore bow ties and had a couple of odd hobbies—odd for Wall Street, that is. In his off-hours, he practiced "magic," in the form of card tricks, and he also loved to whittle pieces of wood and play with a yo-yo.

But what made Greenberg really different was his appreciation for risk. Bear, like all the other firms on Wall Street, was highly leveraged, sometimes three times as high as the competition. But Greenberg considered risk a necessary evil to be contained and managed, not unlike some wild beast. Greenberg's biggest battles with Lewis involved risk, and this was no mere philosophical disagreement. The argument often shook the foundations of the firm.

Partners wanting to take risk went to Lewis to overrule Greenberg's caution, and those who hated Lewis preyed on Greenberg's more cautious nature.

Less senior partners took their cues from the guys at the top. Bear was rife with internal battles and brawls, with partners earning top positions by first making a lot of money and then killing off their in-house competition through corporate gamesmanship. That usually involved proving to Lewis, Greenberg, or both that over time you were more valuable to the firm than the guy you were looking to torch.

Jimmy Cayne wasn't a trader, but he fit right into the bare-knuckle culture at Bear, though there was little in his background that suggested he would become the top executive at one of the biggest firms on Wall Street. The son of a patent attorney, he had grown up just outside Chicago in a middle-class home. Cayne isn't the original family name; his father, Maurice Cohen, an Eastern-European Jew who emigrated from Russia as a young boy, changed it to sound more Irish since the Richard Daley political machine ran Chicago and being Irish, or having a name that sounded Irish, helped with business, particularly if a city contract was at stake.

Cayne was an indifferent student, and that translated to his relationship with his family, particularly his father. Family members say Cayne never thought much of his dad, whom he considered a ne'er-do-well. "I think Jimmy always felt his dad was content being average, and he looked down on him for that," said one family member. Maybe so, but young Jimmy wasn't all that different. He was a mediocre student at Purdue University and dropped out after just a couple of years.

One thing he did love was playing bridge; he learned the card game at an early age, and by college he was skipping class to play bridge with his buddies every chance he got. It's unclear if he ever gave a thought at the time to a possible career on Wall Street. With his quick wit and glib sense of humor, Cayne had many of the attributes of a good salesman. But it was his growing interest in bridge that developed the risk-taking side of his personality. Bridge requires many of the same skills needed in a good trader, such as the ability to weigh the odds, to predict an opponent's next move, and to make decisions with incomplete information. It also takes courage—to take risk in the short term that will provide benefits down the road—and patience, a trader's best friend. Losing positions turn into winners when traders have the stomach not to overreact and to wait out the pain.

Taking risks may not have been in Cayne's father's DNA, but it certainly was in Cayne's. In addition to bridge, he loved to play poker and dreamed of being a bookie. He also hated college, dropped out, and, much to his parents' chagrin, joined the army. "I remember my mother sitting on the sofa when I told her," he said. "She couldn't believe her son was not going to college." Cayne then spent two uneventful years stationed in

Japan. He returned home and became a traveling salesman, married a woman named Maxine Kaplan, and went to work in his father-in-law's sheet metal business as a salesman.

Alas, the family soon met with tragedy. The couple had a daughter and a son, who later died of cancer as a child. The marriage ended in divorce. With everything else falling apart in his life, Cayne's one constant was his love of bridge. Now he wanted to do it full-time. By the time he was in his early thirties, he was winning bridge tournaments and earning the respect of seasoned pros as one of the best players on the circuit.

Cayne was now ready to take possibly the biggest risk in his life. He left Chicago for New York to indulge his dream and become a full-time professional bridge player. He became one of the superstars of the game—not just some guy who likes to sit around with his friends on Friday night but someone who wins the bridge equivalent of the World Series with a home run with two outs in the bottom of the ninth. He won tournaments and accolades in bridge trade publications. Bridge was his life, and through the game he met the woman who would soon become his second wife, Patricia Denner. She was an attractive, smart speech therapist and "bridge groupie," as Cayne put it, a winner of the Miss University of Pennsylvania beauty pageant, and she would soon earn her PhD. But Denner obviously didn't like bridge as much as Cayne did. Cayne wanted to marry her and she wanted to marry him, but under one condition: he would have to put bridge on the back burner and get a real job. Cayne met her halfway.

Cayne might not have graduated from college, but he was smart enough to connect the dots; he used his bridge connections to get business as a part-time municipal bond broker for Lebenthal & Co. and was soon looking to make some really big money. Before long, connecting those dots led him to Bear Stearns, the street-smart brokerage firm that was on the cusp of making it into the Wall Street big leagues, aka "the bulge bracket," where the likes of Goldman Sachs, Morgan Stanley, Lehman Brothers, Kidder Peabody, and increasingly Merrill Lynch dominated the trading and underwriting of stocks and bonds.

Bear was the perfect place for an upstart like Cayne. It had an entrepreneurial culture. Risk taking was rewarded, even if Greenberg was among the most cautious executives on the Street. Moreover, Bear Stearns was a firm where the top executives loved to play bridge—everyone from lowly bankers and traders to Greenberg himself.

Devout bridge players say that the game is like a narcotic; the more you play, the more you want to play. If you're good at it, you almost want to quit your job and do it full-time, as Cayne had. By now Greenberg was not just hooked on bridge, he was making many of the hiring decisions at Bear Stearns. Though Greenberg did not want to quit his job to play bridge, he wanted to bring more bridge players into the firm. The game, he believed, lent itself to what he viewed was calculated risk taking—weighing risk, trying to understand the odds of losing, and limiting your downside—which, unlike Cy Lewis's ultra-aggressive model of trading, was Greenberg's forte.

According to Cayne, he and Greenberg hit it off immediately, despite their personality differences. Cayne said Greenberg seemed impressed with the business Cayne had built up working part-time at Lebenthal, but not as much as he was with Cayne's bridge game. The way Cayne tells the story, Greenberg spent much of the interview talking not about Cayne's abilities as a broker but about his acumen at playing bridge. "How much time and money would I need to spend to be as good a player as you?" he asked at one point. Without missing a beat, Cayne replied, "You could spend the rest of your life and all your money, but you'll never be as good as me."

Greenberg liked what he saw and heard—a confident, smart man who wasn't afraid to speak his mind. The next day, he offered Cayne a job as a broker.

Almost immediately on hearing the offer, Cayne started having second thoughts. He wasn't sure he wanted to work full-time. He was still a bridge bum at heart. His job at Lebenthal had given him latitude to play bridge whenever he wanted and work when he wanted. Another reason: he was a bond broker, he had never sold stocks, and though he might have been a good salesman, he wasn't exactly a financial whiz. He had never even read a balance sheet.

But Greenberg believed he had the DNA to make money, and he wanted Cayne at Bear badly. At one point, he reached out to a common friend, Amos Kaminski, a thoughtful Israeli who worked as an analyst at Bear Stearns and played bridge with Cayne, and asked him to talk some sense into Cayne.

"What do you need to know about stocks?" Kaminski says he asked Cayne. "You pick up the phone and call people." It was that easy, he said. Cayne's wife put the pressure on as well. She saw her husband, a guy who had quit at just about everything, as a star in the making. He might not know stocks and he might not understand much about Wall Street, but he was smart and he knew people and he knew how to gamble. Jimmy Cayne could sell anything to anyone. That's how he'd won her over.

The next day, Cayne took the job.

If Cayne thought he could run Bear someday it was because, like any good salesman, he thought he could sell himself—in this case, to his bosses, Lewis and Greenberg, who were by now deeply divided on nearly every issue regarding the firm, including one of the most important: how much trading risk the firm should be taking on. Lewis wanted more, Greenberg wanted more control of it, and Cayne played them off each other to advance up the ranks. Many longtime Bear partners grimaced when they started to notice that Greenberg and Cayne often arrived at the office together. Cayne made it a morning ritual to drive "Greenie," as he began calling his boss, to work every day, in addition to their regular bridge nights. He used Greenberg as a shield when he butted heads with his colleagues and as an advocate when he intruded on colleagues' turf—including Lewis's. Cayne found himself in hot water when he poached one of the firm's biggest clients, Larry Tisch, who had done his trading with the Bear CEO.

But when Cayne butted heads with Greenberg he wasn't afraid to turn to Lewis, who certainly wasn't bashful about overruling his second in command any chance he got. One such example occurred in the mid-1970s, during the New York City fiscal crisis. It would be Cayne's first big score, the first time he'd do at Bear what he did every day at the bridge table: weigh the odds, take a gamble, and win. But before he could do that, he had to sell Greenberg on the trade.

New York City municipal bonds were held by big banks and wealthy individuals looking for the tax benefits (munis are free of city, state, and federal taxes). As the city began to run out of money, the New York State government passed a moratorium on the payment of interest and principal on the city's short-term debt. The move was later declared unconstitutional, but it didn't matter. The city's major bond buyers, the big banks, had simply had enough of the city's bloated budget and wild spending and boycotted the city's frequent sales of short-term notes to pay for essential services, everything from police and firemen to garbage pickup. The result was a near catastrophe—the city was shutting down. It began to lay off firemen and police. Garbage piled up in Queens. The subways became a haven for criminals. A full-blown fiscal crisis had begun.

New York City Mayor Abe Beame begged the federal government for a bailout. City bond prices sunk to depressed levels, around 30 cents on the dollar, a level that showed that the market believed the city would soon default on all its debt and fall into bankruptcy. No firm would trade the bonds of a bankrupt company or a municipal government except at severely depressed prices, so investors were in limbo—if they held on to the bonds, the bonds might fall into default, costing the investors all their principal, or they could trade them in and sell them for a fraction of the price they had paid.

Like any good trader, Cayne saw opportunity amid the chaos. Being a broker who frequently sold munis to his clients, Cayne soon discovered that no firm was willing to "make a market," or step in and buy city bonds from investors, even at those depressed prices.

Moreover, the city hadn't actually defaulted on its debt. Its creditors—the big banks—had simply stopped buying city bonds, triggering the fiscal crisis. And there was something else that made Cayne think that for all the city's problems, the market was overreacting. Though Cayne had grown up in Chicago, he was slowly but surely becoming a New Yorker, and being a New Yorker he knew some fundamental facts lost on those who live elsewhere. New York was the epicenter of art and culture. It had Broadway and, of course, Wall Street. There was simply too much wealth living in this one place for it to go bankrupt. In short, Cayne believed the city would make a comeback even as that famous headline ran in the *Daily News*—"Ford to City: Drop Dead," describing then president Gerald Ford's response to Beame's plea for a bailout.

New York City was experiencing what would later be known as a credit crunch, which could be broken by having a buyer willing to keep an inventory of city debt and take a chance that even at its bonds' depressed prices, the city wasn't ever going to default on them. That buyer, of course, would be sitting on a pile of cash if the city recovered as he was predicting.

"I couldn't believe that New York was going to just walk away and say 'Game over,'" he said.

Now he had to sell his idea to Bear management. He met with city officials to see if Bear could buy bonds on behalf of the city and hold them in its inventory before they could be sold. The city agreed, but officials at Bear didn't. They thought the plan was fraught with unnecessary risks. No other firm was buying New York City debt; why should Bear be the first?

"Because being the first has advantages," Cayne shot back, looking to get into what he believed was the deal of the lifetime before anyone else could come up with the same idea (Meriwether, it should be noted, made a similar bet). He went to Greenberg, thinking he might jump at the idea of buying up city debt on the cheap with the intention of selling it when the market improved, as it most certainly would. Cayne expected a quick affirmative. Instead he was rejected.

"Too risky," Greenberg responded.

Greenberg was known for his snap decisions, and Cayne wasn't about to beg. So he went above Greenberg's head to Lewis, who he knew would love the city bond idea (and reverse Greenberg), and he was right. With Lewis's blessing, Cayne placed an ad in the *New York Times* announcing that Bear was willing to buy New York City bonds. And investors responded; Cayne ended up buying more than tens of millions of dollars' worth of NYC debt over the next two years. There were some ups and downs to come as the city battled its massive budget deficit; the feds had rejected a complete bailout, but the state came through with a plan to put the city on a stronger financial footing—imposing limits on how much debt it could sell, forcing the city to balance its budget every year in exchange for state funds to plug the city's budget hole, and refinancing the city's costly short-term debt with lower-cost long-term bonds issued by a state agency known as the Municipal Assistance Corporation, or "Big Mac."

By the end of the decade, the city's finances began to strengthen, and Cayne's "risky" gamble turned out to be a wild success as the city's bond prices soared.

Cayne would often say that what he liked the most about Greenberg was that he never held a grudge, and he didn't in this case, even if Cayne had brownnosed his archenemy to get the deal done. In fact, their relationship continued to grow; Cayne was now a partner at Bear and a member of the executive committee, which approved the most important corporate decisions.

Greenberg began delegating more power and authority to his bridge buddy, especially over hiring decisions. Cayne might not have finished college, but he understood how to enhance his power better than most MBAs did, and he began hiring and promoting people who were loyal to him. Cayne's team soon began filling the hallways at Bear. They included people such as "Filthy" Phil Cohen, an old buddy from his days at Lebenthal. Cohen was one of Cayne's closest friends (they would later have a falling-out), and soon rumors spread through the firm about how much they loved to smoke pot and chase women—a lifestyle that Cayne took great pains to keep secret from Greenberg but people at the firm contend was on open display to just about everyone else.

Cohen recalls one such incident of Cayne's free-living lifestyle: Cayne called him to his forty-eighth-floor corner office with its great view of the East River in Lower Manhattan to discuss some firm business. After a couple minutes of small talk, Cohen says Cayne reached down into his desk and pulled out a blue Bromo Seltzer bottle. (Bromo Seltzer is a white powdery antacid.)

"What do you think's in here?" Cayne said, according to Cohen's recollection.

"Bromo Seltzer?" Cohen asked, slightly bewildered.

"No, it's filled with cocaine," Cayne said with a smile.

Cohen never checked to see if that was true, and Cayne in an interview says he has never done coke (he also called Cohen's account "patent bullshit"), but some of the stories about Cayne's partying did make their way back to Greenberg, including one time when a senior executive at the firm, William Montgoris, walked by Cayne's office, detected the scent of marijuana, and reported the incident. Greenberg asked Cayne if what Montgoris was saying was true, but Cayne attributed the marijuana smell to "a new leather couch in my office," and later invited Montgoris in for a whiff. "Does the couch smell like pot or not?" he asked Montgoris, who nervously said it did, and the matter was dropped. Montgoris, in an interview, wouldn't deny the account; Cayne, for his part, maintains the same position he had back then: that he didn't smoke pot in the office.

Greenberg, meanwhile, offers his own explanation. Though he doesn't recall being told about Cayne's alleged in-office pot smoking, he does recall an instance when he discovered Cayne smoking pot at a bridge tournament around that time.

But being busted for smoking pot by your bridge buddy has obvious advantages, and Greenberg could forgive Cayne even that. Cayne's future at the firm, with Greenberg in his pocket, was bright. And now it was about to get even brighter.

Each year, Bear's partners were invited to the annual Partners Dinner at the Harmonie Club in Manhattan. In 1978, the dinner was attended by about seventy-five people. After cocktails the various partners took their seats at their tables. Despite the air of opulence the Harmonie Club radiated, the menu was somewhat less regal: rubbery steak and overcooked vegetables. But obviously, the food wasn't the point. The point was to see and be seen and, if you were a new partner, to revel in your status as an owner in the firm.

As the dinner got under way, Cy Lewis made his way to the podium. In his speech, he reflected on his tenure at the firm, and he told the group that this would be his last annual dinner as CEO. He was ready to hand over the reins to Greenberg. He introduced Ace, noting what a great leader he was. Then, as Greenberg took the podium, Lewis sat down; his face suddenly turned ghostly white and sweat began to pour from his forehead. Within seconds, he keeled over in his seat, clutching his chest, barely able to get a word out, as his body began to convulse in pain. He fell to the floor.

"Someone get a doctor!" screamed one of the attendees. Lewis was suffering from what appeared to be a massive stroke. Medics were rushed in, and he was carried off. He died a few days later.

The crowd was shocked; Cayne couldn't believe what he was seeing. Then things got really bizarre. Greenberg was so impatient about getting the top job—and getting Lewis out of it at last—that even as they were wheeling Lewis out, he proclaimed that he was now firmly in charge of the firm.

"Eat, everyone," he ordered. "Dinner's being served, and Cy would have wanted you to all eat!"

It wasn't long before Jimmy Cayne became just as impatient to replace Greenberg as Ace had been to dethrone his own mentor. What started out as a partnership—Greenberg soon anointed Cayne president of the firm—became a rivalry and, at least to people at Bear, a race to replace Ace as soon as possible.

Jimmy doesn't know anything about finance. But he understands power: how to attain it and how to maintain it. He could teach Machiavelli a thing or two."

That was the assessment of a former Bear partner who witnessed Cayne's rise from salesman to partner firsthand. The partner worked in the firm's municipal bond department, where Cayne had made his name betting on the recovery of New York City. Because of that history, Cayne took a personal interest in the municipal bond dealings, much of which involved wooing public officials with large campaign contributions so they would appoint Bear as an underwriter on bond deals.

Cayne was in the middle of the action, which soon earned the name "pay to play" and drew the attention of regulators and prosecutors worried that government contracts were being exchanged for campaign cash. Bear Stearns was at one point ensnarled in an investigation by then–U.S. Attorney Rudolph Giuliani for providing campaign contributions to the state comptroller in charge of selecting underwriters for New York State debt. After some embarrassing headlines—there were press leaks about possible indictments—the case was dropped and Bear, with Cayne firmly in control, went back to work; winning deal after deal by strategically flooding the coffers of politicians with campaign contributions. Bear Stearns became the major underwriter not just for New York City and New York State but also New Jersey and, in the Midwest, Cayne's hometown of Chicago.

It seemed that Cayne could do almost nothing wrong. Indeed, his rise at Bear was now so fast that many of the old partners of the firm still couldn't believe it had happened. In the course of a decade, Cayne had moved from being a salesman who specialized in selling bonds and didn't understand a balance sheet to being the president of a major bulge-bracket bank, which in 1985 was among the first to become a public company.

Of course Cayne didn't get there on his political skills alone. He was a first-rate salesman; while Greenberg was running the risk committee, worrying about the positions on the firm's balance sheet, Cayne was selling the firm. He met with clients and continued to drive hiring decisions. What made him so good at helping the firm win municipal bond deals was that he had a real interest in politics. He had a genuine liking for politicians and soon developed friendships with those who had powerful positions in Washington overseeing banking and Wall Street issues. He was politically conservative, but he soon had a working relationship with New York liberals who had a say in the financial services business, such as Congressman Charles Rangel and another congressman who would someday become a senator from New York, Charles Schumer.

Maybe more than anything else, Cayne understood that to gain power, you need to win over important people, and from the moment he was made president that's what he set out to do, cultivating one relationship after another with the firm's rising stars, including one with a brilliant young mortgage bond trader named Warren Spector that would have profound implications for Cayne, Bear Stearns, and Wall Street for years to come.

'm looking at this bridge bulletin," Jimmy Cayne said, his voice raspy from too much cigar smoke and an occasional joint, "and it says you won an award for playing bridge and I don't know you."

It was the late 1980s. The guy he was speaking to was a bond trader named Warren Spector. In his seven years since joining Bear, Spector had barely spoken to Jimmy Cayne other than an occasional nod of the head when the two passed each other in the hallway. Cayne by now had a small coterie of friends and advisers. These relationships were based on business acumen and friendship. And now Cayne wanted Spector to join the club.

Bear Stearns was a wild place in the 1980s, particularly for those who hung around with Cayne. It was Wall Street's version of the Delta fraternity in the movie *Animal House*. The people there were outsiders, and they reveled in their outsider status, particularly when competing on deals. But more than that, working at Bear was like one big party, with Cayne as master of ceremonies. There were card games at Cayne's apartment on Park Avenue that lasted well into the morning. Company holiday parties were wild, with some executives disappearing into private rooms with a woman on each arm to snort lines of coke. And, of course, around this time there was Cayne's penchant for smoking pot.

Given Cayne's partying ways, it's easy to understand why he and Warren Spector, the pensive intellectual, had barely spoken. Spector had joined the firm in 1983 as a clerk on its bond desk, after receiving his MBA from the University of Chicago. Given Spector's pedigree, he could have worked anywhere. But he'd chosen Bear because he smelled opportunity—the opportunity to make a lot of money and, like Cayne, run the place someday.

The way Spector tells the story, he was initially befuddled the first time he met with Greenberg for what he described as a "twelve-second job interview" before he was ushered to see the heads of the bond department. But like any good trader, Spector understood the value of intuition and gut, and his gut immediately told him that Bear was a place where an ambitious young man could make a lot of money.

His gut also predicted the coming explosion in the mortgage bond market that Larry Fink and Lew Ranieri would take such advantage of. At Bear, on the other hand, in the early 1980s the mortgage bond department consisted of one trader and one salesman.

Spector wanted to get in on the ground floor in a business that through his training in economics he knew just had to explode. And as it did, so did Spector's career; he, like Ranieri and Fink, was a genius at understanding the complex nature of mortgage-backed bonds, including all the profitable new variations that had taken the market by storm.

In just a few years, Spector had become one of Bear's best traders, luring customers away from Salomon Brothers and First Boston, and was given more responsibility, including oversight of the firm's derivatives unit. He was making so much money that more senior partners ignored his annoying personal traits. Spector would have been more at home, it might seem, in John Meriwether's arbitrage group at Salomon, which cultivated smart, arrogant geeks, than at a firm like Bear, which prided itself on its street fighter culture, and seemed so at odds with a guy who went home after work, read the classics, and rarely cracked a smile.

But Spector's career flourished because at Bear the bottom line is the bottom line. Despite his initial twelve-second interview with Greenberg, by the mid- to late 1980s, Spector was meeting with the firm's CEO on a regular basis at Bear's risk committee meetings, where Spector, like other traders, would be grilled about positions and hedges. Cayne at the time, of course, would barely know a mortgage bond from a bologna sandwich. He rarely attended the risk committee meetings. But he knew how to read the firm's profit-and-loss statements, and he knew that a growing chunk of the firm's profits was coming from the work of this young bond trader whom people at the firm really didn't like.

"So," Cayne asked, "you used to be a bridge player?"

Spector told Cayne he had loved bridge when he was in high school, had won a number of tournaments, and had been considered a "life master" at the tender age of sixteen, and then, he said, "I quit."

But in Cayne's mind, no one quit bridge. "You're kidding me, right?"

Spector explained that he had quit bridge because he "didn't want to be a bridge bum."

"I wanted to go to college. Bridge is addictive, completely addictive," Spector would later explain, "and I wanted a normal life."

At Bear Stearns, there's nothing abnormal about being addicted to bridge. It didn't take long for Spector to get the message. After he'd been at the firm for five years, the mortgage department was cranking out half the profits at Bear Stearns; he himself had a trading book that earned $240 million annually. He was barely thirty, and his take-home salary and bonus amounted to a couple of million dollars a year. Yet he was destined to be a trader, maybe nothing more than the head of the derivatives unit, as he was then, unless he made nice with his boss, and that meant playing bridge.

So that's what he did with Cayne, Greenberg, and their bridge buddies, and just like that, Warren Spector became a man on the rise at Bear Stearns.

3. SEX, DRUGS, AND DEBT

I f the 1980s were like a big party for Wall Street, Jimmy Cayne was hardly the only executive who indulged in a little decadence; in fact, they all did. Along with an embrace of lower taxes and free-market capitalism, a new conservatism swept Middle America with the rise of religious fundamentalism and family values. President Ronald Reagan, hailed the times as "Morning in America," a wonderful way to spin the rapidly expanding economic prosperity and the country's rejection of the 1970s excesses of bad clothes, cheap drugs, and free love. But Wall Street took a different, more reckless turn. Cocaine use was rampant—on trading desks, inside corner offices, everywhere. Call girls made midday trips to offices all over Wall Street. On the floor of the New York Stock Exchange, strippers routinely gave lap dances to commemorate birthdays.

Wall Street of the 1980s was all about sex, expensive drugs, and debt.

Though the stock market made a miraculous comeback from the doldrums of the 1970s, much of the big money being made on Wall Street was in the bond markets—selling, underwriting, and, in the case of mortgage-backed debt, packaging what were essentially IOUs into new securities that investors never seemed to get tired of buying. Indeed, the growth of the debt market in the 1980s was nothing short of amazing: Between 1982, when interest rates reached their height, to 1987, the markets for various forms of debt—everything from government bonds to corporate debt and mortgage-backed securities—exploded.

Wall Street may be amoral, but it reacts to reality, and the reality of the era was simple: with declining interest rates and lower inflation, selling bonds was the least expensive way to raise capital for corporations. There was always a bias against issuing stock to raise money; investors complain that it's "dilutive"; the additional shares put downward pressure on the company's stock price since each share now represents a smaller stake in the company. In the 1980s, issuing stock wasn't just dilutive but also more expensive than selling bonds.

As debt became cheaper, as companies and investors became more comfortable owning IOUs, the Street figured out new ways to capitalize on this sea change in sentiment. Mortgage-backed bonds such as CMOs were just the beginning. Among the biggest innovations on Wall Street during the 1980s were derivatives of the original mortgage-backed securities called "IOs" and "POs," which respectively parceled out the "interest-only" portion of the bond and the "principal-only" portion, thus offering investors various alternatives to the lower yielding plain-vanilla government Treasury bonds or low-risk munis that had become almost passé as yields began to fall and returns stalled.

IOs and POs were not marketed just to investors; they became the favorite of traders because they allowed them to make even bigger bets on the direction of the market. The strips were among the most volatile securities in the mortgage bond arena because they isolated specific parts of the CMO; thus they fluctuated widely with changes in interest rates, so gains made by betting right were more pronounced—as were losses when a trader bet wrong.

Through the early years of the 1980s, wrong-way bets seemed to become a thing of the past. The men (and occasional woman) on Wall Street who created and peddled mortgage debt were now among Wall Street's highest-paid executives. And it wasn't just Larry Fink and Lew Ranieri earning the big bucks. Mortgage traders all over Wall Street were earning $2 million, $3 million, $4 million a year, fortunes that took most corporate executives a lifetime to amass. The merchants of debt were now fully in control at just about every firm. The image of the stockbroker selling shares of IBM to a family looking to participate in the great bull market might have looked good in television ads, but in reality, the money at the big firms was being made in risk taking, and leverage involved debt. By the mid-1980s, debt traders controlled the senior positions at nearly every major firm, except Merrill Lynch, the old-line brokerage firm, which was now desperately trying to expand into the bond market.

Nowhere did the bond traders hold more power than at a once-staid investment firm located in New York City, where most of the money was being made on a trading desk located in Beverly Hills, California.

For most of its fifty years in existence, Drexel Burnham Lambert had never mattered much on Wall Street. It had a strong but hardly dominant commodities business and a solid brokerage arm that specialized in selling stock to rich people. But thanks to the ingenuity of one of its up-and-coming traders, Michael Milken, who led the firm's junk bond department in Beverly Hills, Drexel was no longer mediocre. The firm, in fact, had become the most profitable on the Street because Milken understood the changing attitudes of Wall Street and corporate America when it came to debt possibly better than anyone else in the world. Milken's specialty was the issuance of debt that was rated below investment grade, known interchangeably as "high-yield" or "junk" bonds.

"Investment-grade" bonds are those deemed by the rating agencies to have a low risk for default, meaning they will in all likelihood continue to pay interest until maturity, when the holders of the bond will get back what they originally lent, which is known as the principal. When a rating falls below investment grade, the risk grows significantly that the principal will disappear before the bond's maturity date, hence the sobriquet junk.

But it wasn't junk as far as Milken was concerned. He produced an analysis that purported to show that the chance of companies rated below investment grade ever defaulting on their bonds was exaggerated, and tremendously so. Investors, he argued, were missing a massive opportunity to earn higher returns on debt that for the most part was safe.

Milken and his followers described the market they were interested in growing as a noble mission, capitalism at its best. These were companies with products and employees that were discriminated against because they were new to the markets. Because they couldn't sell bonds, they had to resort to costly bank loans or dilutive stock issuances if they wanted capital to finance their growth.

These were, of course, many of the same arguments used to justify the mortgage and securitized bond market—debt wasn't something to fear but something to embrace because it helped spread wealth. By the mid-1980s, Milken's noble mission was in full bloom as investors got increasingly comfortable with junk as something that wasn't junk but something that produced an annual return of 12 percent. The market for high-yield debt grew to enormous proportions: to $187 billion outstanding in 1987 from a mere $7 billion ten years earlier. And soon the rest of the Street—Merrill Lynch, Bear Stearns, Salomon Brothers, First Boston—all wanted a piece of the action that Milken and Drexel had perfected. And perfected it they had: by the end of the decade, Milken was worth $1 billion; in 1987 alone he earned an incredible $500 million.

The growth of the junk bond market, like that of the mortgage market, was justified by the Street as a win-win for both Main Street and Wall Street. But the history of Wall Street shows that there are no real noble missions, just opportunities to exploit social or economic conditions before reality sets in. The reality of the mid-1980s was that the noble mission of the mortgage bond had changed; deals became ever riskier as leverage reached new and unsettling levels.

Likewise, the junk bond market grew to new and absurd proportions. Wall Street took the innovation to the next level, with Drexel and Milken leading the way. In almost a 180-degree turn from the original intention, instead of creating and growing businesses, junk was now being used to take them over and then dismember them in order to pay off the high-cost debt used in the takeover, a process known as a leveraged buyout, or LBO. Because debt was so cheap, savvy takeover artists such as the billionaire Henry Kravis, Ron Perelman, Carl Icahn, and others could take over almost any company by issuing bonds and offering shareholders a huge premium over current market prices. One by one, some of the nation's biggest companies—TWA, Beatrice Companies, Revco Drug Stores, Safeway, Federated Department Stores, and the granddaddy of them all, RJR Nabisco—were swooped up in the LBO frenzy, their management displaced and control changed from public investors to the men who could secure the most debt.

The LBO artists argued that the debt that was added to an acquired company's balance sheet was actually good for the company and good for the overall economy because it forced downsizing and corporate efficiencies. Wall Street earned massive fees for underwriting the bonds, and average Americans didn't seem to care since the strong economy allowed the companies, at least for the moment, to pay down their postacquisition debt loads.

Everyone was happy except Theodore Forstmann.

What these guys are doing is fucking crazy!" Teddy Forstmann boomed as he sat at his desk in the mid-1980s and read about the latest junk bond–driven exploits performed by his archrival Henry Kravis, with the assistance of Drexel.

Anyone who knew Teddy Forstmann knew he was about as close to an idealist as you will find on Wall Street. He was also enormously wealthy. His takeover firm, Forstmann Little & Co., had been among the most successful of the past decade. Because he was so rich, he had all the accoutrements of the type of wealth that spread across Wall Street during the stock market boom of the 1980s: several homes, a private jet, a condo on the Upper East Side of Manhattan overlooking the East River. He also liked to collect art, so he had a lot of it, in his homes and at his office in Midtown.

Tall and ruggedly handsome, Forstmann looked and behaved like an actor from the 1940s. Donald Trump once described him as a "man's man." In other words, people who knew Forstmann understood he was a tough son of a bitch; not a guy you wanted to mess with.

But Forstmann was also naive, in much the way leaders of nascent political movements often are naive. He believed in the righteousness of his cause, and he couldn't understand why others didn't. And he famously picked fights he couldn't win.

On this day Forstmann was picking a fight with Wall Street itself. He was being outbid on deals because his competitors armed with junk bonds were changing the face of corporate America. KKR was willing to pay three times what he was willing to pay for companies because Henry Kravis knew Drexel could off-load the junk bonds to investors who had no real idea what they were buying. The deals made no economic sense, but more, they were also setting up the entire country for a massive fall; one small economic downturn, and the debt-laden companies would fall into bankruptcy. And because of the size and scope of the deal making in recent years, the entire economy would fall into chaos. Forstmann had no use for Drexel, the devil incarnate of business as far as he was concerned. He had even less use for those making money off the LBO boom, including Kravis, his archrival, whom he believed was getting rich off what amounted to a scam of the public.

But he was shattered when he saw firms such as Salomon, Morgan Stanley, and Goldman Sachs enabling the madness by financing deals that would implode when it became clear to everyone that junk bonds weren't called junk bonds for nothing. As Forstmann put it, they were nothing more than funny money.

Aside from a very occasional piece in the media, no one seemed to listen. Forstmann's friends at the power firms of Wall Street ran the other way when they saw him at some party because they knew what was coming—a terribly boring speech about funny money. The avatars of high finance who had once scoffed at lowly Drexel were now copying its business model.

So Forstmann decided to make his case, up front and in person, to the CEOs of the firms themselves. He set up meetings with people he knew best—the CEOs and lead partners of Goldman Sachs, Morgan Stanley, and Salomon Brothers: John Weinberg, Parker Gilbert, and John Gutfreund, respectively—to get them to reconsider their foray into junk.

They were all friends; in fact, Forstmann was particularly fond of Gutfreund. But they had also become enablers as they began to copy the Drexel/Milken business model, teaming up with takeover artists like Kravis and taking part in what Forstmann believed was one of the most destructive activities in financial history.

One of the obstacles Forstmann faced, of course, was that Wall Street looked at the massive levels of debt and leverage as its gravy train, whereas he saw pending doom. Some of his friends believed his doom-and-gloom forecast was nothing more than sour grapes; after all, he had been beaten to the punch by Kravis on deal after deal because of the latter's strong working relationship with Milken and Drexel. Perhaps Forstmann saw himself losing the takeover battle and so wanted to discredit the entire business.

But Forstmann didn't so much lose the battle as take himself out of the game. He had been walking away from deals not because he couldn't find a willing bank or investment firm to float a massive junk bond issue to make the numbers work but because he understood economic cycles and he could add numbers. Massive levels of debt and leverage make economic cycles more pronounced on the way up and more painful on the way down, when companies can't pay the bills their balance sheets were never designed to take on in the first place.

Forstmann's nuclear winter scenario went something like this: bankruptcies will spread across corporate America, and the biggest companies this country has ever produced—manufacturers, retailers, drugstore chains—that have been caught up in the takeover craze will begin to implode. Junk had become widely accepted as a legitimate investment even when it was financing what Forstmann believed were illegitimate deals. Pension funds, banks, S&Ls, and individuals were holding these securities, which, if he were right, would turn out to be worthless. The ripple of a slowdown in the real economy would thus be magnified into an earthquake of economic chaos.

That's the scenario that Forstmann laid out to the CEOs of Salomon Brothers, Goldman Sachs, and Morgan Stanley. But as he immediately discovered not one of them was going to abandon such a profitable business. Weinberg remarked that Goldman's involvement in junk was merely "defensive," while Gilbert said he would "look into" Forstmann's concerns. "Yeah right," Forstmann thought.

Gutfreund appeared to be the most honest, but that didn't make Forstmann feel any better. "We're doing it," Forstmann recalls him saying, "because we're doing a lot of business here."

To be sure, the money being made in the junk bond market had added countless billions of dollars to the bottom lines of the big Wall Street firms, as had the sale of loans and debt to various Latin American countries and the packaging and selling of mortgage bonds. Forstmann would continue his quest to warn the world about funny money; in 1988 he would lose the epic battle to take over RJR Nabisco—then one of the country's most profitable companies—after a long, bizarre battle among various competing interests, including one team headed by current management and the megabank Shearson Lehman Hutton, a battle recounted in the book *Barbarians at the Gate*. In the end, however, the deal went to KKR with junk bond financing from Drexel.

The year 1987 was the last hurrah of an era. Before long, most of what Forstmann had predicted came true. Drexel came under scrutiny for alleged market manipulation by its fearless leader, Michael Milken. The '87 stock market crash and the looming economic slowdown brought the junk bond craze to a close. The second coming of the Great Depression didn't occur, but businesses loaded with high-yield debt suffered mightily. Most of the LBO targets couldn't pay their bills, and as "funny money" lost its luster among investors, they couldn't refinance their debt.

Dozens of once-profitable outfits—Federated Department Stores, Revco Drug Stores, and others—filed for bankruptcy and began laying off employees and shutting down plants. Investors took huge losses as well. Savings and loans that had been allowed to invest deposits in junk took some of the biggest hits. A wave of bank failures followed, and Wall Street, the financier of the whole mess, began to crumble as well with massive losses and layoffs; just as Forstmann had predicted, the investment firms were forced to finance their own deals, holding on to the debt and the "bridge loans" used to make the deals work until they could find buyers, which, as the markets began to unravel, never came.

"People forgot this was stuff you had to pay back," Forstmann said recently. "And Wall Street would never tell the buyers how difficult it was to pay it all back. You can call this a 'sellout' or say that Wall Street sold out its principles in the 1980s and later on to make a buck. I would tell you as someone who knows these guys, they never had principles to sell out in the first place."

In 1984 the federal government nationalized Continental Illinois, a large bank considered too big to fail because it held deposits of average Americans. No such designation was granted to Drexel Burnham Lambert. By the end of the decade, Drexel had to be liquidated as the government stood ready to indict the firm for fraud, but the debt crisis was spreading. Big banks such as Citibank felt the sting as well; the leverage on the banks' balance sheets included not just bridge loans for LBO deals but loans to Latin American countries that were now falling into default. John Reed, the new CEO of Citibank, was forced to set aside $3 billion, at that time equal to three years of profits, to cover the losses as the firm's share price fell to around $8 and one of the pillars of capitalism appeared ready to implode. Even the RJR Nabisco deal, the largest buyout in history, appeared to be on life support: the company, a former cash cow, which made its money selling products that hold up in even the worst economic times—cookies and cigarettes—was now struggling to repay its bills.

Debt had brought Wall Street nearly to its knees. Like all bubbles, the junk bond bubble of the 1980s had started out as a noble quest. But then, the madness of crowds set in: the good times will never end, people began to think, so the ability to borrow more money at more outrageous terms got easier as the money made went to financing increasingly risky deals. It's a vicious circle: the more money is made from each incrementally risky deal, the more people want to believe the madness won't end. Teddy Forstmann was exactly right: funny money was running Wall Street, and the smartest guys in finance were fooled into thinking the party would last forever.

The funny money of the junk bond market overshadowed, to an extent, the bubble-making funny money being churned out on the mortgage desks of the big firms. Bubbles often burst when markets become focused on some event that is big enough, meaningful enough, to draw attention to the insanity. The junk bubble of the 1980s burst after the 1988 RJR Nabisco deal—a $25 billion deal, then the biggest takeover ever, a milestone that would not be passed until 2006, during a new era of boundless debt that would eventually come crashing down—followed by the demise of Drexel Burnham Lambert the following year.

In the mortgage market, the bubble began to burst when the investors who had suspended disbelief for so long, convinced that CMOs, IOs, and POs were risk free, began focusing on the travails of a man named Howie Rubin.

Rubin, like Larry Fink, John Meriwether, and Lew Ranieri, was particularly good at odds making. He was a trader who had worked at Salomon for seven years and was known as a guy who could turn research from Ranieri's computer geeks into trading gold. He was particularly interested in any and all research on the mortgage prepayment habits of home owners. Mortgage bonds are particularly sensitive to changes in mortgage interest rates.

In one scenario, when interest rates fall, home owners tend to prepay or refinance their mortgages at lower rates. When that happens, the home-owner is saving money, of course, but the bondholder is losing money. Mortgage bonds start to fall because in such cases, the bonds could be "called," or paid off, as the stream of mortgage payments will come to an end. The bondholder essentially gets his principal back, but he's deprived of the higher interest rate payments that would have accrued for the life of the security and must now reinvest in a lower-rate environment.

Being able to sort through this convoluted market and sort out which bonds would end up being prepaid from those that wouldn't was one of the mortgage market's biggest dilemmas in the 1980s, and Rubin had it down to a science, or so it seemed. These predictions became even more important with the innovations of mortgage bonds that were further stripped into portions that paid just interest, the IO, and a portion that paid just principal, the PO. The creation of IOs and POs made predicting prepayments an even bigger headache for traders and potentially even more lucrative when gotten right. If mortgage rates are rising, home owners are less likely to prepay their mortgages, so a bondholder would want to hold on to the interest-only portion, the IO, and earn those high rates of interest over the life of the bond.

Likewise, if rates decline, prepayments will accelerate, and investors will receive their full principal, or "par value," faster. Since principal-only bonds are issued at a substantial discount below par—the amount of money that investors will receive when the bonds mature—the value of the PO will spike. Investors who bought POs before rates declined will receive their full principal at an accelerated pace as home owners prepay, causing the bonds to be called in at par, and the investors get to pocket the difference between the lower purchase price and the full value of the security.

All of this required fairly complicated analysis; researchers at Salomon and First Boston combed through reams of data to figure out which batches of home owners were more likely to prepay their mortgages and which weren't under various interest rate scenarios. In fact, it was Ranieri who spread the word to his customers (and thus to his competitors as well) that Rubin was among the best in the business at understanding this data. There was just one problem, as far as Howie Rubin was concerned: he wasn't being paid like the best. The most Rubin made at Salomon was $300,000, in 1984—at a time the firm was making hundreds of millions of dollars packaging, selling, and trading the various types of mortgage debt—and Ranieri was earning around $5 million a year. Rubin felt he was leaving money on the table—and he was. Merrill Lynch, desperate to compete with Salomon and First Boston, wooed him with a guarantee of $1 million at the age of thirty-four. Like any good trader, Rubin hit the bid.

With Rubin on its trading desk, Merrill began to make a serious run at Salomon and First Boston's dominance of the mortgage market. On Wall Street, trading and underwriting are intrinsically linked. Investors will buy bonds from the Wall Street house most willing to make a market or stand ready to buy, sell, and trade the securities once the deal is sold. That's what Merrill, with the benefit of Rubin's expertise, began doing.

Playing the odds and taking risks was paying off. Having such an active trading desk meant that Merrill won several key underwriting mandates; during the first quarter of 1987, the firm was in striking distance of Salomon and First Boston in the all-important league tables that gauged underwriting activity.

But in time Rubin's luck would run out. Interest rates during the first part of the 1980s remained, in general, remarkably stable. A general consensus had begun to spread among investors who traded long-term Treasury bonds (the ones most sensitive to inflation and upon which many mortgage rates are pegged) that inflation was tamed and, more than that, economic growth had brought the nation's budget deficit, compared to the growing gross domestic product, under control. They began snapping up

30-year Treasury bonds, the most inflation-sensitive of all government debt, which caused interest rates to fall dramatically (with lower inflation and a healthy budget, long-term bonds are safer and the government needs to pay less, in the form of interest, to induce people to buy them). Because mortgage rates are pegged to the rate of 30-year Treasuries, mortgage rates fell as well, causing prepayments to increase as home owners refinanced at the lower rates.

Rubin seemed to have played the market perfectly; he held POs from a deal that Merrill had brought to market. Shortly thereafter, he doubled down on his trade. Remember, POs are profitable if interest rates decline, or at least stay stable, and Merrill was now holding close to $2 billion of the highly volatile PO mortgage strips.

There was just one problem: rates began to rise; in fact, they soon began to rise sharply and aggressively as the Fed, fearing a pickup in inflation (caused by the booming economy, the surging stock market, and the wealth-creating effect of both), started to raise short-term interest rates. Merrill's losses on its huge and volatile PO portfolio began to mount—massively. Rubin suddenly seemed fallible, and worse. When the firm's compliance staff finally sifted through the debris, it discovered that his trades had cost the firm a whopping $250 million. A week later, the losses grew to $275 million. When all the losses were finally counted, the final tally was an incredible $377 million.

Merrill's management was shocked, shocked! that such gambling was taking place inside the halls of the firm whose "thundering herd" of brokers had "brought Wall Street to Main Street," attracting more individual investors than any other brokerage on the planet. They described Rubin to the press as a cowboy and a lone wolf, even though Rubin's hiring and his trading style had been lauded by senior executives at the time he was brought on. Rubin was eventually fired, investigated by regulators, and forced to take a leave of absence from the business. His supervisors were fired as well.

Merrill's CEO, William Schreyer, and president, Daniel Tully, claimed they'd had no idea that so much risk would be taken in such a small corner of the firm's balance sheet. Both were longtime stockbrokers who knew very little of the booming debt markets and the alleged brain surgery that was taking place on the trading desk.

Just weeks earlier, Schreyer, speaking at the firm's annual meeting, had said nothing about trading losses. Before he was briefed on the issue, he hadn't known who Howie Rubin was. But now the losses mounted and analysts questioned his competence: how could a CEO and his senior staff, including Tully, his heir apparent, have been so totally in the dark about something so important to the firm, namely the amount of risk taken with the firm's balance sheet?

Schreyer and Tully soon became Wall Street punch lines. Investors called them "Bartles & Jaymes"—two old guys who were out of touch with the world of risk and leverage that was all around them, just like the over-the-hill actors appearing in the commercials for the once-popular wine cooler. They didn't take the criticism lightly. More heads began to roll. Risk taking was out, and leverage was dramatically reduced. The firm was getting back to basics, focusing on its core business of selling securities to small investors through its brokerage network, the largest on Wall Street. "The ship," a senior official at Merrill told the *New York Times*, "will sail closer to shore." Tully, in an unusual move, took an office near the trading floor to be closer to the trading action and to make sure Rubin's type of trading would never again take place at Merrill, but in a business addicted to risk, it would return, bigger and more deadly than ever.

4. AN EDUCATION IN RISK

Sometimes our technology in creating these securities outpaces our ability to cope with them." That's what Larry Fink told the *New York Times* in May 1987 when asked about Howie Rubin's trading disaster. In the past, Fink would have made that statement to a reporter and then celebrated with his team the fact that one of his competitors, particularly one like Merrill Lynch, which he saw as a pesky upstart in the field he aimed to dominate, was now being nailed with massive losses.

But Fink wasn't celebrating, because, much like Howie Rubin, he had just gotten his first real education in risk. And like Rubin, Fink and his team had thought they had it all figured out. In the second quarter of 1986, Fink had made an opposite bet from Rubin—all of his research pointed to higher interest rates and a reduction of prepayments from mortgage holders, and he was having one of his best runs. With that in mind, Fink began gambling even more with esoteric CMO strips, and, like Rubin, he would end up losing big.

Unlike in Rubin's case, the full extent of the losses wasn't immediately apparent. When the Rubin fiasco became public, Fink and his team thought they had their position hedged—meaning that they utilized a classic technique to reduce losses by making an offsetting trade, thus minimizing the losses that had beset Rubin.

But, as Fink was about to discover, his hedges were losers as well. His former boss Tom Kirch now worked for Fink and was attending an executive committee conference in Carmel, California, enjoying the beautiful northern California weather in the quaint town where Clint Eastwood was mayor.

During a break, Kirch went back to his hotel room and received an urgent message to call the First Boston trading desk in New York immediately. By now the Howie Rubin fiasco at Merrill was huge news; almost every other firm had taken massive losses as well. Fink figured that he was in the hole about $30 million, a manageable number given the countless millions he and his mortgage traders had made for the firm over the past six years.

But Wall Street has a short memory, and the bosses back home were fuming because the losses were bigger than anyone had expected. Apparently there was a glitch in Fink's vaunted computer system, the one set up specifically to crank out data for the First Boston mortgage traders, providing historical data for trades and keeping a record of gains and losses on those bets. In this case, the computers had miscounted the losses—by a lot; "they are bigger, much bigger than we thought," the trader in New York said on the phone.

Kirch felt a knot develop in his gut. "Okay, I'll tell Larry," he said. Fink, who had been travelling, was now back in New York when Kirch gave him the report that their positions had blown up worse than they'd first believed. The losses were close to $100 million, possibly more.

"No fucking way" was Fink's response as he called his team, desperately demanding answers. He got the same report: not only were the trades a bust, but the hedges had failed as well. Even worse, the evidence was clear that Fink's research was wrong; it had miscalculated the direction of the interest rates that were at the heart of Fink's losing trades and faulty hedges.

The old Larry Fink would have called it being "whip chained." But the old Fink wouldn't have lost so much money. He was in fact being whip-sawed, losing money on both sides of his bet, and he couldn't believe it, screaming at one point, "This can't be happening!"

As his staff briefed him on what had gone wrong—the failed hedges, the size of his position—Fink began to realize it *was* happening: the end of his world. Still just in his midthirties, Fink had been a rock star. His desk on First Boston's massive trading floor in midtown Manhattan had been the place to see and be seen. With his wild success, he had been on the fast track to run First Boston—just weeks earlier rumors had been spreading that he and mergers chief Bruce Wasserstein were about to be named co-CEO. The chatter had picked up steam during the first quarter of 1986, when Fink's mortgage group had made a whopping $130 million—its best quarter ever.

But just like that, in little more than the blink of an eye, Fink lost $100 million in one quarter—his worst ever. The firm was bleeding red ink, losing millions, and, combined with other losses, in need of an eventual bailout from the Swiss bank Credit Suisse, which increased its investment in First Boston to a controlling stake. With that came bickering, finger-pointing, and management changes. And the end of Larry Fink's career at First Boston.

In the weeks and months that followed, Fink kept going through it in his mind as he sat at his desk, staring blankly at the numbers on his computer screen. "How the fuck did it happen? How did I guess so wrong?" He couldn't come up with a reason, nor could he come up with an answer as to why he had made so much money in the past. Had it been luck or skill? He'd used to think it was skill—that his knowledge of math and computers and his team of brilliant minds, including Ben Golub, the MIT financial whiz who could crank out differential equations in his sleep, gave him an insight into how complex new bonds and derivatives worked and an edge in understanding odds and beating the markets.

When Fink was on his way up, his ego had grown to massive proportions—being right for so long tends to have that effect. He loved to deride the competition, the risks they took over at Salomon. His archnemesis, Lew Ranieri, was a "cowboy" who didn't have a clue about risk, leveraging up and rolling the dice. Fink had always believed Ranieri was on the verge of a mammoth loss that would put him out of business. But now it appeared that Fink was every bit of a cowboy as Lew Ranieri or Howie Rubin.

At least that's what had become readily apparent as the dust settled on his trades, showing that Fink had been trading in Rubin–like amounts—billions of dollars in volatile mortgage securities. Fink had thought he knew everything there was to know about the mortgage bond—how to trade it, how to structure it to meet the tastes of investors, and, most of all, how to gauge its risk. He had the best risk guy in the business, Ben Golub, and the best computer programs of the time. Golub, a soft-spoken MIT grad, had taken great pains over the past year to make sure the mortgage department had special access to First Boston's mainframe computer. In the past, the mortgage unit had had to share the mainframe and its measurement system with other departments. Golub had cut a special deal with First Boston's head of information technology, explaining to him that the hundreds of millions of dollars the firm had made in the mortgage market could be lost in an instant unless the mortgage department had access to the mainframe on an as-needed basis. IT had agreed, and First Boston was considered to be on the cutting edge in employing technology to gauge the markets and predict its future course. But it wasn't enough.

Fink immediately ordered a postmortem "so this will never happen again," as he told his group. Golub led the effort. His conclusion: as good as the computer systems were, they were obsolete compared to the size and scope of the market. The system used by First Boston had been built to gauge and analyze the mortgage market of the early 1980s, not the trillion-dollar market it was now. It just couldn't produce data on a timely basis about prepayment rates and all the other variables that went into trading analysis. If First Boston was going to defy the odds, it needed a better odds-making system.

Fink had lost considerable standing in the firm, but he still had enough power to order a massive overhaul of the firm's risk assessment system. He assured his supervisors at First Boston that such trading losses would never happen again. But no one seemed concerned about the future. They were worried about the present, as the market for mortgage-backed bonds simply dried up; clients who previously couldn't get enough of them had suddenly had enough. There was almost no price at which Fink could unload his positions. First Boston was stuck, and so was he.

Fink had received not just his first real education in risk but also a lesson in how Wall Street really worked. Computers and mathematical models are only as good as the data that are fed into them; what they don't measure is the unforeseen—sudden and dramatic spikes or dips in interest rates and the once-in-a-lifetime gyrations that the oddsmakers tell us to ignore but history teaches us occur all the time.

Nearly a decade later, the author and mathematician Nassim Taleb would describe what had happened to Fink in his book *The Black Swan: The Impact of the Highly Improbable*. In it Taleb explains why some of the most intelligent people on the planet fail to understand and appreciate the randomness of life as they cling to their statistics and computer programs, which were designed by the very same people. "Alas, we are not manufac-tured, in our current edition of the human race, to understand abstract matters—we need context," Taleb wrote. "Randomness and uncertainty are abstractions. We respect what has happened, ignoring what could have happened. In other words, we are naturally shallow and superficial—and we do not know it. This is not a psychological problem; it comes from the main property of information. The dark side of the moon is harder to see; beaming light on it costs energy. In the same way, beaming light on the unseen is costly in both computational and mental effort."

Randomness was a lesson that Wall Street as a whole would fail to understand over the next two decades—only this time the losses would no longer be measured in "just" hundreds of millions of dollars.

Fink, of course, didn't have the benefit of Taleb's writings, and he was now paying the price. Colleagues who had used to kiss Fink's ass now regarded him as a pariah, the guy who had cost the firm millions and slashed the size of their year-end bonuses, the millions of dollars traders earn at the end of each year that are the payoff for their long hours sitting in front of computer screens. As soon as the bonus spigot was cut off, Fink was cut off from the rest of the firm. The cold shoulder began almost the moment news of the trading losses disseminated widely throughout the firm. Senior executives simply looked the other way when they saw him in the hall, Fink noticed, including Bruce Wasserstein, who had been somewhat of a mentor.

Others treated him worse. Like Fink, Joseph Perella, the flamboyant co-head of mergers at First Boston, often complained about the ruling WASP hierarchy on the Street. But if Fink was looking for Perella's shoulder to cry on, he was mistaken. At least Wasserstein still spoke to him on occasion. Perella acted as if he didn't exist.

Behind the scenes the executive committee wanted Fink out and his trading group dismembered. First Boston was too valuable an enterprise to be left to a bunch of gunslingers. It didn't matter how much money Fink had made in the past. On Wall Street you're only as good as your last trade.

Fink had worked at First Boston for a decade, through good times and bad. He had seen careers upended with one bad move—a lousy trade or a missed deal. He'd just never thought it would happen to him. His chances of running First Boston were over; his main goal now was to survive until he could either go on another winning streak or find another job. But First Boston wasn't the only firm that was reeling.

What Pat Dunlavy loved about working at Salomon Brothers, particularly near the trading desk, was the camaraderie. It reminded him of a frat house, albeit an expensive one, where the frat boys, and the occasional woman, used the firm's expense account to take limos, drink expensive wine, and order lavish meals, much of it for clients but also on occasion to have fun on their own, ringing up such huge bills that Tom Strauss, the firm's number two, was in such a constant state of anxiety that he asked Dunlavy to head an expense committee to keep tabs on the largesse. But it was a frat house nonetheless, with cursing, swearing, and constant sexual and practical jokes (like Mike Mortara's prank call to First Boston concerning the league tables). Things got so crazy that John Gutfreund and Lew Ranieri once ordered the investigation of a managing director who had allegedly received oral sex in one of the conference rooms. But there was also a sense of shared sacrifice. The traders, salesmen, and bankers in the mortgage department were in a constant daily battle, and Ranieri was their platoon leader, Gutfreund the general. Their objective: to defeat the competition and make a lot of money.

The culture of Salomon Brothers was "us against them," which was fine with an old jock like Dunlavy. The firm wasn't quite white shoe; no one would ever confuse a Salomon trader with one of those stiffs at Goldman or Morgan Stanley. And the firm wasn't as "Street" as, say, Bear Stearns or Drexel, places where they pushed the limits by underwriting the crappiest, most speculative bond deals and selling them to investors as if they were solid gold. Salomon Brothers had standards, but it wasn't afraid to take risks, albeit calculated ones, and to fight for business. That's what made it so special; indeed, Dunlavy believed it was unique. For all the infighting among the various partners—the competition between Ranieri and Meriwether over capital, compensation, and power within the firm and the brawls over expenses—they were still a team: the Big Red Machine, the 1977 Yankees, and the 1986 New York Mets wrapped up in one, as it were. Once the game started, egos were put aside for the greater good.

That had been the culture of Salomon Brothers when Dunlavy had arrived at the New York office in the early 1980s, but with each year of the bull market it seemed to dissipate, and now the teamwork was fading, replaced by ego and, to Dunlavy, something far worse: excessive risk. Clients? They were becoming an afterthought. The firm was a "sellout," he concluded, not much different from the rest of Wall Street; it had long since forgotten its original mission of assisting its clients through sound advice. The main objective was to make money, lots of it. And it got to the point, at least in Dunlavy's mind, that it didn't matter enough to many people at Salomon how it was made.

Dunlavy had come to this conclusion over time, but what made it hit home was a conversation he had with Mike Mortara. Mortara was Ranieri's head of trading, and he ran the trading desk like a fiefdom, copying Ranieri's management style. Like Ranieri, he expected absolute loyalty and teamwork. Dunlavy, meanwhile, was suspicious of all traders, but Mortara in particular. Part of the clash involved personalities; Dunlavy was a straightforward midwesterner; Mortara, a New York schmoozer nicknamed "Fat Ankles." Much of it, however, centered on the natural suspicions salesmen have about traders; the sales force is the front line in servicing large clients of the firm, the institutional investors who buy and sell stock and bonds. The traders, meanwhile, wouldn't think twice about stiffing clients if enough money was on the line. Dunlavy heard it firsthand: clients complaining that Salomon's trading desk was trading against their positions, buying bonds that they knew the client had wanted, which would make the trade both more expensive and more profitable for the trader, who would unload the inventory later at a higher price. Dunlavy believed Mortara knew all about these games, and though he had to work with Mortara, he could and would never trust him.

His suspicions about Mortara were reinforced in 1987, when the CMO crisis was sweeping just about every department on Wall Street. One afternoon, one of Dunlavy's guys was working on a fairly large bond sale to a longtime client. The mortgage market was getting hammered; that was the reason Howie Rubin had blown up and Larry Fink was being pushed out the door of First Boston. Yet the prices of the bond Salomon was holding barely reflected the market's travails. They were priced close to 100 cents on the dollar, as if the market were riding high, which it wasn't.

"Mike, what's the market for this stuff?" Dunlavy asked Mortara. Mortara confessed that the pricing wasn't exactly right—it was, in his words, "aggressive," meaning that the bonds were actually priced higher than what the firm could probably have gotten for them on the open market.

To Dunlavy, Mortara had just confessed that he was mismarking a fairly large position on the firm's balance sheet to avoid showing a loss until the market could recover.

Dunlavy believes to this day that Mortara's statement was a slip of the tongue; after he made it he appeared visibly nervous.

"Mike," Dunlavy asked Mortara, "what are you saying?"

Mortara knew he'd said something he shouldn't have and said that the pricing of the bonds was aggressive, not terribly so but aggressive nonetheless.

That was all Dunlavy needed to hear. He was floored, shocked, and bewildered. He'd always thought Mortara was a bullshit artist, but he hadn't figured on his being an out-and-out liar. "I'm telling you I have to tell Lewie, this is bad stuff!" he shot back. Mortara now began to sweat and stammer, trying to explain what he had done and why it wasn't so bad.

But Dunlavy knew it was.

Was what happened illegal? Dunlavy didn't care because he believed "it was an ethical matter." The pricing of mortgage securities was more of an art than a science; it depended on what two traders thought the price *could* be, often communicated over the telephone, rather than being standardized the way stocks are traded on the New York Stock Exchange. "What two traders thought the price could be" could mean just about anything. Even so, what Mortara had done was not just clearly unethical, it was potentially enormously problematic for the mortgage desk to be sitting on huge unreported losses—sooner or later it was going to become apparent that the bonds were not worth nearly what the firm's books said they were.

"We have to tell Lewie," Dunlavy stated once again. Mortara put down the massive stogie he had wedged between his teeth, leaned forward to Dunlavy, and whispered, "Why would you want to do that?"

"Mike, I give a shit about this place, and it's my responsibility to keep Lewie in the loop." Mortara just stared speechlessly at Dunlavy as he walked back to his desk.

Lew Ranieri's office was the size of a small bedroom, with a large mahogany desk, pictures of boats, including Ranieri's yacht, along the walls, and the noxious, ever-present odor of cigar smoke in the air. As Ranieri's success had grown, his hairline had started receding and he had begun smoking pipes along with the cigars. Maybe the biggest change was that he had begun spending more time in the office than he did on the trading floor, leaving Mortara to his own devices.

Aside from the smell of smoke, what stood out was that the office was neat—really neat—in stark contrast to the mortgage-trading desk, which was perpetually littered with pizza boxes and empty Chinese take-out cartons.

Dunlavy scheduled the meeting with Ranieri in advance because he wanted to have time to explain the Mortara situation without interruptions. Ranieri could have a volatile temper, but today he seemed in good spirits, which made Dunlavy feel more at ease given the fact that he was about to rat out one of Ranieri's closest advisers at the firm, someone Lew considered a friend.

Dunlavy, according to his recollection of the event, gave it to him straight: the mortgage trading desk was out of control. The firm was taking on too much risk, and one culprit, possibly the major culprit, was Mortara. Even worse, he believed Mortara might have mismarked the prices of a huge batch of mortgage securities the firm was holding. Those "marks," he said, were "aggressive."

Mortara had all but conceded that the pricing was too high, Dunlavy told Ranieri.

Dunlavy had told no one else about the Mortara matter, just Ranieri. He hadn't gone to Gutfreund, as others might have. If he had, heads would have rolled immediately. As far as Dunlavy was concerned, Mortara's transgression was a family issue and needed to be dealt with by the head of the family.

"I don't understand why you and Mike can't just get along," Ranieri shot back. "Mike knows what he's doing with trading, and you run the sales desk." Dunlavy nearly fell off his chair. Ranieri wasn't a guy who minced words, but those weren't the words Dunlavy was expecting.

"Say what, Lewie?" Dunlavy replied.

Ranieri asked again why he and Mortara were fighting and repeated that the mortgage department needed more teamwork. "Lew, that's not the issue," Dunlavy replied. "I don't trust that Michael is accurately reflecting the value of the book." In other words, he was turning losses into fake gains. (Ranieri says he doesn't recall the incident; Mortara has since died.)

"I'll talk to Mike," Ranieri finally said. The meeting ended and the issue appeared to fade away.

But within a few months, in the summer of 1987, the great Lew Ranieri, the father of the mortgage-backed market, was fired from Salomon Brothers.

Keep in mind that the mortgage business was among the most profitable areas of the firm and Ranieri had been underwriting and selling mortgage-backed securities since the late 1970s and early 1980s. Salomon basically owned the market along with Larry Fink's group at First Boston. The "spreads," or profit margins, on the early deals had been enormous; Dunlavy recalls that Salomon had made 50 percent profit on some deals.

So in running the biggest mortgage department on Wall Street, Ranieri was able to control all aspects of the business: underwriting, sales, trading, but the huge profits that came with them flowed to the firm's bottom line.

But it wasn't enough.

There have been several accounts as to how Ranieri was shown the door, but the one most repeated among Salomon alum (and chronicled in *Liar's Poker*) goes something like this: Gutfreund called Ranieri in for a private meeting, where in a few short minutes he told Ranieri, a man who had spent twenty years at Salomon, that he was out. Ranieri was shocked and devastated. For all the rockiness of the markets and the information about Mortara that Dunlavy had brought to his attention, Ranieri had never seen it coming. He was so blown away that he didn't even return to the office. There were no parties, no speeches. Just a terse memo from Gutfreund saying that Ranieri was gone.

The bigger question is *why* Ranieri was fired. Ranieri had often made the case that he was the primary reason Salomon was so profitable. That, of course, was somewhat of a stretch. Salomon was a bond house, so first and foremost, the chairman of the Federal Reserve deserves most of the credit. Paul Volcker raised interest rates in the late 1970s and early 1980s, so much that the bane of the bond markets—inflation—was significantly reduced for at least the next thirty years. Once inflation was tamed, bonds took off. Inflation erodes the value of fixed-income investments, but with the combination of an inflation rate that was under control and the now-lower interest rates (bond prices rise when interest rates fall), the great bond market rally continued into the mid-1980s and beyond.

But Ranieri had certainly made the most of a good situation. Salomon Brothers' profit margins were off the charts because Ranieri had had the foresight to see how a sliver of the market, the packaging together of pools of mortgages and then selling them as bonds, would take off. For that reason, no one could understand why a visionary who brought in so much money was dumped. Former Salomon Brothers executives explain it by pointing to the enormous paranoia of John Gutfreund. The famously tough CEO was also famously power-hungry. Some people inside Salomon say he allowed the place to degenerate into fiefdoms—Ranieri's mortgage desk, Meriwether's arbitrage group, and the government bond–trading operations—as a way to disperse power and maintain control: with everyone fighting one another for power, compensation, and access to the firm's capital, they had less time to fight with Gutfreund himself.

But Ranieri, unlike Meriwether, had become bigger, more powerful, and more influential inside the firm, given his role in the development of the firm's moneymaker, the mortgage bond department, which had trans-formed not just Salomon but all of Wall Street and the U.S. economy. It wasn't a stretch to attribute the burgeoning of wealth in the 1980s to the housing boom, and it wasn't a stretch to attribute the housing boom to the mortgage bond that Lew Ranieri helped create.

Shortly after Ranieri was ousted from Salomon, Mortara was out as well. Mortara was a little less reticent than Ranieri, chalking up his departure to "a pretty mutual decision" in one newspaper article. But Dunlavy got what he believes is the real story—not just about Mortara's ouster but Ranieri's as well—a few weeks later from Thomas Strauss, Gutfreund's number two. Strauss called him up to his office for a chat about the direction of the mortgage bond department now that it had lost its two most dominant executives and was losing money. In the wake of the firing of Ranieri and Mortara, Salomon had announced losses in the mortgage bond and municipal bond departments that totaled $100 million.

According to Dunlavy's recollection, Strauss began the conversation by asking about the mortgage business. Salomon Brothers had become a free-for-all where traders, obsessed with the fat bonuses that only excessive risk taking can produce, had put their own interests above clients'. Giving the straight skinny to clients became secondary to making money by sticking investors with some crap hidden deep in the firm's inventory. Traders spent so much time studying computer programs and prepayment schedules to trade for their own accounts that they brushed off trading for clients.

Even with all those problems, with Ranieri and Mortara gone, Dunlavy still believed that Salomon's mortgage department was the best in the busi-ness, so he put his concerns out on the table to Tom Strauss.

"Tom," he said, "I told Lewie that I learned that Michael and the boys were not always accurate with the marks."

Strauss, shaking his head, began to laugh. "Now you see why we had to do what we did to those guys," he said, adding, "Let's move forward." Dun-lavy agreed and changed the subject. (Through a spokesman, Tom Strauss said he wouldn't deny Dunlavy's account; Ranieri, through a spokesman, says he was never made aware of Dunlavy's concerns when he was asked to leave the firm, and as far as he was concerned, he was forced out for unrelated reasons.)

What made bond trading so profitable and also so dangerous was that good traders have to have the stomach to ride out big losses, until they turn out to be gains. Joe Giglio, a municipal bond executive at Bear Stearns, was learning that what made Ace Greenberg such an effective trader was that he wasn't willing to ride losses. He knew sometimes, much better than any other trader would admit, that losses don't turn into gains; they remain losses, and they grow and grow. As a result, Greenberg had developed a trading theory centered on making money by limiting how much traders might lose.

Giglio's first encounter with Greenberg's philosophy on risk came one day during a risk committee meeting in the late 1980s. Giglio was now the head of banking in the municipal bond department and one of the most successful bankers dealing with state and city governments in the country. His big clients included New Jersey and New York State, two of the biggest issuers of municipal debt.

Giglio was one of Jimmy Cayne and Ace Greenberg's favorites because of his unique style of investment banking—most investment bankers who worked municipal bond deals were either "quants," that is, numbers guys, or schmoozers who cultivated relationships with politicians and government bureaucrats. Giglio, however, was the complete package: he not only knew how to woo politicians (sometimes with campaign contributions) but could just as easily explain the technicalities of a complex bond offering.

The municipal bond desk had done a deal that wasn't completely sold. The risk committee had agreed to hold the bonds on the firm's balance sheet until the market turned and the securities could be sold at a small gain, or at least not such a large loss. But Greenberg was running out of patience. "I remember him telling me that 'Your first lost is always the hardest one to take,'" Giglio, now a professor of corporate strategy at Northeastern University, recalled. "I wasn't sure what he meant by that until Greenberg explained the following. He said, 'If you're losing money on a trade, just get rid of it. Don't think you're smarter than the market, because you're not and it will prevent you from making another trade.' It was at that point when I realized how Greenberg became such a great trader and damn good CEO."

In short, Ace Greenberg took risks, but he wasn't ready to take just any risk. The firm was leveraged 30 to 1, sometimes as much as 50 to 1, much higher than either Salomon or Fink's group, before they suffered trading losses and cut back on borrowing. But as those firms lost money, Bear Stearns continued to make it because of Greenberg. He took risks, plenty of them, with trades and with people, or else he would never have hired Cayne, who had never sold a stock in his life but was now president of the fifth largest securities firm on Wall Street and in line to become its next CEO. Greenberg took risks, but they were calculated—they fit into his worldview, which meant that risks were taken but with controls.

That was why he hired Howie Rubin, despite his epic trading losses at Merrill.

"We're not saying Rubin did or did not make any mistakes," Greenberg told *BusinessWeek* in July 1988, when the news broke that he'd offered Rubin a job at Bear just eight months after he'd left Merrill and lost all that money. "But we believe in second chances. We have had discussions with him, and we think he can make a real contribution."

Rubin's hiring by Bear was a huge story. This was, after all, one of the all-time risk takers on Wall Street who had lost and lost big, and now he was making a comeback. Rubin, of course, wouldn't have free rein at Bear as he had at Merrill. He would have to answer to the risk committee and Greenberg himself. He wouldn't receive any guaranteed bonuses—just a salary of $100,000 and 10 percent of all profits he generated. But now he was back in the action, doing what he knew how to do best: take calculated risks in the bond market. After the initial amazement of his being hired just months after his firing from Merrill, the press no longer cared about Howie Rubin. But Bear did. He was making countless millions of dollars, with Greenberg looking over his shoulder at every turn.

Despite Greenberg's sometimes bizarre style of management—he wrote memos urging employees to cut down on the use of paper clips and interoffice mail to save money—thanks to his extreme focus on risk, Bear Stearns was an oasis of calm amid the storm that swept Wall Street during the waning years of the 1980s and into the new decade. Through the entire financial crisis, which began in mid-1986 with the mortgage market meltdown, through the 1987 stock market crash and later the slowdown in business stemming from the depressed junk market, and now a recession into the 1990s, Bear continued its record of never having posted a quarterly loss since its creation.

The rest of the Street wasn't so lucky—the combination of risk and leverage that had paved the way for record profits now began to eat away at it. The issuing of new bonds fell dramatically, particularly those that kicked back the biggest profits to Wall Street, that is, mortgage-backed securities and junk bonds. Merrill was bleeding so badly from the slowdown in corporate debt issuance, the Rubin fiasco, and the aftershocks of the stock market crash of 1987 that its total losses over a three-year period added up to $1 billion, nearly a third of its market value.

In the words of the young Larry Fink, the mortgage market was getting "whip chained." The seeds of the implosion had begun with numerous bad bets that mortgage rates would decline and prepayments would continue to mount. In the later part of the 1980s, the Fed began squeezing credit and raising interest rates. The impact was felt across most classes of bonds, where prices move inversely to interest rates, but most strikingly on mortgage bonds and their various volatile derivatives. Deals began to dry up, and firms that had begun by betting that rates would fall and prepayments would rise lost a ton of money. Not even their hedges worked because they were based on the same interest rate predictions.

At Salomon, Ranieri's mortgage bond group never really recovered after his departure, and even though the arbitrage group led by Meriwether was profitable, the firm as a whole began to lose money. Salomon was hardly alone. First Boston was reeling from losses as well, not just in Fink's group but also ones tied to the now-depressed junk bond market.

With the economy slipping into recession and the highly leveraged deals blowing up, Drexel Burnham Lambert was hit the hardest. The "funny money" that had financed the LBO boom and that Teddy Forstmann had predicted would doom Wall Street and the economy began to do just that: companies burdened by junk debt began to slide into bankruptcy while housing prices fell; Wall Street's losses led the firms to lay off tens of thousands of employees; and in February 1990, Drexel Burnham Lambert filed for bankruptcy.

By the end of the decade, the junk bond market was frozen, and mortgage bonds weren't faring much better. As the housing boom that had spurred the economy to such heights during the 1990s came crashing down, real estate foreclosures rose and the issuance of mortgage debt nearly came to a halt. Wall Street's job losses mounted. The damage was spreading to the holders of the imploding debt. Hardest hit were savings and loans, which had specialized in making home loans but during the 1980s, because of deregulation, had begun to speculate and buy mortgage-backed bonds and junk debt. The savings and loans also held depressed real estate, lots of it, and as the mortgage bonds and junk bonds peddled by Drexel imploded, so too did the S&L industry.

The enormity of the disaster was such that the federal government embarked on what was then the biggest rescue of any industry; it created the Resolution Trust Corporation, a new government agency, to bail out the struggling S&Ls by selling the institutions and their assets at bargain-basement prices. Ironically, Lew Ranieri, who by then had been out of the mortgage business for about a year, jumped back in, purchasing a Texas thrift on the cheap that had been forced into receivership because of its bad bets on various forms of debt, including mortgage securities that he might have sold.

There would be no bailout of Drexel, however. The government believed the firm was something close to a criminal organization, while Wall Street simply hated Drexel because of its dominance of the most lucrative market of the past decade, junk bonds. Having been the largest market maker of junk bonds, Drexel was now holding billions worth of them on its balance sheet—bonds it couldn't sell once the market froze up. But more important, Drexel couldn't use the junk bonds as collateral in the so-called repo market, the place most Wall Street firms go to finance their operations. In a repo, firms exchange their own inventory of bonds for cash, but by early 1990, no Wall Street firm would give Drexel credit because the high-yield market had gone, in Wall Street parlance, "toxic," meaning that panic had spread to such an extent that traders believed all junk bonds, not just those issued on speculative deals, were worthless.

And with that, in February 1990 Drexel was forced to liquidate, a precursor that would revisit not just one firm but all of Wall Street in years to come.

It doesn't take Wall Street long to forget the sins of the past and spot an opportunity for the future. Ranieri stayed away from Wall Street and created his own private-equity firm specializing in distressed assets, of which there was now no shortage, while just eight months after leaving Salomon Brothers, Mike Mortara was hired by Goldman Sachs to run its mortgage department. Merrill began by hiring a slew of old Drexel executives, as did Bear Stearns, which poached talent liberally from the now-depleted ranks of Wall Street's bond houses.

Much of the hiring at Bear was directed by Jimmy Cayne. Except for Teddy Forstmann, no one on Wall Street hated Drexel and the competition at Salomon and elsewhere more than Cayne. But that didn't stop him from hiring some of Drexel's bankers in a bid to make Bear the next Drexel if high-yield debt ever revived (which he was certain it would).

Cayne also knew that assembling his own team would be the fastest way to achieve his ultimate goal of becoming CEO of Bear Stearns. By the early 1990s, Cayne had his eye squarely on Greenberg's job. He had been elbowing the competition out of the way for years, assembling a group of loyalists to help him take over, which he was certain would happen very soon. Warren Spector, the firm's most important trader, was now on Cayne's team, and with his blessing the intellectual, aloof, and sometimes unlikable kid with a knack for understanding complex mortgage bonds would soon be the head of fixed income at Bear, pushing aside even the guy who had hired him.

Another Cayne loyalist was the investment banker Alan Schwartz, one of the best deal makers on Wall Street when it came to media companies but not considered one of the firm's deepest thinkers. Around this time, Schwartz and some Bear partners were eating breakfast in the firm's executive dining room. "I got a big problem," Schwartz announced to the table as coffee was served. "Should I get a Mercedes or a BMW?"

It would have been funny, one of the partners thought, except that Schwartz was dead serious.

Cayne was also dead serious about Schwartz, and about the future of Bear and his own future as well. To say Cayne was a rising star at a firm would have been an understatement. By now, he made no secret that he was aiming for Greenberg's job, the sooner the better. In meetings he took credit for the firm's success—Bear had managed to survive and prosper despite the recession that had spread across Wall Street. It was the ultimate contrarianism—Bear had hired people that the Street didn't want, and it made money when competitors were bleeding. It took risks, but, thanks to Greenberg, it knew when to take losses.

Greenberg himself was keeping a fairly low profile, though he wasn't bashful about paying both himself and Cayne what was then seen as outrageous yearly bonuses—they made more than $5 million each in 1990, a year when every other firm was slashing compensation, even at the CEO level.

Cayne, meanwhile, began to take a more public role. He started giving interviews and hired a consulting outfit to do an analysis on what made Bear so great. The rest of the Street, he said, could take a cue from the once-lowly Bear Stearns. The competitors at Merrill were "jokes," he told friends and reporters. He mocked the guys at Morgan Stanley and Goldman Sachs. And Drexel? Unlike Teddy Forstmann, Cayne didn't despise Drexel for the low morality of its business practices. He hated the Drexel firm for another reason: the guys there had taken business away from Bear, and as Drexel fell he rejoiced that the most corrupt firm on the Street was dead. Good riddance.

One day, seated in his office, Cayne was reading the editorial page of the *Wall Street Journal*, which often praised Milken for his role in creating a market that had expanded financing to many companies that otherwise couldn't get it. "You got to be fucking kidding me," he fumed. "Suzette!" he screamed at his secretary. "Get me Bob Bartley, immediately!" Bartley was the editor of the *Journal*'s editorial page, and Cayne wanted a private meeting with the editorial-page chief to set him straight on the people he believed were among the biggest scam artists Wall Street had ever produced. He ranted at Bartley and his staff for about an hour on the evils of Drexel and how much business had gone its way because its traders hadn't thought twice about bending the rules.

"You guys don't know what you're talking about, they were crooks," Cayne fumed. "You don't know what those guys did!" By the time the meeting was over, Cayne thought he had the group, Bartley most of all, convinced of the righteousness of his argument—that is, until he read a subsequent editorial that once again praised Drexel and attacked Bear over one of its regulatory failures.

B y the end of the 1980s, Larry Fink had been persona non grata at First Boston, now known as CS First Boston after the bailout by Credit Suisse following the massive losses from Fink's trades and the company's involvement in a messy LBO deal for Ohio Mattress Company, the maker of Sealy brand mattresses.

Fink had been toying with leaving First Boston for more than a year. But part of the reason he didn't just pack up the minute his role in the firm was diminished was that he loved the place. He had been working there since his midtwenties, and despite all that had gone down, he was just thirty-five and he still couldn't see himself dropping everything and quitting a place he loved. But the place no longer loved him; that much was clear.

It was only a matter of time before he would have enough and would leave to work for the competition or to start his own trading desk. Fink, after all, had been one of the most successful traders on Wall Street during a successful era, and his career was being cut short because of what had turned out to be a single lousy trade.

So in March 1988 Fink left First Boston to create a new money management firm. A former Lehman Brothers investment banking chief, Pete Peterson, and another Lehman alumnus, Stephen Schwartzman, had created their own investment house, known as the Blackstone Group. They had made a standing offer to Fink to set up shop inside the firm. Fink never seriously considered the offer until it became clear he had no future at First Boston: He was still being blamed for the losses at the firm—even losses he had no part in.

By Fink's assessment, the mortgage group had actually made money even after that bad quarter. But that wasn't the spin the company put out to the press. The continued losses at First Boston, they said, were the result of trading losses from the mortgage desk. If there were ever a hint that it was time for Fink to leave First Boston, that was it.

He even had a name for his new firm; it would be known as BlackRock. Peterson and Schwartzman had given him office space and a $5 million line of credit to focus on buying mortgage bonds and other debt now at depressed prices. But more than that, Fink was thinking bigger. His idea was to create a firm designed to understand risk—how to take risks and profit from them without blowing up as he had done

Word of Fink's departure quickly spread through Wall Street. First Boston, which had just lost its banking stars Bruce Wasserstein and Joe Perella, put out a statement that said Fink was effectively fired as co-head of trading and would have had to take a lesser position in the firm if he'd stayed aboard. "He did not have the option of staying in his current job," a spokesman told the *Wall Street Journal*.

Dunlavy read the news the next day. "I wonder what Lewie thinks," he thought to himself as he read the *Journal*'s account of Fink's departure. Then, there it was, a quote from the great Lew Ranieri, who was now said to be setting up his own investment firm. "Larry is a great individual performer," Ranieri said. On its face, the statement could be taken as if it were a compliment. But it wasn't. In fact, Dunlavy knew exactly what Ranieri was really saying: Fink had all the attributes of a great player, but he couldn't work in a team setting like the one he, Ranieri, had established at Salomon Brothers.

But Dunlavy knew better than maybe anyone else the drawbacks of belonging to such a close-knit team that goes to extreme lengths to protect its own no matter what the circumstances.

Dunlavy would stay with Salomon for nearly ten more years, until 1998, to see the firm embrace risk once again and, after again suffering losses, pull back on them. He would see the leadership of the firm change dramatically, particularly after a scandal in 1991 involving Treasury bonds that forced the ouster of Gutfreund, his number two, Tom Strauss, and John Meriwether.

The scandal was odd because on its face it seemed so benign. No single firm can buy more than 30 percent of a U.S. Treasury bond auction. Government officials heard that Salomon's government bond chief, Paul Mozer, was defying the 30 percent rule and finding ways to buy entire lots. They told him to stop. He ignored them. His rationale: Salomon was doing the government a huge favor by helping it finance the growing national deficit. In a few months' time he was arrested, the firm was fined, and its management was fired.

Dunlavy was at first confused as to why Mozer would risk his career on something that seemed so absurd—defying a government edict. Then it dawned on him: what else but money? Mozer had heard of a trader at Salomon who'd earned $23 million in just one year, something that incensed him when he earned a fraction of that amount as the manager of a highly profitable department.

The firm, and of course Mozer's trading desk, was making tens of millions of dollars from the scheme by creating a short squeeze. In a short sale, a trader borrows a bond (or a stock), sells it, and later replaces it with another bond—purchased, he hopes, at a lower price. Traders short Treasuries every time there is an auction because prices are expected to fall due to the glut of new debt. But because Salomon had bought all the bonds in the auction, a short squeeze was created when traders who needed to buy bonds to replace the ones they'd earlier sold short were unable to do so, causing bond prices to soar. Salomon and Mozer's team made millions

Mozer's short-term gain soon translated into long-term pain. When the dust settled, Mozer was fired, of course, and did some jail time. Gutfreund didn't go to jail, but he never ran a major Wall Street firm again and never again achieved the level of status and success he'd had at Salomon. Strauss left as well (he later founded Ramius Capital, a private-equity firm), as did Meriwether, who went on to earn fame and infamy by starting a new and dynamic hedge fund, Long-Term Capital Management, that would capitalize on Wall Street's embrace of risk and leverage in ways that even his arbitrage group at Salomon had never imagined.

Salomon's minority owner, the great investor Warren Buffett, took over and installed new, allegedly more risk-averse management. A new crop of traders began to fill the management ranks. But the one who caught Dunlavy's eye was a young trader on the junk bond desk with a poker face that could rival Meriwether's. His name was Tom Maheras.

Maheras was a Notre Dame graduate with thick black hair and a boyish grin who rose through the ranks quickly. He was bright and polished, and, like any great bond trader, he rarely showed emotion. More than anything else, he seemed to put his new, more risk-wary supervisors at ease. Tom Maheras just didn't seem like a guy who would bet $1 billion on the direction of PO mortgage bonds and lose.

He was also ambitious. If you watch the great ones for a long time, guys like Ranieri and Meriwether, you start to understand how traders think— what makes them get up in the morning and stare at computer screens. The short answer is ambition or, to be more precise, the ambition to make more money than the guy sitting next to you.

But to do that you have to take—and love to take—risk. Risk is part of the trader's DNA, and without it, most traders will tell you, there's no reason to work on Wall Street. And for all Maheras's polish and his easygoing manners, particularly with supervisors, he was a risk taker through and through. Maheras was no different from Meriwether, Ranieri, Mortara, Fink, or for that matter nearly every guy at Salomon and on the rest of the Street. They were all amazed and envious that some trader had made $23 million in just one year. And they all knew there was only one way to be like that guy: they needed to take more risks, borrow more money, indulge in more leverage, and make bigger bets than the guy sitting in the adjacent cubicle.

Wall Street had just suffered massive losses, hundreds of millions of dollars' worth, widespread job losses, and massive management upheaval, yet the firms never broke with their past. They merely took a break, returning to the basics of servicing clients and underwriting stocks and bonds, and waited for the markets to return so they could do it all over again.

PART II

MERCHANTS OF DEBT

5. BIGGER IS BETTER

Wall Street has a short memory, and by the early 1990s, the big losses in the bond market that had ended Larry Fink's career at First Boston, forced Lew Ranieri out of Wall Street, and destroyed Drexel Burnham Lambert seemed like the distant past.

The Federal Reserve, hoping to avoid a massive recession and a banking crisis, began slashing short-term interest rates, sparking another massive rally in the bond market. Once again, it was a giddy time to be a merchant of debt—and to take risks.

Tax-free municipal bonds were thought to be a dying business in the late 1980s after tax reform closed several municipal bond shops, including Salomon Brothers, but by the early 1990s, munis were again on a roll. Junk bonds made a comeback as well. Drexel's absence left a void, but it was soon filled by Merrill Lynch, Smith Barney, Lehman Brothers, and Bear Stearns.

But the biggest bounce came from the mortgage bond market. There were several factors accounting for the mortgage market's renaissance. The economic slowdown that had begun in the late 1980s squeezed the mortgage bond market, though just briefly, thanks in large part to a government bailout of the bankrupt thrifts through the Resolution Trust Corporation, which protected the mortgages within the bonds held by investors. Low interest rates and a mild recovery in the economy around the time of the first Iraq war added to the market's resurgence. But more important was the attitude toward risk: investors were now willing to take more of it and Wall Street was again willing to play with it.

With Treasury bonds rallying (prices rising and yields falling), investors were now demanding bigger returns, and that's what the mortgage bond market offered: yields several percentage points higher than comparable Treasuries because of their greater risk.

One constant of the new Wall Street business model that had developed in the early 1980s was the linkage between borrowing and the growth of the mortgage-backed market, and in the early 1990s, as the mortgage bond market flourished, Wall Street again went on a borrowing spree. According to Standard & Poor's, the use of borrowed funds to make trades and carry mortgage debt once again rose sharply, with Wall Street now borrowing an average of $24 for every dollar it had in reserve to finance positions in various versions of the CMO, including the new classes of mortgage debt being conjured up by the math whizzes at the big firms.

And just as in the 1980s, the mortgage bond market brought Wall Street back to profitability; Bear Stearns earned $150 million in 1993 from mortgage trading. Other firms showed similar results. One key difference was in the players. Fink and Ranieri were gone, both managing money at their own firms, and they were replaced by a new generation of risk takers: the nerdy Warren Spector at Bear Stearns; the amateur weight lifter and math wiz Michael Vranos at Kidder Peabody; and Tom Maheras, a former junk bond trader at Salomon Brothers now moving into the mortgage bond department.

Bonds packed with all types of loans—credit cards, car loans, and most of all mortgages—accounted for some of the biggest deals in the market. In 1991 issuance in mortgage bond departments shot up a whopping 50 percent from the year before, showering profits on the new leaders in the business: Kidder Peabody, Bear Stearns, and a new player looking to make its name in the high-margin, high-risk business of trading and underwriting mortgage debt, Lehman Brothers.

Among its competitors, Lehman Brothers was a firm becoming known for several things, usually in this order: its nasty and insular corporate culture; its willingness to use leverage to make bond market bets; and the firm's rising star, an odd-looking bond trader named Richard Fuld.

want to thank everyone for attending this afternoon, but now the real show is about to begin!" announced a tall, well-dressed investment banker in a dark suit and bright red tie named Michael Madden, Lehman Brothers' investment banking chief.

Madden was speaking to a room filled with five hundred of Lehman's top executives at the firm's annual strategy seminar at the Marriott Hotel in Rye, New York, about twelve miles north of the firm's headquarters in lower Manhattan. The executives had just finished a day filled with meetings, speeches, and panel discussions, all focused on a single purpose: how to make Lehman Brothers the most successful firm on the Street. Now they were ready to party, and Madden was ready to kick off the festivities with a short, and hopefully funny, after-dinner speech.

Madden looked out into the crowd at the tables filled mostly with white men in black suits, including the two men seated right in front of him. They were Lehman's co-presidents, J. Tomilson Hill III, a WASPy investment banker with slicked-back hair à la Gordon Gekko, and Dick Fuld, the guy most Lehman executives considered the real power at the firm.

Madden had been with Lehman for a little more than a year, and it hadn't taken him long to realize that all the stories that swirled around the Street about Lehman were true. Founded in the 1850s by Jewish southern cotton magnates, Lehman had moved into commodities trading before the end of the century and then became a full-service brokerage firm, surviving the Great Depression and later the bear market of the 1970s.

Some firms (Goldman Sachs and Morgan Stanley) are known for their brains and pedigree, others (Merrill Lynch) for democratizing the markets by selling stocks and bonds to small investors. Lehman, for its part, was known for its strongly independent corporate culture and for brutal infighting between traders and bankers, immortalized in Ken Auletta's *Greed and Glory on Wall Street*, which chronicled the power struggle between the firm's white-shoe investment bankers and its streetwise traders, led by Lewis Glucksman and his protégé, a young commercial paper trader, Dick Fuld.

After Glucksman and Fuld took control, Lehman imploded with huge trading losses. Its next move was a shotgun marriage with the credit card giant American Express. By the early 1990s, American Express still owned Lehman, which was now combined with a large brokerage firm and renamed Shearson Lehman Hutton. Glucksman was now out, and Fuld was in charge of Lehman's trading operations, sharing the top job with Hill and reporting to the CEO of American Express, James Robinson.

Risk taking was once again paying off, and though Fuld was nominally only co-president, everyone in the room that night knew the real power at Lehman resided with him and with his bond desk, which produced most of the firm's profits. Indeed, even Hill understood the reality of the situation; in meetings he would let Fuld speak first. When they were together, he would let Fuld walk ahead of him. After Fuld spoke, Hill would tell senior managers why they too should see the world through Fuld's eyes.

Amex had now owned Lehman for ten years, a bitter period both for the firm and for Fuld, who longed for the days when he and Glucksman had called all the shots. The corporate suits at Amex were occupiers who simply didn't understand Wall Street and certainly had no understanding of people like him, the traders who were judged every day based on a simple metric: how much money they made and lost.

By the early 1990s, Fuld was actively planning to wrest control of the firm from Amex and return Lehman to its glory days of independence. His prospects were getting brighter; since the bond market rally had begun in 1990, Lehman had been on a roll, cranking out record profits quarter after quarter. The sources of those profits had been trades, increasingly complex, on mortgage bonds and other types of fixed-income investments that had come into fashion as interest rates declined. Fuld's top executives were all from Lehman's bond desk, and they knew a good bond market when they saw one.

As rates declined, more esoteric securities came into vogue, and they could be easily traded. Liquidity—the juice that keeps the bond market active—was everywhere. People could buy and hold exotic mortgage debt and pocket the heavy returns without worrying that the debt couldn't later be sold. That is, until the market tanked—and they couldn't be sold. The trick was to know when the markets were going to tank, and that's what Fuld and his trading partners were supposed to be good at.

To the rank and file at Lehman, Fuld deserved all the credit for the firm's resurgence. His tough demeanor was already legend both inside and outside Lehman. He instilled an us-against-them corporate culture; at company meetings he sounded like General Patton rallying his troops (in Fuld's case his traders) to kill the competition and "bleed Lehman green" every day on every trade. It was no accident that the firm's official color was the color of money, but as far as Fuld was concerned it could also have been bloodred. He once told a trader who asked him a question he didn't like that he wanted to "break your legs for that." One afternoon while attending his son's hockey game, Fuld got into an argument and went toe to toe with an opposing player's father. Though he nearly got his jaw busted by the bigger man, the incident cemented Fuld's tough-guy rep.

For all of this, and for something else, Fuld was known inside Lehman as "The Gorilla." Now Madden was about to learn what that something else was as he got ready to give his after-dinner remarks and have a little fun at Fuld's expense. "Everybody knows the guys who run our wonderful company," Madden said to a round of snickers from the half-drunk crowd, pointing to Fuld and Hill in the front row. "It's Tom and Dick. Now, I haven't been here that long, but everybody tells me these guys go everywhere together and they're like twins. There's only one problem," he added with a smile, "they don't look like twins. But I know how to fix that. Tom, can you come up here?"

Hill got up, walked nervously to the podium, and stood beside Madden, who whipped out a large, hairy mask of a gorilla and promptly placed it over Hill's head.

"I just want everyone to know who the host of the conference is," Madden said with a hearty laugh. The place went wild. Even Hill thought the joke was funny.

But not Fuld, who shot Madden one of his famous death stares and didn't say another word to him for the entire night and barely a word for the next year—when Madden, despite winning deals and establishing Lehman's investment bank as an increasingly potent competitor on deals, was fired.

What Madden didn't realize until after he was gone was that Fuld hated being called "The Gorilla" because the nickname didn't just come from his legendary temper and aggressive trading style. People at the firm thought he actually *looked* and *sounded* like a gorilla, with his pitch black eyes, his forehead that protruded like a block of cement, and the way he spoke: As a young trader, Fuld had been so intense that he often didn't speak in full sentences. That same intensity led to his running a colleague out of the firm after a well-received joke one drunken evening.

Dick Fuld had clearly come a long way; not only did he now speak fluent English, but he was on the verge of running one of Wall Street's largest firms. In the spring of 1993, he was named Lehman's sole president by Amex (Hill resigned as co-president), setting the stage for Lehman's rebirth as an independent company.

Fuld was jubilant; his dream of running Lehman was nearly a reality, and the firm was now one of the top firms in trading and underwriting debt. Still, people on Wall Street gave Lehman almost no chance of survival without its corporate parent. The Street, after all, was changing, and fast. It had recovered from the 1987 crash thanks to a vast democratization of the markets and low interest rates that had jolted Wall Street's bond-trading desks. Old corporate pension plans were out; people were saving for retirement through 401(k)s and brokerage accounts.

That gave an almost immediate advantage to the big firms such as Merrill Lynch that had brokerage channels big enough to meet the demand. By the early 1990s, the main Wall Street brokerage houses (except for Goldman Sachs and Lazard Frères) had converted to public companies, meaning not only that they were risking shareholders' money rather than the partners' own capital in their risk taking but also that they had more capital to take risks with. Firms of large size and scope, such as Salomon Brothers (with capital from Warren Buffett's Berkshire Hathaway), Morgan Stanley, and Kidder Peabody (thanks to its massive corporate parent General Electric), had the capacity to make big bets in the bond markets, as did Travelers Group with its Smith Barney brokerage firm. Even the relatively small Bear Stearns, thanks to the steady hand of Ace Greenberg, was able to play in markets that demanded size and savvy.

Lehman seemed like the odd man out, particularly since it no longer had access to the nine thousand brokers who had been lost as part of its arrangement with Amex, which had sold the Shearson brokerage unit to Travelers Group in 1991.

Not as far as Fuld was concerned. Even his friends wouldn't consider "The Gorilla" a deep thinker, but when it came to Lehman, Fuld had a clear vision. The firm would be built in the image that had been handed down to him by Glucksman: insular and hypercompetitive, almost paranoiac about the competition. Lehman would be a meritocracy, and most of all, it would embrace risk in the increasingly complex bond market, which Fuld believed was the key to its independence.

Its blind spot may have been in risk—or, to be more precise, the ability to manage the risk that began to define Lehman's business. Fuld was no expert in any of the multitude of new bonds that began to dominate the firm's business, including the complicated derivations of the mortgage-backed securities that surged in issuance in the early 1990s. In fact, he had made his mark trading commercial paper, the safest of all debt issued by corporations. (Commercial paper is basically a short-term IOU used when companies need cash to meet immediate needs.)

What Fuld knew was that there was big money in these newfangled bonds, which is why he geared Lehman's business model to trading some of the most esoteric securities in the market and bask in their higher returns. He also knew how to take the gains of those trades to new heights. As in all trading shops, leverage was the key to success; Wall Street firms don't have enough money on hand to finance their complex trading arrangements, so they borrow. And borrow. Lehman's leverage was now among the largest of any firm on Wall Street, hovering between 35 to 1 and 38 to 1 in 1992 and 1993. Fuld's use of leverage made the firm big money—Lehman earned a then-substantial $250 million in 1992 and nearly as much in 1993. Fuld himself entered the $10 million club, meaning he was now one of the few top executives on the Street to earn an eight-figure salary.

In 1994, American Express agreed to spin off Lehman with an initial public offering (IPO) that would turn Lehman into a public company. Fuld was thrilled: after ten years in the wilderness Lehman would once again be an independent company, with its corporate culture intact.

As the IPO date approached, Fuld constantly reminded his team how far they'd come and what their goal was: to be the best firm on the Street. It was an outlandish claim for a firm that was just barely breaking into the upper tier of Wall Street, and that mostly in bonds. But those who know Fuld say he meant every word of it. Back when Amex had first bought Lehman, he had assembled his senior staff and picked up a pencil from his desk.

"This is what *they* can do if *we* don't stick together," he said and then easily snapped the pencil in two.

He then grabbed a handful of pencils, showed it to the group, and with both hands tried to break the pencils. He couldn't. "If we stick together, they can't break us."

Each of the men in the room—senior executives who would in the next decade build Lehman into a powerhouse—would later receive a package from Fuld. Encased in a small plastic case was an unbroken pencil with the man's name engraved on it.

Not one of the senior executives who received a pencil forgot the incident or removed the item from his desk as long as he worked for Lehman.

In rallying his troops at Lehman to pursue business opportunities aggressively, Dick Fuld told a story about his early days as a commercial paper trader. He had found an opportunity in the market he thought he could exploit, so he'd walked into the office of one of the firm's loan officers to borrow some house money to finance the trade. Fuld, as most people knew, even early in his career, was among the most aggressive traders at the firm, something he cultivated to bully his way through the management ranks. When the loan officer said he "needed to clear" his desk before approving the trade, Fuld took matters into his own hands and cleared the man's desk for him—literally by shoving the papers to the floor.

The officer was stunned, but he approved the trade. And Lehman made money on it.

But by 1994, the same leverage and debt that had boosted profits over the past three years was squeezing Lehman like the rest of the Street as the Federal Reserve began raising interest rates to squeeze yet another speculative bubble out of the market, once again involving mortgage debt and other bonds.

Wall Street, Lehman included, was relying on increasingly complex computer modeling to predict the market. Even without the computers Fuld and his cronies were supposed to be savvy enough to smell the "top of the market," when the Street begins to underprice risk, forcing investors and traders to buy increasingly more exotic bonds to profit from the higher yields and returns they generate.

But Wall Street traders have never been good at sniffing out bubbles— figuring out when prices are too high as trading desks start ignoring obvious risk factors and demanding increasingly lower premiums for increasingly more esoteric bonds. Those who raise the red flags are often ostracized as academics or alarmists who don't understand how the "real" world of Wall Street works. Their warning signs are ignored even if everything they say makes perfect sense.

One of the most prominent alarmists in the crowd this time around was E. Gerald Corrigan, the powerful head of the New York Federal Reserve Bank. Corrigan is a large man who looks and sounds like a tough Irish bartender. He was described once by the *New York Times* as "The Fed's Plumber" because of his vast knowledge of the banking system he regulates.

Corrigan, a disciple of former Fed chairman Paul Volcker, had seen many crises during his long tenure at the Fed. He had aided Volcker during his inflation-fighting campaign in the late 1970s and early 1980s (the campaign that had rocked the bond market but also led to its eventual recovery). He had devised Fed policy during subsequent market tremors, including the 1987 stock market crash.

As Corrigan told the *Times*, he was most proud of having prodded Wall Street banks to improve their "plumbing"—the back-office systems that handle the nuts and bolts of trades—and pushed them into maintaining enough capital to withstand a crisis. Understanding so much about banks' plumbing gave Corrigan insight into just how tenuous the banking system had become in the last few years, especially due to the biggest business of the decade: bonds and all their variations, broadly defined as the derivatives market.

The derivatives market is massive and opaque. Based on whose definition you're using, it involves the transactions known as "interest rate swaps" that Salomon Brothers had pioneered in the 1980s, but also the various new bonds and mortgage-backed securities created by Wall Street trading desks, including IOs, POs, and bonds known as "inverse floaters" or "accrual bonds." Corrigan began sounding alarm bells on the dangers of the growing and increasingly esoteric nature of this market as early as 1992. What concerned him was that these newfangled bonds weren't properly disclosed on the banks' balance sheets. In addition, Corrigan had a natural skepticism about the high-tech risk management systems Wall Street traders were now using. He knew they had their limits and felt that they always seemed to fail when they were needed the most.

"The growth and complexity of off-balance sheet activities and the nature of the credit, price, and settlement risk they entail should give us all cause for concern," Corrigan said in a 1992 speech to the New York State Bankers Association. "High-tech banking and finance has its place, but it's not all that it's cracked up to be. . . . I hope this sounds like a warning, because it is."

Corrigan could see the perils of cheap credit firsthand, as the New York Fed plays a unique role in the financial system. It's the largest of the nation's twelve Federal Reserve Banks, but more than that, it influences the country's monetary policy through something called open-market operations. In other words, the New York Fed controls the nation's money supply by adding credit to or subtracting it from the system. As the head of the New York Fed, Corrigan was like a mechanic who was constantly looking under the hoods of the big banks and investment houses, and he didn't like what he was seeing—or, to be more accurate, wasn't seeing.

The Fed couldn't really control the use of derivatives because most of these bonds and near bonds were hidden off the balance sheets and because the products were so new that no regulatory authority had direct oversight; the derivatives market was the modern-day version of boomtowns in the Wild West, with few sheriffs around to keep order.

The market for swaps and for fancy bonds cobbled together from interest payments had certainly come a long way since Pat Dunlavy had tried to explain "notional value" to the brain trust at Salomon and been told to "be careful." Far from being careful, Wall Street didn't seem to have a care in the world as it created, packaged, and sold derivatives in ways that would have seemed impossible years before. Every firm got into the act; as we'll see later, insurance companies such as American International Group created entire units designed to craft complicated hedging techniques through derivatives, including something known as a "credit default swap," which was an insurance policy on corporate debt. But unlike with regular insurance, AIG didn't have to hold higher levels of capital on the credit default swaps; it simply hedged its exposure to losses with other derivatives.

By the early 1990s, the total value of credit derivatives was approaching $11 trillion, on its way to $100 trillion by the end of the decade. Bond derivatives were everywhere. Wall Street traders used them to make market bets. Companies used them to hedge against quarterly losses. Small towns used derivatives to cut the interest payments on their municipal debt and used the savings to pay for construction of roads, bridges, and schools.

Yet paradoxically, as bond derivatives became more mainstream they also became less transparent.

Loopholes in the accounting laws allowed banks and brokerages to avoid openly disclosing their exposure to the derivatives markets. Corporations could hide their use of interest rate swaps in the footnotes of financial statements in terms that even sophisticated investors could barely understand. The mayors and supervisors of towns and cities that were buying inverse floaters were equally in the dark, relying on Wall Street advisers to explain the financial alchemy.

The growth in the derivatives markets was being greeted on Wall Street as nothing short of revolutionary. Through computers and brainpower, Wall Street was reducing risk and spreading wealth as never before. But as they had in 1986 and 1987, the Wall Street risk takers were missing the randomness of the markets—the possibility that either the computers were wrong or a random event might upend even the most sophisticated models of pricing and probability.

They were also missing something else entirely new. Since 1989, the size of the derivatives market had grown by 371 percent according to industry surveys (it's impossible to know the exact size of a market that has no central regulator). Even though derivatives were designed to reduce risk when properly used, one form of risk, known as "systemic risk," was actually spreading. The swaps and the new bonds whose values were derived from obscure indices created a complex web of interconnected liabilities. The system works great when the markets are calm, when interest rates remain low so traders can borrow cheaply, and when there's plenty of liquidity, meaning that a ready market for the products exists, not that much different from that at the New York Stock Exchange, where sellers know they can find an ample number of buyers for their stocks at prices that aren't distressed.

But Corrigan knew that the markets, particularly for these newfangled bonds and derivative contracts, weren't anything like the NYSE. The derivatives market was completely dispersed. Bonds were traded over the phone between traders or via computer linkups around the globe without any central system of revealing what buyers were paying for the bonds. The bond market can change quickly and violently, particularly in markets that are complex and opaque, such as those for CMOs, which virtually shut down during the IO and PO crisis in 1986 and 1987.

Given the size of the credit market and the credit derivatives markets, Corrigan worried that a sudden spike in interest rates could magnify losses beyond anything the market had previously experienced because the new products were even more sensitive to sudden jolts in the rates and because investors, particularly on the Wall Street trading desks, were using so much more borrowed money than in the past to finance their trades, thus magnifying their gains—and ultimately their losses.

What Corrigan was worried about was the systemic risk: as traders began to lose money on risky bonds, they would have to make up for their losses by selling other assets, possibly stocks and high-quality bonds, thus causing the markets to plunge. A mass selling cycle would be created, starting a panic that could reach across the entire financial system.

Wall Street soon got a taste of systemic risk when Corrigan's boss, Federal Reserve Chairman Alan Greenspan, did what no one on Wall Street seemed to expect and began raising interest rates in early 1994 to squeeze the speculative bubble out of the bond market. The irony of the move was that Greenspan (who replaced Paul Volcker in 1987) wasn't really in Corrigan's camp—he didn't see derivatives as an evil. He never called for broad regulation of derivatives and was regarded as an ally of the increasingly powerful antiregulation brigade on Wall Street. In fact, he believed in the greater good of financial engineering through derivatives, feeling that that it increased market liquidity and its excesses could be contained through actions by the Fed.

Even Greenspan couldn't argue with the sudden and massive speculative bubble that had been created in the bond markets in just a few short years. The Fed raised interest rates five times in 1994, partially to squeeze an economy that was beginning to overheat with higher inflation, but also to squeeze the speculative bubble that had not just caused a rally in bond prices but sparked an increase in risk taking. Wall Street's use of leverage had grown from around 15 to 1 in the mid-1980s to close to 35 to 1 on the high end in less than ten years.

But leverage was only one reason for the losses. The street was leveraged with mortgage debt, which provided a double whammy. Part of the reason was greed—mortgages provided higher returns than other forms of debt, and Wall Street saw the carry trade as easy money. The other part was necessity: the underwriters couldn't sell all of the debt that was now being packaged into mortgage-backed and asset-backed bonds, which itself hit a record issuance of $66 billion in 1993. So they had just kept them on their balance sheets, earning the carried interest while interest rates remained stable and the market calm.

Either way, that's where all these bonds were sitting when the markets began to collapse and losses piled up. The problem was that mortgage-backed securities were among the most interest rate–sensitive bonds ever created because they were packed with loans to home owners who had borrowed heavily to buy their homes in the first place. Therefore a sharp decline in interest rates often led to prepayment as home owners rushed to refinance their loans. This is good for home owners but bad for mortgage bond holders since the return on the bonds resets to the new, lower interest rate. Likewise, when interest rates spike sharply, defaults sometimes rise, especially for home owners who have adjustable-rate mortgages.

Firms like Lehman that specialized in bonds were particularly hard hit. The Fed's action caused trading desks to begin to reprice risk; the spreads between Treasury yields and those of mortgage bonds—once razor thin— now began to grow as mortgage bond prices fell and investors began snapping up supersafe Treasuries. The only players looking to buy mortgage debt were the so-called vulture funds, which began buying at steeply discounted prices, magnifying the trading desk losses.

The 1994 bond market crash had many victims besides Wall Street traders. Orange County, California, was one of them, filing for bankruptcy after using its pension account money to bet big on fixed-income derivatives that folded once interest rates began to rise. Likewise, the coffers of a school district in the small Pennsylvania mill town of Tyrone were wiped out. Corporate America suffered as well, underscoring the level of systemic risk that had been attained, when the consumer giant Procter & Gamble lost $180 million in an interest rate swap.

Most of Wall Street would avoid the systemic risk Corrigan was worried about, albeit barely. In addition to the illiquid bonds, the derivative contracts that Wall Street had to take losses on, the big firms had hedged their bets to a certain degree; the Street's collective balance sheet carried a sufficient level of Treasury bonds, which investors bid up in a flight to quality when the mortgage bonds imploded.

Still, 1994 was a terrible year both for Wall Street and for investors nationwide. Bear Stearns and Morgan Stanley remained profitable but just barely, as their earnings sank dramatically due to mortgage bond–related losses. Hedge funds that specialized in mortgage debt, such as Askin Capital Management, collapsed. Salomon once again lost money, more than $300 million over two quarters, while Lehman eked out a marginal $22 million profit.

Hit worst of all was Kidder Peabody. GE's deep pockets notwithstanding, the firm lost $85 million in one quarter thanks to the losses in the mortgage group run by Michael Vranos and on the Treasury bond desk because of a trading scandal. After the dust settled, GE sold Kidder, leveraged at its height as much as 100 to 1, to the hyperconservative brokerage firm PaineWebber, which promptly set out to downsize the firm's operations.

In the wake of the 1994 bond market rout, the Street vowed to reform itself. Leverage would come down—and it did for a while. The Street would rein in its excesses—and it did, at least for a while.

Lehman wasn't about to go out of business, but the question remained whether as a firm it was still relevant. Its investment bank went back to being mediocre at best (Mike Madden, who had made the "gorilla" joke that had so incensed Dick Fuld, was now gone), its research department had suffered massive defections. Moody's Investors Service, the bond-rating agency, cut the firm's long-term rating, alerting investors that Lehman's capital base was becoming too thin to support its business model of using high leverage to make huge bets in the bond market.

Now that it was independent of Amex, with little if any capital on which to grow, Lehman was learning what it was like to be a small fish in the ocean surrounded by much bigger predators. In order to survive, Lehman would have to take even bigger bets and borrow even more money, even if it put its survival in jeopardy.

Lehman's relevance came to be questioned even more as Wall Street continued to evolve. It wasn't just the "bigger is better" philosophy; it was the way business was being done. Morgan Stanley, the famed investment bank, merged with the brokerage giant Dean Witter under the theory that Dean Witter's eight thousand brokers would help Morgan's famed investment bankers place IPOs and stock deals with many more small investors, something every corporate stock issuer lusted for because individual, or "retail," investors usually bid higher for stocks than large, more sophisticated investors do.

Sandy Weill was busy one-upping the Morgan Stanley Dean Witter deal by creating a firm with nearly the same type of structure that Fuld couldn't wait to get out of at Amex: Weill's Travelers Group would combine insurance, stock and bond underwriting, brokerage operations, and soon a commercial bank into the megagiant Citigroup.

Fuld, at fifty-one, suddenly seemed like a dinosaur compared to the sixty-three-year-old visionary Sandy Weill. Analysts predicted that Lehman would be bought or merged into another bank—rumors swirled that JP-Morgan was ready to buy the firm, pushing Lehman's stock to a then all-time high.

Fuld's response was what might have been expected: he told Wall Street to go to hell; Lehman was there to stay and stay independent (although Fuld no doubt used many more four-letter words when communicating his position to his potential acquirers).

Though Fuld periodically met with potential merger partners such as Donaldson, Lufkin & Jenrette, he vowed that Lehman would finally diversify—and although over time, as the firm's stock price rose, he hired more bankers, rebuilt the firm's research staff, and dipped his toe into other businesses, those changes were mere window dressing.

But the real Lehman never changed. Power struggles returned: his new number two, Christopher Pettit, who wanted more of a say in what happened at Lehman, abruptly left the firm (Pettit tragically died soon thereafter in a snowmobile accident). Even as the firm's long-term bond ratings were cut, Fuld refused to slash borrowing; the amount of leverage barely budged, and bond trading and risk taking were as big as ever. His senior staff remained composed of all bond guys. He even appointed a bond trader to run the equities department.

To be sure, there was much to admire about Dick Fuld's relentless drive to make Lehman into something bigger and grander than it had been before the merger with Amex. He instilled a corporate culture that didn't rely on pedigree or Ivy League degrees (he himself had graduated from the University of Colorado). Senior managers all received the same paycheck for a time, and no one seemed to work harder than Fuld, which set a firmwide standard.

But there was also something distressing about the manner in which Fuld ran Lehman Brothers. In many ways, the firm was managed more like a cult than a Wall Street firm, with Fuld as the guru. "I'd run through a brick wall for Dick Fuld!" was a common refrain of many of his top people.

What was so scary about the cult of personality surrounding Fuld's rule manifested itself as the need for dissent and discussion grew. Leverage was growing, and financial engineering was now the dominant business model on Wall Street and beyond—commercial banks, such as NationsBank in Charlotte, North Carolina, and its crosstown rival, First Union, were peddling mutual funds that invested in derivatives of mortgage-backed securities to their customers, sometimes at the teller window. All this innovation demanded disclosure and debate. Yet at Lehman there was little if any.

Fuld had such control over Lehman that he was able for years to avoid appointing a president—the official number two that is required at just about every major corporation—until he named his longtime friend and fellow bond trader Joseph Gregory to the post.

His board of directors—in theory, the ultimate guardian of corporate governance—appeared equally compliant to his wishes. On Wall Street boards, members are appointed by the CEOs, which means they owe their jobs and their six-figure stipends to the men they're supposed to be watching. But at Lehman, the board was unusually eager to please. Lehman's board stood out for being a sinecure for a retired actress (Dina Merrill), a former Wall Street executive who'd given up high finance years earlier to produce Broadway shows (Roger Berlind), and others who weren't savvy to the new leveraged environment that Lehman had plunged itself into.

In the aftermath of the interest rate hikes of 1994, the obviously annoyed Fuld told analysts that trading had once produced 10 percent of the firm's net revenue, but now that figure had fallen to just 6 percent. "Clearly the conservative hedging of our balance sheet caused our cash positions in fixed income to underperform the market," he added.

Translated into English, he was saying that any firm that was prudent and conservative when it came to risk taking was destined for failure. Nor was he alone in that point of view.

6. FINANCIAL RENAISSANCE

He's a superstar," Jimmy Cayne said as he lounged back in his black leather chair, his telephone in one hand and one of his trademark cigars in the other, "an absolute superstar."

Cayne wasn't talking about the mortgage bond wiz Warren Spector, whose department was responsible for much of the profits Bear Stearns was now producing, but a trader in Spector's mortgage group, Howie Rubin. Rubin had once been the poster boy for excessive risk taking after he'd lost a fortune at Merrill Lynch during the mid-1980s trading the principal-only strips of CMOs. But now, in the mid-1990s, he was making Bear Stearns a fortune.

Lehman and the rest of the Street were reeling from their bond market losses, but not Bear—it was experiencing a financial renaissance. Rubin's trading group raked up winnings with massive trades in the esoteric areas of the bond market and became possibly the most profitable unit of the firm. Beginning from Rubin's first year of employment, 1991, and continuing through the middle part of the 1990s, Rubin had made more than enough money to make Cayne, Greenberg, and most of the rest of Wall Street forget those early losses had ever happened.

Rubin's success also almost made Cayne forget just how much he was starting to hate Warren Spector.

Cayne and Spector had forged their relationship over a shared sense of ambition, the desire to make money, and, of course, playing bridge. Cayne loved the money Spector's expertise in mortgage bonds brought in, and, in 1992, at just thirty-five years of age, Spector was rewarded with one of the highest salaries on Wall Street, $11.5 million, and a position on the firm's powerful executive committee.

At a brutal meritocracy like Bear Stearns, power flows where the profit does, so the betting on the trading desks had Spector as the odds-on favorite to eventually be CEO of Bear. Spector, according to people there, thought he should be running the firm already.

But Cayne wasn't about to turn over the keys to the kingdom, at least not yet and possibly never.

That was because for all the money Spector made the firm and Cayne, whose lofty salary was tied to the profits that were disproportionately generated by the bond unit, over the next few years, Cayne would come to the conclusion that he really didn't like Spector. Longtime executives said Spector acted as if he were smarter than everyone else at the firm. He was probably right, but people at Bear were sick of having his high intellect thrown in their face. Even Greenberg, who now seemed to disagree with Cayne on everything, agreed with him when it came to Spector: both considered Spector a "prick."

Spector was demeaning during meetings; he began ignoring Cayne and Greenberg in the hallways. People thought he was wearing his University of Chicago master's degree on his sleeve while reminding the rest of the guys that they had barely gotten through City College. (Cayne himself, you'll recall, was a dropout from Purdue.) "He doesn't know how to talk to people," Cayne said at the time. "He's not exactly Mr. Nice Guy."

No one's ever accused Jimmy Cayne of being indirect when describing how he feels about people. Friends are "solid citizens" or "studs." But when you fall out of favor with Cayne, you're an "asshole," a "snake," or simply "full of shit." (If you're a woman and you get on Cayne's bad side, he refers to you as "a cunt.")

All were names that Cayne would ultimately use to describe Warren Spector with increasing frequency. Over time, he would mock just about everything about him, from the way he wore his hair to his wife, Margaret Whitton, an actress who was famous for her roles in the *Major League* movies, to Spector's politics: he was a liberal Democrat, while Cayne was a conservative Republican. The political differences became a particular sticking point for Cayne, who couldn't stand the fact that as Spector made more money, he began donating more to left-of-center politicians.

Of course, Cayne had no problem with giving money to Democrats for business purposes—he did it all the time, and he was good at it. He often used campaign contributions to entice both state and local government officials to appoint Bear to underwriting positions on the sale of municipal bonds. Cayne would almost brag how Charlie Rangel, the left-leaning congressman from Harlem, paid regular visits to him, hat in hand, as did, much later on, Chuck Schumer, the liberal senator from New York, and how they never walked away unhappy.

But this was nothing more than business—Rangel sat on the powerful House Ways and Means Committee, while Schumer would eventually become a ranking member of the Senate Banking Committee.

As much as Cayne was starting to hate Spector, there was really nothing he could do to him, not now and, given the firm's business model, maybe never. Among the prospects for Bear's future leadership, there was simply no rival to Spector, except possibly the investment banker Alan Schwartz, who, despite his vast web of contacts in the media business, had neither the brains nor the balance sheet to compete with Spector.

The Bear bond department was simply the most productive part of the firm, far outpacing any other area, and it was Spector's knowledge of bonds, particularly mortgage-backed securities, that showered so many riches on the firm and its senior management.

So Cayne, the master of Machiavellian corporate politics, made his peace with Spector, appointing him to higher positions in the firm and signaling that at a place where money mattered the most, the guy making the most of it would someday have a chance at running the show, even if he really didn't mean it.

That's because Jimmy Cayne, at least for the moment, had bigger fish to fry.

Y ou must fuck like you play bridge! That's right, you fucking jerk-off! You don't know what the fuck you're talking about! You're full of shit!" Cayne screamed into the receiver after getting a call from his boss, Ace Greenberg, according to a former Bear investment banker who witnessed the conversation.

Cayne may have hated and despised Spector because he was arrogant and because Spector wanted his job, but he had grown to hate and despise Greenberg even more. People at Bear used to joke about the moment that Cayne's relationship with Greenberg started to deteriorate. It was in 1987, when Cayne became president of Bear, the odds-on favorite to be the firm's next CEO, and began actively plotting Greenberg's overthrow.

That's when all the brownnosing came to a halt; Cayne stopped picking Greenberg up at his Park Avenue condo and driving him downtown to work, and he started using the English language in a way only Jimmy Cayne knew how.

Cayne also began trying to convince board members about the need for new, younger leadership (he was ten years younger than Greenberg) and aligning key people on the firm's executive committee on the need for a new leader who understood that the firm had grown beyond the Greenberg model—a myopically focused trading house where the CEO sat at the trading desk barking out orders. Bear was in the "bulge bracket" of the Wall Street hierarchy, and it needed a CEO who understood that the firm had arrived.

In other words, Cayne was setting the stage for Ace's eventual ouster. Such moves are usually made when firms are losing money; it's then that boards of directors tend to believe change at the top is needed and necessary. But Bear was always different, both in the people it hired, such as Howie Rubin, and when management upheavals occurred. Cayne made his move as Bear was becoming one of the most profitable of all Wall Street firms, even if much of that success had nothing to do with him.

Jimmy Cayne believed that Bear's good fortune was the result of his good management. "Do you know how difficult it is to deal with these people?" he would say. In Cayne's mind, his management style, which involved stroking the egos of big producers such as Howie Rubin and even Spector (which made him sick), was the secret to Bear's success. It motivated people to work harder and make more money.

But Bear was also benefiting from trends that Cayne's ability as a corporate schmoozer had little to do with. The bond markets, the mortgage market in particular, were booming in no small part because of the federal government. Bear was earning a fortune underwriting mortgage debt issued by the bond market's new nine-hundred-pound gorillas: Fannie Mae and Freddie Mac, the two quasi-governmental agencies known as government-sponsored enterprises.

By the mid-1990s, the GSEs had morphed into something that Larry Fink and Lew Ranieri could have never imagined when they were pitching early mortgage bond deals to them in the early 1990s. They were now massive institutions issuing billions of dollars in debt each year—backed by an implicit guarantee of repayment by the federal government in case of default—to finance their congressional mandate of promoting home ownership. They were also large public companies with boards of directors and shareholders who demanded growth and stock appreciation. They were both clients of brokerage houses (Bear both structured the GSEs' mortgage bonds and underwrote the debt issued by the agencies to finance their operations) and competitors of the big Wall Street firms, which were vying to purchase and sometimes hold the same mortgages the GSEs were packaging and selling to investors as mortgage-backed securities.

More than anything else, they were increasingly the tools of a political establishment that believed the GSEs should be expanding home ownership well beyond their original mandate of making housing affordable to the middle class.

Speculative bubbles, of course, are never caused by any one factor; it's more a blend of multiple instances of irrationality that causes the prices of everything from tulips (in seventeenth-century Netherlands) to real estate (many years later) to explode beyond their intrinsic value. The great housing bubble that would ultimately explode in 2007 and 2008 had many causes: the greed of Wall Street investment bankers, who packed mortgage bonds with increasingly risky loans; historically low interest rates, which allowed banks to lend money freely and to almost anyone; and government policy, which pushed Fannie Mae and Freddie Mac to expand home ownership as a social entitlement rather than something that had to be earned.

The GSEs' contribution to the housing bubble may well be traced to a meeting in Arkansas in early December 1992 between President-elect Bill Clinton and Henry Cisneros, the man who would later run the Department of Housing and Urban Development.

In some ways, the meeting marked the culmination of a triumphant year-and-a-half-long campaign and an even longer friendship dating back to the mid-1980s, when Clinton was the charismatic Arkansas governor who could seemingly pack convention centers just by reading the phone book and Cisneros was the mayor of San Antonio, Texas, considered one of the country's best urban officials for forging close ties with the city's business community.

Both were known as "New Democrats" who believed public-private partnerships could work—that the business community could help achieve the aims of government and spread wealth around.

Cisneros in particular believed that one way to spread wealth and bring the poor into the middle class was to give people homes of their own, so when Clinton offered him a spot in his cabinet, he knew exactly where he wanted to go: HUD, the Department of Housing and Urban Development.

HUD had been a mess, the poster child for government bureaucracy and mismanagement; in the late 1980s, Ronald Reagan's HUD chief, Samuel Pierce, was investigated for political favoritism after leaving office.

During the administration of George H. W. Bush, Pierce's successor, Jack Kemp, an advocate of supply-side economics or tax cuts to spur economic growth, attempted to expand HUD's mandate by calling for "empowerment zones," which involved granting tax breaks to inner-city neighborhoods to spur business activity and letting the poor tenants of housing projects purchase their apartments as a way to prod them into the middle class, but his success was limited. Cisneros wanted to pick up where Kemp had left off, but with a difference. It wasn't just empowerment zones that were needed, he felt. People needed to own their homes, and he believed government could and should help through the ultimate in public-private partnerships, the GSEs

At the meeting, Cisneros went through his list of top priorities: reforming HUD, revitalizing inner cities, and leading a focused effort to decrease homeless rates. The homeless issue hit hard with Clinton. He remarked how he was always amazed at the number of homeless people in the cities. But most of all, Cisneros wanted to increase home ownership, which he believed was a gateway for the poor into the middle class; as the grandson of Mexican immigrants, Cisneros believed home ownership was what had pulled his own family out of poverty and that the entrenched poverty of the inner cities that had plagued minority communities for decades couldn't be addressed in a broad way solely by relying on the free market or the fortitude of striving immigrants.

No, eradicating poverty through home ownership would take big government, working with the banking business and mandating laws to force banks to lend to those most in need and with the least ability to pay. Clinton agreed.

Cisneros believed that by moving the home ownership rate from its current level of 60 percent of the population to 70 percent, the wealth creation possibilities would be tremendous. It was a key sign of the widening gap between rich and poor that lower-income families were being priced out of the housing market. Home ownership penetration into minority and low-income neighborhoods, he said, would be noticeable and real. But to do so, he needed buy-in—he needed the GSEs and private mortgage lenders to be prodded into expanding mortgage lending to sectors of the population that in the past had largely been ignored. *Why* they had been ignored varied depending on whom you asked. According to the banks, loan denials resulted from poor credit and low incomes. According to a 1992 study by the Federal Reserve Bank of Boston, race was the issue, with banks discriminating against minorities applying for mortgage applications.

Whatever the root cause of poor people's not getting loans, to move forward with this plan, Cisneros needed not just an advocate at his side but an enforcer.

know this guy who you will absolutely love," said a voice in a distinct New York brogue. "It's my son Andrew."

Cisneros had been housing secretary for less than two hours when he got an impromptu call from New York Governor Mario Cuomo. Cisneros had known Cuomo since the 1984 convention, where Cuomo had given a rousing keynote speech and Cisneros had been a viable candidate to be the Democratic nominee for vice president. At the time Andrew Cuomo had been working as his father's chief of staff, but he had grand aspirations of following in his father's footsteps as governor of New York State—and possibly rising even higher.

Andrew Cuomo was incredibly well connected, and not just because of his powerful father. In addition to being a Cuomo, he was then married to Robert Kennedy's daughter Kerry. In the late 1980s, he was a partner at Blutrich, Falcone & Miller, a prominent law firm that feasted off its ties to New York State's Democratic Party and its control of state and local government. More than that, he had earned a degree of fame both as a housing advocate (he had created a controversial nonprofit organization called Housing Enterprise for the Less Privileged, or HELP, which built upscale homeless shelters around New York City and in its suburbs) and as a realist when it came to explaining the root causes of homelessness.

Cuomo's report on homelessness in New York City, issued at the behest of then New York City Mayor David Dinkins, earned applause even from conservatives for stating in straightforward terms that homelessness wasn't just the result of bad luck: the homeless were often mentally ill or addicted to drugs. They needed treatment and a decent place to live. Shelters should be run not by the city government but through partnerships with the private sector and nonprofits such as HELP, which were the most capable of providing the necessary services in a cost-effective manner.

Cuomo was completely on board with the administration's ambitious new goal to increase home ownership to 70 percent of the population. That was Cisneros's primary goal, and with Cuomo as his right hand, he would use government policy to spur home ownership in ways not seen in years.

Above all, Cisneros and Cuomo took the first major step of making sure that Fannie Mae and Freddie Mac began serving the poor as never before.

Fannie and Freddie were loosely regulated agencies, created by Congress to increase home ownership. They did this mainly by placing guarantees on mortgages doled out by banks for middle-income borrowers, thus lowering mortgage rates, and through purchasing the mortgages directly from banks, pooling them, and then selling the pools to investors. The process was supposed to give the banks more latitude to make home loans to working-class families.

Complicating matters was the fact that the GSEs weren't normal federal agencies. By the early 1970s, both were public companies, and by the 1990s, their shareholders included some of the biggest institutional investors in the financial business. Because they had public shareholders, the GSEs' mission wasn't just to spread the wealth through home ownership, but also to make money.

The tension between these two missions had haunted the GSEs for years. Who should receive affordable housing was always a matter of debate. Housing advocates criticized the agencies for not doing enough for the poor by extending loans to riskier borrowers. The agencies themselves countered that since they were public companies, they had to focus on helping families who could afford to repay the mortgages they were guaranteeing.

Ramping up the pressure was the 1992 study by the Federal Reserve Bank of Boston that had concluded there was widespread racial and income discrimination in the mortgage-lending process; in other words, people were being denied mortgages based purely on their race and income level, even if they had the means to repay those mortgages. The study purported to prove statistically that many minorities might have had bad credit scores, but they also held jobs and were being denied mortgages even though they had the ability to repay them.

The study argued that the traditional ways of weighing creditworthiness—credit scores and loan-to-value ratios—were discriminatory, and it became a rallying cry for housing advocates, including some members of Congress, such as Barney Frank, the chairman of the House Finance Committee; Maxine Waters, the California congresswoman who made home ownership for minorities one of her key crusades; and a senator from Connecticut, Christopher Dodd.

Though conservatives argued that the Boston Fed study presented faulty data and ignored how the deep levels of poverty in minority communities made home ownership difficult, its main conclusion was daunting and heralded in the media as proof of racial disparity in mortgage-lending practices.

The Clinton administration viewed the study and the controversy it sparked as political dynamite and immediately began to address its concerns through an expansive housing policy. The Community Reinvestment Act, which traced bank lending policies in poor and minority neighborhoods, would be enforced to its fullest. The Federal Housing Administration ramped up its insuring of home loans going to the poor, making mortgages more affordable. Cisneros banded together with regulators at the Treasury Department and Federal Deposit Insurance Corporation to prod—the banks would say "force"—them to cut transaction fees and down payment costs to make mortgages more affordable.

Even more, the GSEs, Fannie Mae and Freddie Mac, would be forced to use their massive and growing financial clout to guarantee and buy loans from banks that served minority and other lower-income home owners.

In 1995, Cisneros began his first major effort to influence Fannie and Freddie to dedicate more of their resources to providing mortgages for low-income families. His new measure directed the GSEs to set aside 42 percent of all their mortgage guarantees to serve what the government classified as low- to moderate-income borrowers.

When Andrew Cuomo succeeded Cisneros as HUD secretary in 1997, he increased that number to 50 percent and began to pressure the GSEs to buy up mortgages of people classified as "very low income."

If this move was an important step toward the democratization of the housing market, it was also an important step toward the expansion of risk in the financial markets. For the first time, it opened the GSEs to the part of the housing market that dealt with so-called subprime borrowers.

Subprime loans are the riskiest of the risky mortgages. They were initially limited to people with spotty credit ratings—maybe a missed credit card payment or two—but were actually dependable borrowers when given limited amounts of credit. Later, as the housing market picked up steam, the definition expanded and the pool of subprime borrowers now included people with little or no credit history, some who couldn't document their incomes, many others who didn't even have a regular job.

Of course, giving low-income families and minorities that had been shut out of the housing market assistance to purchase their own home is a laudable goal, though it totally ignored the fact that the surge in lending to the poor would push housing prices up beyond what average people could afford without government help or gimmicky adjustable-rate mortgages.

As the GSEs continued their expansion into areas of the housing market long dominated by private lenders and later Wall Street, their power continued to grow. Bankers still gripe about the "report cards" the GSEs handed out that graded the firms on how much information they supplied Fannie and Freddie. They complained that the GSEs' implicit backing by the government amounted to unfair competition because it allowed Fannie and Freddie to borrow cheaply to finance their purchases of loans.

But what Wall Street could not complain about was the enormous amount of money it made from the bubble Fannie and Freddie were helping to create. For nearly twenty years, Wall Street sucked in billions of dollars in fees and investment gains from pools of mortgage bonds that Fannie and Freddie guaranteed. With the GSEs ensuring the ever-growing loan pools, the housing boom seemed destined to continue forever, and, as a result, so would the fees and profits the big brokerage houses made by securitizing those loans into mortgage bonds.

It was as close to a "sure thing" as Wall Street could ever experience. But if there is a single given in the history of Wall Street from its founding under a buttonwood tree in 1792 to the present, it's that "sure things" don't last forever.

At Bear, the one sure thing appeared to be Jimmy Cayne's relentless drive to become CEO.

Bear earned record profits in 1992 and again in 1993. Greenberg and Cayne were now earning around $15 million each year, the highest salaries of anyone at a major Wall Street firm at the time. On the surface, it seemed that the Greenberg-Cayne team was working better than ever.

Cayne and Ace shared duties, each doing what he liked and was good at. Cayne schmoozed with big clients and set the general direction of the firm. He hired brokers and worked the political end of municipal bond deals. Ace was the straight man who traded for clients such as Donald Trump and managed the firm's risk position. He hired people he called "ferrets" to snoop on traders trying to cover up losses and back-office help stealing paper clips. More than anything, Greenberg relished his role as chief risk officer, and it didn't matter who came into his crosshairs. In fact he was so unfazed by even Warren Spector's intelligence that during one meeting he snapped, "You've got fifty million dollars in bonds that are four months old—that's terrible. It's out of control!" Spector, known for his sharp-edged retorts, just sat in silence.

In reality, Cayne was tightening his control over the firm—and changing its tone. His hiring went beyond brokers to include other producers in other areas of the firm. The firm's brutal meritocracy was contained by Greenberg's high ethical standards; he was proud of the fact that under his watch the firm was rarely involved in any major scandals. Cayne didn't go out and tell his traders and bankers to break securities laws, but people internally saw a different tone in business practices as Cayne tightened his grip. "It came to the point where we were expected to go right up to the line of legality if it meant making money," said one executive who worked at Bear during the 1990s.

Cayne changed the firm in other ways: Bear began to expand into investment banking, secure in its spot as the leading firm in the highly profitable business of processing, or "clearing," trades for hedge funds and brokerage firms. For a guy accused of smoking joints at the office, he became so concerned, even obsessed, with the firm's image that one day he issued an edict: the old security guards, many of them former cops, who escorted clients to the executive suites would be replaced by what he called "nice, clean, pretty women." Those pretty women—dressed in provocative, low-cut outfits—were soon known inside Bear as the "geisha girls," and they caused an uproar among some of the senior female executives, not just at the firm but across Wall Street when the move created some press attention. Cayne's explanation that "we had to get rid of those guys because they looked like crap and smelled like they just drank a fifth of old McMullen" didn't work; women's groups on Wall Street protested. It wasn't long before the notoriously stubborn Cayne capitulated, somewhat. The women would stay, but he made sure they dressed more conservatively.

The geisha girl incident added to Cayne's reputation among women as a sexist pig—in a magazine article Cayne was once quoted as saying that women are both lousy brokers and traders because when facing pressure, they would "probably have to go to the ladies room and dab [their] eyes"—but the geisha girls took his stature to an even higher level on the less-than-PC Street: Bear was a place that lived up to its reputation as a no-holds-barred meritocracy that put up with wild personalities, and no one could be wilder than the cigar-chomping, pot-smoking president of the firm, who was on the verge of becoming its CEO.

By the mid-1990s, most of the top people at the firm had either been hired by Cayne, owed their jobs to him, or were simply Cayne loyalists, having seen the power at the firm shift away from Greenberg, who cared so little about the politics of running a big firm that he let Jimmy Cayne, a man consumed with office politics, have free rein. And now it would cost him.

Over a period of weeks, Cayne had been lobbying key members of the firm's executive committee and had lined up their support for Greenberg's ouster and his elevation to CEO. He went to Greenberg with the news. At first, Greenberg said he was ready to fight, but that was before Cayne laid out the odds of him winning—"zero," he said. The executive committee was on his side, and so was just about everyone else who mattered. Cayne said it would be the worst trade of Greenberg's life.

Greenberg slept on it a night and came back and told Cayne the job was his. That day, Cayne called an executive committee meeting and proposed a new management team: he would be named CEO, Greenberg would remain as chairman. The vote was unanimous, and just like that the Ace Greenberg era at Bear Stearns was over and the era of Jimmy Cayne had begun.

Okay, this is what I want," Cayne said one morning not long after becoming CEO, as he handed his PR woman, Hannah Burns, a report he'd commissioned from PricewaterhouseCoopers that described Bear's corporate culture. "I want this to be a story in the *Harvard Business Review.*"

The "study," which Bear had paid for, spoke in glowing terms about how Cayne had used his skills at bridge to build one of the great American investment banks. It mentioned how he had been able to ingratiate himself with key officials in the Chinese government because they were also bridge players, to win a lucrative financing deal. It spoke about his big, successful New York City bond bet and how he used his "instinct and judgment" to bet big and win, much as he'd done during bridge tournaments.

It spoke about Cayne's hiring philosophy, which the new CEO articulated during a St. John's University commencement address: "Education, title or rank no longer confer power. . . . People are expected to create their own opportunities. . . . Those who wait for someone to tell them what to do quickly find themselves on the sidelines, or out of work."

A few bones were thrown to Greenberg, now Bear's chairman, about how he still conducted his legendary "cold sweat" risk meetings. But mostly the study was an encomium to Cayne and his alleged brilliance as a manager—and ironically, as a *risk manager*. The study described how Cayne had been confronted at a recent industry conference about how Wall Street firms were taking on more risk than ever before when trading complex bonds and derivatives. He had been questioned about the possibility of a catastrophic event imploding one or more of the big Wall Street brokerage houses.

"Jimmy's response was immediate and confident: 'If the industry goes so will Bear Stearns. However we will be the last standing. Our deliberate approach to growing such businesses and our controls ensure our continued stability.'"

What struck people at Bear as odd about the statement was not the fact that Cayne was prophetically predicting the doom that would ultimately engulf Bear and the rest of the Street but the irony of his talking about risk. He liked to take it, but he never demonstrated a deep understanding of how to manage it and chose to leave that part of it to others. The few times he showed up at Greenberg's risk management meetings, he said and did almost nothing apart from chomping on his cigar.

Risk meetings just weren't that important to Jimmy Cayne. What was becoming important to Jimmy Cayne was Jimmy Cayne, as people in the firm were now finding out.

Unlike Cayne, Greenberg had never really cared much about press attention; the only thing that got him riled up was the notion that the firm was loaded with gunslinging traders who thrived in an entrepreneurial culture like Bear's because they ate what they killed. "Goddamn it," Greenberg snapped one morning at Burns after reading a story in the *Wall Street Journal* about Bear's risk-taking ways, "tell them we don't trade, we invest!"

But Cayne had grander plans for Bear's image, because Bear's image was increasingly his own, as he continued to take control of the firm and, at least in Greenberg's view, marginalize the man who had built the company: blocking his suggestions at meetings, ignoring him at lunch, cursing him out, and making it known that the buck stopped at the CEO's office, not the chairman's.

Bear would hit a relatively small pothole in 1994, compared to the massive implosion elsewhere, as the Wall Street business model of leverage and risk faced its first major threat. It wasn't just Gerry Corrigan at the New York Fed sounding the alarm bells. There were calls in Congress for regulation of the trillions of dollars in complex bonds, mortgages, and swaps all loosely labeled as "derivatives," which suddenly became a word that fascinated the political class. According to *Institutional Investor* magazine, even the fringe presidential candidate Lyndon LaRouche likened derivatives to "unemployment, hunger, and the spread of AIDS." LaRouche's solution to the excessive risk taking didn't seem so fringe: a massive tax should be imposed on certain trades.

Then House Banking Committee Chairman and noted populist Henry Gonzalez called derivatives a "monstrously global electronic Ponzi scheme." Corrigan, still at the New York Fed, brought together the top dealers in swaps and bond derivatives to study the use and misuse of derivatives and how they could lessen the likelihood of systemic risk causing a global trading meltdown. More mainstream politicians joined the debate as well. Congressman Edward Markey, who had chaired hearings on the Salomon Brothers bond scandal, told *Institutional Investor*, "The market appears to have sprinted beyond the vision of the umpires and the referee."

Then, just like that, all was forgotten. The Fed's policy of raising interest rates ended, and profitability returned to Wall Street, as did increased risk and leverage. Corrigan left the Fed and went to work for Goldman Sachs, becoming Wall Street's master risk taker. He also changed his tune quite a bit about the need for new legislation and regulations to clamp down on risk taking.

"When I say I don't think legislation is needed," he explained, "I'm not saying that I'm satisfied with the status quo. But the things that need to be done can be done under existing legislative authority."

By 1996 the topic of conversation had turned away from excessive risk to excessive profits. Wall Street was swimming in money, thanks in part to the dot-com boom but also because the average leverage of the big firms had risen to 27 to 1 after momentarily dipping after the 1994 bond market rout. The asset mix that the firms embraced with that leverage had changed as well: bonds that provided the highest returns were in as interest rates began to fall, making low-risk trades less desirable.

Meanwhile, Jimmy Cayne and Dick Fuld, who ran the two smallest firms on the Street, had learned a valuable lesson in their quest to survive on Wall Street. Their firms had about half the equity capital of either Merrill Lynch or Morgan Stanley (Goldman Sachs was still a private partnership), yet both firms were surviving and thriving for the same reason: they shared the same penchant for taking risk—and it was working. In 1996 Lehman and Bear were Wall Street's most highly leveraged firms, borrowing an identical $32 for every $1 the firms had in capital.

It's easy to forget the perils of risk and leverage when you're making so much money. Well into the latter part of the decade, the debt merchants were on a roll; mortgage bonds at Bear, Lehman, Kidder Peabody (when it was still in business), Merrill Lynch, and others brought in colossal fees, so colossal that Bear's mortgage chief, Warren Spector, earned $12 million in 1992 and $20 million by 1999—many times what Larry Fink and Lew Ranieri had pulled in during their heyday in the 1980s. Michael Vranos, when he was at Kidder, wasn't far behind, and after the firm blew up, Vranos started his own hedge fund, Ellington Capital, where he made sums equal to those of any of his peers on the Street by trading mortgage debt.

The types of mortgage bonds—the CMOs, IOs, and POs—that had been industry standards in the 1980s now seemed almost quaint. The mid-1990s brought new variations, "the sticky-jump Z," other variations on the CMO, and especially the new collateralized debt obligation, or CDO, a mortgage bond that is actually a bond within a bond; a CDO is made up of various types of asset-backed securities, from bonds comprising home loans to those containing credit card receivables.

It wasn't long before Wall Street created two instruments, the "synthetic CDO" and the "CDO squared," derivatives of the original security that pushed the investor even further away from the ultimate sources of payment, namely home mortgages.

Why would anyone want to buy a CDO squared? One reason was the higher returns. Another was their accounting treatment. There was a perverse relationship between risk and disclosure under the accounting laws: the more exotic the security, the more it could be defined not as a bond but as a derivative that could be hidden in off-balance-sheet structures and corporations.

It was unclear if Wall Street's risk models would be able to keep pace with the innovations that began to crop up almost by the day in the bond markets, but some people were trying. By the early to mid-1990s, Larry Fink, who had lost so much money at First Boston, was busy building his money management firm, BlackRock, into a powerhouse. He had spun BlackRock out of the private equity firm Blackstone and would take it public before the end of the decade, setting him on his way to billionaire status.

But more than that, Fink's reputation had done a 180—he had become an industry leader and an expert in risk and the bond markets, particularly the market he had helped create for mortgage securities.

Fink now had a firm of a hundred money managers and traders and an entire floor in midtown Manhattan. In addition to managing $20 billion in bonds, much of it in the same mortgage debt he had gambled and lost with at First Boston, BlackRock had quickly developed a reputation of creating some of the best risk analytics. In 1993 Fink's risk models began to spit out data that showed that the bond markets were heading for disaster. The mortgage bond Fink had helped create—the CMO—wasn't quite obsolete, but it had been supplemented by all the new, complicated derivatives. The new mortgage bonds now were everywhere, held by banks, the big firms themselves, pension funds, and even school districts. Low interest rates had fueled their growth, and Fink knew that a sudden change in rates would make the reversal of fortune he and others had suffered in the 1980s look like child's play.

When the markets imploded in 1994, Fink had his opening to both make a lot of money and take his reputation to new heights. The opening occurred when Kidder Peabody collapsed, and Fink knew exactly what to do. One of his first calls was to GE Chairman and CEO Jack Welch, who was looking to unload the firm from GE's vast stable of businesses, but to do so without taking massive losses on the soured mortgage investments that had been made by the Vranos team.

Kidder's mortgage portfolio had a face value of about $7 billion; the mortgage market was locked down, so if Kidder sold it now, it would be worth far less, just pennies on the dollar. Welch's plan was to put the portfolio out to bid; the firm with the best analytics and capability to come up with a way to hedge the portfolio against future losses and figure out a way to unload the securities, possibly over time and hopefully at a profit, would win.

The bidding was fierce, and Fink had a lot on the line, namely his rehabilitation from his humiliating exit from First Boston, which still elicited occasional snickers from the Wall Street elite he dined with at San Pietro. Fink had met Welch years earlier, and the two had a cordial relationship; moreover, BlackRock managed GE's pension fund assets, no small selling point in BlackRock's pitch. But what sold Welch was Fink's risk management system: how he had valued the Kidder portfolio and how he planned to unwind Kidder's positions in a way that not just minimized the pain but also provided a possibility of some gain.

Fink won the bid, and his risk management systems—he called them "hedging systems"—helped GE avoid massive losses, and with that, he was again a star. Gone was his reputation as a wild trader who had nearly toppled First Boston. He now had the imprimatur of having helped Jack Welch, the best CEO in corporate America, save money and avert disaster.

He also had the image of a risk manager, even if his "hedging systems" were amazing after-the-fact inventions used to price securities when they blew up and sell them off in an orderly fashion when the markets revived, as they were now doing.

Larry Fink had become Wall Street's cleanup guy, and there would be far bigger cleanups to come.

7. PAYING THE PRICE

Wall Street had sold out before, performing the usual dance of ignoring its past mistakes and ramping up the risk level in order to make money. But each time, as we've seen, the risk had gotten higher; the "moral hazard" created when the government sought to lessen the pain of risk taking gone bad had increased, giving added incentive to Wall Street's reckless bets at a time when the complexity of the bonds and other securities bought, sold, and held on the balance sheets of the big firms increased.

In 1995, after five consecutive interest rate hikes, the Fed began to change course, bringing interest rates down almost as fast as it had raised them. With that decline a flood of credit came rushing into the banking system as banks began lowering the interest rates they charged everyone from home owners to corporations to Wall Street traders who tapped lines of credit to finance their return to the bond market.

Mortgage bond desks started hiring again, and investors' appetite for the risky securities returned. A collective amnesia was now sweeping the Street. Leverage returned with a vengeance. Lehman, which had briefly cut back on borrowing to below 30 to 1, now began to ratchet it up again. For every $1 Lehman had in capital, it now borrowed a whopping $32 more. Merrill Lynch, which had vowed to permanently cut its leverage and risk taking after the Howie Rubin fiasco in the 1980s, was now trading and borrowing even more, $33 for every $1 in capital. Bear Stearns' leverage approached 40 to 1.

Traders were once again kings. At Goldman Sachs, power shifts according to the source of profitability, and the bankers who had taken charge in the late 1980s and early 1990s were replaced by the people making the money, the bond traders. Jon Corzine, the former head of the government bond desk, was now in charge of the entire firm. Dick Fuld was fully in command at Lehman and alone at the top, and he did what he knew best, ordering his army to begin trading—massively.

To be sure, Lehman in the second half of the 1990s had attempted to diversify from its post-IPO status as a bond shop that merely dabbled in other businesses. Fuld had since built up the firm's ranks of investment bankers to make it appear that it was now in the league of Goldman Sachs and Morgan Stanley as a full-service, diversified brokerage firm. Its research department was among the better ones on the Street. It had a growing brokerage department that focused on wealthy investors.

But its real power came from trading and underwriting bonds, which were increasingly tied to commercial and residential mortgages. The Lehman bond department generated two-thirds of the firm's profits. And the trading had become more complex. Fuld began cranking up its leverage to nearly where it had been before the 1994 market crash; Lehman was now looking to squeeze every dollar it could out of its thin capital base so it could make bets in bond markets all over the world, including the once-obscure but now-booming market for Russian bonds as the former Soviet Union embraced capitalism.

Leverage reached a crescendo in early 1998 and with grand results. Wall Street capped off nearly four years of strong growth with its best quarter in years. Lehman posted a 30 percent rise in profits in March 1998; Bear, Morgan Stanley, Goldman, and Merrill weren't far behind.

Fuld, who had been considered a dead CEO walking just four years before, was rewarded for the string of strong results, receiving a bonus of $13.3 million.

Then, just as it had in 1994, in the summer of 1998 the markets hit a pothole that turned into something much worse. It began with Russia devaluing its currency and then defaulting on part of its debt, setting off a global bond market panic.

In such cases, a familiar pattern can be seen. Bond desks around the globe suddenly realize that they have undercharged to take risk. Traders and investors begin unloading the most esoteric bonds in the market—first mortgage-backed securities, then high-yielding corporate bonds, or junk—and begin what's known in the markets as "a flight to quality" to Treasury bonds, long the safest security in the world because they are backed by the full faith and credit of the U.S. government. Often scorned because of their relatively low returns, Treasuries are embraced at such times.

The markets begin to focus on systemic risk, specifically, the risk that the vast global system of borrowing by issuing debt will implode amid widespread losses. As this fear picks up momentum, panic selling spreads and losses mount. In the worst case, financial Armageddon occurs: banks are hit with so many losses that they stop lending; traders are bleeding so badly that they no longer make markets in stocks; and the vast global system of lending and borrowing grinds to a halt. The yield spreads between the supersafe Treasuries and the esoteric mortgage bonds tell the story. Yields move in the opposite direction from prices—the higher the price, the less a trader demands in yield or return. The spreads were once narrow, just a percentage point or two, reflecting the market's belief that even the most bizarre mortgage bonds created were nearly as safe as Treasuries, and thus traders didn't demand extra yield to compensate for their risk.

But with reality settling in, market sentiment changed; suddenly the prices of Treasuries skyrocketed, while the prices of mortgages and junk bonds plummeted; spreads widened to levels not seen since 1994. A truly global bond market panic had begun, and Wall Street was now getting ready to pay for its partying with risk and leverage with hundreds of millions of dollars of losses that would last through the end of the year.

By 1998 the average leverage of the top five Wall Street firms, Goldman Sachs, Morgan Stanley, Bear Stearns, Merrill Lynch, and Lehman Brothers, was close to 29 to 1; eight years earlier they had been leveraged around 20 to 1. Such borrowing demands a huge level of trust; if for some reason fears spread that a bank might be taking huge losses, that trust could dissipate instantly and the bank's funding could dry up literally overnight. Even worse, in terms of risk, was what Wall Street was leveraged *with*: an analysis by SNL Financial, a market research firm, showed a doubling, tripling, and for some of the firms quadrupling of the volatile mortgage-backed securities held on their balance sheets.

Many of these bonds held Triple-A ratings, but bond market panics, whether they're sparked by higher interest rates, as in 1994, or the Russian default of 1998, do nasty things to mortgage debt. These bonds are usually the first victims of the carnage because they are so esoteric and so highly leveraged *internally*—mortgage bonds are, after all, packed with thousands of loans where the borrower is putting little to nothing down. Put all this together, and you have a recipe for disaster. Traders run for the safety of Treasury securities while selling everything else—corporate debt, junk bonds, even debt issued by the GSEs, Fannie Mae and Freddie Mac, but mostly mortgage-backed securities.

The same type of run on the bank that had doomed S&Ls during the Great Depression as people raced to pull their deposits from banks feared to be on the edge of collapse, thus bringing about that very collapse, was threatening a return. Only this time, instead of commercial banks, the run threatened Wall Street as creditors began to panic about which firm had lost the most money and wouldn't be able to pay its bills.

Compounding the fear was that those bills were increasing short-term loans from the "repo" market. By 1998, such borrowings—as opposed to safer types of funding such as issuing long-term debt and securing lines of credit—accounted for more than half of all of Wall Street's funding. When times are good, repos are fast, cheap, and efficient. A firm could raise billions of dollars overnight.

But repos are also a risky way to raise money; creditors can refuse to take anything but the highest form of collateral such as Treasury bills, or simply refuse to lend, as Dick Fuld and Lehman Brothers were finding out.

Because of its small size and low capital base, the rumor mill fingered Lehman for the biggest losses, and a run on this bank, as creditors became increasingly skittish about repoing cash to Lehman, took its first tentative steps. Dick Fuld, an old commercial paper trader, saw the situation first-hand; his financing desk gave him the word that creditors were demanding additional capital on repos. But more than that, investors were demanding additional yield on Lehman's commercial paper to make up for the risk that the firm might not make good on its obligations because of its losses. The word from the trading desk was that rumors were spreading among counterparties that Lehman was taking such heavy losses it might not survive.

By now Fuld and his small coterie of advisers had been together long enough, survived enough brushes with death, that they were battle-hardened, almost literally. For years now, they had used war metaphors to describe doing business on the Street, whether trading or fighting for investment banking deals.

Fuld reveled in the fact that the firm had survived past brushes with death and vowed it would fight and survive this one. "It's us against them," he began telling his senior staff as they girded for what would be the firm's biggest battle for survival. "We're in a war that we've got to win!"

Fuld's bluster and warlike attitude were regarded by some as an attempt to instill loyalty and esprit de corps among the troops. As Lehman's funding problems evolved into a funding crisis, he was in full battle mode. He ordered his senior advisers to assure friends and contacts on the Street that the firm was fine—not even close to insolvent—although according to several people who worked at the firm at the time, Lehman had huge exposure to commercial real estate, which had been hammered by the fallout from the Russian default, as well as other bonds that were now underwater.

Lehman had been the leading underwriter of mortgage bonds tied to commercial real estate, which before the markets imploded had grown to $200 billion a year. These deals had powered the firm's first-quarter earnings, but now they were powering a potential run on the bank.

Lehman was stuck in a vicious circle. Its losses were made even worse by the leverage it had taken, and because it was so highly leveraged, it didn't have the capital on hand to weather a prolonged crisis, particularly one that would eat away at the confidence of the counterparties who were needed to keep lending the firm money in the first place.

Nonetheless, Fuld called up rival CEOs, such as Jon Corzine at Goldman Sachs, where he believed the rumors about the firm's heavy losses had started, and demanded that Corzine order his traders to stop spreading them. Fuld at one point accused Goldman of leading the charge against Lehman so it could steal away large institutional investors—clients that would naturally be wary of trading through or lending Lehman money if it were about to be liquidated because of massive bond losses.

Corzine denied the charge, as did the others, but the rumors didn't stop; in fact, speculation about Lehman's health continued. To this day, Wall Street executives are split over Fuld's motives in making the calls. Some say he was truly outraged because the rumors caused some of Lehman's creditors to think twice before lending it money. Others say he was bluffing; he knew Lehman was in trouble and was doing anything and everything he could to keep the firm alive.

Whatever the reason, there was certainly a degree of desperation in Fuld's attempts to prop up his firm. While he was attacking competitors, he sought relief from anyone who could help out. One call he made was to Richard Grasso, the chairman of the New York Stock Exchange. Grasso's job was twofold: he ran the world's largest stock market, enticing companies to list their shares to be traded on the Big Board, but he also served as a regulator of Wall Street firms that were members of the stock exchange.

Under federal securities laws, the NYSE, along with the other major U.S. stock market, the NASDAQ, not only serve as markets but also provide frontline regulation, making sure that Wall Street firms, such as Lehman, that trade on them comply with securities rules and meet solvency standards.

It was an arrangement rife with conflict; on the one hand, Grasso was regulating firms like Lehman, but on the other, it was his job to protect the member firms that conducted the lion's share of business on his exchange. Fuld and Grasso had bonded over the years, and they were frequent dinner partners, occasionally eating at the exclusive Rao's restaurant in East Harlem. Grasso had appointed Fuld to the NYSE board, which had later awarded Grasso one of the highest salaries on Wall Street—more than $15 million a year.

Grasso now launched an examination of the books of the biggest brokerage firms, and in pretty short order, he called Fuld to tell him the NYSE had taken a snapshot of the firm's finances and capital levels and given Lehman a clean bill of health.

In a subsequent interview, Grasso said he had given Lehman nothing more than a tentative thumbs-up so it could fight another day. Maybe so, but the effort worked; creditors backed off, and Lehman survived.

For now.

When it comes to making bets in exotic bond markets—whether mortgage debt or Russian bonds—things can always get worse. And so they did in early September 1998, when John Meriwether, one of the most successful debt merchants of the past two decades, who had left Salomon Brothers earlier in the decade and launched perhaps the most successful hedge fund of all time, Long-Term Capital Management, suddenly made a disturbing announcement: his hedge fund was on the verge of losing billions of dollars. Its risk models, thought to be the best in the world, were wrong, and because much of the rest of Wall Street had copied his trades, the entire financial system was on the verge of collapse.

The unraveling of LTCM, as told best in Roger Lowenstein's book *When Genius Failed: The Rise and Fall of Long-Term Capital Management*, can be traced to the impetus that had led to Meriwether's rise and fall at Salomon and had led him to start a new career as a hedge fund manager: his desire to take risk. But Meriwether was less of an addict than a student of risk. LTCM's premise was to manage risk for *long-term* results. The hedge fund relied on the best risk management techniques of its time, including the calculation known as value at risk, or VAR, which had been popularized in the early 1990s at J.P. Morgan and had since become a staple of Wall Street risk managers because it could assign a dollar value to the amount of money that could be lost, given a set of risk parameters.

The problem with VAR, however, is that it measures what will likely happen 99 percent of the time—which sounds good unless those odds are compared to others. The odds on winning the California SuperLotto jackpot are 1 in 18 million, and people win all the time. The odds of being struck by lightning are 567,000 to 1; the odds of being killed in a plane crash are around 52 million to 1.

Would anyone in his right mind board a plane knowing that there was a 1-in-100 chance of its crashing?

That's what LTCM (and Wall Street) had just done, and it was paying the price in the late summer and early fall of 1998. Measuring what happens 99 percent of the time doesn't do much when the Russian government suddenly and unexpectedly defaults on some debt and a global domino effect begins affecting trades that nominally have little to do with whether the Russians will make good on their loans but are stigmatized nonetheless as investors recalibrate risk on a global basis, selling products deemed too risky and esoteric and hiding in the safety of U.S. Treasury bonds.

LTCM's losses approached $5 billion, but that was just the start of it. Most of the big Wall Street firms had money in the fund. Some of the Street's leading executives had pieces of their personal fortunes tied up in the hedge fund—Jimmy Cayne and David Komansky, Merrill's CEO, to name two.

Even worse, many of the firms had piggybacked on LTCM's trades, meaning they stood to lose hundreds of millions, possibly billions of dollars. The trades were in some of the most bizarre debt instruments ever created; one of LTCM's favorite trades wasn't even in U.S. mortgage debt, as complicated as those bonds had been, but in Danish mortgage bonds, of all things.

The potential damage went still further as bond investors ran for cover, fleeing to the safety of Treasury bonds and selling or trying to sell everything else. In such a panic, mortgage bonds often take the biggest hit because of their heightened volatility in a crashing market. Hedge funds that had feasted on mortgage debt and produced huge returns during the past couple of years, such as the Ellington Fund run by former Kidder Peabody mortgage chief Mike Vranos, nearly imploded as well. Ellington was saved by an arrangement worked out by a consortium of the fund's creditors, including, significantly, traders at Bear Stearns, who felt that Vranos was too big a client to be left for dead. Others weren't so lucky. A subprime mortgage lender named Cityscape Financial filed for bankruptcy during the crisis, foreshadowing problems to come in this still small but growing part of the mortgage business.

Wall Street was now getting an education in the notion of systemic risk as the financial panic began to pick up steam once the Wall Street establishment discovered the true extent of its exposure to LTCM. Until 1998, the term had rarely been used in financial circles except in academic settings because the notion that big losses generated by one player, such as a hedge fund, would create a mass panic of selling was considered unthinkable.

But as news of LTCM's problems began making their way past Fuld's corner office and into the executive suites of Morgan, Goldman, Merrill, and the rest of the Street, the talk was all about the vast and systemic damage LTCM's demise might cause.

As it turned out, not only had the big firms copied the hedge fund's losing bets, particularly in the mortgage bond market, but they had also relied on LTCM's risk management systems. LTCM's calculations of value at risk had never taken into account the low-probability disaster that had just occurred, nor the possibility that LTCM's hedging strategy would collapse. The fund had shorted Treasuries, making a bet that Treasury bond prices would decline—which proved disastrous when Treasury prices rose dramatically as traders made a flight to quality. Wall Street had copied those trades as well. The concern about systemic risk had spread to where it mattered most: the Federal Reserve and the Treasury Department, where top officials, including Fed Chairman Alan Greenspan, a staunch supporter of the use of derivatives and the controversial trades that were at the heart of the crisis, decided that LTCM was "too big to fail." The fund was so big and its trades so complicated and touching so many players in the global financial markets that if John Meriwether and the LTCM brain trust were forced to liquidate their positions, the losses would grow and spread.

And spread was just what they were doing. Goldman Sachs was regarded as the best trading firm on the Street. But its LTCM-related losses were so high that Goldman had to drop its plans to convert from a private partnership—the last major one on the Street—and join the rest of the financial industry as a public company. Merrill lost nearly as much as Goldman—$150 million or more in just one quarter—and was forced to lay off thousands of employees while drastically scaling back trading in the face of its market losses.

And Lehman, having just barely survived the Russian debt crisis, was now fighting for its survival once again.

O kay, so what do you want me to do?" Richard Dickey asked with more than a hint of sarcasm as he listened to his supervisors reports that, on the one hand, despite all the rumors, all the stories in the newspapers about Lehman's desperate situation, the firm was actually fine, but that, on the other, he also needed to begin making calls to his best contacts on the Street and ask them if they would be willing to lend Lehman money on an emergency basis.

Dickey had just joined Lehman as head of equity sales after working several years at Morgan Stanley. He had left his job at Morgan for the reasons most people on Wall Street leave their jobs: money and career advancement. Lehman had given him both, but now he was questioning whether it was worth it. Lehman was a bizarre place, he had learned. It was more akin to a cult of personality surrounding its CEO, Dick Fuld, than a firm that encouraged creativity through debate, like Morgan and Goldman or even Bear Stearns. At Lehman, decisions were handed down from the top—namely, Fuld and his team of advisers, which included, most importantly, Joe Gregory, Lehman's president and number two, who was also one of Fuld's closest friends on the Street. Dickey resigned himself to get with the program, but now management was asking him to do something so outrageous he couldn't bring himself to carry it out. They wanted him to call up his best clients—Dickey was close to Bruce Calvert, the CEO of Alliance Capital, a big institutional investor—and spin the best story possible. "Are we running out of money?" Dickey asked. "Are we having trouble rolling over our commercial paper? If everything is fine, why am I making these calls in the first place?" The answers he received were tentative at best. He was ordered to talk around those questions and highlight the positives of the firm—its strong management in Fuld and the rest of his crew. "Hey, these guys you want me to call are my best clients, and I can't fuck them," Dickey shot back. "I need to tell them *something.*"

"Trust us," was the answer he got, "everything is fine."

What concerned Dickey the most about Lehman was the odd way Fuld and his team responded to the rumors: they said almost nothing to employees, except for making an occasional statement along of the lines of what he had just been told. Though Fuld was threatening other CEOs, inside Lehman management acted as if there were something wrong; Dickey could see it on the faces of the people on the bond desk. They looked as if they were scared, and they said almost nothing. So Dickey did some digging on his own, hitting up every source he knew at the firm, and what he found nearly brought his world crashing down: Lehman wasn't insolvent, but it was clearly moving in that direction.

Creditors formerly worried about Lehman's exposure to Russian debt now turned their concern to Lehman's exposure to LTCM's losing trades. They were demanding billions of dollars in collateral on repo; investors who normally bought the firm's commercial paper were starting to walk away.

Dickey, like any other good soldier in Fuld's army, did make the calls that were asked of him. But he began every call with a standard disclaimer—"I have to do this"—before making his pitch. Dickey found few of his sources willing to help out, as Lehman remained in the crosshairs for weeks. Though the attention of the press and the rest of the Street remained focused on Lehman, at least one other large firm was feeling the heat as well, setting the stage for an implosion the likes of which the Street had never seen before.

8. ONE BIG HAPPY FAMILY

Mighty Merrill Lynch, whose massive brokerage operation of fifteen thousand financial advisers who sold stocks and bonds to investors across the country was known as the "thundering herd," wasn't in Lehman territory yet, but it was coming alarmingly close, so much so that Stan O'Neal, the CFO (and eventual CEO) of Merrill Lynch, began to worry about a bank run as well.

Apart from the CEO, the CFO holds the most important job at any Wall Street firm; part company flack, pushing the firm's stock to investors, and part finance minister, making sure the firm is well funded and has access to capital and funding to support its operations. Since taking over earlier in 1998, O'Neal had noticed something odd about Merrill's finances. The firm's leverage had now hit $33 for every $1 it had in capital—a jump of 65 percent in just a few short years to one of the highest levels on Wall Street. Moreover, Merrill was increasingly dependent on short-term funding in the risky repo market, where its very existence was increasingly hostage to investors' willingness to extend the firm a line of credit.

The LTCM collapse took the bond market turmoil sparked by the Russian debt crisis to a new and more dangerous level. Merrill and Lehman might have been the most exposed to leverage and borrowing, but the rest of Wall Street wasn't far behind. That's why the nightmare scenario of systemic risk that had rarely been whispered in public before most people heard of the hedge fund Long-Term Capital Management became a distinct possibility as LTCM's losses grew, efforts to buy up its assets floundered, and fear gripped the markets. The credit markets were essentially shut down—there were still sales, an occasional trade of junk or triple-A mortgage debt, but only at depressed prices as fear spread that LTCM was toast. The lesson of leverage is that it magnifies losses more than it magnifies gains, as Wall Street was discovering in late 1998. Since not just Lehman Brothers but all of Wall Street had copied the hedge fund's trades and had paid for those trades through massive amounts of borrowing, LTCM's failure could have a devastating impact on the financial system, with firms taking huge losses and those most dependent on short-term borrowing (Lehman and Merrill) facing a funding crisis.

The virus would then spread from Wall Street to Main Street. There would be panic selling of stocks, causing the already jittery markets to fall further and decimating the savings of Middle America, which was now more closely linked to Wall Street than ever before through 401(k) retirement plans and brokerage accounts that were tied to the fortunes of Wall Street.

The definition of systemic risk would no longer be something spoken about in academic settings or at the Federal Reserve Board meetings; for the first time Middle America would understand what it meant as well.

None of this doom prediction seemed to faze Jimmy Cayne as he sat down in his usual spot in the Bear Stearns dining room, surrounded by several of his chief lieutenants. Cayne ate breakfast every day at 7:45 a.m. His table was located in the middle of the room, in full view of the rest of the firm's managing directors, who occasionally would come by and kiss his ring (if they couldn't stand cigar smoke—Cayne lit up his first stogie of the day during breakfast—they would wait until later in the day to say hello).

Not everyone was allowed to sit with the boss—certainly not "Filthy" Phil Cohen, who by now had had such a severe falling-out with Cayne the two weren't on speaking terms, and certainly not Ace Greenberg, whom Cayne considered such a relic he was taking the first tentative steps toward stripping him of the chairman's title.

Cayne's coterie of advisers now included, most prominently, a man named Vincent Tese, a former top aide to New York State governor Mario Cuomo and billionaire commodities trader whom Cayne had met while pitching a municipal bond deal but was now his best friend and consigliere. Cayne trusted Tese so much that he not only seated him at the breakfast table, but he'd given him a seat on the board and an office down the hall from his own, a move that raised eyebrows since Tese was the lead outside director and was supposed to be independent of management to protect interests of shareholders.

Various brokers who were the firm's biggest producers would sit at Cayne's breakfast table as well, a testament to the amount of money they made the firm. One, Richard Sachs, had the reputation of being the highest-producing broker on Wall Street; he earned a salary of $10 million a year thanks to his well-heeled client list, many of them Hollywood stars such as the actor Edward Burns. Another frequent tablemate was Kurt Butenhoff, whose client list included one of the world's richest men, the reclusive commodities trader Joseph Lewis.

But at Bear, particularly at breakfast, Cayne was the star. As he did every day, this morning Cayne did most of the talking, occasionally puffing on his cigar (he was the only person allowed to smoke in the Bear Stearns dining room) and from time to time chewing a bagel with a light smear of cream cheese.

Today the conversation surrounded the financial crisis and LTCM, run by the group of financiers Cayne referred to as "Meriwether and his Merry Men."

As news of LTCM's problems made their way from the Wall Street trading desks to the popular press, Cayne spoke about his various meetings with the legendary John Meriwether. He said they had a lot in common. They both played bridge (Cayne assured everyone that he was better). They both liked to take risks; Cayne and Meriwether had made their bones on Wall Street with the same trade, buying New York City debt on the cheap during the 1970s fiscal crisis.

But that's where the similarities ended, according to Cayne. To be sure, Cayne had been an early investor in LTCM and Bear was the hedge fund's clearing broker, meaning it acted as the middleman, settling trades between LTCM and its multitude of counterparties. But Meriwether and his merry men wanted more; when they had launched their fund, they had wanted a bizarre assurance: that Bear would agree to step up to the plate and clear all of LTCM's trades, even in the unlikely occurrence that the firm fell into deep trouble and began to implode. It was as if they had been preparing for the day that had just arrived.

"I told them, 'No fucking way.'"

Then Cayne explained "the bailout": the financial Armageddon that had been threatening to take down Lehman, possibly Merrill, and God knew who else had been avoided because LTCM didn't have to liquidate.

A couple of weeks earlier the Federal Reserve and the Treasury Department, in a historic move, had convinced Wall Street to do something it was unaccustomed to doing, namely cooperate for the common good. Under the plan, every Wall Street firm would put up money to assume LTCM's liabilities.

Every firm, that is, except Bear Stearns.

Cayne then explained just how Bear Stearns had been able to say "Fuck you."

Bear might have had the rep as the Animal House of Wall Street, willing to take huge risks, particularly in esoteric markets—it was after all, one of the biggest players in the mortgage bond market thanks to the expertise of its bond chief, Warren Spector, and it was leveraged at times close to 40 to 1, the highest of any Wall Street firm—but as a firm, its exposure to LTCM had been minimal (it cleared trades for the fund), and Cayne said he wanted to keep it that way.

"So they call us in to the Fed," Cayne said, as he eyed his big, dark stogie, "and they're looking for big money." The "us" he was referring to was the heads of the big investment banks, including Cayne. "Komansky was there," he said, referring to Merrill Lynch CEO Dave Komansky. Also in attendance were Jon Corzine, the CEO of Goldman Sachs, and the most powerful CEO of them all, Sandy Weill, the takeover artist who had just completed his biggest trade, merging his Travelers Group investment bank with the commercial banking powerhouse Citicorp to create the world's largest bank, Citigroup.

"They all were there." And, based on what he was hearing, they were all ready to put up the money to bail out the hedge fund, essentially buying the bad debt from the fund and unwinding the trades in an orderly fashion, because they were bailing out themselves.

But not Cayne. Why should the tiny Bear Stearns bail out the mighty "Goldman Sucks"—or anyone else, for that matter? The Fed had taken a roll call vote on how much money each firm would pitch in, to which he had responded, "'Don't go in alphabetical order,' because if they did, I said, they would be disappointed."

The room at the Fed was taut with tension. Officials from the Fed and Treasury were shocked. His friend, the large, rotund Merrill CEO Dave Komansky, nearly put Cayne in a headlock to convince him to chip in a couple hundred million dollars. Cayne explained how there was a lot of rumbling from the crowd about Bear staying out of the bailout. Komansky, for one, was particularly livid and told Cayne so, but the others just steamed in silence.

Cayne knew that what he was doing was controversial, to say the least. He argued that Bear had always been a team player, that the firm's reputation for making good on its obligations had never been questioned on trades or when it came to the business of Wall Street.

But asking Bear to put up money to bail out guys who didn't deserve a bailout—Meriwether and his Merry Men had failed and should be allowed to fail—was a step that he as CEO wasn't willing to take. If Bear needed a bailout, would any of the CEOs sitting around the table have offered him one? Cayne said he bet no, and in time he would be proved right.

Others in the room just shot him dirty looks. Bear wasn't part of the team, and they let Cayne know it.

Cayne, however, was proud to be an outcast, and now he wanted the people he was closest to at the firm—his breakfast partners—to know it.

Despite the gravity of the situation, there were some funny moments during the meeting. When they begin to go around the table, the CEOs and top executives stated they were in, ready to contribute $300 million each.

Komansky stood up and told the group how great it was that all of Wall Street was sticking together. They were a "financial family." But one of their family members had some problems. Lehman Brothers didn't have the cash to put up $300 million, so it was his opinion that Lehman, its CEO, Dick Fuld, seated a few feet away, his dark eyes looking straight at him, should contribute just $100 million. No one said a word except for one of Morgan Stanley's representatives, capital markets chief Peter Karches. Karches was known as a guy who pulled no punches. His most famous line inside Morgan was "Ask me a question, and I'll give you a straight answer." And he always did.

"Okay, Dave," Karches responded, "we're a financial family, sitting at the financial family's dinner table. But if we're a family and Lehman is only chipping in $100 million, why aren't they sitting at the children's table?"

The place erupted in laughter, including Cayne but not Fuld, who stewed silently, staring coldly at Karches, who couldn't have cared less.

Cayne's "Fuck you" during the LTCM bailout is now Wall Street legend, partially because he has told the story so many times to so many different people, but also because it says a lot about how Cayne was remaking Bear in his own image. Ace Greenberg's Bear Stearns was quirky and contrarian; he hired people who didn't fit in at other firms and focused on businesses that the rest of the Street ignored. Jimmy Cayne's Bear Stearns was a renegade outfit that thumbed its nose not just at competitors but at regulators as well.

By the late 1990s, Cayne could take credit for building Bear into a full-scale investment bank—something much bigger than anything achieved under Greenberg, who would have been content to have Bear remain a niche trading shop. But it was also becoming a highly leveraged casino.

Even after the LTCM crisis, when most of Wall Street gave leverage and risk taking a breather, Bear remained the most highly leveraged of the big firms, still borrowing nearly $30 for every dollar it had in cash and capital.

Even more, the old controls were now gone—or largely so. By the end of the decade, Greenberg was out as company chairman, and though he still showed up for work every day on the trading desk, he was hardly a meaningful presence. Each Monday after the market close, the heads of each trading group at Bear would still meet in a bland conference room to discuss the firm's risk positions, but without Greenberg's active presence, things had changed. Greenberg still attended, but now Warren Spector ran the show. Spector believed Greenberg's risk-managing techniques to be nothing more than anachronisms as Bear matured into a modern firm. With that, Greenberg's "three-week rule," under which all losing trades needed to be unloaded after three weeks, the "cold sweat" encounters with traders over their holdings, and his famous badgering of others to sell off losing positions became more and more infrequent.

"Ace," said one senior executive who regularly attended the risk management meetings, "just seemed to check out over time."

Bear's limited exposure to LTCM was a testament to Greenberg's straightforward approach to risk, which emphasized profits and sought to minimize losses. As much as Jimmy Cayne saw Bear as being different from its competitors, it really wasn't anymore. Bear's approach to risk had morphed into the same approach that was sweeping the Street: risk was a manageable concept that could be measured with VAR and other models. Trades in mortgage bonds and other investments were not part of a strategy of making money—they were the entire strategy. Cold sweats were slowly replaced by risk analysis and dissertations about various market scenarios.

The demise of LTCM, the losses that followed, and the near death of one, possibly two firms failed to have any long-term corrective effect. There was, as there had been in the past, a brief respite from risk taking and leverage. Merrill fired some of its traders and scaled back on risk for a few months. Lehman did the same, albeit briefly. Goldman pushed out Corzine and delayed its IPO until the spring of 1999. Leverage came down to earth a bit, but only temporarily.

Profits returned just about as quickly as the crisis had hit the trading desks of the firms. The losses were papered over by revenues from IPO underwritings from the dot-com boom and a return of the bond market once the LTCM situation worked itself out. Lehman's share price, now a bellwether for the health of the market, recovered sharply as the firm's profits returned.

Then leverage and risk taking were back among the big Wall Street firms, as if nothing bad had ever happened. Because of the LTCM bailout, because Fed Chairman Alan Greenspan responded to the crisis by lowering interest rates dramatically, flooding the financial system with money that could be borrowed cheaply, the acute pain of systemic risk was certainly avoided, but so was a valuable lesson: that risk taking should have consequences, and without consequences, Wall Street was destined to repeat its mistakes again, and with larger stakes.

Pat Dunlavy had spent nearly twenty-three years at one of the epicenters of risk, Salomon Brothers, when the news broke that it wasn't a scandal or massive trading losses, or even a mismarking of the prices of mortgage debt along the lines of what he had told Tom Strauss some ten years earlier, that would end the firm's independence. Salomon Brothers ended not with a bang but a merger.

Dunlavy vividly remembers sitting in a large conference room at Salomon Brothers headquarters in Lower Manhattan in 1997. The firm's employees had received a memo from Sir Deryck Maughan, the British banker who had been appointed CEO after the bond market scandal had claimed John Gutfreund as a victim.

It was billed as an "informational meeting," the type of gathering in which senior management discloses the location of the next executive retreat, but it was actually much bigger than that. Maughan got right to the point: Salomon Brothers, the greatest bond house of all time, had now lost its independence. It had been sold to Travelers Group, the brokerage and insurance conglomerate run by Sandy Weill and his number two, James "Jamie" Dimon.

"This will be good for us," Maughan said. But no one else in the room seemed to share that view. Although Salomon had never quite recovered from the Treasury scandal and had lost around $1 billion during the mortgage implosion in 1994, the firm was still considered the gold standard of the bond-trading business, and it was now being gobbled up by what many in the room considered the Wal-Mart of Wall Street.

Weill and his protégé Dimon were focused on the creation of a "financial supermarket," the notion that the modern financial firm could be all things to all people, cross-selling its various products: mutual funds and bank accounts to small investors, bridge loans and bond underwritings to large ones. In a few months' time they would announce their biggest deal ever, merging Travelers with the commercial bank Citicorp to create Citigroup.

They were also known for running a boring operation; Weill hated bond trading because he hated risk taking, and he imbued the same distrust of risk and leverage in his top lieutenants, chief among them Dimon, who was the odds-on favorite to run Weill's empire when Weill was done building it.

For all those reasons and more, Dunlavy believed, the Salomon trading culture he had resented for so long—the risk taking, the leverage, the thirst for personal profit—was toast. That would be a good thing. The bad thing was that Dunlavy also believed that Salomon housed some of the smartest people on Wall Street. For every gunslinger on the trading desk, there were people who cared about their clients, and they would be toast as well.

But Dunlavy was only half right, which is why 1998, as Wall Street was first bleeding and then recovering from the LTCM crisis, was maybe the worst year of Dunlavy's career. First off, there was no room for Dunlavy at the new firm, renamed the Salomon Smith Barney unit of Travelers Group, combining Salomon Brothers with the brokerage firm Smith Barney, purchased by Weill some seven years earlier.

Even worse, the gunslingers weren't out of a job; in fact, they were now running the place.

Salomon's bond desk would be run by Thomas Maheras, whom Dunlavy viewed with a mixture of envy and suspicion for his meteoric rise through the senior ranks of the firm. Maheras had begun as a junk bond trader but had made his mark by making the firm billions of dollars by allegedly repairing the mortgage desk in 1994 following massive losses. He had been well compensated for that feat; at the time of the Travelers deal, Maheras was running all of fixed income at Salomon.

Dunlavy thought Maheras was more lucky than good, even if that was all it took to be a successful trader. The way Dunlavy saw it, Maheras's big blowout year in 1994 had come by his performing one of the oldest tricks in the trader's book: he had written down the losses as much as he could, chalking up $1 billion in losses that he blamed on his predecessors, and then had taken credit for all the upside when the mortgage market recovered. More troubling to Dunlavy was Maheras's dark side when it came to trading and risk—all the hours he believed Maheras had spent trading bonds for his own personal account when he should have been handling client matters and his self-described "love of leverage." Dunlavy believed that someday the leverage would catch up to Maheras as it had to Ranieri, Meriwether, and the other Salomon gunslingers. "And this is the guy to put in charge of every trade?" Dunlavy scoffed. "And I'm out? Something had to be wrong." (Maheras, in a brief interview, said his personal trading was not unusual.)

That's what Dunlavy began telling the new Travelers transition team as they began meeting with Salomon's traders to get a handle on their new acquisition and how to integrate it into the larger firm. Dunlavy was out of a job, but he was a treasure trove of Salomon-related information given his quarter century at the firm. Dunlavy didn't hold back, telling people in the new firm who had worked for Dimon at Smith Barney that they had better be careful; the traders at Salomon Brothers who were getting the senior jobs, including the seniormost job, had excessive risk taking in their blood.

He didn't mention names, but everyone knew who he was talking about.

Dunlavy's misgivings made their way up the ranks to Jamie Dimon, who demanded a meeting. By Wall Street standards, Dimon had a Spartan office in the new Salomon Smith Barney headquarters on Greenwich Street in Lower Manhattan. There wasn't any fancy art, just some bookcases and lots of annual reports lying around.

Dimon's boss, Sandy Weill, was known to love the trappings of power—he had a fireplace in his office and allowed other top executives to do the same. He was possibly the least detail-oriented CEO on Wall Street, unless it involved details of the firm's stock price, which he watched relentlessly on a computer screen on his desk. He saw himself as a visionary, and his vision was to create the first true financial services conglomerate.

Dimon, on the other hand, was a self-described "geek." He loved to pore over annual reports and deal with issues such as the firm's trading leverage and whether its traders took too much risk. Also unlike his boss, Dimon loved the nitty-gritty of the business. He constantly badgered traders and bankers about their business. His senior staff was on call at all hours. His favorite way of finishing a long day was to grab a handful of his top guys, open a couple of bottles of expensive wine, and sit in his office talking about work.

Dunlavy didn't know it at the time, but Dimon had already heard tales about the risk-taking ways of Maheras and his bond team. Though he liked Maheras personally—they were both of Greek ancestry, which added to their bond—Dimon was concerned about Maheras's excesses, something he vowed to put an end to quickly.

Dimon believed he could "manage Tommy," take him away from the "goombas," as he called Maheras's buddies from the old Salomon Brothers, mainly mortgage bond chief Randy Barker and chief risk officer Dave Bushnell.

Others, including Pat Dunlavy, weren't so sure.

When Dimon welcomed Dunlavy into his office, he told him that he'd heard Dunlavy had "some concerns about the business" and he wanted to hear them. Dunlavy asked how much time he had. After a brief smile, Dimon indicated that he wasn't there for small talk and asked him to begin.

"The traders are just out of control," he explained, making nearly the same argument he had made in Lew Ranieri's office in the mid-1980s and in Tom Strauss's office a few months later.

Dimon took copious notes, every now and then interjecting a point or asking a follow-up question. After about a half hour, he'd had enough. He thanked Dunlavy for his time, shook his hand, and said he would be missed at the new firm. Dunlavy smiled, said thank you, and noted that though he'd given the same warnings to Ranieri and Strauss, at least this guy had taken notes.

Later, when Dunlavy was packing his belongings, he reminisced about his early days on the Street. "It was a gentleman's business," he thought. "We were in the business not just to make money but to serve clients." Then he thought back to when things had begun to change. "It was the 1980s," he said to himself, that's when the model had changed, when advice really didn't matter as much anymore, when everyone had been paid to take risks and use leverage to take risks. Each time the system had blown up—in 1986, 1994, and then 1998—the damage had been greater, but the overly risk-based business model had returned shortly thereafter, bigger and riskier than ever.

How would it all end—or would it ever? Dunlavy didn't know, and at this point he didn't care. He was leaving the business. He was thinking about a career change, possibly getting a master's degree in something he loved, such as education, something he could be proud of.

Dimon, meanwhile, was proud of something else: the effect his management style was having on Maheras. Dimon now considered Maheras his friend as he continued to make him his special project. Dimon has said that he took steps to get Maheras to scale back on what Dimon thought was Maheras's excessive personal trading. (In a brief conversation, Maheras said he didn't recall such an admonition.) Maheras soon learned what every executive who ever worked for Jamie Dimon quickly discovered was standard operating procedure—he was in your face every day and, for some, just about every hour of the working day.

B y the time Jamie Dimon, then forty-one, was overseeing Tom Maheras, he was possibly the most experienced manager on Wall Street. Of course, he couldn't match Weill's long years on the Street, particularly as a deal maker, Cayne's ability to schmooze politicians, or Dick Fuld's trading desk experience.

But they all lacked the skills that Dimon had picked up thanks to his unusual tutelage under Weill.

Jamie Dimon never worked as an investment banker, a trader, or a broker, the way most senior executives begin their careers. Dimon began his career by building companies, or helping someone else build companies. Dimon had joined Weill as his assistant back in the 1980s after Weill had created the Shearson brokerage empire, sold it to American Express, and then was fired after coming inches away from running the company. Weill had then made his comeback with Dimon at his side as he built the biggest financial empire in Wall Street history through a series of high-profile acquisitions, first Commercial Credit; then the Shearson brokerage department, Smith Barney; and then Salomon Brothers. Weill came up with the big ideas, and increasingly over time, Dimon carried out the details: slashing staff, containing costs, integrating units. No one on Wall Street had put together as many companies, slashed more redundant businesses, or managed as much top talent as Jamie Dimon.

Weill and Dimon's partnership became Wall Street legend. The prominent financial columnist Alan Sloan described their relationship this way: "Wall Street had always thought of the Weill-Dimon relationship as father-son. Especially since Weill has no sons." (Weill actually has one son, Marc, who worked for him as well but left the company after a brush with drug use.) But as the empire Dimon put together on Weill's behalf grew, it also left Dimon resentful and bitter. Everyone seemed to agree that Dimon should soon be running the firm except Weill, who hung on to power and lusted for more, making it clear that he had no plans to leave Dimon the keys to the kingdom anytime soon, even as, at sixty-five, he completed his greatest feat yet, creating Citigroup.

The deal was a milestone for many reasons: the new company was the largest financial services organization in the world, operating in 107 countries with 22,000 offices worldwide. It was valued at $140 billion, the largest merger during an era of megamergers. Maybe most important, the deal was illegal. A Depression-era law, the Glass-Steagall Act, still prevented the combination of investment banks and commercial banks, a law Weill quickly worked to unwind.

According to the press accounts, the megadeal had been a brainstorm of Weill's a few months earlier. In fact, he had been thinking of such a deal, of creating a financial supermarket, since his days at American Express. Weill was never considered a deep thinker, even according to people who worked for him. He was and always had been myopically focused on boosting share prices through his mergers by having Dimon squeeze out inefficiencies and redundant costs.

But one area into which Weill seemed to have insight was his theory of what the modern financial services company should look like. There was no reason why a bank customer shouldn't be able to buy a mutual fund while depositing a check. Likewise, there was no reason a corporation should have to get bridge financing from a bank but float a bond deal through a securities firm. The modern financial services firm should be a massive supermarket.

Dimon shared his boss's theory and enthusiasm for the all-in-one concept, but he now believed he was the person best qualified to run the empire he and Weill had created, and he never missed an opportunity to let his boss know his feelings.

By the time the Citigroup deal was announced, their relationship had already turned toxic. The two could barely spend a few moments speaking to each other over lunch at the Four Seasons, dinner at the downtown restaurant Ponte's, or in the executive suites of Citigroup without erupting into a screaming match. When Dimon failed to appoint Weill's daughter, Jessica Bibliowicz, to a senior position at Salomon Smith Barney, Weill was livid with rage.

Advisers to Dimon say it was at that time he believed his days at the firm might be numbered, but in the meantime, his education of Tom Maheras continued.

"Billy, let me ask you a question," Maheras said, pulling aside one of his managers, a former bond trader named William Heinzerling, one afternoon, according to people with knowledge of the conversation. "Why the fuck does Jamie keep asking me all these questions?"

Heinzerling had worked for Dimon for about a decade and learned to accept and respect the tough love that Dimon meted out—because he meted it out to everyone. He showed no favorites, and it was effective.

"Tommy, that's Jamie," Heinzerling answered with a laugh. "Just deal with it, and don't take it personally."

Maheras said he wouldn't, but maybe he should have. The part of Dimon that saw Maheras as a friend wanted to make him someone Dimon could trust as a part of his inner circle. The other part worried about Maheras's tendency to take risk. Jamie Dimon was way ahead of his time in being profoundly concerned about Wall Street's addiction to borrowing huge sums of money and then making increasingly esoteric bets in markets he didn't really understand, nor did he think his traders understood. One of those addicts, he had heard as he continued integrating Salomon into Travelers and then into Citigroup in the waning months of 1998, was Tom Maheras, as well as the traders he surrounded himself with.

That was the reason why, not long after announcing the Salomon deal, Dimon shut down Salomon's famed arbitrage desk, the place where John Meriwether had gotten his start, and it was the reason why Travelers' losses tied to LTCM were minimized. Having put so many brokerage firms together and squeezing out costs, Dimon had come to believe that the aura of banks as being epicenters of wealth and capital, and too big to fail, was a facade.

The LTCM implosion and the huge losses it spread across Wall Street (including at Salomon Brothers, which he'd had to pick up the tab for) only reinforced this view. He could read the leverage numbers as well as anyone else; with most firms borrowing many more dollars than they had in capital, the business model was fragile, too fragile to be left to a bunch of hyperaggressive bond traders, even one he really liked.

But Dimon's education of Tom Maheras came to an abrupt end in late 1998, just a few months after the Citigroup deal, when maybe the most effective risk manager on all of Wall Street was fired by Weill. The move, of course, wasn't a complete surprise, given their deteriorating relationship, but the timing was. It occurred during a weekend of executive meetings at a retreat when Dimon was led to believe that he would have an enhanced role in the new firm. A story to that effect even ran in the early editions of the *Wall Street Journal*, as Dimon was called on Sunday afternoon by Weill to meet him at the executive retreat in Westchester County as Dimon entertained more than one hundred company brokers at a brunch at his Manhattan apartment. Despite his deteriorating relationship with Weill, getting fired was the last thing on Dimon's mind. He had done his fair share of it in his career, and it generally doesn't happen during an executive retreat, or so he thought. With that he began spreading the news to his top aides that a promotion was pending.

But when he was ushered into a conference room, Dimon could tell something was wrong. He was seated in front of the somber-looking Weill and John Reed, the CEO of Citicorp and now the co-CEO of Citigroup. Weill did most of the talking. He stated matter-of-factly that it wasn't he, but others, namely Deryck Maughan and another senior executive named Mike Carpenter, who would be getting promotions.

Dimon, meanwhile, was getting fired. The firm's general counsel (and future CEO), Charles Prince, would call him to negotiate his severance.

Dimon was shocked and hurt but, at least for the moment, showed no emotion. "I'll do what you need to do to make this work," he said. But that lack of emotion wouldn't last. He told friends that the firing had more to do with Weill's lust for power than anything else and that Reed would be the next to go. Dimon's criticism made its way back to Weill, who a little later asked to see Dimon.

Dimon, now unemployed for the first time since he'd left college, showed up in Weill's office, only this time he was visibly angered and hurt. As Dimon walked in, Weill told Dimon he was sorry the relationship hadn't worked out and tried to hug Dimon, who pulled away, and said, "Sandy, I'm not going there." After a brief conversation the meeting ended, but the animosity between the two men continued.

As did Weill's power grab. A year later, Dimon's prediction about Reed proved accurate. Weill forced out Reed in a boardroom showdown, enlisting none other than Robert Rubin, the former Treasury secretary, now a Citigroup board member, to convince the rest of the Citigroup board that there should be only one CEO of Citigroup, and that CEO should be Sandy Weill.

With Reed's ouster, Citigroup was now Weill's to run as he saw fit. But in reality, Weill never really ran anything. He was a visionary, to be sure, but one whose vision was so myopically focused on building the empire he had lusted for for so long and on its share price that he ignored just about everything else. In the past it wouldn't have mattered because he'd had Dimon and his team to do the heavy lifting. But with Dimon gone, the others began leaving as well, one after another, leaving Weill with a far less accomplished team of assistants to run one of the largest banks in the world.

They were also left in charge of some of the biggest risk takers on Wall Street. Without Dimon around, Maheras and his "goombas" were left to their own devices. Of course, Maheras, now the head of the newly combined firm's bond division, had to answer to compliance officers, lawyers, the head of the Salomon Smith Barney division, Michael Carpenter, his successor and longtime Weill protégé Robert Druskin, and of course the CEO himself. But there was no one left to badger Maheras about his trading positions, his leverage, and his personal trading, and no one who understood that any trader's natural inclination to take risk must be curtailed when the survival of the firm is on the line.

In fact, with Dimon gone, just the opposite was happening. Maheras and his men were given a larger canvas on which to perfect their leveraged artistry, Citigroup's massive balance sheet; the size of its deposits and cash on hand coming close to $3 trillion, the size of a small country's GDP. It was much bigger than anything Maheras had seen before or even than traders such as Mortara or Meriwether had had to work with at Salomon Brothers. It was so large that the leverage and attendant risk of the nature that had been seen at Salomon and the rest of the Street during the 1990s now seemed insignificant.

eill, his power consolidated, was now focused on improving Citigroup's stock price, and so were his lieutenants. One of these was Robert Rubin, who had joined Citigroup just weeks after the big merger was made official.

Rubin made headlines for his high-profile support of Weill's plan to oust Reed. Meanwhile, he did something else that didn't make big headlines but should have: he pushed the risk-adverse Citigroup CEO to take more trading risk.

Rubin hadn't been planning to become a workaholic when he arrived. Those days were over for the former bond trader turned Goldman Sachs chairman turned Treasury secretary. Many people credited Rubin for the reduction of the nation's budget deficit during the 1990s and the subsequent economic boom. Just before he took the Citigroup job, Rubin met with Maurice "Hank" Greenberg, the notoriously tough, hardworking CEO of the insurance giant AIG, about possibly joining his firm. Greenberg was beside himself when he heard what type of job and ballpark salary Rubin was looking for. "This guy wanted to make eight million dollars a year to just travel around the world," he later said.

Greenberg wasn't impressed with Rubin, but Weill was. His dream of creating a mega–financial conglomerate was within reach except for one small detail—the Depression-era Glass-Steagall Act, which formally separated investment and commercial banking activities. Rubin became an important behind-the-scenes advocate for Weill and Citigroup in Washington to reverse the law. And it worked: Glass-Steagall was abolished little more than a year after Citigroup was created, and a few days later Rubin was rewarded with his dream job: a seat on the board and, maybe most importantly to Rubin, a position without responsibility as chairman of the firm's "executive committee," where he earned millions of dollars a year to merely meet clients and dispense advice. His timing couldn't have been better. Once Dimon was ousted, Rubin was the second most powerful man at Citigroup, even though he continued to fill a job without any operating responsibility. But he had something else going for him: he had Weill's ear on just about every major issue affecting the firm, including the concept of taking on additional risk.

Taking risk was a natural for Rubin. Like John Meriwether, he had earned his stripes as an arbitrageur at Goldman Sachs. Goldman, of course, had earned headlines for its prowess in winning high-profile investment banking deals, but over the years, its trading profits had exploded, and Rubin wanted Citigroup's to do the same. So Rubin began holding meetings with Weill and traders in various areas of the firm, giving them pointers on how Citigroup could compete with Goldman Sachs in amassing big profits by taking more risk. Of course, Rubin said, the trading should have proper oversight. Even so, to the Salomon Brothers traders such as Maheras, Rubin's advocacy of risk was music to their ears, but to the old Smith Barney traders, schooled in the Jamie Dimon model of risk taking, the changes were appalling.

A re you sure that's what Rubin said?" asked Bill Heinzerling when he first heard that Rubin was making the rounds of the various departments and talking about taking more trading risk, according to people with knowledge of the matter.

The firm's risk profile was important to Heinzerling because he was the head of Citi's bond finance department, meaning if Citi's traders wanted to borrow the firm's money to make a bet on the direction of the market, they had to get Heinzerling's approval.

Now he was getting a briefing from one of his underlings about a meeting he had just sat through with Rubin over the firm's risk profile. Heinzerling had worked for Weill and Dimon for around ten years. He had been there when Weill and Dimon had slashed the firm's risk dramatically, disbanding Salomon Brothers' famed bond arbitrage department during the LTCM-induced bond market crisis.

When asked by subordinates to describe his trading philosophy, Heinzerling called it "blocking and tackling." As a working-class kid from Brooklyn's tough Bay Ridge section, Heinzerling had seen his share of degenerate gamblers—people who rolled the dice on everything and almost always lost everything. No way was that going to be him.

Heinzerling didn't think it was his job—or the firm's—to do anything too exotic; he worked for clients, he traded on small increments with the company's money and made quick and relatively safe profits. He didn't believe in VAR or the other risk measurements that were all the rage on trading desks across Wall Street. He went with his gut, and with that came a disdain for excessive leverage and risk.

But now, with Rubin pushing for more risk by increasing positions in certain markets it traditionally avoided, Citigroup was morphing into Salomon Brothers. Even when Dimon had been watching the show, Heinzerling could see the tenor of the firm shifting to more risk and leverage. Up until now it had been only around the edges. If he was hearing it right, Rubin was asking for a fundamental shift in strategy by trading in more exotic bonds and holding on to those trades much longer than ever before. So Heinzerling wanted more proof that increased risk taking was indeed the future of the firm. It was, he was told, and it wasn't just Rubin pushing for it but Tom Maheras as well.

Like Dimon, Heinzerling personally liked Maheras. "He's a great guy to be around *outside* the office," Heinzerling told coworkers, emphasizing "outside." But inside the office, Heinzerling said, Maheras was a "riverboat gambler" who was infecting Citigroup with the wild and reckless ways of Salomon Brothers.

He was even tempted on one occasion to let Weill know that Maheras and his men were busy remaking the firm's risk profile in ways that Dimon and Weill would never have allowed. One afternoon Weill called Heinzerling into his office to get a status report on the bond department, where rumors were now circulating that there was a culture clash between the Salomon and Smith Barney traders.

Those rumors had made their way up to Weill, who decided to confront Heinzerling with them directly. The meeting was typical of those Weill conducted with his staff: he barely paid attention to the people he was speaking with.

As Heinzerling later told people at the firm, he stood near Weill's desk and noticed something odd: his boss was staring intently at his computer screen, not reading but watching the firm's share price change with each and every trade.

After a couple of grunts, Weill told Heinzerling he wanted to talk to him about "Tommy," as Maheras was known in the firm.

"He's good, and I want you guys to work together," Weill said.

"I'll do whatever it takes to make this work," Heinzerling said. Weill said, "Great," and then resumed watching his screen as if Heinzerling weren't even in the room.

But making it work would prove more difficult than Weill realized. Maheras had his first off-site meeting with his new management team not at some corporate training center in New Jersey or even at some golf club in Westchester County but in Las Vegas. He used the company's jet, a Falcon 800. It was a weekend of big steaks, expensive wine, and lots of gambling. Maheras spent his nights at the crap table, rolling the dice. It's not a stretch to see that the setting of the meeting proved insightful about how Maheras wanted to run things.

In fact, most of the Smith Barney bond traders couldn't figure out exactly why Weill or, for that matter, Dimon had liked Maheras so much, given that their positions were so different on the issues that mattered most when it came to risk taking and the use of leverage. Not long after he took control, Maheras openly professed that he "loved leverage," which was odd because Weill and Dimon had done all they could to squeeze leverage out of Salomon Brothers after the purchase.

But Maheras's star continued to rise. During another meeting, Weill described Maheras as among the best risk managers on Wall Street. It's unclear how Weill had come to that conclusion. There was some talk among the Smith Barney traders that Salomon Brothers executives had approved the sale of the firm only on the condition that one of their own, namely Maheras, run the bond department. Others say that Maheras's friendship with Dimon, their common Greek background, and their genuine liking for each other had played a key role in his resurgence inside the newly combined firm.

A better explanation might have included Weill's growing blind spot: his overinflated ego.

Maheras, one of the smoothest people at Salomon Brothers and maybe on all of Wall Street, knew how to stroke it, which he did every chance he got. As their relationship grew closer, the two began spending time together on Weill's yacht or at his mansion in Greenwich, Connecticut. Maheras's power within the firm skyrocketed. At bottom, Maheras knew how to *trade* Weill, as he had his bosses at Salomon Brothers. It was a different mentality from the one Weill was used to in Dimon, who had challenged his decisions and thinking on a regular basis. That's how Maheras was able to make the best trade of his career and convince Sandy Weill, a skeptic of legendary proportions when it came to risk, that he should take on more of it, with "Tommy" in charge.

Others, of course, saw a different side of the man Weill considered so measured and so secure. This was the Tom Maheras with ambition as big as the thick mop of black hair on his head, and it centered on making a lot of money in order to become a billionaire. The only way to make $1 billion as a bond trader would be to turn the Sandy Weill risk model 180 degrees in the opposite direction, and that's exactly what Maheras, with the help of people like Bob Rubin and the acquiescence of senior executives who didn't know the first thing about trading, did in just a few short years.

As head of the finance desk, Heinzerling was the first to see the changing nature of the firm's risk profile and, according to former colleagues, one of the few people inside the firm to raise issue with the way Maheras was running things. One morning, a fairly large account at the firm that had made investments on borrowed money, something known as margin, was losing money and facing large "margin calls," or demands from lenders, in this case Citi itself, for cash to cover the losses. But in order to meet the margin calls, the firm was now financing the losing position, meaning that Citigroup was lending the account additional money until the losses could be covered directly by the investor.

Given the heavy trading of the account, people at the firm became concerned; it had all the earmarks of an account held by a large institutional investor that traded big bucks and was important to the firm because of the business the institution brought in. They alerted Heinzerling of the losses and the frequency of the margin calls. "Let me check it out," these people recall Heinzerling saying, and when he did, he found out it was not an institutional account at all but the personal trading account run by his boss, Tom Maheras.

While the trading wasn't improper—in fact, Citigroup rules allowed executives to trade for their personal accounts within certain guidelines— the activity appeared unusual to Heinzerling, a longtime trader himself, for a couple of reasons, as Heinzerling would later recount to associates: for one thing, it was incredibly active for an individual investor, much less someone whose primary responsibility was to tend to the firm's business, not his own. And for another, the firm was "financing" the account, meaning it was trading with borrowed money at favorable rates, those usually reserved for large institutional investors, not individuals.

Michael Carpenter, who was Maheras's direct supervisor as head of Citigroup's Salomon Smith Barney brokerage unit for several years, was certainly aware of his head bond trader's private activities. "Tom traded actively for his own account," Carpenter would later say. "As far as I am aware, he was in compliance with Citi's" policies.

Still, that didn't stop people inside the firm from raising additional questions. Another executive on the finance desk, Tim Douglas, pointed out that Maheras at times wasn't meeting his "margin calls," or demands to repay borrowed money to cover losing positions. He brought the matter to the attention of an attorney who worked in the fixed-income department, Marcy Engel, who, according to people who worked at the firm, in turn discussed the matter with Maheras. (Neither Douglas nor Engel disputes this account.) Heinzerling, according to former colleagues, also explained this incident to senior management, and his relationship with Maheras, which had never been easy, was about to get worse as they continued to battle over the direction of the bond department, specifically, its use of leverage and risk. In Maheras's view, the Citigroup bond department would be the most productive at the company; given the way interest rates were heading and the boom in the market for turbocharged bonds such as mortgage debt, the firm was on the precipice of making it big. And it would do so through leverage.

During one of their many conversations and confrontations about the direction of the bond department, Maheras appeared to sum up his trading philosophy, Heinzerling would later tell colleagues: "I love leverage," Maheras said.

"How much leverage would you like?" he then asked Heinzerling, who replied with more than a touch of defiance: "Zero."

9. OPENING THE FLOODGATES

itigroup certainly wasn't alone. After a brief period of reflection following the blowup of LTCM, the love of leverage was spreading fast and Wall Street fell in love, once again, with the mortgage bond. It was an odd romance. Wall Street had flirted with excess leverage tied to mortgage securities markets and survived three rather large bond market implosions when the complicated activity of trading mortgage debt had been either at the center of each upheaval or pretty damn close.

Yet the love affair lived on, even as the risk of trading those securities grew with every episode.

And the concept of trading had evolved over time as well. The textbook definition of the craft is one of rapid-fire buying and selling to take advantage of changes in pricing and interest rates. Though a few of the most successful traders, such as Steven Cohen, the head of the powerful hedge fund SAC Capital, were now becoming famous for making huge amounts of money by exploiting "teenies," or small increments, in the prices of bonds, Wall Street firms had now redefined trading to mean not just buying and selling but also holding securities on their own books for a period of time. Back in vogue was the carry trade, where firms would buy and hold mountains of debt on their balance sheet, borrowing at low short-term interest rates to finance the purchase of much higher-yielding bonds. Carrying bonds was nothing new, of course; it was what had landed Larry Fink and Howie Rubin in so much trouble in the 1980s, Mike Vranos in 1994, and just about everyone else around the time of LTCM in 1998. What was new was the size. In the 1980s, a bond desk was considered aggressive if it carried $1 billion on its books; in the new era that amount would be considered chump change.

The continued fascination with mortgage-backed securities bears some explanation. Mortgage-backed securities had always been particularly enticing for the carry trade because they were high-yielding yet most were highly rated because of the diverse portfolio of mortgages that went into each bond. But the ratings belied their real risk.

Through the 1990s and into the next decade, those characteristics were compounded by ever more complex structures, with tranches, or sections, sold individually to investors and the bonds including an ever-growing number of mortgages held by borrowers who in the past could never have qualified for a home loan. In the past, when Lew Ranieri and Larry Fink had pioneered the market, the risk of default had been almost an afterthought. The risk in the mortgage market then involved prepayments; if interest rates fell and home owners began to prepay their mortgages, the mortgage bonds, particularly the interest-only strips (IOs), which have no intrinsic value beyond the expectation of the interest payments they would generate, could decline rapidly in value. When many mortgages are paid off, the interest payments to the IOs shrink, so their value plunges.

But now, as the mortgage market continued to grow, default risk became an issue as well. One big reason was the sharp rise in housing values as low interest rates spurred massive buying of homes. The housing boom began to price families out of the market, and the mortgage industry responded with new loan types designed to make the unaffordable affordable—at least in the short run—with so-called adjustable-rate mortgages, in which interest rates start out low and affordable but adjust later on to higher rates that are more difficult to pay.

Present Wall Street with a problem and it will create a bond as a solution. That solution in the late 1990s and beyond was the collateralized debt obligation. The advent of default risk as a consideration in the mortgage market also correlated with the explosion in issuance of the CDO.

That's because CDOs are supposed to be the ultimate in risk reduction. They are bonds packed with other bonds made up of mortgages, car loans, credit cards, high-yield securities, and anything that allows the risk of defaults in one class of debt to be offset by others.

That was the theory. Reality was different. The concept of the CDO was so enticing that issuance grew by more than 700 percent over the next decade, and with that explosion of issuance came an explosion in risk—for both the bond itself and those buying the bonds. With interest rates stable, the firms borrowed huge amounts of money to finance the various CDO structures, which of course magnified their gains—and their losses if rates started to rise, though few traders and underwriters of CDOs considered the possibility of losses. That's because they were brainwashed by the magic of "structured finance," a process involving the complex computer modeling of risk, the diversification of borrowers on the basis of geography, income levels, and the types of loans handed out, which in theory was supposed to prevent the bonds from going into default.

The rating agencies continued to slap their gold-plated triple-A ratings on the CDOs' senior, or most diversified, tranches, which made CDOs a growing favorite among the big Wall Street firms looking to collect underwriting fees and earn interest income on the carry trade. Indeed, the leverage used to create these bonds, combined with the leverage used to trade them and carry the bonds on the books of the Wall Street firms, became the engine of earnings growth at the big investment banks. Even as the dot-com bubble exploded in early 2000 and the associated billions of dollars in underwriting profits vanished, the mortgage departments that churned out new "structured finance" products like CDOs kept chugging along.

The risk takers were the firms' saviors. At Goldman Sachs, the leading underwriter and banker of stocks, more money than ever before was being made on the trading desk. As the profits tilted in their favor, the men who controlled Goldman's trading desk, people like Lloyd Blankfein and Gary Cohn, were secretly plotting to remove Henry "Hank" Paulson, the former head of investment banking who was now the firm's CEO. Zoe Cruz, a savvy trader at Morgan Stanley, emerged as one of the top women executives on Wall Street as her risk taking paid off. Jimmy Cayne hated Warren Spector more than ever, but he couldn't part with the man behind Bear's continued strong earnings.

If "money is the mother's milk of politics," taking risk through leverage or esoteric investments and trades had become the mother's milk of Wall Street. Leverage wasn't new to Wall Street. In fact, as we've seen, it's exactly what got Larry Fink, Lew Ranieri, and a whole host of traders in trouble during the 1980s and in other previous market routs. But as the Street kept avoiding disaster, the deadly brew of borrowing and holding on to risky positions grew in importance to the big Wall Street firms, and increasingly the commercial banks as well, as they expanded their trading operations at the end of the 1990s.

The expansion hit full throttle in 1999, after the separation between investment banks and commercial banks was shattered. That year, spurred by Sandy Weill and Robert Rubin's lobbying after the Citigroup megamerger, the Republican Congress and the Clinton administration enacted the Gramm-Leach-Bliley Financial Services Modernization Act, which "modernized" the banking industry by killing the Glass-Steagall Act, the formal separation of investment banks and commercial banks that had been created following the Pecora Commission hearings into the causes of the Great Depression, and in particular the way bankers had been manipulating the markets.

Glass-Steagall was created to protect bank deposits from the risk-taking activities of investment bankers, though for years banks had pushed the limits of that separation by convincing regulators to allow them to keep increasingly larger brokerage and trading operations. Now Gramm-Leach-Bliley opened the floodgates for the banks to be more like Wall Street. Once-staid balance sheets changed overnight; Citi had been among the least leveraged firms on the Street, borrowing no more than $5 for every $1 it had in capital. But that quickly changed. Rubin and Weill were moving toward leverage of 20 to 1. Traders like Tommy Maheras who had fought for scarce capital when they were working at Salomon Brothers now had a massive balance sheet to play with. Adding to their comfort was the fact that Citigroup depositors would always be protected if the traders bet big and lost. The federal government, through the Federal Deposit Insurance Corporation (FDIC), insured bank deposits up to $100,000 (and later $250,000), which was part of the house money being spent increasingly on large bond market bets. In effect, the federal government, and hence taxpayers, was subsidizing the risk-taking activities of the big banks such as Citigroup.

With the competition from commercial banks growing, the investment banks ramped up their trading in response. Investment houses such as Goldman Sachs, rebounding from their losses in the wake of LTCM, were no longer content with "stable" earnings from their core businesses of asset management and financial advice.

Investors demanded more, and as Wall Street looked for ways to jazz things up, to keep cranking out profits, traders continued leveraging one of the most volatile parts of the bond market, the mortgage bond, with a little help in Washington from Fed chairman Alan Greenspan.

In late 2001 the economy was reeling, battered by the one-two punch of the end of the dot-com bubble and the September 11 terrorist attacks, which brought Wall Street and the business community to a sudden halt.

By now Greenspan was considered possibly the greatest Fed chairman in history. He was an avid free marketeer and a disciple of the libertarian philosopher Ayn Rand. He was a supporter of financial innovations like interest rate swaps, which he believed enhanced market liquidity by limiting risk to corporations that bought these assurances, and mortgage debt, which through Wall Street rocket science gave the poor access to the lending markets so they could buy a home and enhance their standard of living. He was also an advocate of deregulation like the Gramm-Leach-Bliley Act, which cleared the way for the creation of Citigroup and mixing risk taking with bank deposits, even as he famously called for the markets and investors to show some temperance when he warned of "irrational exuberance" overshadowing risk.

But now he was faced with a financial crisis of unparalleled dimensions. The people who had lost money in the dot-com bubble's burst weren't just Wall Street traders but average Americans who had ignored his irrational exuberance speech and gambled their life savings in now-cratering Internet stocks. Trillions of dollars in wealth had been destroyed when the NASDAQ—the market where most of the securities had traded—fell from over 5,000 points in March 2000 to just 1,500 a year later. The economy was sputtering, and then the second shoe fell, the 9/11 terrorist attacks, which closed down Wall Street for a week. Fear of a total collapse of the economy spread.

So Greenspan did what Fed chairman have always done in troubled times: he began slashing interest rates.

The Fed controls the nation's money supply through the Fed funds rate. By cutting it, the Fed lowers the rates at which banks can borrow from the Fed if they need additional cash. That starts a chain reaction; since money is available to banks on the cheap, the banks can provide more loans to businesses and mortgages to people who want to buy homes.

Cutting the Fed funds rate is a time-honored tool to spark economic growth, but the cuts Greenspan made reflected a degree of panic. For starters, the rate wasn't very high to begin with—just 3.5 percent when

Greenspan began cutting in 2001 (by comparison, it was hovering around 13 percent in 1981, 9 percent in 1989, and 6 percent in 2000). In a few months, Greenspan reduced the rate to 1 percent, its lowest level ever at the time.

In effect, the Fed created "free money." The banks, flush with cash, increased their lending to businesses and to people looking to buy homes, causing home values to rise well beyond the already high levels of the late 1990s. Eventually, the rising home values papered over the losses in the dot-com stocks; average Americans might have lost money on their stock portfolios, but they were making it up on the appreciation of the value of their homes.

But booms often create bubbles, and with historically low interest rates that lasted until 2004, the housing bubble continued to grow.

Banks, of course, make money by lending money, and they were now having trouble finding people to lend to. In order to keep lending money to home owners, they needed to dig deeper into the pool of potential buyers. The family with a stable income willing to put down 20 percent of a home's purchase price as a down payment became an anomaly as banks opened their coffers to real estate speculators and individuals with little or no credit history.

The so-called subprime loan—once a backwater in the banking industry—became a standard tool to lend money to this new, more speculative class of buyer. Banks could charge higher interest rates on those mortgages, which was good news for the banks and good news for Wall Street, which packed the subprime loans into mortgage bonds that offered higher interest rates.

Under pressure from HUD (Andrew Cuomo, who had succeeded Henry Cisneros, was now at the helm), Fannie Mae and Freddie Mac, which had begun to dip into this risky market a few years back, for the first time embraced it wholeheartedly by ramping up their guarantees of subprime mortgage loans.

Over time, subprime grew to be one of the biggest parts of the GSEs' business; by 2000, Fannie Mae had guaranteed some $1.6 trillion of mortgages, but increasingly subprime loans, compared with about $830 billion just five years earlier. This policy continued through the Bush administration as well. The GSEs' foray into the subprime market was a bold move, one that raised the eyebrows of some in Congress, but was applauded by others, including the two most vocal advocates in Congress who had been pushing the agencies for years to help more low-income people afford homes, Representative Barney Frank and Senator Christopher Dodd.

The magic of securitization—the slicing and dicing of various types of loans, be they credit cards, auto loans, or mortgages, and repackaging them into bonds—allowed banks to keep lending at levels they never could in the past. Mortgage lenders such as Countrywide Financial were now emboldened to take even more risk and lend to anyone. And why not? They could simply take their loans and sell them to Wall Street, which would later take them and pack them into a mortgage bond that would be sold to other investors.

It was a Wall Street version of the game "hot potato" that was about to get even hotter.

We don't service these loans, we just sell them to Wall Street," said the manager of the marketing department at Chase Mortgage, a subsidiary of the big bank, when asked why the bank was lowering the credit requirement of the home buyers it was targeting in a new marketing campaign.

For years, the bank had refused to market loans through advertisements such as direct mail or telephone calls to subprime borrowers. That all changed in 2001, when officials of the bank relaxed lending standards and the pool of borrowers suddenly expanded to include "just about anyone with a warm body," according to one marketing executive who spoke on the condition of anonymity.

Chase, like many other lenders, didn't wait for this new class of borrower simply to show up at one of its branches to fill out a mortgage application. Those days were long gone. Chase worked with credit bureaus to find people who might qualify for subprime mortgages (and later home equity loans) and targeted them in marketing campaigns.

The marketing team would spend hours combing through lists of targets; the executive recalls checking the credit rankings of those people and finding distressing signs: the targeted mortgage applicants were no longer making even small down payments on a home (just a decade earlier, 20 percent had been the minimum to buy a home). During routine checks of how people spent their home equity loans, he noticed another disturbing trend: people were not just using the loans as they had in the past, to build additions to their homes, say, or to pay for a college education. Because the borrowers were so strapped for cash, they were now using home equity loans for living expenses.

More than that, people were now borrowing more than the value of their homes to pay for closing costs and other expenses, jacking up the size of their mortgage to as high as 120 percent of the value of the home.

It was obvious to people at the bank that many of these high-risk borrowers couldn't afford the price of the homes or the high interest rates they were charged.

When one of the marketing executives brought all this up to her supervisor, he reiterated the standard line: don't worry about the details because the bank doesn't hold the loan. In other words, it was now Wall Street's problem.

And Wall Street wasn't complaining.

Wall Street may have gotten killed during the dot-com bust, but the mortgage bond desk proved to be its savior. According to Standard & Poor's securitization (both mortgage and asset backed issuance) exploded 230 percent between 2000 and 2006 to $2.7 trillion, after the Fed rate cuts spurred a massive issuance of mortgage debt. Profits soared on the mortgage bond desks of every firm because the deals were so complicated and thus commanded higher fees. Those bonds that couldn't be sold off to investors were simply held by the firms that did the underwriting, earning the "carried interest" that the mortgage debt generated.

As securitization grew, banks could lend more, and with banks flush with so much capital, lending standards started to fall even more. People didn't even have to show documentation that they had jobs to qualify for a loan. People on welfare showed phony documentation of income and savings. Many brokers, content to make a quick buck and pressured to find more and more customers, simply looked the other way.

The mortgage originator First Franklin rented conference rooms in about two dozen cities around the country for a sharply dressed, well-mannered fellow from a credit-rating company to describe how to turn credit lemons into lemonade so people with low credit scores could qualify for mortgages. "Make sure you get these people to pay their utility bills," he said. "That automatically increases their credit score." Having credit card balances, no matter how high, is a good thing as well, he pointed out, because it shows that the potential borrower can pay a bill. Multiple cars are good things, he said. No one gave much thought to the fact that the credit score system could, with just a few adjustments, transform a near deadbeat into someone who deserved a new and expensive home.

And why should they? The brokers believed they were doing God's work giving people access to the American dream, and, more important, too much money was on the line.

There were some brokers who both didn't think they were doing God's work and didn't believe the mania would last. One was a woman named Marilyn Barber, a licensed mortgage "solicitor" in New Jersey who had been assisting brokers to sell loans for more than twenty years.

"Another liar loan," Barber snickered as she heard the latest tale of excess from a colleague who had just placed a mortgage for a minimansion with a family who couldn't afford a down payment.

It was the dirty secret of the housing boom that brokers and the so-called wholesalers, or the mortgage originators, made the most money by selling mortgages to those least able to repay them. They were called "liar loans" by Barber and others in her office, even ones (unlike Barber) who made them. This new class of loan was made through tricks that made the mortgage "affordable."

These tricks included a slew of new products such as adjustable-rate mortgages (ARMs), where the interest rates and principal payments were low at first but later grew to immense proportions when the grace period ended and the bank needed to make enough money to compensate for the increased credit risk of the borrower.

Another variation was the interest-only ARM, where borrowers had to pay only the interest for the first couple of years until (in theory) their economic status had changed enough that they could afford not just the interest payments but the principal that must now be paid as well. During the second half of 2004 a whopping two-thirds of all mortgages were of either the adjustable-rate or interest-only varieties.

The beauty of these loans, at least on paper, was that everyone won: the loans demanded no down payment; in fact, interest and principal payments didn't begin for a year or possibly even later, so for the first year, the mortgage holder lived for free. In the meantime, housing prices continued to soar, giving the mortgage holders a sense of security that they could sell their homes at a profit.

The wholesalers didn't mind extending credit to those least able to pay them back because they were off-loading so many of their mortgages to Wall Street, which through securitizing them could diversify away the risk of the subprime loan, or so it seemed. Whether a family could afford to make the payments when they came due didn't matter—the broker who had made the sale had earned the fee. The wholesaler didn't care because it had sold the mortgage either to Wall Street or, increasingly, to Fannie Mae or Freddie Mac, to be packaged in a mortgage-backed security.

No one gave a second thought to this gravy train ending, particularly on the ground floor of the bubble, where Marilyn Barber worked. Every day, Barber saw people with no jobs and poor credit ratings being told they could afford virtual mansions. The mortgage borrowers believed everything they read about the housing market—it was the American dream, and according to the textbooks and brokers it would never end—so why worry if they could afford a $500,000 house on a $50,000 annual salary?

Barber didn't buy the theory; that's why she called them "liar loans." In her mind, there was no way they could be repaid, and she wasn't alone in thinking this. Brokers would openly mock the loosened credit standards now available and how people with not even a job could now qualify for the home of their dreams. The mortgage brokers, for the most part, laughed about the lower standards, but who were they to argue with the wholesalers, who provided every incentive to find home buyers?

Barber heard stories all the time that brokers had created fictitious incomes or glossed over credit problems to push through applications. She believed the wholesalers knew the games that were being played but didn't care; in fact, she couldn't remember a time when they had ever brought up the possibility of fraud and abuse in the mortgage application process.

All the big wholesalers came by her office on a regular basis. The wholesalers seemed out of place in her little town of South Marlton, New Jersey—the men in dark suits and starched white shirts; the women in black power suits. They looked and sounded as if they should be working for a Wall Street firm, and as it turned out they were, since their firms off-loaded so many of the mortgages to Wall Street to be packaged in mortgage-backed securities.

As the market continued to explode, the wholesalers offered all types of incentives to brokers to find live bodies to take out mortgages—expensive gifts, fancy dinners, clothing, books, and promises of bigger paydays down the road. Barber didn't need liar loans to make a decent living. "I couldn't live with myself," she says.

Others, however, could.

The market for selling mortgages to the masses couldn't be accomplished, of course, without a leader, someone who could crystallize the need for such largesse, distill it in an easily understood set of principles. That leader was Angelo Mozilo, the flamboyant CEO of Countrywide Financial, the California-based mortgage lender.

"Do whatever it takes to win," Mozilo exhorted three of his top executives. Money would be no object, and losing was not an option. "If you guys aren't one, two, or three in five years, you're out," he snapped

It was a standard speech for Mozilo, who could be alternatively daring and insightful—he pushed Countrywide in the direction of the lucrative subprime mortgage market at the perfect time—and inspirational, as the executives in his office had just witnessed.

Said one, "We would run through a wall for Angelo."

With his year-round tan and custom-made suits, Mozilo transformed Countrywide from a sleepy mortgage-lending company into a Wall Street darling. By 2000, his founding partner, David Loeb, had retired, and Mozilo ran the show, earning accolades and honors as one of the nation's best CEOs as the stock price soared.

But Mozilo wanted more. He started his own commercial real estate group, pushing the bankers he had hired to "crush," "demolish," and "dominate" the competition. He started his own financial advisory firm that focused on the working-class borrowers who were the bedrock of Countrywide's market but had been ignored by Wall Street brokers who wanted to manage money from the rich. And he focused on the subprime borrower.

Mozilo's drive to make Countrywide the best in everything was now extending into the lucrative subprime lending market. For years, Countrywide had focused on the higher end of the market. Under pressure to produce profits, Mozilo began cracking the whip; the firm began offering ARMs and interest-only mortgages, and subprime lending more than doubled in just one year between 2003 and 2004, to make up nearly 12 percent of all the loans the firm made.

And with it, Countrywide's stock shot up. By 2004, Countrywide had a market value of $200 billion. Mozilo was now "Banker of the Year," according to *American Banker*, a respected trade publication.

Mozilo recognized the vast riches in the subprime end of the market, but friends say he wasn't in it purely for the money. The son of a butcher from the Bronx, Mozilo believed that working-class families, many of them minorities and immigrants, were truly shafted by Wall Street, which usually ignored people with less than $100,000 in savings. With that in mind, Mozilo created his own mutual fund company to help the typical Countrywide client to invest in stocks and bonds.

Mozilo's push to expand the pie of home owners into the subprime arena began to change the face of the housing market. With more money chasing fewer homes, prices soared even in working-class neighborhoods. The boom in housing prices wasn't just a New York or an East Coast phenomenon. Edge cities began to spread throughout metropolitan areas in Nevada, southern California, and Florida.

And the boom would continue, in large part because of Mozilo. Obsessed with the notion of securitization and its ability to paper over risks and take his business to new heights, Mozilo used his political clout to push Fannie and Freddie to guarantee more subprime loans. He also made a series of speeches citing the need to expand home ownership to the poor and minorities, as the rationale to expand subprime lending.

In one speech, he called for the elimination of down payments, which he believed made it impossible for the poor to own a home. "Let's look for every possible reason to approve applicants, not reject them," he said.

And that's exactly what Countrywide did. With Mozilo leading the charge, subprime lending became the fastest-growing part of the lending market, rising a whopping 26 percent between 2001 and 2002 and at similar rates thereafter, compared to a 13 percent increase for the rest of the housing market.

Wall Street was toasting subprime lending as well. The draw of the subprime market for the lenders was the fees—lending to high-risk borrowers meant the banks could ramp up fees and interest rates on their loans. Those higher interest rates had another effect: they increased the returns of the mortgage-backed securities that were created from these high-yielding subprime loans.

In effect, Wall Street was nearly demanding that originators lower their credit standards because that would result in higher interest rates and higher returns on the bonds. Just as it had before the 1998 LTCM crisis, Wall Street was once again undercharging for risk. The interest rate spread between Treasuries and other bonds tightened. Investors began turning away from mortgage debt that didn't pack enough yield, and with that Wall Street looked for ways to produce bonds with still higher returns.

The only way to do that was to get the lenders to push mortgages to less creditworthy borrowers who could be charged higher interest rates or to simply buy the lenders themselves.

And that's exactly what Wall Street did.

10. NO MORE
MOTHER MERRILL

n the middle of Merrill Lynch's high-tech trading floor on the ninth floor of its lower Manhattan headquarters, Vincent Mora leaned into his phone and asked one simple question: "Bill, don't you want to make money?"

It was a question William Dallas had heard many times before, but now the answer was far more complex. As the mortgage boom continued, Dallas, the CEO of a small but fast-growing subprime lending shop called Ownit, could see the subprime market was starting to become a bubble, if it wasn't there already.

Dallas had seen the market hit rough spots before, with a spike in delinquencies that had left mortgage lenders with huge losses, but this time would be greater. Dallas could read the data as well as anyone; since 1990, despite occasional pullbacks, housing starts had almost doubled, and now in 2003 they were rising incrementally again. When would the demand end? He couldn't predict the exact date, but a run of more than ten years had "bubble" written all over it.

Given the size and length of the current housing boom, the losses this time could be catastrophic.

That's why Dallas had begun telling his wholesalers to cut back on the riskiest type of loans, both so-called alt-A lending, which involved mortgages sold to people with intermediate credit scores, and the riskiest of all, subprime mortgages. But now he was getting heat from above. Ownit was 20 percent owned by Merrill Lynch, now the leading underwriter of mortgage debt. The problem for Merrill: it needed loans, particularly subprime loans with high interest rates—"coupons," in the language of Wall Street—

so it could pack them into mortgage bonds and CDOs with enough return to entice investors to buy the securities, at a time when Dallas wanted to scale back on this part of the business.

"I'll ask again, Bill, do you want to make money? Because if you do, we're going to need a higher coupon," Mora repeated. Vincent Mora ran the mortgage desk at Merrill Lynch, and he was making a fortune. A slightly pudgy man in his late thirties, Mora was in charge of the desk that would buy and package billions of dollars worth of loans from various lenders and then sell them off to investors. Each month, Merrill would buy about two-thirds of Ownit's production—to the tune of some $6 billion a year—and then sell it to investors, big pension funds, mutual funds, and increasingly to itself. Merrill had begun to increase its own holdings of mortgage-backed bonds, particularly high-yielding CDOs, pocketing the carried interest to boost earnings.

And that was why Mora was pushing Dallas to ramp up lending to subprime borrowers, which would in turn boost the coupons on mortgage-backed securities. The higher the coupon, the more money Merrill could make when its mortgage desk got through repackaging those loans into other securities.

Dallas listened to Mora's request and rolled his eyes in disgust. "You're missing the point," he told Mora. "You can't look at it that way. You've got to look at each loan and its credit quality."

Over the past few months, Dallas and Mora had been having increasingly contentious conversations over the types of mortgages being underwritten by Ownit. According to Dallas, Mora, like many of his peers on Wall Street, had a monolithic view of mortgage lending. He didn't care about the measurements of risk, such as the credit score of the borrower or whether the borrower had a job.

As Dallas saw it, all the bankers really cared about was "product" and "yield"—or, more accurately, just how much product Dallas could deliver at a high enough yield.

In Wall Street parlance, Mora was looking for something called a "weighted average coupon," the total blended return that a pool of mortgages distributes, in the range of 8.25 percent, about a full percentage point higher than what he had been asking for nearly a year before.

It was a vicious circle of risk and return. In order to sell more bonds to investors, the yields would have to get larger, and the only way to do that was to pack the bonds with riskier mortgages, granted to people who because of their low credit scores or shoddy work histories were willing to pay exorbitant fees and interest rates for a subprime loan.

Merrill Lynch didn't seem to care about the consequences of packing bonds with more risky loans, even as some people in the business such as Bill Dallas were now predicting a correction; Dallas, on the other hand, realized that such yields came at a price. He knew that an 8.25 percent weighted average coupon would mean cutting corners.

In order to deliver an 8.25 percent coupon to Merrill and keep doing business with it, Dallas would have to make bad loans in lousy housing markets, where the chance of a home owner's defaulting grew to astronomical levels.

"Are you stupid?" Dallas responded during one of those conversations. "This is when you want to be doing better business. The devil is in the details." In Dallas's mind, Merrill could set itself apart from the other Wall Street firms if it did higher-quality deals; it would reap the benefits if and when the housing market faltered, as it had always periodically done.

Mora responded that he needed higher coupons to make deals work; in any event, it wasn't his call. There was a firmwide mandate for these mortgages, and it came right from the top.

Eventually Dallas stopped arguing; Merrill after all, was an investor in the firm and his best customer. Even so, as Dallas hung up the phone, he was confronted by a familiar but disturbing thought about doing business with Wall Street's mortgage bond desks, as the business for mortgage debt grew beyond even the wildest dreams of Larry Fink and Lew Ranieri twenty years before.

"These guys have no idea what they're doing."

For a business it apparently had no idea about, Wall Street was more involved in the market for securitizing assets, including mortgages, than ever before. In 2005, the total issuance of all asset-backed securities had surpassed $1 trillion for the first time on its way to $1.3 trillion in 2006 (by comparison, the national debt of the entire country at the time was $8 trillion). The fees kept growing not just because the size of the pie grew but because the big firms created more innovative ways to package and sell the increased risk they were asking people like Bill Dallas to create. The CDO itself was now being eclipsed by something called the "CDO squared," in which the firms sold not the CDO itself but interest in a pool of CDOs.

Because of the increased complexity in securitizing subprime debt—spreading out the risk in the bond so it could win approval from rating agencies—the Wall Street firms that underwrote these securities were paid higher fees to package them for resale to investors. And there was always the carry trade—borrowing money at low interest rates and buying mortgage bonds that now offered better yields and the alleged safety of securitization. It was the easiest money around.

What bankers soon discovered was that there was another way to make money from the housing boom and its evolution into the market for risky loans. If you buy into the notion—as many economists did—that the housing boom would continue for years, fueled by immigration, rising global prosperity, and other macro trends—the business of originating loans was ripe for the taking as well.

As the housing bubble continued to grow into the new millennium, Wall Street was seeking a trifecta—not only would it carry mortgage bonds on its books or package and underwrite them, but the big Wall Street firms had begun to buy up mortgage originators that supplied the product necessary to keep the entire process going. Why have to cajole the Bill Dallases of the world into selling more subprime loans when you can buy a mortgage originator, issue as many subprime mortgages as you like, and sell those mortgages to yourself?

Lehman Brothers was first out of the box in 2003, purchasing a small mortgage lender, BNC Mortgage, which specialized in subprime lending; later Lehman finished purchasing the entire company. Executives at Bear Stearns, innovators of mortgage debt that they were, soon started a subprime originator of their own, Bear Stearns Residential Mortgage Corporation, while Morgan Stanley and Deutsche Bank both acquired small lenders as well.

But the biggest and most audacious move into the origination business came from Merrill Lynch. Merrill had been relatively late to the mortgage origination business with its mere 20 percent interest in Ownit. Still, its underwriting desk was clearly one of the most successful on the Street, winning the number one spot in the league tables or coming in a close second to Citigroup quarter after quarter. It was something the company's new CEO, Stan O'Neal, touted as one of his major achievements—Merrill was now beating firms like Goldman Sachs and Morgan Stanley in some of the most profitable business on Wall Street. Likewise, the Merrill bond chief, Dow Kim, one of O'Neal's most trusted lieutenants (and O'Neal didn't trust many people), made expanding the CDO business one of his top priorities.

But to do that, he needed product or, as Vincent Mora would say, higher-coupon product, to take the business to the next level. And if Bill Dallas refused to supply it, Merrill would get it someplace else.

"This is what *he* wants," explained Michael Blum, the head of Merrill Lynch's structured finance department. The "he" Blum was referring to was Stan O'Neal, the autocratic CEO of Merrill Lynch.

A little more than a year after Merrill purchased its stake in Ownit, the firm—or, to be more precise, the only person at the firm who really mattered—was hungry for more. O'Neal was prepared to announce the acquisition of the subprime lender First Franklin, the second largest subprime lender after Countrywide Financial, and Blum was giving the people at Ownit a heads-up.

Blum received a cool reception from Ownit's board as he relayed O'Neal's intentions. The meeting was held in the company's two-story office on Agoura Road in Calabasas, California, a sort of Wall Street for subprime lenders, where a large number, including Countrywide, were headquartered.

Leading the discussion for Ownit, of course, was Bill Dallas, Ownit's CEO. Dallas might have been wary about recent trends in the subprime market, but among his colleagues, he was highly regarded as a pioneer in the subprime industry, starting back in 1981, when subprime had been an underdeveloped business—at that time not even Countrywide would have considered venturing into that area of the mortgage business.

Dallas had done so because he believed people with a few blemishes on their credit history not only deserved the chance to own a home but also, more often than not, were good customers. The subprime business had been built in many ways on the same premise as the junk bond market: finding creditors who were considered junk but whose ability to pay was actually much higher than average, thus making people who fell into this category one of the last untapped areas of the mortgage business.

But now Merrill wanted to buy First Franklin, a company Dallas had actually cofounded back in the early 1980s and one that he knew well. Merrill didn't want to be burdened by Dallas's credit standards and his hand-wringing over granting loans to people with lower credit scores. Dallas, of course, had seen it coming, given the constant battles he was having with Mora. Still, when he heard that the firm was moving away from Ownit and getting ready to buy a competitor, it hurt.

"You got to be kidding me," Dallas shot back to Blum. "Those guys are our competitors!"

Blum said there was nothing he could do. O'Neal and his senior management were on a mission to purchase a subprime lender in order to compete with Lehman Brothers and Bear Stearns. Buying First Franklin was the only way Merrill could get there quickly, which was the way O'Neal liked to do things.

Though Dallas considered Blum a smart guy who understood how to securitize mortgages, "like a lot of guys on Wall Street, he knew nothing about mortgage origination." And there was a lot to know. Mortgage origination was and always had been a cyclical business. The firm had had periods of huge earnings, but when interest rates rose, they could barely make ends meet. In 2006 the housing market was so overheated that many analysts began predicting a crash.

"You realize you are getting into the market at the very top, right?" Dallas told Blum. "This is absolutely the wrong time to be buying them." He went on to warn Blum that the price Merrill was looking to pay, $1.3 billion, was just absurd, much greater than any of Merrill's competitors (twice what Lehman was bidding), and he warned that First Franklin was heavily involved in subprime lending, a part of the market that was now showing strain.

He also said the rumor mill was overloaded with speculation that the company was not being forthcoming about potential losses on its books. His prediction to Blum: "You'll have huge losses if you buy them now. "

But the folks at Merrill took Dallas's gripes with a grain of salt. The housing market was booming; a correction might be in the cards, but all of Merrill's indicators showed it would be short-lived. Meanwhile, Merrill considered Ownit to be inferior to First Franklin. "I understand what you're saying," Blum answered after hearing Dallas's objections. But there was nothing he could do. The decision was coming from the highest levels of Merrill, from the last guy he would ever argue with.

The purchase of First Franklin by Merrill Lynch in late 2006 was typical of Stan O'Neal's leadership at Merrill: almost no one, other than a few advisers he plucked out of the ranks of Merrill's vast executive corps (based on a combination of friendship and trust that they posed no threat to his leadership), could tell him what to do. In addition to Dallas,

O'Neal's market strategist, David Rosenberg, had been predicting for years that the housing market had reached bubble proportions and that the inevitable bursting would leave firms leveraged to the mortgage market battered and bloody. So did a battery of analysts at the firm who saw the warning signs of the housing bubble and were ignored.

O'Neal was unimpressed by them all, in fact by just about anything, that is, except for himself, which made his tenure as CEO of Merrill Lynch, the storied brokerage franchise that had vowed to bring Wall Street to Main Street, one of the strangest, most volatile, and ultimately most disastrous that Wall Street had ever seen.

For all intents and purposes, E. Stanley O'Neal (the "E." stands for Ernest) had no business being the CEO of Merrill Lynch, the nation's largest brokerage firm, and it had nothing to do with his intelligence or qualifications, and everything to do with the culture and traditions of the firm he would lead. The grandson of a slave, O'Neal had grown up in segregated Georgia during the 1960s, the son of a farmer who eventually moved his family to Atlanta to take a job at a GM assembly plant. O'Neal survived racism and segregation as a child and excelled as a student. He went to college at the GM Institute, which was at the time a highly selective engineering college that trained a generation of senior executives at the big automobile manufacturer, and then to Harvard for his MBA.

After working in various positions at GM, lastly in its finance department, he joined Merrill in 1986, recruited into its investment banking division.

In the mid-1980s, Merrill's investment banking department wasn't exactly a training ground for future company leaders, which had traditionally been plucked from the firm's brokerage sales force, which hawked stocks to Middle America. O'Neal, by contrast, had never sold a single share of stock in his life.

But from nearly the moment he started, O'Neal was regarded as a force to be reckoned with inside the firm. He was smart and aggressive, and he could turn on the charm—that is, when he wanted to and when he wasn't brooding about not being promoted fast enough as he moved up the firm's management ranks, heading up the junk bond desk, becoming CFO, and later running the brokerage business.

A lean, cerebral, and intense man, O'Neal was known for his aggressive intelligence and at times odd behavior. He could be stunningly ambitious; when his boss, Merrill's then chief Dave Komansky, set up a horse race among senior executives to become his replacement, O'Neal was the only candidate to state publicly and emphatically that he wanted the job. Komansky promptly upbraided him for the comments that appeared in the *Wall Street Journal*.

There were also times when O'Neal, without any warning, would simply not show up at the office; at least once he was away so long that underlings worried about his safety. Several times during his twenty-one-year career at Merrill, O'Neal threatened to quit because he believed he was either working for inferior supervisors or being undercompensated. Each time he stayed and was given more money.

O'Neal, for all his occasional charm and despite his working-class roots, could be nasty, condescending, and elitist, the exact opposite from the personalities of Merrill's past CEOs, most of them gregarious former stockbrokers such as Bill Schreyer, Dan Tully, and David Komansky. They were men who made money but also took pride in caring for their people—an attitude that led to the firm's being known as "Mother Merrill."

By contrast, O'Neal stated early and often that he believed their governing philosophy was leading to corporate bloat—by 2001 Merrill had a workforce of about seventy-five thousand, the largest on Wall Street aside from the giant Citigroup, and its earnings growth had stalled. In every management position he held, O'Neal did everything he could to do away with this part of the firm's culture, firing staffers without a second thought.

For those reasons, the oddsmakers inside Merrill never gave O'Neal much of a chance of becoming CEO. And there was more: Merrill was the one Wall Street firm that appeared to celebrate its whiteness. It had been run by Irish Catholic brokers, some of whom had taken pride in learning the business without college, certainly without attending Harvard Business School, since the firm's founding in 1914 by the legendary investor Charles Merrill.

How would it ever be run by a black man with a Harvard MBA?

When asked that question in 2000 as he announced a round of firings in the brokerage departments, firings that no doubt upset the culture and his standing within the firm, O'Neal confidently stated, "The list of people who doubted me is very long, and I'm still here." He became CEO three years later.

When Dave Komansky had been named CEO of Merrill Lynch in 1997, the firm had gone out of its way to show that Komansky was himself a trailblazer of sorts: He was the first Jew to run

Merrill Lynch in its history, a change from the Irish Catholics who had run the place for generations. He'd grown up working class in the Bronx; his dad was a postal worker. "I was always proud that I gave a poor Jewish kid from the Bronx a shot to run Merrill," Dan Tully, the firm's beloved former CEO, once said. This from a guy who'd grown up in Woodside, Queens, a neighborhood of working-class Archie Bunker types who once also said that the proudest moment of his career had come during his retirement party when a band complete with bagpipes played "Danny Boy."

At Merrill, although the board of directors has to approve any selection, the choice of a successor is left up to the retiring CEO. According to Tully, when he was ready to retire, he believed Komansky was perfect for the job. He had been a broker, meaning he understood the heart and soul of the firm; he had worked his way up Merrill's management ladder and had the right temperament for the job.

He also believed in Mother Merrill.

When you ask people like Tully and Komansky what Mother Merrill means, the answer is always the same: family. Merrill, like all Wall Street firms, has its fair share of power struggles and pettiness, but what made the firm different from the rest of the Street was that in many ways, it was a big family. The losers of those power struggles didn't always have to leave the firm; they simply found other places to work within the comforting embrace of Mother Merrill. Executives who came under suspicion for wrongdoing weren't immediately fired; the firm's first response was to protect and defend.

Mother Merrill extended up and down the corporate ladder. Douglas "Sandy" Warner, the former CEO of J.P. Morgan, held extensive discussions with Komansky about a merger. Komansky was one of the leading Wall Street executives who pushed for the repeal of Glass-Steagall, so he believed the combination of investment and commercial banks was the future of Wall Street. But such mergers, even in the best of times, lead to downsizing, and that's why, after weeks of negotiations, Komansky decided to scuttle the deal. "He worried about the elevator people," Warner said during an interview. Komansky told Warner that he didn't have the stomach to make such drastic cuts, particularly of average people: the secretaries, receptionists, and back-office types who feel the brunt of the downsizing and postmerger cost-saving moves, the people, as he put it, "who go up and down the elevators each day."

Komansky announced O'Neal's status as his heir apparent in the summer of 2001, when O'Neal was named Merrill's president, a stepping-stone to the CEO job that he was now all but assured. "Through his broad experience across many facets of our business, Stan O'Neal has demonstrated keen strategic vision and a great ability to inspire and lead people," Komansky said in a press release.

What he left out was the battle behind the scenes that had propelled O'Neal to power. O'Neal hadn't been his first choice, his second, or even his third. O'Neal had by now proved himself to be, at least on paper, the firm's best and perhaps most ruthless manager. He cut costs in the brokerage department, increased productivity, and all but abandoned an effort to offer cheap computerized stock trading to brokerage customers. That was the good part. The bad part, to Komansky, was that O'Neal had made it known that Mother Merrill—a culture he considered more racist than paternalist—would be out the minute he was named CEO.

As one O'Neal loyalist put it, "Stan believed Mother Merrill bred mediocrity. He thought it was racist because it was a self-fulfilling prophecy; people who were at the firm because of connections or because they were drinking buddies of some guy on a trading desk stayed and were promoted, while newcomers were screwed."

This executive felt that much of O'Neal's animus to the maternalism of Merrill's culture stemmed from his belief that not only did he not benefit from it, it had held him back, and it would influence nearly every decision he made going forward.

"Stan believed he had to rise above this culture to get to where he was, and he always had a chip on his shoulder about it. As a result, he never saw the good side of Mother Merrill, how the firm did what other firms didn't do, and that was take care of its own."

Komansky, intent on controlling the succession process, had told the four executives in the horse race, including O'Neal, not to lobby board members on their own. Every man had complied except O'Neal, who had wooed top directors with plans to reshape Merrill, boost profits, and cut costs firmwide, much as he had done during his relatively short stint as the head of the firm's brokerage business.

O'Neal's end run angered Komansky, who like most insiders at Merrill had never developed a relationship with O'Neal. Komansky, a rotund man with a congenital bad back, was the prototype of the Merrill Lynch broker:

he hadn't graduated from college, but, he knew how to sell stocks. Gregarious and at times profane, Komansky wears a pinky ring and a gold wrist chain. O'Neal, meanwhile, was almost never without a finely pressed suit and starched white shirt. Around the office, he rarely cracked a smile and almost never cursed.

Though their styles were clearly different, others say Komansky was most of all put off by O'Neal's overwhelming ambition, particularly when it came to competing for the top job. Either way, people close to Komansky say he was prepared to name Merrill Lynch asset management chief Jeffrey Peek as his president and his successor, but in the end, the board, impressed with O'Neal's cost-cutting strategy, aggressive lobbying, and plans to slash costs and boost profitability, ignored the traditional prerogative of the CEO to select his successor and pushed Komansky to pick O'Neal.

Komansky was sixty years old when he named O'Neal president, which meant he had four more years to linger around the executive suites before he retired. But O'Neal wanted Komansky out immediately; he swiftly consolidated his power by pushing out Peek and marginalizing the other losing competitors for the top job. The opportunity to complete his takeover presented itself one morning in mid-September 2001, when two jets flew into the World Trade Center's twin towers. The attacks devastated Wall Street, but Merrill in particular because of the proximity of its headquarters, just yards away from Ground Zero.

The firm was immediately evacuated, and Komansky, in ill health with a bad back, barely made it out alive. He ran north along the Hudson River as fast as he could, just like thousands of others who escaped the attacks, and nearly passed out in exhaustion as he made his way to safety.

O'Neal reached safety as well. He made it to a colleague's town house in the Greenwich Village neighborhood about ten blocks north of the attack site, along with other top Merrill executives, eventually including Komansky, who was visibly shaken—in a "state of shock" according to one colleague—after witnessing the carnage that included people jumping to their deaths from the burning twin towers of the Trade Center.

As the group huddled, Komansky sat silently smoking a cigarette while O'Neal filled the void by declaring that the management team would meet daily and demanding he be given frequent updates about the firm's relocation efforts. A few weeks later, O'Neal pushed for and received nearly total operating authority at Merrill; the firm's powerful executive committee, which used to report only to the CEO, would now report to him as well.

The Stan O'Neal era at Merrill Lynch had officially begun.

The changes brought about by O'Neal were swift and jarring to long-time Merrill executives. As brokerage chief years earlier, the first thing he did was announce a major downsizing, not of brokers but of the people who go up and down the elevators every day, the administrative staff and secretaries.

Komansky had a heart and so did Merrill, but O'Neal believed this heart was weighing on the firm's success, making it vulnerable to competition and, more than that, hurting its chances to remain independent. When O'Neal was made president of Merrill, he produced an analysis showing that its operations were so bloated that it trailed just about every other major firm in return on equity. Komansky still resisted downsizing the firm, so as O'Neal assumed more power, he launched plans to do it on his own, dismantling operations around the globe and cutting jobs at home. When he was done, Merrill had eliminated some twenty thousand jobs.

O'Neal believed that Mother Merrill was not just idealized and ill suited to modern times but bred a certain degree of racism at the firm and hampered the firm from being what it needed to be in order to survive: a truly modern, international, meritocratic company.

As a result, O'Neal believed Merrill's senior ranks needed to reflect the international nature of the modern financial services business. An Indian-born investment banker was soon running all of the capital markets, while Dow Kim, a bond trader from Korea, was appointed to run the firm's bond department. The biggest change may have been the ascension of an Egyptian-born numbers cruncher named Ahmass Fakahany, who would soon become O'Neal's closest adviser.

Fakahany's appointment shocked the dwindling ranks of the old guard at Merrill, not because he was from Egypt but because his résumé was so thin to be among the top executives running the firm. A former executive from Exxon, Fakahany joined Merrill around the same time as O'Neal in 1987 and held a series of accounting-related positions, mostly overseas, until he returned to the United States in 1998 as an aide to O'Neal, then the company's CFO during the LTCM crisis that O'Neal worried might cause the firm's collapse.

Merrill, of course, survived LTCM's demise, as did the relationship between O'Neal and Fakahany, which not only survived but thrived from that point on.

Unlike other Merrill executives, O'Neal rarely socialized with his co-workers, except for Fakahany, someone O'Neal concedes he truly liked. Fakahany liked O'Neal as well. They shared a love of wine and fine food (Fakahany owned a couple of restaurants) and a zealousness to make Merrill a more international firm. People close to O'Neal credit Fakahany as the person who more than anyone else at the firm prodded O'Neal to compete for the CEO job and to move the firm's culture and business model away from Mother Merrill.

The friendship grew even tighter in mid-2003, not long after O'Neal took total control of the firm from Komansky, when O'Neal faced his first real managerial test. A brilliant investment banker, Arshad Zakaria, now one of O'Neal's top aides, secretly began lobbying for the number two spot at the company, a stepping-stone to the ultimate prize.

As he made his move, Zakaria thought he had powerful allies on his side. His closest friend at Merrill was CFO Thomas Patrick, who also had helped O'Neal unseat Komansky, and the board's most prominent member, Robert Luciano, the former CEO of Schering-Plough.

Patrick had a vast and detailed knowledge of Merrill's balance sheet and moonlighted as an investment banker; in the early to mid-1990s, he and Zakaria developed a new type of bond deal—a tax shelter called "contingent payment installment sales"—that saved some of Merrill's biggest corporate clients, including Schering-Plough, hundreds of millions of dollars in taxes. These deals were among the most lucrative Merrill had ever produced, and they made Zakaria and Patrick enormously rich and powerful.

That's how they were able to bounce Komansky and elevate O'Neal, and they attacked the task of elevating Zakaria with the same zeal, once again approaching their old friend Luciano with the proposal and asking him to convince the larger board to approve Zakaria's promotion. But Zakaria and Patrick severely overplayed their hands. Luciano was taken aback by what he considered an end run. "Shouldn't you guys be going to the new CEO about this first?" he asked. Soon word about the power play leaked to Fakahany, who promptly reported the machinations to O'Neal.

With lightning speed, O'Neal forced out both Zakaria and Patrick before they could sell the idea to the larger board. A few months later he forced Luciano off the board as well.

Fakahany was amply rewarded for his loyalty as he became chief operating officer of the firm, in charge of every department, including the company's use of risk and leverage. More than that, Fakahany cemented his place as O'Neal's eyes and ears for years to come.

Few inside Merrill could argue with the firings of Patrick and Zakaria after such a brazen attempt to end-run the CEO. But their removal would have long-term implications for Merrill Lynch. First, the firings would send a message to other executives that O'Neal was willing to let go of just about anyone, even two men who had made the firm a lot of money and to whom he owed the fact that he was CEO in the first place, to maintain his grasp on power.

But more important, the firings consolidated power and oversight of the firm's soon-to-burgeon balance sheet with Ahmass Fakahany, who had never run a business, much less a balance sheet this large, in his entire career.

Merrill Lynch had never been very good in the risk-taking department. The firm's business model had been built not on taking risk but on giving advice. Merrill had the nation's largest brokerage firm, its thousands of financial advisers peddling stocks to investors in cities across the country.

The brokerage division was Merrill's heart and soul. Nearly every CEO of the firm had been a broker. It's the reason O'Neal had pushed to manage the brokerage department in 2000 on his way up the ranks. He'd viewed it as a stepping-stone to the big time. And he was right. The problem is that while brokers are trained to sell stocks, they tend to make lousy risk managers.

Each time Merrill rolled the dice in the bond markets, it lost big. In the late 1980s it had almost fallen into insolvency when Howie Rubin lost hundreds of millions of dollars trading mortgage bonds. In 1994 it took big losses trading mortgage bonds, not to mention the legal costs associated with selling mortgage bonds to Orange County, California, which filed for bankruptcy. Merrill then lost hundreds of millions of dollars during the one-two punch of the Russian debt crisis and the implosion of LTCM.

In all these cases, the firm's response was to drastically scale back on risk and miss out on the subsequent bond market rally, only to get back into the market near its top, when the bubble was about to burst once again.

It's unclear if O'Neal understood that he was about to repeat the mistakes of the past. He announced his new risk initiative not long after he became CEO. In meetings with his executive team, he produced an analysis of how Merrill was falling behind both Morgan Stanley and Goldman Sachs, which had now made leverage and trading risk a centerpiece of their business. O'Neal was particularly keen on copying Goldman. While most of the Street hated Goldman, envious of its success, its culture of recruiting only the best people from the best schools, and its snooty attitude of being the best, O'Neal envied and obsessed over Goldman.

He studied how Goldman, following the dot-com underwriting crash in 2001, had made a huge push into proprietary trading, using house money and leverage to become, by 2003, the most profitable firm on the Street. The success of the strategy would even lead Goldman's CEO, Hank Paulson, to quit in 2006 and become George W. Bush's Treasury secretary. Paulson's successor, Lloyd Blankfein, was behind Goldman's successful strategy of taking calculated market bets. When Blankfein became CEO, Jimmy Cayne finally had something to like about Goldman. "The Jews are making a comeback over there," he said. Blankfein was the first Jew in years to run Goldman, which traces its roots to German-Jewish immigrants.

O'Neal wasn't so much enamored with Goldman's ethnic makeover as he was with its profit margins and business model based on taking risk. In 2003 Merrill was leveraged just 16 to 1, the lowest of any of the major Wall Street firms, but O'Neal's edict to copy Goldman and take more risk unleashed a massive increase in leverage, with the company making maybe the biggest uptick in borrowing and risk taking in the firm's history. In just a few years, Merrill increased the size of its balance sheet—the sum total of trades and investments—from $500 billion to around $1 trillion quickly and abruptly. Merrill ramped up its activities in the bond market, pushing to be the top underwriter in every category. More than that, Merrill didn't simply buy and then sell the bonds at a profit; it embraced the "carry trade" in a major way as traders began hoarding billions of dollars of bonds, many of them high-yielding mortgage bonds, CDOs, and their particularly volatile offspring known as the CDO squared on its books to earn the higher returns.

Stan O'Neal was considered a decent investment banker and a good CFO, but he didn't have Blankfein's background in risk management and trading, and even worse, he lacked the team that Blankfein possessed at Goldman (Blankfein's number two, Gary Cohn, was considered among the best traders on Wall Street), but that didn't stop him from trying to copy Blankfein's strategy. People who worked with O'Neal recall countless conversations in which Goldman's strategies and market presence were discussed. Newspaper articles were often clipped and sent out to his senior staff, with O'Neal circling in red the sections that discussed how much money Goldman was making from its trading operations. O'Neal laid down a profitability target: a 30 percent return on equity. It wasn't a coincidence that that was Goldman's ROE as well.

O'Neal said it was what investors were justifiably demanding. "Why would any investor want to put money in an investment bank if it produces the earnings of a commercial bank?" O'Neal asked his senior staff. His point was a valid one: Investment bank revenues are typically more volatile than commercial banks' because they zig and zag with the periodic implosions of the markets, particularly the debt markets, which drive the risk-based trading model.

Just a year after O'Neal took over as CEO, his strategy was showing progress. Mortgage underwriting had become the most profitable underwriting business on the Street, and Merrill was number one or close to it every quarter, underwriting just about every deal it could, even those it couldn't profit from. Not only that, the firm was trading more, using its leverage to buy and hold mortgage debt more than it had since it had been burned in the LTCM crisis.

But what made Goldman dangerous as a competitor and Blankfein a trading genius was that he had transformed risk and leverage to an art form. Goldman wasn't simply hording vast amounts of mortgage debt in a carry trade; it was making savvy market bets while staying nimble and ready to change directions, as it often did at a moment's notice.

There may have been a more important reason for Goldman's success. Goldman was able to stay nimble because it was, at bottom, a meritocracy. Dissent and debate were part of the fabric of the firm because everyone there understood the fickle nature of the markets.

O'Neal, on the other hand, was increasingly running an oligopoly. Underlings were afraid to question his moves as he began weeding out longtime executives who had been on the losing side in his battle to become CEO. A chill engulfed the firm once known as Mother Merrill, and O'Neal couldn't have cared less.

That chill extended to the CEO's office. O'Neal was proud of his accomplishments: his rise from banker to CEO and the pit stops he'd taken along the way: junk bond chief, brokerage head, and CFO. But he had also forgotten his own roots. As CFO, he had taken steps to cut risk at the firm as he saw the bond markets getting too inflated. During the Russian bond crisis and the LTCM imbroglio, he had worried openly to colleagues about the firm's risk profile: Merrill's reliance on short-term borrowings and the people it used for funding its trading operations, the banks, and other investors who purchased Merrill's commercial paper and who could in a heartbeat turn on the firm and place it in great jeopardy.

But now O'Neal was pushing Merrill further and deeper into the hands of those same entities than ever before. Why the change of heart? O'Neal's many detractors say any executive brave enough to confront O'Neal would have been fired—as we'll see. O'Neal later said the realities of the business had left him no choice but to ramp up risk and leverage: fees were being squeezed due to greater competition, and investors were demanding that he crank out Goldman-type earnings. The data back him up. Over the past ten years, underwriting fees had declined by 26 percent. Merrill had been hit particularly hard by the commoditization of the brokerage business, in which fees declined by more than 80 percent because of increased competition and the advent of cheap online trading.

Goldman had been the first to figure out that the only way to combat the decrease in fees was to fight it through risk-taking activities on the trading desk, and its earnings had soared, as had those of others that embraced the Goldman philosophy—which now included just about every firm on Wall Street.

In O'Neal's mind, Merrill couldn't survive without taking more risk, even if taking risk put its very survival in jeopardy.

11. THE MONEY MACHINE

By the middle of the new millennium's first decade, risk wasn't a way to boost profits on the Street, along with investment banking and asset management. It *was* the business of Wall Street.

That's what John Mack of Morgan Stanley was telling clients in 2005, as he met with hedge funds that were clients of Morgan's fast-growing prime brokerage business. One of his meetings was with Stanley Druckenmiller, who ran the hedge fund Duquesne Capital.

The prime brokerage units at the big Wall Street firms aren't well known to the general public or, for that matter, to many shareholders, but in the investment banks they had become hugely profitable, a veritable fee machine. The investment houses earn fees by clearing or processing trades for their prime brokerage customers. They also earn considerable amounts of fee income by lending their prime brokerage clients stock and cash to carry out their investment strategies.

But the real importance of the prime brokerage accounts derives from the fact that Wall Street is one big lending bazaar—investors lend money to the investment banks so they can finance their trades and market bets, and the investment banks lend money to investors, namely hedge funds like Druckenmiller's, so they can trade and invest.

By 2005, one of the largest pools of money for the security firms to lend out and earn fee income from was found in the prime brokerage business, as the hedge fund industry caught on fire and new hedge funds cropped up almost daily and then turned to Wall Street to set up these accounts.

It seemed like such easy money, but only because the markets were so strong. There is also a dark side to the prime brokerage business, and Wall

Street's increasing reliance on these funds for lending purposes. Hedge funds can pull their money out of these accounts at a moment's notice, causing a cash-drain at the firm and, depending on the level of withdrawals, sending the investment bank into insolvency, as Wall Street would soon discover.

But no one was even considering any of this now, particularly not John Mack, who explained to Druckenmiller that he was about to join the risk-taking party with the rest of Wall Street in a major way.

A longtime executive of Morgan Stanley, where he'd earned the nickname "Mack the Knife" for his cost-cutting prowess, Mack had pushed for its 1997 merger with Dean Witter, which combined Morgan's white-shoe investment bank with Dean Witter's large brokerage sales force. Dean Witter CEO Philip Purcell ran the show, with Mack as his number two. Mack was a charismatic leader; he had been with Morgan Stanley for his entire career and had built a loyal following. He had started his career as a municipal bond salesman, but what got him going every morning was deals—investment banking deals. Mack's travel schedule was filled with trips around the world to help the firm win banking deals. To the former college football standout at Dartmouth, winning a merger was like scoring a touchdown: every day in every way, he wanted Morgan Stanley to crush the competition, particularly its long-standing rival Goldman Sachs.

Mack also wanted to be leading that fight. Though he agreed to take the number two spot for the sake of the merger, he believed he had squeezed a handshake deal out of Purcell, in which Purcell agreed to turn over the reins of the firm five years after the merger. Mack may not have traded a bond in his life, but he was a risk taker, and without a second thought, he signed the merger agreement and waited his turn. When Purcell later refused to turn the firm over to Mack as agreed, Mack attempted a coup, lost, and was forced from the firm. Defeated, he joined Credit Suisse and later the hedge fund Pequot Capital, one of the largest in the business, as its chairman.

Purcell ran a decidedly low-risk operation focused on brokerage fees and underwriting—known on Wall Street as the "agency" model, since the firm was nothing more than an agency or conduit for risk taken by others.

But Morgan's returns lagged far behind those of rivals that had begun to embrace risk, including so-called proprietary trading, where firms use house money, expanded through leverage, to make market bets. This, combined with a nasty culture war between the Dean Witter and Morgan Stanley executives, forced Purcell out and brought Mack back to run Morgan Stanley.

Now that he was back, Mack immediately recognized that Purcell's business model showed all the signs of stress that had pushed Stan O'Neal, and increasingly the rest of the Street, to expand into risk and trading. "The old agency model is gone, and it's never coming back," Mack said to Druckenmiller. "The proprietary model is here to stay."

Druckenmiller was one of the most experienced traders on Wall Street, having made his mark working for George Soros's Quantum Fund. After more than a decade with Soros, he had gone out on his own after losing $200 million trading technology stocks and built one of the top hedge funds in the business. He had since become one of the richest men in America, worth more than $3 billion.

Mack began discussing the success of Goldman and its trading culture, and how he had every intention of taking Morgan there as well. He told Druckenmiller he had all the players in place to be able to match Goldman and the other big bond houses in the trading department, including a brilliant former bond trader named Zoe Cruz to run the show. Druckenmiller wasn't so sure. Not about Cruz, whom he liked, but about Mack, who only a few years earlier gambled and lost the CEO job.

Now Mack was going to the gambling table once again, and the stakes were even larger. Morgan Stanley was planning a major foray into bond trading, mortgages in particular, and their derivatives, he said. Morgan was returning to its roots as an organization that looked to dominate the competition in every part of the securities business, and the part of the business to be dominant in in those days wasn't the brokerage arm that Purcell had cherished from his Dean Witter days but the bond-trading department that was making Goldman Sachs, Lehman Brothers, Merrill Lynch, and Bear Stearns so much money.

Druckenmiller listened to Mack's plans for the future. But he thought back to a meeting he had just had with an analyst from Bear Stearns who with a series of charts and graphs had predicted the imminent demise of the mortgage bond market—the very place where John Mack now wanted to gamble. The analyst had spoken about the deteriorating quality of loans and the rise in housing prices, and how they would lead to the demise of the mortgage bonds packed with so many of those loans. It had been an odd meeting, Druckenmiller thought, because mortgage bonds were at the heart of Bear Stearns' business model, and here was one of the firm's analysts all but predicting the market's demise.

Did John Mack know anything of this? Nothing from his conversation with Druckenmiller suggested he did, yet he was now going to roll the dice in a market that seemed ripe for a fall. As far as Druckenmiller was concerned, risk and leverage were now, in John Mack's hands, a potential nuclear bomb.

And he was far from alone.

Jamie Dimon leaned back into his chair and began thumbing past the standard disclaimers and statements in the annual report to get to the "good stuff," as he called it: the statistics, charts, and graphs that are used by experienced numbers crunchers to determine the real story of a firm's financial health. While CEOs like John Mack could think of nothing better than traveling to the Middle East to win a banking assignment or, in the case of Jimmy Cayne, holding court with politicians to manage a large municipal bond offering while inhaling one of his cigars, Dimon considered it fun to read financial statements and balance sheets, and the balance sheet he'd been most obsessed with lately wasn't his own. It was Citigroup's.

By 2005, it was seven years since Dimon had been bounced from Citigroup by Sandy Weill, and now, after an arduous journey, he was in the middle of his professional resurrection. It had been in 2000, when he had been appointed CEO of Bank One, a large but troubled regional bank located in Chicago. He had convinced top executives from his Travelers/Citigroup days who were tired of Weill's egomaniacal behavior to join him in the journey.

Dimon was also creating a tremendous buzz on Wall Street, particularly at Citigroup, that he was really plotting his return to Wall Street: once he improved the results at Bank One, his next move would be to take on Citigroup through a megadeal creating the "Son of Sandy," a large financial services supermarket that could compete directly with Weill's $3 trillion gorilla.

Dimon might have matured—he was closing in on fifty, less inclined to scream at people in the office, as he often had during the Weill years, or to pester them with dozens of questions about deals and trading positions—but he was as competitive as ever, and everyone knew it.

The move that Citigroup feared came in 2004, when Dimon merged Bank One with the banking giant JPMorgan Chase. It wasn't quite a combination on the size and scale of Citigroup, but even so, the press heralded the deal as Dimon's comeback—a chance to replicate and surpass the success of his old boss, former mentor, and now sworn enemy at a time when Weill and the Citigroup empire he had created appeared most vulnerable. Scandal had shaken the foundations of the firm—Weill himself became a target in a wide-ranging investigation by New York Attorney General Eliot Spitzer into Wall Street sleaze and would eventually step aside, leaving the top job to the firm's attorney, Chuck Prince. (Citi paid a large fine; Weill escaped charges.) Meanwhile, Citigroup's earnings growth had slowed; the firm threw off $25 billion a year in net cash, but all that money seemed to get lost in the morass of the many operations created by its massive size and by the costs needed to complete the integration of the various subsidiaries Weill had purchased all over the world.

What Dimon couldn't understand as he reviewed Citigroup's finances in 2005 was that amid all this—the management turmoil, the massive costs needed to integrate its operations—a corner of Citi's balance sheet was kicking ass, and he couldn't figure out why.

That corner was the firm's fixed-income department, the department that deals with underwriting and trading bonds of all types. Citigroup was consistently at or near the top in the various underwriting categories, but many of the bond deals were known on the Street as "loss leaders," meaning that the underwriting firm wasn't making a lot of money on them—or any money at all. Even so, Citigroup's fixed-income department was now generating massive revenues—Citi's $25 billion in net cash from its various and far-flung operations was the result of billions of dollars in revenues being generated by its bond department, roughly $9 billion a year, nearly one-quarter of the firm's $44 billion in revenues—and Dimon couldn't figure it out.

To be sure, the sources of the booming trading profits were mostly hidden in opaque disclosures and footnotes in financial statements that were nearly impossible to figure out. Under then-existing accounting rules, financial firms didn't have to come clean with a detailed accounting of their growing investments in mortgage debt and other bonds. All they had to do was price the securities on their books based on financial models they themselves created, a process known as "mark to model." It incorporated what the market believed the bonds were worth, but also what the financial formulas created by the firms' risk managers believed as well.

The problem with mark to model was the element of human intervention; the models themselves could be tweaked in such a way as to make losses of nearly any magnitude disappear. Many people believed Lehman had been able to survive the 1998 trading debacle that had caused losses in bond positions that had nearly doomed the firm because its models priced its bonds and other holdings at generous levels (Mike Mortara would have called their pricing "aggressive"), not at what they would have sold for on the open market.

Even so, few people on Wall Street were better at figuring out numbers, as opaque as they might be, than Dimon, and as he studied the firm's financial documents, he still couldn't understand how they were adding up. The fixed-income balance sheet—the revenues derived from trading and holding bonds—was exploding. That was obvious from the disclosures. But far less obvious was *how* those revenues were growing. The amount of capital Citigroup was holding against those assets was remarkably small and steady, given the surge in revenues.

In order to make that kind of money, the bond department must have made some incredibly smart trades (his guys had heard nothing about that), or else it was starting to take much higher risk—carrying massive amounts of interest-bearing securities such as mortgage-backed bonds off its books, in some Enron-like maze of limited partnerships that were hidden from normal disclosures, much as the former energy conglomerate giant had hidden its risky trades and investments before its ultimate demise.

"I can't understand how they're doing this," Dimon said.

W hat the fuck is this crap?" Billy Heinzerling asked one afternoon during a risk management meeting. The department's boss of bosses, Tom Maheras, wasn't there, but some of his risk managers were. Heinzerling's gripes with Maheras and his team had been building for years, but what he heard now took his issues with his boss to new levels.

Heinzerling, as he would later tell colleagues, said he had found something odd, only this time it had nothing to do with Maheras's personal trades but some "legacy trades" that were still on the firm's books. These were bonds bought nearly a decade earlier, dating to the time when the trio of Bushnell, Barker, and Maheras had been at Salomon Brothers. Heinzerling wasn't exactly an advocate of Ace Greenberg's "three-week" rule (by now largely ignored at Bear), which said that losing carry trades must be disposed of after three weeks, but this was the mother of all carry trades.

Heinzerling had a simple question: how could bonds remain on the firm's balance sheet for almost a decade?

The risk managers were at first flummoxed; they didn't appear to know that the positions existed in the first place. Later they said that the trades had been there for so long because they were still making money. Heinzerling wasn't convinced. How and when had they been purchased? he asked, and at what levels? And how come no one seemed to know the trades were even there until he raised the issue?

No one seemed to know, or care, and the matter was quickly dropped as Citigroup embarked on one of the greatest binges of risk and leverage in Wall Street history.

The firm's timing couldn't have been worse. In 2003 Weill announced he was about to step down as CEO as the research scandal initiated by Spitzer snared its highest-profile target, Weill himself. He was saved the indignity of formal charges over pressuring the analyst Jack Grubman to upgrade a stock in exchange for getting Grubman's kids into the exclusive 92nd Street Y preschool in Manhattan. That he had escaped at all was thanks largely to the deft lawyering of the man who had emerged as his closest adviser, Chuck Prince.

For nearly two decades, Prince had worked in various administrative capacities for Sandy Weill, including as the general counsel of Weill's growing financial empire. He had little experience running a business, but he was a smart and dependable attorney. As the Spitzer investigation was building steam, Weill's orders to Prince were twofold: protect me, and make this go away for the sake of the firm.

Prince did both. Weill was questioned by Spitzer's investigators and embarrassed by his role in the scandal as it leaked out to the press—sexually laced e-mails from Grubman to a female money manager discussed Weill's role in helping him get his children into the school. The scandal was the last thing Weill needed at the time; Citigroup was still engrossed in its massive consolidation of various operations. Weill's dream was to cross-sell investment banking services and commercial banking services. But that took time and money and management talent—talent that was in short supply as the exodus of senior managers leaving the firm continued as Dimon took control of JPMorgan Chase.

Meanwhile, Weill had to explain those damning Grubman e-mails. In addition to the banter about Grubman's sexual prowess, the e-mails suggested that Weill had prodded Grubman to upgrade AT&T as a favor to one of his board members, AT&T CEO Michael Armstrong. By upgrading AT&T, Weill had been able to curry favor with Armstrong, who had then voted with other Weill supporters on the Citigroup board to oust John Reed as Weill's co-CEO a couple of years earlier. Weill had then rewarded Grubman by helping his children get into the preschool, which was "harder than Harvard" to gain admission to, or so Grubman wrote.

It was a sordid affair, one that, Prince argued, was nothing more than a misunderstanding; Weill, he said, had showed no intent in committing fraud when he had pushed Grubman to upgrade AT&T. He was, after all, a company shareholder, board member, and believer in the stock. In the end, Spitzer let Weill off the hook, but not Citigroup. The firm paid $400 million in fines and had to agree with the rest of the securities business on a set of reforms to the process of issuing stock research, reforms that were so weak that firms began violating them with impunity nearly from the minute the deal was struck.

However, in exchange for not charging Weill, Spitzer's staff wanted change at the top, former investigators for Spitzer say, and they wanted Weill to announce a successor and retire. It was a bitter pill to swallow. People who know Weill say he'd had no plans to step down until he was dead or incapacitated.

With the announcement that Weill would give up the CEO spot by 2004, the company announced that Chuck Prince, the man Weill and the company owed so much to, would become the next CEO of Citigroup. Weill would remain as chairman, though his management role would diminish from nearly the moment the announcement was made.

"What the fuck does Chuck know about running a firm?" Jamie Dimon scoffed when Prince was named for the post. A better question might have been, what did Chuck Prince know about mortgage bonds, risk, and leverage, which by now had replaced the financial supermarket as the firm's business model?

In Prince's defense, he also didn't have much time to learn. When he became CEO, his first order of business was to fix the regulatory mess he had inherited from Weill. In addition to Spitzer, other regulators were questioning the firm's business practices, including its role in the demise of Enron and WorldCom. Fines and settlements piled up well into the billions of dollars. "I give you my word we will make a concerted effort to clean up our act," Prince told SEC Chairman William Donaldson in 2004.

Donaldson said he was eagerly waiting for results. So were the people at the Federal Reserve, one of the many regulators in charge of watching Citigroup's operations, who were so concerned about the firm's regulatory problems that they put a hold on the firm's acquisitions until it could get its legal house in order.

Citigroup's problems should have given Dimon a chance to rejoice, but he was feeling the heat in a different way. Dimon's problems weren't regulators or corporate sleaze but investors' complaints about JPMorgan Chase's stalled stock price. His first year was hardly the comeback story investors had expected—major investors questioned him repeatedly and pointedly about the wild zigzags in the stock price, which reflected the volatility of the newly formed firm's earnings. The stock price could barely crack $45 a share.

By now investors had grown weary of the model that Dimon and Weill had forged at Citigroup, Prince had inherited, and Dimon was trying to re-create at JPMorgan Chase: the universal investment bank, which was proving no match for the earning power of the more nimble Goldman Sachs and its mimics Morgan Stanley and Merrill Lynch, all of which had lean operations focused on trading and risk taking, much of it in mortgage debt, to crank out huge profits.

So how did Dimon respond? Not by taking more risk but by taking less. If Dimon had learned anything from his former mentor, Sandy Weill, it was that risk taking should be confined to those who truly understood it. Dimon and Weill understood "business risk," or the risk embodied in bringing two businesses together and making the new combined company into one functioning company. As a team they were pretty good at handling business risk. What neither of them really understood was trading risk, and that's why both had avoided it like the plague.

During Dimon's first year at JPMorgan, he reduced leverage some 40 percent. He ordered his staff to cut back on derivatives trading, once a hallmark at the bank. It was, after all, JPMorgan's team of financial rocket scientists that had developed risk management techniques such as VAR, enabling the massive growth of the derivatives market.

Citigroup took the opposite track.

"The only undervalued asset is risk" was the comment to Robert Rubin one afternoon by an investor during one of Rubin's many trips around the world. Rubin's job at Citigroup was the envy of his colleagues and all of Wall Street. The former Treasury secretary was paid handsomely, some $15 million a year; he could (and did) travel all over the world doing what he liked, namely schmoozing with clients; and he had no operating responsibility at all.

Not one employee actually reported to Rubin, yet aside from his responsibilities as a member of Citigroup's board, Rubin, in his role as "chairman of the executive committee," was the most powerful single person at the world's largest financial institution. The new risk strategy Citigroup was employing was his baby, and now, based on what he was hearing from investors, his instincts were right.

The investors he met with all but demanded that Citigroup take more risk—borrow more to engage in increasingly complex trades, the types of activities Goldman was undertaking. In a low-interest-rate environment, where the bond markets are calm and the liquidity of complex bonds is strong, a firm can make a lot of money—with, of course, the right people in charge—and Rubin assured every investor he spoke with that Maheras and his team were the best in the business.

Rubin might have been an ambivalent manager when it came to most of Citigroup's many businesses, but he took a particular interest in the firm's dance with risk. It's unclear exactly why; some have suggested that risk was in his blood from his days at Goldman Sachs. A more plausible explanation may be that he saw risk as Citi's only savior. Citigroup was in many ways his brainchild as it was Sandy Weill's, since it had been Rubin, through his lobbying efforts in Congress and the White House, who had possibly been most responsible for dismantling the Glass-Steagall Act. Rubin saw risk—in trading, investing, and leverage—as the way to pay for everything Citigroup needed to survive.

Whatever the reason, Rubin had begun prodding Prince much as he did with Weill, to use risk as a tool to grow profits. Rubin's advocacy prompted Prince to hire a consulting firm to take the firm's risk profile to another level. Its first task was to determine how Citigroup measured up to the rest of the Street in the use of its balance sheet to trade. Even with Maheras and the Salomon Brothers gang in charge, it was still less than the competition's, particularly Goldman's, the most successful firm on the Street. It confirmed what Rubin had thought and said for the past two years. Once again, in a meeting with Prince and his senior staff, Rubin reiterated the notion that Citigroup should expand its risk taking to new levels but do it intelligently and with oversight.

Rubin says he was assured time and again that Citigroup had not only the right oversight but the right people in place. Maheras and his team, he says he and the Citigroup board were told, were "the best people in the bond business."

With that, Rubin gave his blessing to one of the biggest sprees in borrowing and risk taking Wall Street had seen in years.

A t Citigroup they called it "the machine." It was the process, honed and perfected, that Citigroup used to create mortgage-backed debt. It worked like this: Citigroup traders reached out to mortgage wholesalers for product, namely home loans. Those loans were sorted and categorized by zip codes, credit scores, and borrowers' incomes. The information was fed into Citigroup's risk models, and presto, out came a mortgage-backed security. The last step: a trip to the rating agencies for one of their triple-A ratings.

By 2005 the machine was running strong, particularly in the high-margin market for creating collateralized debt obligations. CDOs are among the most complex of all mortgage-backed bonds because they consist of bonds inside of bonds; a CDO is packed with various types of mortgage bonds and other asset-backed securities such as credit card receivables. It was created by financial rocket scientists at Drexel Burnham Lambert and perfected after a brilliant academic, David X. Li, developed a model to weigh and distribute risk among its various tranches.

The "machine" reduced CDO creation to the efficiency of a car assembly line. Traders on the underwriting desk constantly warehoused various types of mortgage- and asset-backed securities for the CDO group, headed by Shalabh Mehrish, a smart, affable quant who made the group the leading CDO underwriter, along with Merrill Lynch, beginning in 2005, at a time when Citigroup needed the money the most.

Increasingly in 2005 and early 2006, there was a waiting list of investors looking for CDOs to feast on their high yields and their (alleged) safety through diversification. Citigroup was telling large investors they'd have to wait months before their orders were filled—and they were happy to wait. The CDOs were a gold mine: the firm was making tens of millions of dollars in a business that was a fraction of the size of its brokerage sales force or investment banking division, for creating and selling CDOs, not to mention countless millions more from hoarding the securities on its books and earning the interest.

It was a glorious time to be working on the mortgage bond trading desks at Citigroup. At its height, the trading, underwriting, and warehousing of mortgage bonds were earning Citigroup profits of close to $1 billion a year. Citigroup had snared or shared all the big buyers of CDOs in the financial business, including an investor that had emerged as possibly the biggest of them all: the Bear Stearns mortgage bond hedge funds, run by Ralph Cioffi.

In a business known for its blowhards and braggarts, the affable Ralph Cioffi came as a breath of fresh air. He was always calm and composed— it's one reason clients loved him so much—even when explaining the most intricate bond-trading strategy, and in his twenty years at Bear, he had developed a reputation as possibly the best bond salesman on Wall Street. Before he was a CNBC star anchor Lawrence Kudlow was an economist at Bear, and trotted out to large clients because of his star status as a former Reagan administration official. But Kudlow would beg to have Cioffi accompany him during client meetings because Cioffi could explain the most complex bond market concepts in simple English.

Cioffi had also developed a voracious appetite for mortgage debt, a specialty he took up in the 1980s as the market was in its developmental stages, and his success as a salesman in the mortgage market made him one of the highest-paid people at Bear and a close associate of Warren Spector, the Bear bond chief. In 2003 Spector did something that raised more than a few eyebrows on the Street—he gave the green light for Cioffi, the salesman, to convert into a money manager. He was ultimately put in charge of creating two hedge funds, the tongue-twisting High-Grade Structured Credit Strategies Fund and later the High-Grade Structured Credit Strategies Enhanced Leverage Fund, which would exploit the growth of the mortgage market and Cioffi's expertise in the new types of bonds that were exploding in value, the CDO and the CDO squared.

The creation of the funds marked a new and important phase in the development of the mortgage bond market. A mere handful of hedge funds invested in mortgage derivatives, primarily because individual investors were wary of the complicated nature of mortgage bonds and the wild swings in returns that a fund investing in them can experience.

Cioffi and Bear were betting that times were changing. Much like Wall Street itself, individual investors were less wary of risk than before, and they yearned for bonds with higher yields than Treasuries and corporate bonds could offer. The funds themselves made numerous disclosures about their level of risk and leverage; Bear would market them only to "qualified investors," meaning people with at least $1 million in net worth who signed waivers stating that they understood the risk involved.

Cioffi's funds soon became among the hottest items peddled through Bear's brokerage arm, and it made the firm's asset management division, an earnings laggard, into a profit center. Cioffi's success was immediate. Brokers who were pushing the investments referred to the funds not by their convoluted official names, but by the name of the man most responsible for their creation. They were known as "Ralphie's funds," as Cioffi's operation grew into one of the largest in the financial business focused purely on mortgage debt. By 2005 he was managing more than $1 billion in clients' money in one of his funds, but its size and importance to firms like Citigroup that supplied mortgage products were magnified by the leverage he undertook to magnify his returns.

Cioffi, one of the most down-to-earth people in the securities business, loved leverage as much as the volatile Tommy Maheras. Cioffi and Bear sold the funds as basically safe investments, according to some of his former investors, and even the name of the fund, "High-Grade," suggested its safety. Cioffi denied that the fund invested in subprime mortgages and said the securities consisted mostly of triple-A securities.

He was technically right. The bonds were triple-A because they were culled from the "super-senior" tranches of the CDO. He didn't invest in subprime debt directly, because he purchased CDOs, which are made up of various securities, including and increasingly subprime bonds. But none of this means the funds were in any way "safe." Aside from the pressure put on rating agencies to place triple-As on just about everything, Cioffi's High-Grade Structured Credit Strategies Fund, with its emphasis on buying and trading CDOs, gave new meaning to the concept of leverage, since Cioffi not only borrowed to take his positions, but he invested in securities that relied on heavy borrowing, namely CDOs with exposure to subprime mortgages.

The second fund Cioffi created would use even more leverage, as reflected in the "Enhanced Leverage" part of the fund's name—and investors so far were loving it. Cioffi's personal wealth soared. Like many Wall Street executives with too much money on their hands, he produced a movie (*Just Like the Son*, starring Rosie Perez); indulged in his passion for expensive cars (Cioffi loved Ferraris); and had homes in New Jersey, Florida, Vermont, and Rhode Island.

And who could blame him? He was Wall Street's best mortgage client: "Everything stopped when Cioffi called," said one Citigroup mortgage trader. At Citigroup, underwriters trying to unload CDOs packed with gobs of risky subprime mortgage bonds that couldn't be sold elsewhere knew they had a receptive ear in Cioffi.

More important, investors loved Cioffi because his funds seemed so much like Cioffi himself, producing a string of strong but steady positive returns since their inception.

One of the odd things about the CDO market, and the Wall Street mortgage market in general, was how far removed it was from the real action—the loans and mortgages being handed out to people in Florida, southern California, Texas, and Nevada, where sleazy loan officers were pushing mortgages, known on Wall Street as "product," on people who had no business owning homes they couldn't afford.

Every now and then, there would be talk among the traders about how lending standards had slipped or how the housing market in Las Vegas was overheating. But no one seemed to care. First, the traders' computer models and economic forecasting showed the housing boom continuing ad infinitum. The diversification of the mortgage bonds made them virtually risk free. And anyway, the mortgage debt merchants were having too much fun and making too much money. As talk of a housing bubble busting began to pick up steam, as the dire warnings of a pending housing crash by Nouriel Roubini, an economist at New York University, began to spread from academia to the mainstream, the traders took it all in stride.

"Live for today" was the motto of the Citigroup trading desk.

One afternoon in early 2006 when CNBC ran a report about a slow-down and a correction in housing, a longtime Citigroup analyst wandered by the bank's trading desk to gauge its reaction.

"So whaddaya think?" the analyst asked one trader.

"What's there to think?" was the answer. "What's the worst that can happen? We make $200 million and then we get fired."

12. PERVERSE INCENTIVES

f anything goes wrong," said Harvey Goldschmid in his thick New York accent, "it's going to be an awfully big mess." It was April 2004, and Goldschmid, a longtime official at the Securities and Exchange Commission, had a knot in his stomach. Throughout the 1990s, Goldschmid had served as the general counsel of the commission, Wall Street's top regulator, under Arthur Levitt, Jr., its former chairman. Levitt loved Goldschmid, a law professor and old friend, and believed he was one of the smartest people around when it came to finance law. When George W. Bush was elected, Levitt, a Clinton appointee, was forced to leave, but he made sure Goldschmid and his institutional knowledge stayed; in one of his last acts as SEC chairman, Levitt appointed Goldschmid a commissioner, where he remained through Bush's two terms.

During Goldschmid's years at the SEC, the commission talked a good game—Levitt had launched high-profile investigations into fraud in the municipal bond market and at the NASDAQ stock exchange—but in reality, the SEC had taken a hands-off approach to regulating Wall Street. The dot-com bubble fueled huge profits for Wall Street firms that sold public offerings of Internet companies to small investors through fraudulent research reports (the companies, it turned out, were paying for the positive research), but the SEC under Levitt was nowhere to be found.

Glass-Steagall's demise, smack in the middle of Levitt's tenure as SEC chairman, was a massive act of deregulation that took risk taking among banks that were supposed to be safekeeping savings accounts of average American to new levels. Among the vast changes it allowed was the ability of these new financial behemoths like Citigroup to combine commercial and investment bank operations under one roof, transforming once-staid commercial banks into gambling dens of trading increasingly exotic bonds. Even worse, the massive amount of borrowing or leverage that had spread across Wall Street during the past two decades escaped serious scrutiny by the SEC despite the periodic losses, including the 1998 bond market fiasco that had nearly led to Lehman's destruction.

But now, with the election of a Republican president and his selection of a business-friendly Republican to run the agency, the commission was looking to take deregulation to a new level. Levitt was out, of course, and William Donaldson, a former Wall Street financier and founder of Donaldson, Lufkin & Jenrette was in as chairman. Goldschmid, meanwhile, was one of five SEC commissioners gathered in the basement hearing room of the agency's headquarters in Washington to vote on an amendment to the so-called net capital rule, an obscure but important provision that governed how broker dealers (i.e., the big investment banks) set aside capital to cover potential losses on trades on their balance sheets.

The event received little media attention. There were no cameras at the hearing, no reporters waiting outside for an interview with the commissioners, as is often the case on important votes. But the changes were vast, important, and potentially lethal, which had Goldschmid worried.

In the past, investment banks had had a set model for capital requirements for the assets they kept on their books: the less risky an asset, the less capital they had had to set aside to hold against that asset. The capital requirements for Treasuries, for example, were much smaller than those for corporate or asset-backed securities. This was because the risk of a Treasury default was minimal (backed as the bonds are by the full faith and credit of the U.S. government), while corporate bonds and mortgage-backed bonds carried more risk of default. The "haircut" system, as it was called, was designed to dissuade brokers from taking on too much risk—the added capital necessary to hold more risky debt was costly, and it ate into the firm's bottom line.

But under the amendment the SEC was set to vote on, everything would change. First, the big firms were now supposed to *self-monitor* their risk levels and report those risk levels to the SEC, but, most important, they could use any risk measurement tool they wanted, including VAR. VAR was as controversial on Wall Street as it was universally accepted. Every quant worth his weight understood its limitations, the biggest being that VAR could be incredibly misleading. Its risk measurements were based on historical trends, and every risk manager knows you can't count on history to repeat itself forever. Yet every firm held up its VAR as if the measurement were the final word on the matter.

The next major change was even more daunting: all triple-A securities would be treated about the same, meaning that firms could now hold just about the same amount of capital for a triple-A mortgage-backed bond, a CDO squared, for example, as they did for ultrasafe Treasuries. If the risk takers on Wall Street had ever needed an incentive to carry CDOs on their books, this was it. Risk and leverage, made cheap by low interest rates from the Federal Reserve, was now made even cheaper by the SEC.

Despite his gut telling him otherwise, Goldschmid supported the measure, and with Goldschmid's support, tentative as it was, the commission's vote was unanimous and the rule was adopted. As he recalls, there was some discussion, a little arguing, but not much and none of it particularly enlightening given the far-reaching consequences of the decision.

The SEC commissioners, of course, weren't walking away empty-handed, or so they believed. The conventional wisdom is that the SEC, as Wall Street's "top cop," has broad and unlimited authority. It doesn't. The SEC's authority is strictly prescribed in the Securities Acts of 1933 and 1934, which limited the commission's focus to the brokerage activities of the big firms. For years, the securities industry blocked the SEC's attempts to expand its legal and regulatory mandate to cover new business lines such as private equity and hedge funds. Now the SEC's mandate covered the entire breadth of the Wall Street business model.

SEC examiners would be able to delve into firms' private-equity arms, units that were making investments in corporations and were growing in size and complexity, as well as the hedge funds that the firms were now opening. Wall Street firms could no longer hide the positions in off-balance-sheet corporations and claim the units weren't part of the SEC's regulatory mandate. Nor could the Street claim that the SEC had no authority to monitor derivative trades, which had become a huge business over the past decade, as had off-balance investment funds, known as structured investment vehicles, which were packed with mortgage debt and that now fell under the new mandate.

Which might have made the decision a good trade for the SEC if the commission had had the bodies and brains to carry out its new mandate. The SEC was essentially putting its stamp of approval on risk taking of historic proportions. The repeal of Glass-Steagall had made brokerage firms and commercial banks almost indistinguishable, making regulation even more important as commercial banks such as Citigroup adopted the risk-taking strategies of Wall Street firms. Yet instead of enhanced oversight, the new capital rules were met with less regulation, not more. The Fed, which had been the eyes and ears of systemic risk for years, had now abdicated that responsibility to the SEC, which despite its enhanced powers had no idea what it was doing, its ranks depleted by years of defections of longtime officials to higher-paying jobs on Wall Street.

With that, the commission missed scandal after scandal, including a brewing $50 billion Ponzi scheme perpetrated by Bernard Madoff, who was examined by the SEC nine times and each time given a clean bill of health.

In retrospect, Goldschmid, now a law professor at Columbia University, concedes that his vote was a mistake, primarily because the SEC's staff didn't really know what to do with its new powers to monitor the Street and its evolving business model. "We didn't have a sophisticated enough staff," he said. "The sad part is, we should have been getting that information on the risk they were taking, but we didn't because we didn't know what to ask for."

James Nadler rated his first mortgage bond deal in 1988 as an analyst at Standard & Poor's. His first assignment as a young analyst was to tell Salomon Brothers what it would take to earn a triple-A rating on one of its deals, a CMO pool made up of home loans from a company called Ryland Mortgage, which operated in all fifty states and handed out mortgages to the best borrowers, known as prime borrowers.

But that didn't mean the bonds would automatically receive the triple-A rating. Nadler's boss gave him the rating criteria; they reflected little of the real estate boom of the 1980s. In fact, Nadler was shocked to learn that the rating's methodology was one that based default rates on the soaring levels of the Great Depression. If Ryland wanted a triple A, it would have to back up the pool with additional collateral. Nadler recalls the bankers' shock when they learned that they would have to come up with cash representing 12 percent of the entire pool of loans just in case another Great Depression was about to happen.

Fast-forward a little more than a decade later, and times had certainly changed. Not only had the mortgage bond market changed—it was far bigger and more complex—but so had the way the raters assessed the risk of mortgage-backed securities. They no longer used Depression-era worst-case scenarios when judging whether a bond had enough collateral and safeguards to earn their coveted AAA.

In fact, somewhere along the line they had stopped worrying about the possibility of falling housing prices altogether. By 2005 triple-A ratings were being handed out like candy; underwriters could nearly demand the rating they wanted on a deal and did.

The question is why.

T hat can't be true!" Neil Baron gasped when he first heard from a friend of his at Fitch Investors Service just how much money the analysts who were rating structured deals were now making. The year was 2003, and Baron had retired from Fitch three years earlier after having created the agency's structured finance practice back in the early 1990s.

Baron is a tall, aggressive man who worked his whole life in the rating business. His dad was one of the founders of Standard & Poor's, where Baron served as general counsel during the 1970s. Later he helped make Fitch a competitive threat in a business that had been dominated by Moody's and Standard & Poor's.

They were businesses that earned stable money, considered cash cows by their corporate parents. Their objective had always been to weigh the risk of a bond's defaulting through a rigorous analysis of the bond issuer's ability to repay its debt. They did this through a rating system, Triple-A being the highest rating, D for those in default. No bond could be issued in the public markets without a rating; this was due both to investors' reluctance to purchase unrated bonds and to the government. Though it didn't directly regulate the raters as it did brokerage firms through inspections and examinations, the SEC issued a kind of Good Housekeeping Seal of Approval to Moody's Investors Service, Standard & Poor's, and Fitch Investors Service. They were "nationally recognized" by the federal government as the best in breed.

But many in the market recognized the raters for their appalling record at protecting investors from some of the biggest credit market crashes in history, often keeping bonds at or near triple-A levels until just before they fell into default. Among their big misses that had led to massive losses for bondholders: New York City's near bankruptcy as well as Orange County's; the accounting-related implosion of Enron and WorldCom; and the brewing mortgage crisis.

Now they were about to become even more powerful—and profitable—high-growth companies, and they had the government and the mortgage bond market to thank. The new SEC rules on capital combined with the international banking regulations known as Basel II made ratings even more important to Wall Street and the capital markets. Banks were now allowed to treat all those suspicious triple As slapped on mortgage debt—debt increasingly held on their own books—as having the same level of safety as the U.S. Treasury bond, the ultimate in triple A (since it had the full faith and credit of the government behind it) when calculating their capital requirements. It was a powerful incentive for the Wall Street firms to hoard as much high-yielding mortgage debt on their books as possible whenever investors took a pass or the firms wanted to earn interest income on a carry trade.

It was also a powerful incentive to keep cranking out mortgage-backed bonds, which the raters saw as a gold mine. The raters could charge higher fees for rating a CDO, claiming it required more work to understand a derivative's complex structure than that of a simple municipal bond, and as the structured finance market exploded, Moody's profits quadrupled; between 2000 and 2007, it had the highest profit margin of any of the companies in the Standard & Poor's 500 Index of big-company stocks. Standard & Poor's (no relation to the S&P Index in this regard) wasn't far behind, more than making up for having a new competitor in the privately held Fitch, which also had seen its profits quadruple.

When Baron had started the department in the early 1990s, analysts there had been earning a decent six-figure salary. Now, to his amazement, the salaries had exploded; some of the raters in the structured finance group earned around $1 million, and those considered barely above average not that much less. These were Wall Street–level salaries that for the first time were doled out by the rating agencies because of the boom in mortgage-backed issuance, which had made rich just about everyone in its path, from the home owner who purchased a house with no money down and a subprime mortgage, to the bank that issued the mortgage, pocketed the fees, and sold it to an investment bank, to the investment banker who packaged that mortgage into debt, to the investor who feasted off of the bond's high interest rates, and now to the formerly geeky bond rater who became a millionaire rating the debt.

No other part of the rating agency business was bigger than, or even close to as profitable as, mortgage bonds and other structured products. In

1999 Moody's booked $172 million in revenues from rating deals in structure finance. By 2007, rating structured finance bonds produced $900 million in revenues, nearly four times what Moody's made in rating municipal debt, all of which set up strong incentives for the raters to cut corners in order to rate even more of these lucrative deals. And they did.

Obtaining a triple-A rating for the top tranche of a CDO or a CMO is a near necessity for the issue to be sold in the public markets. With that in mind, one might think the raters were in the driver's seat, that they could set standards as high as they wanted. Not quite. First, bankers had long since figured out how to get around each rater's "black box," its analytical methodology that computed the various ratios and credit scores of the underlying mortgages, including the types of mortgages, whether they were prime, subprime, or the in-between step on the credit ladder known as "alt-A," and whether the loans were issued in geographically diverse regions of the country. Making it easy to game the system was the simple fact that the black boxes inexplicably never took into account the possibility of declining home values. Throw in the cutthroat competition among the big three raters—two of which, Moody's and S&P, were now public corporations and needed to grow their revenues to keep their shareholders happy—and the groundwork was laid for disaster.

And that disaster started the minute bankers figured out how raters were paid. By the end of the 1990s, rating agencies were still paying their analysts a salary and year-end bonus based on a range of factors, including how well their ratings had fared. But increasingly, the salaries, particularly for raters in the structured finance department, were also based on *production*, or the volume of deals they rated.

The raters' pay now had less to do with the quality of their ratings or even the complexity of deals. Instead raters were paid not much differently from bankers, traders, or brokers: the more business brought into the firm, the more money they made.

It was a recipe for huge profits but also disaster. The bankers began pitting analyst against analyst. Because they were paid based on revenues, the analysts began cutting corners, ignoring red flags and handing out triple As on even the most risky mortgage deals.

The declining standards at the rating agencies in the mortgage area didn't make big headlines—at least not yet. Nor did they provoke much thought at the Securities and Exchange Commission, which took the position that it didn't have the authority to regulate the raters, despite the power these firms wielded in the bond markets. But they were an open secret inside the agencies themselves.

In 2001, Frank Raiter, a managing director for residential mortgage-backed securities at S&P, at first refused to rate a CDO without access to what he considered crucial information on the loans in a pool of mortgages. Raiter's boss, Richard Gugliada, told him in an e-mail that such a request was "totally unreasonable!!!" and he should render a rating without even looking at the pertinent data, according to subsequent congressional testimony.

Raiter wasn't alone in questioning the lax rating standards. "It could be structured by a cow and we could rate it," read an e-mail sent by another executive at Moody's regarding a mortgage-backed security.

The response to the e-mail said it all: "Let's hope we are all wealthy and retired by the time this house of cards falters."

Becoming wealthy may have been the overriding concern on Wall Street and at the rating agencies, but at AIG executives had another objective. In 2005 Martin Sullivan, the new CEO of the world's biggest insurance company, told senior executives that in addition to bringing in new business they should go out and have "some fun."

It was, to say the least, an odd request for a CEO to make, particularly the CEO of the company defined for so long by Hank Greenberg. Over a thirty-four-year career, Hank Greenberg had built AIG into the world's largest insurance company and one of the Dow's best-performing stocks. He had done it through hard work—senior executives sometimes received calls from Greenberg at 3 a.m.—and a relentless desire to make AIG the largest company in the world.

And he did it in part through the mortgage bond business—or, to be more precise, an insurance policy that covered mortgage bonds, known as a credit default swap.

By 2005, the credit default swap, essentially a guarantee to cover the cost of a bond if it fell into default, had become the glue that held together the mortgage-backed securities markets. Merrill Lynch, Citigroup, and nearly all the major underwriters were able to brag that they were holding triple-A mortgage securities on their books because of the AIG insurance. It allowed the big firms to underwrite bigger pools of bonds and hold more of them on their books.

Greenberg had begun the financial products group that sold all those credit default swaps in the late 1980s, just as the market for interest rate swaps and other derivatives had begun to take shape. The initial intention of the group was to help corporate America hedge, or manage, its risks. In the simplest form, a company that has issued debt at a floating interest rate might want to hedge the risk of rising rates by swapping into some fixed-rate debt, or vice versa.

Greenberg had a love-hate relationship with the group and its various leaders. He hated the risks they took and the independence the top people in the group sought. But he loved the profits the financial products group, known inside AIG simply as "FP," produced.

Since its inception, FP had grown from a $323 million operation with just a handful of people to one that produced billions of dollars in revenues annually. In recent years, the impetus of this growth had been the credit default swap, which AIG had introduced in the wake of the LTCM collapse in 1998. It was a derivative tailor-made for the growing risk in the bond market, and especially the boom in the mortgage-backed securities market, particularly as the market began to embrace more risky types of securitizations, such as CDOs.

With that in mind, AIG became the largest credit default swap seller in the world and the premiums charged flowed right to its bottom line.

Yet Eliot Spitzer, the New York State attorney general and by now Wall Street's most famous enforcer, believed Greenberg ran a company that regularly committed accounting fraud and created fictional profits through a series of sham transactions that had nothing to do with credit default swaps. Greenberg considered those possible transgressions so trivial that he called them "foot faults," as in a minor foul in tennis.

Greenberg, now eighty, denied the charges and vowed to fight them until the day he died (as this book goes to press, he's still fighting them). Put simply, Greenberg believed that no one at the firm could manage its complicated array of businesses, and with him gone, AIG would self-destruct. But that didn't stop Spitzer from turning up the heat even more— he wanted Greenberg out. Soon, leaks about the investigation began appearing in the press (a frequent Spitzer tactic) as the attorney general began threatening criminal charges against AIG and possibly Greenberg himself if he didn't resign. Greenberg eventually did resign, and his replacement was Martin Sullivan, an AIG lifer and someone who had the approval of Spitzer.

But not Greenberg, who considered Sullivan reasonably competent but never CEO material. "He's Irish and good with insurance brokers," Greenberg once said of Sullivan. "He has little education . . . I thought he had street smarts."

Sullivan was English, not Irish, but Greenberg's assessment of his abilities, particularly as they related to risk, appeared spot on, as did his assessment of what Spitzer's actions might do to AIG. The day after Greenberg left AIG, risk taking surged to unprecedented levels. The company lost its coveted triple-A bond rating from all three major rating agencies, yet Sullivan urged his managers to have "fun" in pursuing new business aggressively, and that's exactly what the people in financial products did.

For AIG, the triple-A rating was more than a badge of honor or a selling point to investors. It had real bottom-line ramifications. Immediately, AIG had to post more than $1 billion in collateral on deals it was now insuring, and the mortgage securities market was still strong. If the market hit a pothole, the amount of collateral AIG would have to post could explode, given all the credit default swaps it had sold to Wall Street firms and investors.

By now the swap business was booming alongside the continued growth of mortgage-backed securities, including collateralized debt obligations, the most complex and risky of all mortgage bonds. In just two years' time, CDO issuance had grown by 250 percent to more than $500 billion in 2006. AIG responded by ramping up its swap business to unprecedented levels; by the time Greenberg left in March 2005, it had issued swaps on $40 billion of CDOs. In just nine months' time, it had doubled its exposure.

Talk about fun.

13. TOP OF THE WORLD

We go to battle every day. . . . This is nothing short of hand-to-hand combat—and we're here to win!" Dick Fuld shouted as he gave what some executives called his "War Speech." On this night in 2005, Fuld was speaking to eighty senior directors, the crème de la crème of Lehman's employees, who were gathered for an annual dinner in the swanky executive dining room at the firm's lower Manhattan headquarters.

Much had changed at Lehman since the dark days of 1998 during the LTCM crisis, when it had nearly failed. Fuld was achieving his dream: to remake Lehman in its pre–American Express image: strong, vibrant, independent, and hugely reliant on taking risk.

The Lehman that Fuld now bragged about was no longer the little trading shop that had nearly blown up; it was hugely profitable, and its stock price was soaring. It had a competitive investment bank, a private client arm that dealt mainly with affluent investors, stakes in hedge funds, and a private-equity unit. Mostly, though, it was real estate that powered Lehman Brothers. Fuld wasn't an expert in real estate debt, which had become the company's core business, but he had people who were, particularly Mark Walsh, a leader in the area of commercial real estate.

Walsh was the perfect Fuld disciple: smart, loyal, and above all aggressive. He had become what Lew Ranieri and Larry Fink were to the residential mortgage market, a pioneer in the use of commercial mortgage-backed securities, or CMBS, as a way to finance the vast explosion of properties from luxury high-rises to office complexes that sprouted during the economic boom. As Lehman's profits approached $4 billion a year, Walsh's operations could take credit for nearly 20 percent of the bounty.

Lehman was basking in those riches. It had barely survived the 9/11 terrorist attacks, its offices near Merrill's across from the World Trade Center, but now it had a ritzy office in Midtown, its trademark green wrapped around the exterior. Lehman's success had also made it a particularly enticing catch; foreign banks yearned for a equity stake, while others would have paid mightily to buy the entire franchise, if, of course, they could have afforded it at $150 a share.

Fuld, meanwhile, wasn't looking to cash in but to grow, as the longtime investment banker Andrew Malik learned one afternoon when he picked up the phone and shared an interesting idea with his boss. Malik had worked at Lehman long enough that he could pick up the phone at any time and make a suggestion or two, and Fuld would listen.

"Dick," Malik said breathlessly, "I think we should buy Bear Stearns."

Lehman and Bear shared an interesting distinction in the Wall Street hierarchy: most of their competitors over the years hadn't thought either would make it through the massive consolidation in the brokerage industry or survive the current competition with financial conglomerates such as Citigroup. Yet as the bull market in bonds continued, both were among the most profitable firms on Wall Street. Jimmy Cayne, Bear's CEO, was now doing a victory lap, making headlines with his new building in Midtown, a posh headquarters a few blocks from Grand Central Station, as a symbol of his success.

"Dick, they've got a new building, and we need to build a clearing operation and they have one of the best. We can double the number of brokers we have."

"And one more thing," Malik added, "you personally get to fire that prick Jimmy Cayne!"

Fuld just laughed. "I don't know Jimmy Cayne personally," he said, before adding that "we looked at them and the price they want is too expensive."

In reality Fuld felt he didn't have to buy anyone, and he was going to make sure that no one was going to buy him. Fuld was Wall Street's comeback kid, and he was basking in all the glory of being right that risk would set Lehman free.

"It's us against them," he said, his dark eyes fixed intently on the crowd, "and we're winning!" Long gone were the days when Lehman's shares had traded at a mere $6, which meant that many in the room were now incredibly wealthy thanks to Fuld and his management style. During the speech he spoke about "the war" that Lehman was waging for business, despite its success, and reminded the wives who attended to "support their husbands" as they went off to "battle" because they were doing great things in making Lehman the best on the Street.

Most of the crowd loved the speech. Fuld received a standing ovation, but while some saw heroism, others in the crowd, albeit just a few, saw hubris. Halfway through the speech, Richard Dickey, the head of equity sales, felt a sharp kick to his shin. It was his wife, who simply rolled her eyes at Fuld's repeated war analogies. Dickey just smiled and clapped politely when the speech ended.

"Is this guy out of his mind?" Dickey's wife snapped on the ride home to Connecticut.

Dickey's response: "You have no idea."

Richard Dickey wasn't part of Fuld's inner circle, but he was close enough to it. He reported to one of the handful of top executives who served on Lehman's management committee and ran the firm with Fuld. Dickey had been on the trading desk during the LTCM crisis and on the orders of his supervisors had used his own Rolodex to drum up support to get Lehman short-term financing; he had seen just how close the firm had come to closing down at that time.

Since then he had seen Lehman come back to health, but in his mind, it had had less to do with Fuld than it had with timing and luck. Lehman had simply benefited from the continuation of the bull market and the decline in interest rates, which helped its bond department continue to crank out record profits.

Dickey had witnessed something else: contrary to Fuld's reputation as a hands-on manager, which brought with it the growing respect of his peers on the Street, Fuld had became more isolated and arrogant, often downright nasty, in meetings with colleagues and clients. "You deserve to have your legs cut off!" was the answer Fuld gave to a tough question from one employee during a town hall meeting when asked about a balance sheet issue.

But at least he was at the meeting. More often when he was in town, Fuld rarely left his office and made even rarer appearances on the trading floor, where the actual risk was being taken as Lehman's use of leverage began to soar to even more lofty levels; it was now borrowing $30 for every $1 in capital, nearly a 30 percent rise between 2003 and 2006, and its balance sheet of trades and investments had grown to around $600 billion.

The soaring stock price glossed over the massive risks Lehman was taking and Fuld's increasingly detached management style. The firm had now jumped into just about every new business line Wall Street had created, from high-octane private equity deals to volatile hedge funds and investments in mammoth commercial real estate projects. One such deal was known as Archstone-Smith, a massive $22 billion financing that closed in 2007 that Lehman would solely finance at the top of the market.

Dickey recalls thinking that the situation at Lehman was just the opposite of what he had heard was taking place at Goldman Sachs, where friends at the firm told him that as the Wall Street debt leverage frenzy grew, CEO Lloyd Blankfein made increasingly regular visits to the trading desk to discuss risk positions with his traders.

But Fuld was known to cancel meetings that bankers had scheduled with potential clients. And woe to any employee who dared ask Fuld a question during one of the appointments he decided to keep. Dickey did once, questioning Fuld about the firm's return on equity during a managing directors' meeting. Fuld responded with a gruff, almost inaudible nonanswer.

"Don't do that ever again," Dickey's boss warned him as soon as the meeting concluded, and he didn't. Nor did anyone else.

As the firm's leverage increased, Fuld's grip on his management and board grew. He was revered by so many people in his circle of senior advisers that almost no one dared to speak out about the firm's risk and leverage, and almost never to Fuld himself. Everyone else was so scared to be cursed at in public or even fired that they simply kept their mouths shut.

Fuld's leadership was more like that of a cult leader than even that of an imperial CEO. His board of directors remained silent as his risk taking grew, and he surrounded himself with so many sycophants and true believers in leverage, such as his president, Joe Gregory, that no one in the executive ranks seemed to notice the risk taking reaching such absurd levels; if they did, they just kept their mouths shut as they watched Lehman's stock price soar.

There were, however, a few exceptions.

Lawrence Lindsey, a former Fed governor who had predicted the Internet stock market bubble before joining George W. Bush's economic team, was now a $200,000-a-year consultant to Lehman Brothers. His presentation to the firm's risk committee focused on the business of alt-A mortgages. Lehman had become a leading underwriter of alt-A mortgage bonds because it owned the mortgage originator, and that meant that it was holding plenty of the alt-A bonds on its balance sheet, earning the carried interest before trying to sell them off.

Lindsey is a heavyset man who during meetings rarely takes off his suit jacket. On this day in the summer of 2006, people in the room wished he had because he was sweating bullets as he explained that the housing boom was moving into bubble territory. In other words, Lehman, given its exposure, should be cutting back.

Fuld wasn't at the meeting, although some of his advisers were; the heads of nearly every department showed up, according to one person who was in attendance, so presumably he received a briefing on what Lindsey had said. But would it have mattered?

People who worked with Fuld doubt it. By the middle of the decade Fuld was not just increasingly isolated but also amazingly rich thanks to a business model built on risk. He earned $40 million in 2006 and had a net worth of around $500 million, which afforded him a lifestyle that included a $21 million Park Avenue apartment, a $13 million vacation home in Florida, and a $9 million mansion in Greenwich, Connecticut. He owned thirty acres of land in Sun Valley, Idaho; Fuld was the area's largest landowner and angered the locals because he wouldn't let them fish on his property. Perhaps the reason was because that's where he was planning to keep his pet lion—he had been going on safaris in Africa and telling people he had made arrangements to bring one back. Later he decided against the move, but that didn't stop him from adorning his office with photos of the big cats and talking about his admiration of them.

His top people didn't do badly either. Joe Gregory was now taking a company helicopter to work every Monday from his home in Long Island and, according to one senior Lehman executive, was racking up huge expenses. Gregory appeared to take extreme pleasure in showing off his growing wealth. By 2006 he was earning more than $30 million annually. He owned a Jaguar and bought a new boat for $100,000. He was now divorced and remarried, and living the life that only excessive risk can get you, and apparently not afraid to flaunt it. Colleagues were blown away when they heard that Gregory had sold his new Range Rover for about half the price he paid because he wasn't satisfied with the color.

Like Fuld, Gregory was a commercial paper trader. They had been together from the start, during the dark days when Lehman had been ruled by American Express and later during the trying times when the bond markets had imploded in 1994 and 1998. Surviving that long together had created a certain level of trust, and over time Fuld began to trust Gregory with more and more of the day-to-day responsibilities of the firm, including the level of risk Lehman took.

"The Gorilla," once one of the most hands-on managers in the business, was now coasting, and Gregory was increasingly setting the terms. Gregory's view of risk and leverage was no different from Fuld's—both saw it as a necessary ingredient in Lehman's success and independence, though some people at Lehman say Gregory indulged in risk taking even more than Fuld did.

It was Gregory who made two decisions that would haunt Lehman for years to come. The first was to oust Michael Gelband, the head of fixed income who was Mark Walsh's boss but also one of the few people at the firm to question its highly leveraged business model. Gelband took his complaints to Fuld, telling the CEO during discussions of his bonus that the mortgage market was getting too frothy and that the Lehman business model that relied on making leveraged market bets would lead to disaster. Fuld thought Gelband was talking heresy and questioned whether he had the stomach to remain at the firm.

For Gelband, speaking his mind to Fuld was the kiss of death. In late 2006, with Lehman stock at an all-time high, Gelband received word that either he had to be part of the team or he would have to go. A few months later Gregory broke the news to him: he had to leave.

The second decision wasn't a firing but a hiring. Gregory believed a senior executive named Erin Callan was the complete package: smart, loquacious, and very attractive. She mixed well with clients and she was also a woman, something Gregory saw as an asset in terms of diversity given the male-dominated world of Wall Street.

More than anything she bled Lehman green, meaning she shared Fuld's and Gregory's enthusiasm for risk, and Gregory trusted her. Gregory let it be known that Callan, currently the head of hedge fund sales, was going places. Within a few months, she was named CFO of Lehman.

Callan was going places, but Fuld, Gregory, and Callan all believed that so was Lehman. With the SEC weakening capital standards, the regulators demanding little accountability on disclosing balance sheet assets, and Fuld's complete and total control of Lehman, risk taking by the firm continued to soar with no end in sight.

In 2003 Lehman Brothers held about $35 billion in Treasury bonds on its books, compared to about $39 billion in mortgage- and asset-backed securities. By 2006 Lehman grew its mortgage- and asset-backed bond holdings to a little more than $111 billion, an increase of nearly 185 percent, while its Treasury holdings grew by just 14 percent.

The massive leveraging of Lehman with risky assets didn't scare off investors, who appeared to be hypnotized by Fuld's stature and his grandiose plans to boost Lehman's share price even more. That guarantee came in public statements, comments to analysts, and also a little gift Fuld gave some key executives and board members.

In the spring of 2006, Lehman stock was trading at around $140 a share; that's when senior executives and board members received a gift courtesy of Dick Fuld. It was a T-shirt with the words "LEHMAN BROTHERS: DRIVE FOR $150" emblazoned across the back.

Lehman didn't get that high, but not because its stock price didn't grow and grow. On April 6, 2006, shares of Lehman broke the $150-a-share level when the firm announced a two-for-one stock split, meaning that for every share owned, investors would get two in return, thus cutting the firm's stock price in half. Such a move is designed to make it easier for investors to buy a firm's stock.

When Fuld promised that even after the split he was aiming to reach $150 again, no one argued with him.

Lehman wasn't the only brokerage stock on a tear in 2006. The share prices of the big Wall Street firms had soared by 64 percent since 2001 as trading accounted for two-thirds of the record revenues. Trading desks were expanding everywhere; UBS, which employed a thousand traders, more than any other firm, wanted to add even more.

Nearly every major firm had or was in the process of creating in-house hedge funds or groups of traders that used the firm's capital and leverage to bet in various markets. Morgan Stanley's VAR increased dramatically in 2006, but it was now coming close to Goldman in earnings growth, so no one really cared. Between 1998 and 2006, Bear Stearns' inventory of risky asset and mortgage debt had grown by more than 300 percent as its leverage soared to around 40 to 1. Few, if any, seemed to think twice about the increased risk at Bear either, probably because its stock price rose to a Lehman-like $140 a share on its way to an all-time high of $170.

Dick Fuld was incredibly rich, worth about $500 million; Hank Paulson of Goldman Sachs and John Mack of Morgan Stanley had personal fortunes in about that range. But Bear Stearns' Jimmy Cayne was now richer than them all, becoming the first and only CEO on Wall Street to be worth more than $1 billion.

How did the former bridge bum and sheet metal salesman suddenly become a billionaire?

Members of the press (myself included) attributed the supercharged earnings now being produced at Bear to Cayne's deft leadership, with some justification. Cayne had kept Bear Stearns from indulging in the dot-com mania that had swept Wall Street, in which major firms, in a bid to underwrite any stock that had anything to do with computers or the Internet, had all but guaranteed buy recommendations in research reports to entice dot-com companies to hire their firm for investment banking work.

And like Fuld, he had brought Bear back from the brink, or so it seemed at the time. By the end of the decade, and despite continued strong earnings from its bond department, Bear's stock price had suffered, falling to around $18 a share because it couldn't keep up with Morgan, Goldman, or Merrill in taking new Internet companies public. "Bear wasn't even considered a takeover candidate. Who would want them?" said one high-ranking banker at Goldman Sachs at the time.

That didn't mean Cayne didn't try to make a deal. He had some conversations around this time with Fuld to merge their respective firms and create one large gambling den that specialized in mortgage debt; the talks broke down because neither Cayne nor Fuld seemed willing to share power. Cayne approached some European banks as well, and once again came up empty. One problem was the firm's corporate culture—the risk-taking ways of its bond desk and its outsider image as the bad boys of Wall Street—wasn't a great selling point to banks, including the cash-rich European banks looking to gain a foothold in the United States. Bear's frequent skirmishes with securities regulators over clearing and processing trades for several bucket-shop brokerage firms didn't help either.

Cayne began to panic. He began talking up the stock, literally. He told Wall Street securities analysts that the firm was in play, meaning he would consider a sale. It was a bold, unusual, and some would say desperate move for any Wall Street CEO, much less Cayne. Most CEOs talk up their stock by explaining how well the firm is performing, not announcing that it's in play. But Jimmy Cayne has proved time and again that he's not like most other CEOs. The stock soared on the move; the SEC contacted Bear and explained that such comments, particularly if intended to do nothing more than goose the stock, would be a problem. Cayne backed off but not before shares of Bear reached a fifty-two-week high.

Bear's stock would dip again after the 9/11 terrorist attacks, along with the rest of Wall Street, but with the help of Alan Greenspan and lower interest rates pushing the mortgage bond markets to new heights, its resurgence began. In certain categories of mortgage-backed securities, Bear even beat out its much larger rivals Citigroup and Merrill Lynch in underwriting fees. It held large pools of the stuff on its sheet. Analysts kept questioning whether the firm's winning streak would continue—but it did, as the Fed continued to cut rates through 2004 and kept them low for the two years thereafter.

By the middle of the decade, Jimmy Cayne was now the CEO of a firm with a stock price of more than $170 a share. He regularly reminded reporters of this fact from his office in Bear's newly built headquarters, which had a private bathroom and showers, an audiophile's dream of a sound system, a ventilation system that could circulate the smoke from his foul cigars, and an unusual deal trophy—a motorcycle from a Chinese company he had helped Bear win business from.

Cayne was on top of the world—he was rich and now respected by his peers. The stock price of $170 a share made everyone forget his "Fuck you" to the Wall Street establishment during the LTCM crisis; even Hank Paulson from the hated "Goldman Sucks," now running the Treasury Department, called Cayne to tell him what a great job he had done.

Cayne and Dick Fuld didn't know each other; they probably wouldn't have liked each other if they had. But they had much in common. Their firms were now standouts on Wall Street; both men were incredibly rich from riding the wave of speculation that had lifted all boats, particularly those at Bear and Lehman, which focused so intensely on taking risk; and most of all, both went unchallenged by any of the normal checks and balances that are supposed to be in place to prevent excessive risk.

Cayne's inner circle was much like Fuld's—full of yes-men—while Warren Spector, the firm's bond chief, was increasingly on the outs with Cayne, isolated and inattentive to detail, and, according to friends, praying for the day his boss would retire so he could run the firm. Also like Fuld, Cayne had become increasingly isolated amid the firm's success. He spent hours in his office playing bridge on his computer or on the golf course at his country club.

And maybe most of all, Cayne had a firm grip over his board of directors. Several of Cayne's board members were friends and dinner partners. Vincent Tese, for example, considered the board's lead director, was a frequent golfing partner of Cayne. This friendship and admiration meant that the board barely debated the firm's risk taking or the firm's latest invention, Ralph Cioffi's highly leveraged hedge fund that made huge bets on CDOs. The fund's steady returns were a great selling point to Bear's brokers, and finally Bear's asset management unit, long a backwater at the firm, had something to cheer about. In fact, it was making so much money that Cioffi created another hedge fund in 2006, as the mortgage market reached dizzying heights. This fund would use even more leverage to crank out returns, vastly expanding Bear's exposure to one of the most volatile segments of the mortgage market.

Why didn't the board step in to put a stop to any of this? "In retrospect we should have," one board member said. "But the firm was doing so well."

Cayne provided an even better answer to me. It had come in response to the news a few years earlier that the Merrill Lynch board had ordered Dave Komansky to pick Stan O'Neal as his heir apparent, despite Komansky's reservations.

"Jimmy, would your board make you do that?" I asked Cayne.

His response, with a dry laugh: "My board is *my* board."

Cayne may have had full support from his board, but he had almost no support from the company's second most powerful executive. By 2006 Cayne and his bond chief, Warren Spector, were barely on speaking terms. Their conversations, when they had them, were short and hardly sweet. When they shared an elevator or passed each other in the hallways, they barely acknowledged each other's presence.

At least for now, it didn't matter whether or not Cayne and Spector liked each other. Profits had soared in large part because of the firm's bets on mortgage debt, so even if Cayne wanted to get rid of Spector, he wouldn't.

Over the years, Cayne was nothing if not a true believer in the firm. While other CEOs, most notoriously Stan O'Neal of Merrill, had cashed in during the long bull market and sold their shares, Cayne kept buying more. Because he had taken most of his salary—all those $20 million bonus awards—in Bear Stearns stock, he had become a billionaire, one of the four hundred richest people in the world listed by *Forbes* magazine.

But when the story citing his wealth first broke in the *Wall Street Journal*, Cayne feigned annoyance over the news. "I really didn't need that," he said at the time. But in reality he was beaming. The news was recognition that the former bridge bum had done what the old-school Wall Street establishment couldn't do, that he and Bear had hit the big time.

Jamie Dimon, meanwhile, was sitting back digesting the news of Cayne's success. Dimon's JPMorgan Chase was nowhere near a "Wall Street darling" like Bear Stearns, and that had Dimon steaming, particularly because Cayne and he had had merger discussions and Cayne, in a move fraught with arrogance, had set the price of any merger so high that there was no way Dimon could meet it.

For Cayne, snubbing Dimon was almost as delicious as snubbing the Fed, the Treasury, and all those Wall Street executives during the LTCM bailout. It showed he had arrived.

Dimon took it a different way. He thought Cayne was arrogant and shortsighted; Bear, he believed, needed a balance sheet if it was going to survive. Over the past couple of months, the two had gotten to know each other pretty well. Dimon genuinely liked Cayne despite his arrogance; his bizarre sense of humor, his affection for cigars, his easygoing mannerisms were appealing. "This is a guy I can work with," Dimon thought.

But he was also perplexed: He couldn't understand how Cayne ran the company. Dimon knew what it took to run a major Wall Street firm— eighteen-hour days, countless meetings, constant traveling. But Cayne always seemed to be playing bridge on the computer (or away at tournaments), golfing, or palling around with his friends at his favorite restaurant, San Pietro.

Even worse, Cayne didn't really understand his balance sheet. Dimon came to that conclusion pretty quickly after discussing the firm's business model and its reliance on risk and leverage. It derived huge profits from the mortgage desk, prime brokers, and now increasingly the asset management unit as well, because of Cioffi's two hedge funds packed with high-risk mortgage debt. Though highly profitable, the hedge funds were leveraged so highly with the riskiest CDOs in the market that they risked becoming a runaway train.

"Jimmy looked like he knew what was happening," Dimon would later say, "but he really didn't."

Meanwhile, at San Pietro, the Midtown restaurant that serves as a hangout for the Wall Street elite, no one questioned Cayne's business acumen. He was now treated like a king as news spread about his wealth and his success running Bear Stearns.

Cayne himself seemed to be as high as his paycheck.

"More wine, more wine, more wine!" he loudly chanted one night while sitting at his favorite table, eating dinner with some friends.

San Pietro's owner and maître d', Gerardo Bruno, who serves the Wall Street brass on a daily basis, knows how to appease his high-maintenance guests, even unruly ones. When they want more wine, he gets them more wine.

Tom Patrick used to joke that Merrill shouldn't take any risk because it had no clue how to trade bonds. He should know. Patrick was appointed CFO in the late 1980s to clean up the mortgage trading debacle left by Howie Rubin. He fixed that mess and went back to his native Chicago as the lead investment banker specializing in complex derivative transactions. Then Dave Komansky brought him back to New York in 1998 after Merrill copied LTCM's bond trades and lost big. After helping O'Neal take control of the firm, Patrick helped the new regime cut costs and manage its risk—that is, until he was fired by O'Neal in 2003.

Patrick is a large, intense man with thick brown eyebrows and a volatile temper. He is also brilliant, an expert in tax shelters, balance sheets, and derivatives, the complex financial tools that had become all the rage on Wall Street. He and his closest friend, Arshad Zakaria, who you'll recall was also fired by O'Neal after their abortive plan to elevate Zakaria to president of Merrill Lynch, had now set up their own growing and successful hedge fund to invest in foreign companies, many of them high-tech outfits in India.

But Patrick, a thirty-year veteran of the firm, couldn't get Merrill out of his system. Even before the Zakaria plot, he had hatched a plan to pay a documentary producer to run an unflattering profile of Merrill's nemesis, New York State Attorney General Eliot Spitzer.

The latest he was hearing from Merrill was that the firm was cranking out such huge profits because O'Neal was embracing risk and leverage as never before in the firm's history. Moreover, O'Neal had recently become enamored with "structured products," the plethora of mortgage- and asset-backed securities that were all the rage, including the most popular of them all, the CDO.

"This isn't the O'Neal I know," Patrick began telling his circle of friends and contacts from the Merrill days. He recalled several conversations he had had with O'Neal over the years, particularly when O'Neal had been CFO and saw firsthand how holding such risky products could decimate a firm's balance sheet, as it nearly had in 1998. "I hate structured products," O'Neal had said at the time.

But that was Stan O'Neal before he became CEO; the new Stan O'Neal was dramatically different. O'Neal was now privately complaining that as CEO he had to appease board members and shareholders with quarterly returns and a consistently higher stock price to match the success of Goldman Sachs, Lehman Brothers, and Bear Stearns, the firms whose risk taking had made their stocks investors' darlings.

"If you want thirty percent ROE [return on equity], you have to take risk," O'Neal said. As such, Merrill launched a massive and unprecedented borrowing spree to finance trades that kicked out huge returns. Leverage grew, incrementally at first, from $16 borrowed for every $1 of capital to around 20 to 1. Then it exploded as Merrill's short-term loans from banks and its hedge fund clients jumped radically in size. Merrill was now returning to its pre-LTCM leverage levels. The firm would be leveraged close to 30 to 1, behind only Bear Stearns, which at times approached 40-to-1 leverage.

What made the leverage even more dangerous was the types of assets being purchased with the borrowed funds—the so-called structured products that O'Neal had once hated.

If Stan O'Neal had disliked CDOs in the past, he certainly loved them now in all their variations. The complex securities generated enormous fees, and since they offered returns 2 to 3 points higher than corporate bonds, or around 6 points above Treasuries, just carrying them on the books was hugely profitable. He ordered his bond department to ramp up the firm's CDO operations and make Merrill a top underwriter of CDOs.

It wasn't long before Merrill became the leading underwriter of CDOs. Just how much O'Neal liked CDOs and the profits they generated was immediately apparent. At Merrill, O'Neal socialized with no one except perhaps his one close friend at the firm, his deputy, Ahmass Fakahany; he even often chose to play his favorite sport, golf, alone. But when Christopher Ricciardi, Merrill's head of CDO operations, held an off-site with his staff at his favorite golf course, O'Neal jumped into the Merrill Lynch helicopter and made the short trip a couple miles north of New York City so he could play golf with Ricciardi at his side.

With structured finance boosting Merrill's earnings, O'Neal, like Jimmy Cayne, was now the toast of Wall Street. Since O'Neal had taken over the firm in 2003, the story of Merrill Lynch had primarily been about cutting costs to achieve profits. But suddenly the firm began to churn out operating profits as well.

The firm's success was best summed up by David Trone, a senior brokerage analyst at Fox-Pitt Kelton, who said at the end of 2005 that "this was a tough quarter to find anything wrong with Merrill's earnings."

And they would get even better, earning O'Neal even more accolades.

But lost in all the hoopla was the way Merrill was able to crank out those big returns. The dirty little secret was that Wall Street was now eating its own cooking as the housing market started to show the first signs of strain. Greenspan was largely oblivious to any of this and to what his post-9/11 cheap money policy had done to Wall Street and the entire financial system. The cheap money supplied by the Fed created a bonanza of risk taking never seen before. Banks had created more types of debt than even the smartest rocket scientists on the Street could keep track of. And with interest rates low, banks chose not to sell all those bonds to investors; they were hoarded on balance sheets at full value because of their triple-A ratings the firm's could pocket the higher interest rates the securities threw off.

What Greenspan wasn't oblivious to was the threat of inflation. After slashing the Fed funds rate to just 1 percent by 2004, Greenspan began to raise it gently, trying to cool off the overheated housing market. Credit suddenly became a little more expensive, and because so many of the weakest borrowers for homes had taken out ARMs and other mortgages tied to short-term rates, defaults began to rise.

By early 2006, CDOs and other bonds packed with subprime mortgages were starting to receive a chilly reception from investors. For years, analysts had predicted a slowdown in the mortgage market, a much-needed correction. No one knew how much of a correction would occur, though some, like the economist Nouriel Roubini and the analyst Josh Rosner, were now predicting a correction of massive proportions.

Amid all this concern, what did the man who had once said he hated structured products do?

His firm loaded up on even more of them.

t's pretty simple," Jeffrey Kronthal deadpanned. "If the markets change, we're stuck with this stuff, so let's find a way to get rid of it."

Kronthal wasn't the most dynamic executive inside Merrill Lynch. In fact, he was probably the least dynamic. A friend once described listening to Kronthal speak as the auditory equivalent of watching paint dry. Kronthal was such a boring man that another friend, Peter Kelly, a compliance attorney in Merrill's finance group, jokingly gave Kronthal some legal advice before he went into a meeting with O'Neal—"Get to your points fast and move on"—because he was worried that O'Neal might lose his patience and throw him out of the firm.

Kronthal may have been boring and long-winded, but he was also one of the best in the business at understanding the mortgage bond market. He had begun his career at the epicenter of mortgage bonds, working with Lew Ranieri at Salomon Brothers, and since then had held about every job there is in the mortgage business.

Kronthal lived and breathed mortgage debt. He was a disciple of securitization—one of the many on Wall Street who believed that packaging loans and mortgages into bonds and selling them to investors wasn't just good for Wall Street but good for the country as well. Structured finance, after all, was behind the long economic boom that had begun in the early 1980s—banks were able to make additional loans for homes and expand credit to people other than the wealthy because Wall Street had figured out how to spread the risk.

He was also once in favor of Merrill buying a subprime lender so the firm could have access to its own mortgages to package into bonds. But Kronthal was also a realist. Having lived through so many implosions of the mortgage market—from the downturn in the late 1980s that had doomed Ranieri to the LTCM-induced implosion of 1998, he understood just how toxic mortgage-backed securities and their numerous derivatives could be when interest rates change, when prepayments of mortgages rise, or when people start to default.

Now he was worried. The data were pretty stark. In 2005 the median price of a home nearly doubled and home ownership rates were close to the

70 percent target that Henry Cisneros had set years before—both signs that a bubble was about to burst.

And that could spell trouble for Merrill. Kronthal had been sounding alarm bells since 2003, when he had seen how O'Neal's relentless push for profits was beginning to change the firm's leverage and risk profile. Merrill was no longer leveraged with Treasury bonds or even corporate debt; it was now increasingly leveraged with the stuff he knew best, mortgage-backed securities.

That's what had Kronthal concerned now as the mortgage bond market continued to roll and Merrill's leverage with them grew. In memos to executives and in personal conversations, he implored them to keep the warehousing of CDOs at a minimum.

But O'Neal had had enough, according to people who worked with both men. He was sick of Kronthal's long-winded sermons, sick of his management style. "He can't make a decision," he said about Kronthal. So O'Neal decided to make the decision himself.

O'Neal had not only slimmed down the firm and ramped up profits, he had also eliminated every potential rival from within the executive ranks and most of the voices of opposition on the board. Gone were seasoned members, including those who had helped put him into power, longtime board members such as Schering-Plough CEO Bob Luciano and John Phelan, a former head of the New York Stock Exchange. They were replaced by business associates—Alberto Cribiore, a private-equity investor who had been doing business with O'Neal for years, and other people O'Neal believed he could bully.

With all that control came massive amounts of personal freedom. O'Neal would disappear for days at a time without contact with his senior team to play golf, which had become almost an obsession for him. He used the company helicopter and jet to fly to his favorite golf courses across the country.

Most of all, he had near-complete freedom to run the company as he saw fit. His friend Ahmass Fakahany's official title was chief operating officer, but in reality the Egyptian-born accountant was a shadow CEO; O'Neal delegated almost all major decisions to him. Department heads understood that every move they made needed Fakahany's approval, not just because it was getting increasingly difficult to track down O'Neal but because when O'Neal was around he made it clear who was his number two.

Fakahany's power was enhanced not just because he was O'Neal's eyes and ears but also because not long after taking over, O'Neal had appointed him to be in charge of risk management. At most Wall Street firms, the chief risk management officer is the CFO. One major reason Goldman Sachs had turned risk taking into an art form was that its primary risk management officer, CFO David Viniar, had survived several management changes during his long tenure at Goldman and now answered to no one at the firm apart from Goldman's CEO.

But not at Merrill during the O'Neal era; CFO Jeffrey Edwards had limited control over the firm's risk profile because all the risk officers who monitored the various trading desks reported to Fakahany, as did Edwards himself.

And O'Neal made it clear that Fakahany's word was golden, in just about every respect, even if what he was doing had little to do with his most important job as the firm's chief risk manager.

What the fuck is this?" demanded a senior Merrill Lynch investment banker who noticed something strange as he entered Merrill's executive dining room, a place where white-gloved waiters served the company's top executives and clients in a classic, decidedly American setting.

At Fakahany's orders, all that had changed. The dining room took on an Asian tone, with artwork imported from China and the Far East and a new menu of Asian cuisine.

The changes were daunting for longtime Merrill executives who had bragged to clients that the best steak in New York was served there. Now they could barely find any meat. In a way, the resistance to this change epitomized much of what O'Neal wanted to change about the firm's culture: clubbish, closed to outsiders, somewhat stuck in the past, and decidedly not reflective of the modern world of international business.

Forcing some of the old guard to come to terms with the reality of changing times by eating something other than beef was one thing, but Fakahany wasn't content to settle for changing the menu and decor to become more international. What the old guard could now find besides the new eats of sushi and other dishes most of them couldn't recognize or even pronounce was wine—very expensive wine, for O'Neal and his inner circle.

Fakahany built a high-end wine cellar for the firm's executive committee—which consisted of O'Neal's top advisers, including himself—and thousand-dollar bottles were regularly served to guests of the top executives in the dining room or on Merrill's fleet of corporate jets.

In late 2006, former executives recalled lectures from Fakahany on ballooning expenses when they fêted clients at dinners, while he ordered the placement of large vases, about three feet in height, near his and O'Neal's offices. They were filled with ornate flower arrangements that were changed each week at a cost of $200,000 a year.

Fakahany even extended his power to include the firm's philanthropic department, which had always been headed by someone outside the executive committee out of fairness; the board didn't want the company to be seen as favoring the top brass's pet projects. The more power Fakahany got, the more he wanted people inside the firm to know who was boss when O'Neal wasn't around. He was making more speeches, particularly in front of senior executives and clients, which he became surprisingly good at because unlike O'Neal, who was stiff and awkward in public, Fakahany knew how to laugh at himself. During one memorable speech he told people he was "so popular" inside Merrill these days that he "received lots of gifts. Some of them are good and some of them aren't," he said, holding up a pair of boxer shorts that could have fit a buffalo.

Now, three years into O'Neal's term as CEO, many things had changed at Merrill. The old guard continued to shrink; Mother Merrill was increasingly a distant memory; profits kept rising; and risk taking continued to grow, as did Fakahany's power inside the firm. O'Neal loved having someone like Fakahany around. He handled all the heavy lifting of dealing with the firm's massive bureaucracy and the power struggles between departments that naturally come with running a company the size of Merrill, and he never complained. He pushed and prodded senior managers to delve more deeply into the mortgage bond market. Fakahany was behind Merrill's acquisition of its own originator of mortgages, particularly the subprime variety that could be stuffed into high-yielding CDOs.

And why would he complain? In addition to his almost limitless power, O'Neal paid his friend between $10 million and $30 million a year. Fakahany was so good that when a civil complaint was filed that he had fathered a son out of wedlock and allegedly tried to stiff the mother on child support payments (after receiving a massive bonus)—a move that might have gotten someone fired during the Mother Merrill days—O'Neal did nothing. (Fakahany paid the child support, he says, after learning the child was his.)

But the story only served to confirm what many top executives who dealt with Fakahany had come to believe: that he had all his boss's bad qualities (glib in public, he could be demeaning and egotistical during strategy meetings) and none of his good ones (O'Neal's intelligence).

Investment banking chief Gregory Fleming couldn't stand to be in the same room as Fakahany. One reason brokerage chief Robert McCann left the firm was his distaste for Fakahany (he was later wooed back by O'Neal). When traders witnessed Fakahany kissing up to his boss one afternoon, they began to refer to Fakahany as "Fuck Me in the Heinie."

The only person at the firm they had more contempt for was the man who had given Fakahany so much power, the man they called "Stan bin Laden."

Surprisingly, at least one of O'Neal's competitors didn't feel the same way.

"He's the real thing," Jimmy Cayne said when asked why on earth he was having lunch with Stan O'Neal. Cayne's lunches were legendary because Cayne spent so much time bragging about what celebrity he was breaking bread with in the private dining room he had installed in Bear's new headquarters in midtown Manhattan. The $600 million, forty-five-story edifice was a testament to Bear's success, and Cayne loved to show it off, particularly to his luncheon guests; at the time he had feelers out to Supreme Court Justice Antonin Scalia, and he had already broken bread with *Wall Street Journal* columnist and former presidential speechwriter Peggy Noonan, whom he later called "boring."

Now he was fêting O'Neal. Cayne usually had nothing but disdain for most of his fellow CEOs on Wall Street. He had once referred to former New York Stock Exchange chairman Dick Grasso as a "pig" for his attempt to cash out his $140 million retirement package even after Cayne, then a NYSE board member, approved the move. Morgan Stanley CEO John Mack was a "bullshit artist"; Hank Paulson, when he was the CEO of Goldman, was simply a "snake." Cayne had even less tolerance for Sandy Weill, and even as he considered selling Bear to JPMorgan Chase chief Jamie Dimon, he told friends that he wasn't crazy about Dimon because he was a "screamer."

But O'Neal was a "solid citizen" and a "real player," even though he had virtually stolen the CEO job from Dave Komansky, whom Cayne still considered a friend.

For the next two years, the meetings would continue with some degree of regularity. It was an odd relationship on the surface because the two men couldn't have been more different: the foulmouthed, cigar-chomping Cayne and O'Neal, who almost never cursed and never smoked.

But dig deeper, and O'Neal and Cayne had much in common. Both were CEOs. Both loved to play golf when they should have been working. And both ran firms increasingly addicted to leverage and risk without the slightest idea what was in store for them.

14. CASH STOPS FLOWING

I n early 2006, AIG's financial products division chief, Joseph Cassano, received an urgent warning from one of his risk managers, who had been studying the rapidly expanding composition of subprime mortgages in the CDOs that AIG had been insuring through its credit default swap business. Since Hank Greenberg had left AIG in March 2005, and at the urging of the new CEO, Martin Sullivan, Cassano had been on a tear. The credit default swap business had doubled in just nine months' time to meet the demand from Wall Street underwriters, which had been cranking out CDOs as never before.

But the risk managers were getting nervous about the subprime part of the CDOs, particularly the 2005 "vintage." What gave the CDO so much allure for investors was its diversification—it was packed with bonds of many classes, including mortgage-backed and asset-backed securities. In 2005 and 2006 that had begun to change. More and more of each CDO consisted of mortgage-backed securities, and the most risky type at that, those packed with the hottest mortgages now in the market, subprime loans.

All of which came at a time when the real estate market by many estimates was overheating. William Gross, the bond guru at Pacific Investment Management Company, was now openly warning about falling real estate prices and the resulting decline in the prices of mortgage bonds. Many other analysts on the Street were also issuing warnings that housing prices would experience a pullback, and if that happened, the subprime part of the market would take the biggest hit.

In late 2005, after being presented with evidence about lax standards in subprime lending, Cassano announced a major pullback in his credit default swap business; no longer would the company underwrite swaps on

CDOs underwritten by U.S. financial institutions. It was a wake-up call for AIG and a bitter pill for Cassano to swallow, since he had built the business from scratch less than ten years earlier.

By the time Cassano put a hold on the credit default swaps, AIG was exposed to $80 billion in CDOs. The financial product group's balance sheet had grown to a "notional value" of $480 billion. AIG couched its decision in the least sensationalistic terms: it told clients it believed it was being prudent and protecting itself from a downturn that Cassano believed had a minimal chance of ever occurring, one that even if it did occur would be short and mild.

News of AIG's decision soon reached Jeff Kronthal at Merrill, who, nervous about Merrill's expanding CDO portfolio, had been hedging his risk with swaps from AIG. He ordered his traders to put a halt to the CDO buying binge until they could figure out another hedging technique. News of Kronthal's decision filtered through the firm. Merrill had made much of its role as the top CDO underwriter on Wall Street. Bond chief Dow Kim wore the distinction as a badge of honor, bragging about it in one magazine article, and COO Ahmass Fakahany was busy implementing O'Neal's broad vision that the firm should be increasing, not cutting back on, risk, and he was zeroing in on buying a mortgage originator to aid in that effort.

Most of all, Stan O'Neal thought Kronthal was missing the logic of a market where risk was being rewarded with massive profits.

With that, Jeff Kronthal, the single most knowledgeable person at Merrill about the mortgage bond markets, became the latest victim of the Stan bin Laden regime.

"Stan, no way is Osman the right guy for the job!" is the way Peter Kelly described to colleagues his reaction when he heard the news of Kronthal's ouster and who would be replacing him. Kelly was a longtime Merrill Lynch executive, now the general counsel to the mortgage bond department, meaning he approved the legality of trades and products that the department had been creating and selling to investors. He had worked with O'Neal for years and felt he could level with him even after O'Neal became the CEO, particularly when he did something so destructive it put the firm in danger.

What Kelly heard O'Neal had done was about as destructive a move as Kelly had seen at the firm in a long time. Kronthal's replacement as the head of the firm's entire fixed-income department wasn't a seasoned veteran of risk and mortgage debt (an outside candidate had turned down the job) but a thirty-five-year-old bond salesman named Osman Semerci, whom O'Neal considered "brilliant," the prototype of a modern Merrill executive: young, highly educated, and international (he was of Turkish descent).

But others saw Semerci differently. Kelly had seen him in action and thought he didn't know the first thing about how to manage a large business, much less how to manage risk. Real estate prices are shooting through the roof, he thought. Bill Gross, one of the experts in fixed income, was calling for a pullback in mortgage debt. Even Joseph Cassano, the AIG executive who had become rich by issuing credit default swaps, was now predicting at least a lull in the market.

And in Kelly's mind, here was O'Neal allowing Kronthal, who had begun his career at the knee of Lew Ranieri, to be replaced with a salesman, a bullshit artist and "knucklehead," whose expertise was in wining and dining clients.

"Are you fucking crazy?" Kelly says he told O'Neal when informed of the Semerci promotion. O'Neal told Kelly that Semerci was right for the job—or, to be more precise, the right guy to execute his risk strategy.

Kronthal couldn't do that, O'Neal said. "He doesn't know how to make a decision." Then in an uncharacteristic use of profanity, Kelly has told people that O'Neal told him to "shut his fucking mouth."

The encounter over Semerci's appointment soon got even more heated. "This fucking guy is an idiot! He can't manage a balance sheet . . . he's the exact wrong guy for the job!" the six-foot-two, two-hundred-plus-pound Kelly shot back, with the veins in his neck popping out and his face turning red.

According to Kelly, O'Neal now ordered him to "Get out of my office now, or I'm calling security!" and a security guard did in fact show up.

Kelly stomped back to his office. When his friend and former colleague Tom Patrick heard the news of Kronthal's ousting he gasped. "You've got to be kidding me!" A little later, Patrick discovered that it wasn't just Kronthal who was out. O'Neal had approved a dismantling of the firm's mortgage-backed unit, including Patrick's friend Mac Taylor, an expert in CDO trading and risk, who was axed as well.

The news of the departures didn't make huge news on Wall Street; in fact, it wasn't even mentioned in the next morning's newspapers. But inside Merrill, its significance was large. Kelly's description of his encounter made its way back to O'Neal, who simply brushed it off saying Kelly was too much of a wimp to confront him face-to-face, but among the rank and file, it cemented O'Neal's reputation as "Stan bin Laden," the man who couldn't handle even an ounce of dissent. More important, O'Neal had just removed a huge amount of institutional knowledge from a firm in desperate need of it. Kronthal and his team were pushed out just hours after Merrill's board of directors made one of its few trips to the firm's profit center—its mortgage trading desk—where board members met Kronthal personally. But now almost a half century of experience in the intricacies of the mortgage market that O'Neal had bet on in a big way had been removed.

O'Neal, of course, saw it differently. He was on a mission to modernize Merrill Lynch, and Kronthal and Taylor represented the old guard that needed to be cleansed from the firm. Both were in their fifties, and he had big plans, he said, for Semerci. He believed Semerci possessed the attributes to one day run Merrill Lynch, which in O'Neal's mind, was still five years away from being at a level where he believed it could fully compete with the best of the best, Goldman Sachs.

"Brilliant" was not a term many others would use for Semerci, a bright salesman and a highly successful one but not someone to manage a balance sheet of complex mortgage securities. What O'Neal failed to understand was that Semerci, Dow Kim, the head of capital markets, and Fakahany—the people most in charge of managing risk—were all missing a fundamental element needed to understand it: the humbling experience of losing money from taking risk with possibly the most volatile and complex bond ever created. Jeff Kronthal, in contrast, had lived through the last two mortgage meltdowns and seen careers destroyed and the best risk models upended.

Under no circumstances are we in this deal," boomed Craig Overlander, the head of Bear Stearns' mortgage unit, to his underwriting desk. "If we can't sell it, we're not in it."

Under Cayne and Warren Spector, Bear Stearns had more than made its mark in the mortgage-backed market—it had bet the ranch. Two-thirds of its profits came from mortgage bonds in one way or another. It ran two of the most successful hedge funds on Wall Street that focused on making bets on mortgage-backed bonds and subprime credits. It was always one of the top underwriters of mortgage-backed debt.

But Overlander, a mortgage-backed bond veteran, had lots on his mind in late 2006, and none of it good. He had some of the same conclusions as Kronthal had over at Merrill, only in his case he didn't lose his job over it. Spector agreed with taking a more cautious approach, so Overlander began to cut back—no underwritings unless the entire CDO could be sold to investors. Bear already had too much money in CDOs on its balance sheet. If mortgage bonds were volatile, Overlander believed, CDOs could be nuclear. No one really understood what they were, given the stew of mortgages, including those of the subprime variety and loans that had found their way into the most recently issued CDOs.

The history of bond market panics is pretty simple: when markets begin to tank, the most complex bonds can't be priced and can't be sold. And there was nothing more complex than the CDO.

It was an odd time for Overlander or any other senior executive at Bear to become so negative. Bear's stock price was surging; its management was widely admired; Jimmy Cayne was a Wall Street legend based on the firm's prowess in underwriting mortgage debt and other bonds that Overlander was now turning away.

But Overlander and other senior executives at the firm saw numerous signs of trouble despite Cayne's and the firm's exalted status. For one thing, they thought Cayne was coasting, playing too much bridge and golf, while other CEOs were making friends in China to create business or strategic partnerships.

"Hank Paulson made over forty trips to China," Warren Spector once remarked to Overlander about the former Goldman CEO during one of their frequent conversations about the direction of the firm. "Jimmy's made one." Overlander couldn't believe what he was hearing.

Even worse, Bear's highly leveraged business model had left it more vulnerable than ever before to forces outside its control. Overlander, Spector, and other senior executives believed Bear simply needed a partner to help grow the firm's balance sheet in order to compete effectively with Merrill, Goldman, and Morgan Stanley, all of which had balance sheets as high as $1 trillion compared to the roughly $400 billion that Bear had. Without that capital, Bear was forced to borrow massive amounts of money from banks, which had now become Bear's competitors since the demise of the Glass-Steagall Act had given them the power to run investment banks themselves—and investors, including hedge funds, which would turn on Bear in a New York minute, pulling money from their prime brokerage accounts and draining Bear of cash if they could make a quick buck by also shorting Bear's stock.

The creditors provided much of this funding to Bear in the form of short-term loans in the risky repo market, which by now had grown to astronomical levels. Wall Street firms' reliance on repos had grown by 130 percent since 1999, to more than $1 trillion by 2006. Repos may have been the most efficient way to raise cash and the most popular because Wall Street firms exchange cash for collateral, such as all those mortgage-backed securities they had been holding on their books, but the risk of relying on repos can't be overstated. That's why some firms, Goldman in particular, had taken steps to reduce reliance on repos for their funding.

Bear, meanwhile, had increased it. The result was that at times repos now accounted for 70 percent of all of the firm's funding.

The problem for Bear was simply one of size. In order to compete with larger players, not just Goldman and Merrill but Citigroup, which could borrow deposits to finance its operations, Bear had to borrow and borrow big. At one point in 2006, Overlander approached his boss, Warren Spector, with his concerns. The firm was playing with fire, and it needed to be sold to a bigger bank to survive. Spector told him he too worried about Bear's size and now the size of the firm's borrowing, but there was nothing he could do; Cayne wasn't in a listening mood.

One reason: Cayne distrusted Spector's motives. Cayne knew Spector wanted more than anything in the world to be CEO of Bear Stearns, and the seventy-year-old Cayne wanted to keep the job as long as he could smoke his cigar. Cayne began telling people at the firm, "*I'm* going to be the last CEO of Bear," suggesting two things: at some point, no one knew when, he would sell the firm, but also he planned to do all he could to deprive Spector of his shot at the top job.

The statement made its way back to Spector, who, according to people at Bear, now became even less interested in the day-to-day activities of managing the bond department, often disappearing for his own private activities (he had a large home on Martha's Vineyard). In the meantime, Cayne was happy to have Spector out of his way so he could sit in his office and play bridge on his computer while he smoked his cigars and counted his many millions.

Despite the few voices of caution, risk and leverage had become a national fixation, embraced both on Wall Street and in government. The SEC and the Fed, the main regulators in charge of monitoring the buildup of risky assets on the banks' books, together with the rating agencies, were the modern-day equivalents of Nero fiddling as Rome burned. The fire in this case was the massive and rapid buildup of mortgage debt on the balance sheets of the banks; by 2006 it was approaching $1 trillion and heading higher without so much as a peep from the traditional watchdogs.

Still, the risk taking and leverage went beyond the brokerage houses and the banks. The GSEs, Fannie Mae and Freddie Mac, were in the game as well. By now, Fannie and Freddie had fully and completely conceded their original mandates to the whims of the Washington political class, which demanded "affordable" housing for all, even those who couldn't afford it. The politicians were giddy with Fannie and Freddie's conversion from staid mortgage banks to subprime lenders that would make Angelo Mozilo, the CEO of the largest subprime lender in the markets, Countrywide Financial, envious.

It was an evolution that took years in the making. As HUD secretary, Andrew Cuomo boasted in one report in the late 1990s that the new mandates he was imposing on Fannie and Freddie to ramp up subprime lending "could be of significant benefit to lower-income families, minorities, and families living in underserved areas."

With prodding from HUD, including Cuomo's successor, Mel Martinez, appointed by George W. Bush, the GSEs would become even larger catalysts of the mortgage market that dealt with subprime borrowers and one of the big reasons that housing prices continued to soar well into the new millennium. The combined balance sheets of the GSEs grew by an average of 15 percent a year, from $1.4 trillion in 1995 to $4.9 trillion in 2007, about $1 trillion of which was subprime.

One of the ironies of the bubble Fannie and Freddie helped create through their guarantees and purchase of subprime loans is that it made housing less affordable, not more so. To own a home, working-class and poor families were now more reliant than ever before on the various gimmicks the mortgage business offered—the adjustable-rate mortgages and "no-money-down" loans that allowed families to live in their homes at minimal initial cost, only to have their mortgage payments skyrocket later.

The strain on the system was becoming apparent in mid- to late 2006 as subprime mortgage delinquencies and defaults started to spike in a meaningful way, with the GSEs picking up the cost. At first, the top executives at Fannie and Freddie didn't seem to notice or care; the losses and the additional risk the agencies were taking on were papered over by the theory that housing was something that almost never went down in value, even as it was showing the first signs of doing just that.

The housing boom had done many things, including papering over the accounting scandals that hit both agencies. During this time, the top executives at both agencies earned salaries that could be found only on Wall Street. Fannie Mae chief Franklin Raines earned a whopping $90 million between 1998 and 2004, when he was forced to leave amid an accounting scandal not much different from what had occurred at Enron or World-Com.

There was just one difference: Fannie and Freddie were doling out the American dream to the poor, and consequently the outcry, particularly from the press, was muted.

There were, of course, a few skeptics. Richard Baker, a Republican congressman from Louisiana, was a longtime critic of the GSEs, but his warnings were ignored even as he discovered that the agencies weren't just in the business of facilitating mortgage lending through guarantees or through buying mortgages with cheap money, bundling them in mortgage securities, and selling off to investors; they were now keeping the mortgage securities on their own books and earning the interest in the same carry trade that had become popular on Wall Street and led to massive losses in 1986, 1994, and 1998.

Baker had no direct authority to regulate Fannie and Freddie, other than calling hearings after the accounting irregularities first surfaced and later proposing reformist legislation, which he did, much to the dismay of the powerful congressman and housing advocate Barney Frank, who dismissed the idea of either Fannie or Freddie being out of control as "overblown." In addition to Frank, Fannie and Freddie had other powerful allies in Congress, especially Chris Dodd, the senator from Connecticut who was not just a ranking member of the Senate Banking Committee but also a "Friend of Angelo," meaning he was on a special list of VIPs who had received low-cost mortgages from CEO Angelo Mozilo of Countrywide, the largest purveyor of subprime lending.

Frank, it should be noted, finally agreed to rein in Fannie and Freddie in 2007, but only with the provision that agencies must provide even more low-income housing—the very practice that was leading the GSEs into insolvency. (The effort eventually failed.)

It might seem easy to blame the accelerating train wreck of the GSEs on the Democrats: Cuomo at HUD, Frank and Dodd in Congress, and of course, the Clinton administration, beginning with HUD Secretary Henry Cisneros's desire to expand home ownership to up to 70 percent of the population.

But they weren't alone. Free-market types such as Fed chairman Alan Greenspan both warned against Fannie and Freddie's irrational exuberance and extolled the GSEs' subprime lending as they expanded home ownership to the poor, as did Clinton's successor, George W. Bush.

For the president, and for Republican leaders looking to soften their image with minorities, Fannie and Freddie's efforts to expand home ownership seemed to be the perfect vehicle to broaden their appeal to groups often hostile to Republicans. Fannie Mae's exposure to subprime mortgages tripled from 2005 to 2008, when Republicans ran both Congress and the White House. George W. Bush may not have been as fervent a supporter of the expansion of housing to risky borrowers as Barney Frank or Chris Dodd was, but he was hardly a bystander. During one speech at the height of the bubble, the president said:

[I've] issued a challenge to everyone involved in the housing industry to help increase the number of minority families to be homeowners. . . . I'm talking about your bankers and your brokers and developers, as well as members of faith-based community and community programs. And the response to the home owners challenge has been very strong and very gratifying. Twenty-two public and private partners have signed up to help meet our national goal. Partners in the mortgage finance industry are encouraging homeownership by purchasing more loans made by banks to African Americans, Hispanics, and other minorities.

Representatives of the real estate and homebuilding industries, through their nationwide networks or affiliates, are committed to broadening homeownership. They made the commitment to help meet the national goal we set. Fannie Mae and Freddie Mac . . . have committed to provide more money for lenders. They've committed to help meet the shortage of capital available for minority home buyers. . . . There's all kinds of ways that we can work together to meet the goal. Corporate America has a responsibility to work to make America a compassionate place.

Angelo Mozilo couldn't have said it better.

Nor for that matter, could Stan O'Neal, who now pinned the firm's future, and his own, on the continued boom in the mortgage market. Within nine months of Kronthal's departure from Merrill, Osman Semerci helped keep Merrill the top underwriter of mortgage-backed securities, including the highly profitable CDOs.

But more than that, the $4 billion CDO portfolio that Kronthal left behind grew more than tenfold, to around $50 billion. Before the year was up, Merrill finally purchased its own mortgage originator, First Franklin Financial, for $1.3 billion to ensure it had a steady flow of mortgages to pack into the CDOs it was underwriting at a record pace. As far as insuring those CDOs went, Merrill could no longer rely on AIG, so it went elsewhere, including to an undercapitalized insurance company called ACA, which had once provided insurance on high-risk municipal bonds but had recently dived into the structured finance business, insuring billions of dollars' worth of CDOs despite the company's low, single-A rating.

It was a spree that made anything that Howie Rubin had done at Merrill in the 1980s, or even the positions the firm had taken during the LTCM crisis, look puny. With the new insurance from ACA and other sources, Dow Kim, the bond department's chief executive as the head of global capital markets, and the bond department's direct supervisor, Osman Semerci, told their supervisors the portfolio was safe and hedged. It was made up largely of the least risky, most senior tranches of the CDO—the triple-A portions.

It's unclear how much of this or of the growing size of Merrill's exposure to CDOs Stan O'Neal knew, understood, or even cared about as Merrill's profits soared, which affirmed his broad strategy of taking more risk and his selection of the team of advisers who carried out this strategy. The Merrill Lynch board agreed: for 2006 O'Neal would get his biggest paycheck yet, earning $48 million. Fakahany and Dow Kim were the next-highest-paid people at the firm, earning $30 million and $37 million, respectively, that year. Only the senior management team at Goldman, led by Lloyd Blankfein's nearly $70 million bonus, made more than the top guys at Merrill.

O'Neal and his team viewed the generosity (which was easily approved by the board) as proof of a job well done. Sometime in 2006, O'Neal had hit his target, reaching a 30 percent return on equity, while shares of Merrill soared 40 percent that year, and he reminded the board that he had the team, the talent, and now all the pieces in place with the firm's acquisition of First Franklin, to repeat this success in 2007.

Merrill looked like a firm on the move, but beyond its board and investors, so focused on the firm's bottom line of quarter-by-quarter success, it was seething internally. Whether he knew it or not, O'Neal had created a climate that could only lead to disaster, where risk could fester and grow out of proportion without checks and balances. His board was clueless; dissent was outlawed; and power had become centralized among people who either didn't know enough about risk management or could benefit the most from reckless risk taking.

When word of the massive bonuses hit the press, the climate turned even more ugly. Longtime executives, who chafed under the new regime, considered leaving; Bob McCann, the head of the brokerage unit, did leave, only to be wooed back by O'Neal because he feared growing discontent among the sixteen thousand brokers, Merrill's "thundering herd." The herd may not have fit neatly into O'Neal's plan for the modern Merrill Lynch but were in the meantime providing the firm with tons of money through the simple buying and selling of stocks on behalf of small investors.

And where was Stan O'Neal as Merrill simmered? Playing golf and taking credit for the firm's revival while relative novices were left running the show.

Amid all of this, Larry Fink worried about what he had gotten himself into. By the end of 2006, BlackRock, his money management firm, was managing about $1.4 trillion and earning more than $230 million a year in profit. He had sold 49 percent of it to Merrill, in a deal he had crafted with O'Neal personally over glasses of expensive wine at Fink's apartment and a modest breakfast at the Three Guys diner on Manhattan's Upper East Side. But now he was starting to worry. He remembered one of the many discussions with O'Neal during their deal talks.

"Stan," he said, "we have these great risk models," talking about BlackRock's mortgage bond analytics, carefully crafted since the time Fink had lost his shirt and nearly his career in the mortgage market, which would now be available to Merrill after the deal went through.

O'Neal, uninterested, just changed the subject.

Chuck Prince was wishing he could change the subject as well, away from all the problems he was facing as Citigroup's new CEO. While O'Neal spent far too much time playing golf and Jimmy Cayne too much playing bridge, people in Citigroup knew where Chuck Prince was nearly all of the time—either in his office, attempting to get his arms around the massive, disjointed set of businesses he had to run, or begging regulators to go easy on Citigroup, assuring them that the firm would behave. It was no easy task. The Federal Reserve had basically prevented Citigroup from doing another merger until Prince could stop the flow of investigations and lawsuits charging the firm with misleading investors, conducting shady trades, and general malfeasance.

More than that, Citigroup's far-flung operations had never been integrated. Sandy Weill's vision of Citigroup was as a vast integrated platform of services for individual consumers and corporations, one-stop shopping for finance services. That was theory; reality was much different. To be a truly global financial supermarket, Weill had gone on a spending spree—Citi owned commercial banks in Mexico, Argentina, India, and Brazil. It had an insurance company and a large brokerage firm.

But the parts were not aligned, not even in the basic ways. A commercial banking client in Germany, for instance, would be unknown to the bankers in New York, who could be offering different products and services. Likewise, on the retail level, Weill's dream was for the retail bank to become a source of business for the Smith Barney brokers, but because the units worked on different computer platforms, they couldn't share even the most basic pieces of information.

The biggest problem was management, starting at the top. Chuck Prince had little experience running a business, and now he was in charge of running scores of them and, more than that, making sure they ran well together. The costs of doing this seemed insurmountable. One longtime Citi board member recalls speaking with Weill after he left as CEO but was still receiving regular briefings in his role as chairman.

Weill couldn't believe how big the firm had gotten in just a few years. "Sandy said, 'I left Chuck a balance sheet of a trillion dollars, and it had grown to $1.5 trillion.'" What Weill didn't say was that it was quickly heading to $2 trillion and above.

The reason for the growth was the amount of risk that Citigroup was now taking. As its costs to integrate its operations shot through the roof, while it was simultaneously prevented by regulators from growing its bottom line through acquisitions, Prince had no choice but to make up the difference through risk and leverage.

The man leading this effort was, of course, Tom Maheras, whose power and prestige only grew as the checks and balances at the firm seemed to diminish by the day. The board of directors believed Maheras was the best, and the profits seemed to back them up. With Prince working on other issues and key managers in charge of risk (such as former brokerage chief and now COO Robert Druskin and a musical chair of CFOs—from former General Electric executive Todd Thomson and his replacement to Sallie Krawcheck, installed by Weill despite her lack of experience, and her eventual replacement, Gary Crittenden, a longtime American Express CFO who landed the job long after the Citigroup machine had been created) putting their faith in Maheras, the bond department had free rein to do what it wanted, including partaking in a little (literal) gambling.

"Tell your buddies to stop being so loud," came the stern warning from the large man at a nearby card table to Maheras, who had taken his team on another trip to Las Vegas, this time to the MGM Grand. The way one person with knowledge of the scene later described it, they were all standing around a card table. Maheras and some other Citi people were getting a little rowdy, agitating some of the other players, including the pit boss, who now shut the table down.

Maheras became enraged. "Do you know who I am?" he snapped.

The pit boss said he didn't know and didn't care. The game was over; either leave nicely, or get thrown out of the hotel.

Maheras and his entourage left for a show, but the incident soon made its way back to the trading desk. Several traders said it spoke volumes about Maheras's growing ego, which appeared to match the growth of his power inside Citi, where he and his top managers ran the place with an iron fist and loved to gamble, literally, while his boss, Chuck Prince, basically left Maheras to do what he pleased, including growing the bank's inventory of risky debt.

Maheras didn't seem to have a care in the world, but he should have. During his tenure at Citigroup the size of the mortgage bond market had exploded and so did Citi's exposure to it. But in the fall of 2006, the first signs of trouble began to emerge in a then-obscure and little-followed credit index called the ABX. The ABX Index, or "*asset backed securities index*," serves as a benchmark for subprime mortgage-backed securities, and for much of 2006 the ABX remained stable. But around October 2006 it started to fall, though it was largely dismissed by Wall Street as an aberration even as evidence started to build that the entire system of bonds and derivatives, underwriting fees, and profits from carry trades built around mortgages was beginning to crumble.

The concerns expressed by AIG in late 2005 about the increase of subprime loans in mortgage bonds, the doom and gloom of housing articulated by Roubini, were now hitting investors. One problem was risk or, to be more precise, how to price risk. The yield on super-senior CDOs had narrowed to just 3.15 percent—that means that an investor was getting just a 3.15 percent return for buying into a bundle of subprime loans, which, if you listened to the skeptics, might begin defaulting on a daily basis. By the end of 2006 investors were not just shying away from CDOs, they were turning their backs on many of the big issues brought by Merrill and Citigroup, now including the least risky super-senior tranches of the CDOs.

Yet no one on Citigroup's mortgage-backed bond desk seemed overly worried. "We'll just warehouse the bonds," said one trader at the time. "The problem will correct itself."

While Wall Street ignored the growing threat, a very worried Josh Rosner walked into his local diner on Broadway in the TriBeCa neighborhood of Manhattan. He was there to meet the economist Nouriel Roubini for lunch. Rosner, a financial analyst with a small firm named Graham-Fisher & Company, and Roubini, an economics professor at New York University, had never met, but they had been fellow travelers in the dismal science of predicting doom and gloom.

Both had been among the few—the very few—who had believed the housing boom at its height back in 2005 was nothing more than a speculative bubble that was about to sink the U.S. economy, though they had come at it from different directions. Roubini had approached his research in macroeconomic terms: the wages of average Americans had stagnated for decades while they were able to build false wealth through borrowing and buying homes. The irrational exuberance of housing prices wouldn't—couldn't—last forever, because no bubble ever does. Since around 2005, he had been warning that when reality hit, the massive unwinding of housing prices that most Americans now based their wealth on would flatten the economy as never before.

Rosner, on the other hand, was focused on the Wall Street side of the equation. He had just written a paper titled "How Resilient Are Mortgage-Backed Securities to Collateralized Debt Obligation Market Disruptions?"

Before getting a phone call from Roubini's assistant, Rosner had actually never heard of the professor or his economic consulting firm, RGE Monitor. But Roubini had heard of Rosner, particularly as his paper detailing how the massive securitization market for mortgages was on the cusp of taking down the entire economy circulated within some academic circles and the media (*New York Times* writer Gretchen Morgenson had written about it in one of her columns in February 2007).

As Rosner entered the diner, Roubini stood and extended his right hand. "It's a pleasure to meet you," he said in his slight Turkish accent. "Please, have a seat."

Roubini said that he was impressed with Rosner's paper but wanted to know more about the securitization process—how exactly the packaging of all those mortgages into bonds had given a false sense of security that the banking system was strong and stable, when in reality it was just the opposite.

Rosner explained that securitization led to increasingly more risky lending practices. People who couldn't afford homes were given alt-A and subprime mortgages because they could be packaged into bonds, which, through securitization, were considered risk free by the traditional gatekeepers of the markets, the underwriters, bond insurers, and rating agencies.

But Rosner's premise was that there was no way many of the borrowers could afford the high rates and fees associated with such mortgages, much less the cost of paying the principal on the loans. The ratings were phony, issued by analysts who were incompetent or corrupt and were bullied by underwriters to grant triple As to earn fees. The insurance companies were just as incompetent and corrupt, Rosner said.

And while those risky securities had once been held by large investors such as pension funds and mutual funds, Rosner also believed that banks and brokerages would soon be choking on a toxic stew of mortgage debt warehoused because they had either underwritten the debt and couldn't sell it all or chose to hold on to the securities and pocket the carried interest.

All of which works when the economy is strong, when interest rates are low, but if the housing market tumbles, Rosner said, the economic contagion will spread to Wall Street and the banking system that's holding at least a portion of this debt.

When did Rosner start to suspect the system had become so vulnerable? Roubini wanted to know.

Rosner said he had never had a single "aha" moment when he suddenly realized the financial system was basically a house of cards. Instead, the realization had come to him slowly, after several years of observing Wall Street interact with the housing market.

Securitization allowed the banks to continue lending because they could simply sell mortgages off their balance sheets. People bought homes they couldn't afford, based on the evidence he was able to find. Families pulling in $50,000 were buying homes worth $500,000 or more. "It was mostly speculative buyers who thought, 'Hell, home prices never go down,'" he said.

The Fed was now raising interest rates, but home prices were still defying gravity. At the beginning of 2006, Rosner came across a report put out by the FBI that said mortgage fraud was one of the fastest-growing areas of fraud in the United States. This was particularly troubling. Rosner had predicted in previous papers that the housing market could long outperform GDP as long as unemployment and interest rates stayed low. Now interest rates were rising, and the integrity of many of the loans was in question.

Already some investors were balking at buying CDOs, which is why banks were beginning to hold more of the bonds on their books, bonds that could soon fall into default if the housing market stalled and default rates continued to accelerate. It was a somewhat technical but terrifying prediction. Without securitization, banks wouldn't be able to make additional loans. Without new product, the major Wall Street firms would stop packaging mortgages into securities, and the housing market would collapse.

If what Roubini was predicting—a massive and unrelenting retraction in housing prices—was correct, all those CDOs held by Wall Street would start defaulting, setting off massive balance sheet chaos. The banking system would implode on a level not seen since the Great Depression.

The question neither Roubini nor Rosner could answer is what would be the spark, the inflection point, that would cause these terrifying towers of debt to launch into free fall? Wall Street, and the rest of the world along with it, was about to find out.

PART III

WHAT'S THE WORST THAT COULD HAPPEN?

15. NO CLUE

f you keep predicting for long enough that something is going to happen, eventually it will. That's probably how Nouriel Roubini and Josh Rosner felt in early 2007, as the housing market began to unravel and the shock waves both had predicted—Roubini from the standpoint of the economy and Rosner from the standpoint of the securities markets— began to pick up in intensity.

Bubbles never burst all at once. It's the cumulative effect of information, proving that the naysayers were right and warning signs were ignored, that pushes a speculative frenzy in the opposite direction.

The warning signs of significance became most pronounced about midway through 2006, when the riskiest of mortgage bond deals couldn't be sold to outside investors and were being done at all only because Merrill Lynch and Citigroup began to warehouse most of their deals on their balance sheets. Investors were getting increasingly skittish about the housing market and the bonds that market spawned, with their increasing concentration of subprime debt.

By late 2006 market players estimated that nearly 80 percent of even the safest pieces of each CDO, the triple-A super-senior tranches, were packed with debt directly tied to the increasingly credit-challenged subprime mortgages; meanwhile, default rates, stable for the past five years, began to rise, at first gradually in mid-2006 but then with great ferocity in the new year and beyond.

If the mortgage bond market was about to relive the horrors experienced by Larry Fink and Howie Rubin in the 1980s, Michael Vranos and Kidder Peabody in 1994, and Lehman Brothers and the firms that piggybacked the trades of LTCM in 1998, it was easy to see how this time it would be worse. By 2006 mortgage-backed issuance reached $1 trillion, and that didn't count the $500 billion of CDOs sold that year and into 2007. Many of those bonds couldn't have been sold unless Wall Street firms had gobbled them up and hidden them on their balance sheets because of tepid investor demand through the second half of 2006, meaning that they were more exposed than ever to the market once if and when it began to fully unwind.

The question in early 2007 wasn't really if and when it would unwind but how much. In the old days, the full extent of a growing problem in a market as opaque as that for mortgage bonds wasn't known until the Wall Street firms were forced to disclose their losses, and because of meek disclosure requirements, the big firms rarely fess up to losses until the bleeding is severe.

This time the market's initial unraveling could be ascertained fairly early, not because of enhanced disclosure requirements (that would come later) but because of the ABX Index, which charted the health of the mortgage bond market. The index actually charted the costs of insuring mortgage-backed bonds, since many of the risky mortgage bonds, such as CDOs, now carried those guarantees. The insurance had come first from AIG, but after the big insurer had stopped writing mortgage bond insurance, other companies had stepped in, including Ambac, MBIA, and the thinly capitalized ACA.

The index was created so that if the costs associated with insuring mortgage debt rose, the index would fall in value, reflecting the increased risk, as perceived by the insurers, for the now trillions of dollars in bonds tied to housing.

And that's exactly what happened as the index reflected the growing default rates first among subprime loans but then increasingly in the rest of the housing market. During the first six weeks of 2007, the ABX went into free fall, declining by a whopping 30 percent, reflecting something more than the mere tremors of a housing correction that had gone largely ignored by Wall Street through 2006. There was little if any debate about a housing correction in the executive offices at Merrill as O'Neal and his team rushed to complete the First Franklin purchase or at most of the other firms—though now there would be as the unraveling predicted by Roubini and Rosner had begun.

In May 2005 Stan Druckenmiller had given a speech at the Tomorrows Children's Fund conference, a fund-raising event that prominent investors use to promote their market calls. Druckenmiller told friends he had "ripped off" briefings he had been given earlier in the year by a Bear Stearns analyst and one from Lehman who had predicted a massive decline in housing sales and its devastating impact on the mortgage bond market and the economy in general. Druckenmiller remembers the speech primarily for two reasons: first because of the Bear Stearns and Lehman analysts, whose positions seemed odd given their firms' business models, which were highly dependent on mortgage bonds, but also because of another speech at the conference, from David Einhorn, a thirtysomething whiz kid who ran Greenlight Capital, a hedge fund that generated some of the biggest returns in finance.

Einhorn was a bookish Cornell graduate and expert poker player who had placed eighteenth in the 2006 World Series of Poker. He was generally considered a short seller, meaning he traded in a way that allowed him to profit when stocks declined in value. One of his most famous short trades, of Allied Capital, had led to an SEC investigation of his trading techniques after he publicly announced his short position and accused the company of fraud. Neither Einhorn nor Allied Capital was ever charged.

Druckenmiller was an old-school money manager who had never been particularly impressed with Einhorn's flamboyance, so he decided to leave the conference early. He later wished he hadn't. Friends told him Einhorn had basically "crapped all over" his position in the market (in a few months Druckenmiller would begin to "short" mortgage bonds) as Einhorn touted his long position in a home-building company, M.D.C. Holdings, and the sustainability of the housing market despite the trend lines just the opposite. "For deterioration, you would need prices to actually fall," Einhorn said. "While it is quite possible for an individual market to suffer price declines, there has not been a material decline in housing pricing on a national basis in seventy years." Einhorn also had a large stake and a board seat in a company called New Century Financial, one of the largest originators of subprime mortgages, which Einhorn believed would do great things along with M.D.C. as the housing market rolled to new heights.

But those great things—New Century was now the second largest originator of subprime mortgages nationwide, behind Countrywide Financial—were coming to an abrupt end.

In fact, mortgage originators were now falling like dominoes. Just after Merrill Lynch's Stan O'Neal had put the finishing touches on his purchase of the subprime originator First Franklin, Ownit, the mortgage originator founded by Bill Dallas in which Merrill held a 20 percent stake, had filed for Chapter 11 bankruptcy and was on its way toward liquidation because of subprime delinquencies. People at Merrill say O'Neal didn't seem to care; he believed First Franklin had a better loan portfolio than Ownit, even though Ownit's chief executive, Bill Dallas, had warned his bankers that just the opposite was true.

Dallas was out of a job, but his assessment of First Franklin was right; the losses at the mortgage originator stemming from delinquent subprime mortgages began to mount just a few weeks after Merrill paid $1.3 billion to purchase the company, and the losses continued into the New Year.

Einhorn, meanwhile, was losing his shirt on M.D.C., but that was nothing compared to his investment in New Century Financial, one of the era's success stories. Its loan origination had jumped to $60 billion in 2006 from just $5 billion in 2001, the result of its automated loan system, which, through algorithmic modeling, allowed the company to grant a mortgage in as little as twelve seconds.

New Century may have been fast, but that didn't translate into quality. Brokers figured out how to manipulate the system by tweaking credit scores and pushing loans on people without the means to repay them, because as long as there was a demand for product from Wall Street, the mortgages were no longer the bank's problem. New Century issued most of its mortgages through independent brokers, who had no stake in making sure the borrowers were creditworthy, and neither did New Century for that matter, because of Wall Street's voracious appetite for packing its mortgages into bonds.

But as Wall Street underwriting slowed, New Century was forced to hold more and more subprime debt on its own books. Unlike First Franklin, New Century didn't have a big new corporate parent to shield it from the losses. As subprime defaults picked up steam into 2007, New Century found itself the first casualty of the now-expanding mortgage crisis. Einhorn wasn't singing the mortgage business' praises now; he resigned from the board just before New Century filed for Chapter 11 bankruptcy protection in early March 2007.

So far, however, nothing could shake Wall Street and its top executives out of their bubble-induced haze—not the declining ABX Index, the bankruptcy of New Century, or the $10 billion bad-debt charge the banking giant HSBC surprised the markets with one morning, the direct result of defaults and delinquencies stemming from subprime mortgage lending. Shares of Lehman, Bear Stearns, Merrill, Citigroup, and even JPMorgan

Chase were off their highs in early 2007, but just barely, as the firms began to release year-end bonuses to top executives based on their record profits in 2006. Goldman Sachs, Morgan Stanley, Merrill Lynch, Lehman Brothers, and Bear Stearns would dole out $60 billion in bonuses. And that didn't account for the billions more handed out by the big banks, Citigroup, JPMorgan Chase, and others that feasted on mortgage debt through their brokerage units.

The good times would continue to roll, Wall Street assured the markets. Shares of AIG, the nation's largest insurer, whose fortunes were so tied to the housing market, held up as well. At AIG's financial products group, Joseph Cassano and CEO Martin Sullivan weren't fretting about their massive exposure to subprime through the issuance of $80 billion of credit default swap guarantees on mortgage debt. Instead they were patting themselves on the back for their decision to stop covering CDOs with the guarantees just in time to miss much of the subprime mortgage bonds issued in 2006, regarded as the market's most risky. "We were breathing a sigh of relief," said one former AIG executive. "As the market was blowing up, we thought we dodged a bullet."

The veteran securities analyst Guy Moszkowski also believed that Wall Street had deftly sidestepped the subprime crisis as well. While acknowledging that the Wall Street "firms have expanded aggressively into the mortgage business in recent years," he said in a research report, "these banks have been careful. . . . Securities firms and big banks don't retain much sub-prime exposure, or not enough to be significant," adding that "securities firms quickly move off their books the loans they underwrite or buy."

Others weren't so sure. Larry Fink had lived through three bond market disasters, one in which he had lost lots of money and nearly his career and two when he had made a fortune and rebuilt his reputation as one of the market's smartest people. His firm, BlackRock, had been on a roll since the Merrill Lynch deal he had made in 2006, when BlackRock's stock price had been at an all-time high, which made him suspicious to begin with. The subprime crisis made him worry even more.

Fink knew the mortgage business better than anyone else on the Street, and he knew how the game was played. Like the analysts who covered Wall Street for either the rating agencies or the big firms themselves, no one knew exactly how much mortgage debt was held by the firms because of the toothless disclosure rules.

The exact size of the holdings didn't need to be disclosed (not yet, anyway) under accounting rules until losses began to mount in the firms' pricing models—and the firms took another step to limit their need to disclose their holdings. Banks such as Citigroup created something known as a structured investment vehicle, or SIV, which was basically a large bond fund, where short-term paper was sold to investors to finance the purchase of higher-yielding long-term debt, including CDOs and mortgage bonds.

The SIVs were hidden, massive, and totally legal under accounting rules. Fink and others on Wall Street had heard rumblings about the scale and growth of the SIVs over the years, but those rumors didn't even begin to reflect the reality. Citigroup and other banks had hundreds of billions of dollars of debt hidden in these funds, much of it mortgage debt they couldn't sell anywhere else—so they sold it to the SIV.

This shell game of hidden assets was highly profitable, as accounting rules didn't require the issuing bank to set aside the same amount of required capital against the assets in the SIV as it would have to if it held the debt on its balance sheet. Thus, by creating SIVs and moving the debt off their balance sheets, the banks feasted on a regulatory gimmick that allowed them to take on enormous risk without a hint of disclosure or the expenditure of capital to cushion possible losses.

In fact, under SEC rules, disclosure perversely becomes better when things get worse—the banks issuing the SIVs began to disclose their existence only when the mortgage markets began to tank and losses needed to be taken. Citigroup was among the first out of the gate when in early 2007 it revealed money-losing SIVs in public disclosures. But that would be only the start, not just for losses tied to the SIVs but for losses from every risky investment that Wall Street had made during the great mortgage bubble. The markets were unraveling, and the unraveling couldn't be hidden and massaged as it had in the past. That's because slowly making its way through the SEC's approval process was a new rule, which stated unequivocally that bond prices must be marked down to the prevailing market prices, not ones derived by some obtuse, easily manipulable computer model.

The rule, known as "mark-to-market accounting" would go into effect before the end of 2007, and Larry Fink knew that would spell trouble for the Street. From his perch at BlackRock, his dinners with top executives, Fink knew Wall Street had embraced a business model that was tied to more risk and leverage than ever before, and that had the old risk taker scared. "I don't think Wall Street understands what it's in for," he said at the time. "These guys are fucked."

And having experienced being "fucked" at least once in his career, Fink knew what it meant all too well.

We're looking at somewhat immature markets that are going through a growth phase" was how Bear Stearns' Ralph Cioffi put it that February. "There is a catharsis and a cleaning-out process."

Since 2003, Cioffi's hedge fund investments had been at the center of the subprime market. They had grown to a size of around $2 billion in just that short time, the result of massive leverage and investors' appetite for the big returns generated by his portfolio. Though he told investors he didn't directly invest in subprime mortgages but in mostly supersafe triple-A-rated bonds, the reality was somewhat different. Cioffi, despite his laid-back manner, was among the biggest and most aggressive buyers of CDOs, albeit the super-senior tranches, which carried triple-A ratings despite being packed with high-risk subprime mortgages.

By 2006 Cioffi wasn't someone who was known to the general public in a way Jimmy Cayne was, particularly after making the *Forbes* billionaire's list, but he was a legend in the world of mortgage finance. At Citigroup all the action stopped when a call from Cioffi came in, and not just because he was golfing buddies with bond chief Tom Maheras. Cioffi had become the firm's best customer because he would purchase CDOs that other investors avoided, including the 2006 "vintage," which carried the riskiest subprime mortgages. In fact, as the CDO market sputtered in late 2006, Cioffi had become the *only* buyer of those products, people at Bear now concede.

Cioffi was regarded as the gorilla in the market; the guy who bought big—$200 million chunks of bonds were his norm—and bet big. He was so big in the mortgage market that he wasn't just a buyer; when Citigroup's bond desk needed product—the mortgage-backed securities that made up CDOs and the CDOs used to make the bonds known as CDO squared—they called on Cioffi for help.

Cioffi was now equally respected inside Bear, now a smoldering inferno of corporate warfare between Jimmy Cayne and Warren Spector, where "Jimmy guys" and "Warren guys" battled for control. (Cayne had recently made headlines for upbraiding Spector in a memo to employees for Spector's vocal support of presidential candidate John Kerry.) Though Cioffi and Spector had grown close during their long working relationship—Spector, after all, had given him the go-ahead to manage the funds—Cioffi was one of the few at Bear who got along with both him and Cayne, and for good reason. Cioffi had made the Bear bond department a lot of money over the years, and his hedge fund, with one of the longest and possibly the most confusing names in the fund business—The High-Grade Structured Credit Strategies Fund—had become a cash cow.

The fund was part of Bear's asset management division, known internally as BSAM, and the fund, with its high degree of leverage, produced the strong returns Cioffi had promised, anywhere from 10 to 12 percent a year since its inception. Bear pumped the funds to its best clients through its brokerage department, which dealt exclusively with high-net-worth investors.

For a lifelong bond salesman, Cioffi had, at least initially, turned out to be a damn good money manager. Cioffi now lived large, with fancy cars and several homes, but he had started small. The son of a restaurant owner in Burlington, Vermont, he attended a small liberal arts school, Saint Michael's College, and majored in business and economics. His professors didn't think he had the ability to make it big on Wall Street, but now he was such a large benefactor that he sponsored a scholarship at the school.

During his twenty-plus years at Bear, working mostly as a salesman, Cioffi developed a reputation as someone who could simplify the seemingly endless array of complex new mortgage products and hedging strategies for Bear's clients. Cioffi didn't play the company game, bridge (he loved to golf), but he and Warren Spector became close friends. "He was Warren's best salesmen," said one former Bear executive. "Ralph wasn't just making money for himself but for the entire firm." In a place like Bear, where the saying "you eat what you kill" was the standard, Cioffi ate very well, earning more money than he had ever imagined. He told people that life couldn't be better.

But he wanted still more. In 2003, he went to Spector with a plan to do something new—not just sell mortgage debt but manage it. It was, after all, only fitting that Bear Stearns, the expert in mortgage debt, develop one of the few hedge funds in the mortgage-backed market. Cioffi saw an immense profit potential. As opposed to mutual funds, hedge funds are targeted to wealthy investors. They can take massive amounts of risk to produce bigger returns than funds that simply track the market. They can also charge high fees, and management gets to keep a large chunk of the winnings, meaning that if Cioffi were successful he could make more money than ever before.

Given his close relationship with Spector and how much money he had made Bear over the years, the firm gave him the green light to start a fund that mixed the two ingredients that had showered profits on Bear and the rest of Wall Street: leverage and investments in risky mortgage-backed securities.

No one seemed to think twice about whether a great salesman could make the transition to becoming a money manager, whose main job isn't selling bonds but investing in them and managing their risk.

Now, in 2007, Cioffi's long run of successes, which had earned him a salary of more than $10 million a year, was coming to an abrupt end. First, he was only technically right when he claimed that his funds stayed away from subprime mortgages; by 2006, the super-senior tranches of CDOs were packed with more subprime than ever before, and of the weakest credit quality the market had ever seen. Wall Street, meanwhile, gave new meaning to the term "diversification," that mortgage bonds, the CDO in particular, were supposed to represent in its fullest. "Diversification came to mean jamming subprime mortgages from different parts of the country into a bond," said one Wall Street investment banker. "The theory was that not all the housing markets could crater at once. But this meaning of diversification had nothing to do with the borrower's ability to repay the loan."

Cioffi's problems would be compounded for another reason; despite the convoluted names of his funds, he marketed them as rather safe investments that shied away from subprime mortgages and, of course, focused on gold-plated triple-A-rated securities, which were at bottom, a function of the rating agencies' illiteracy about a market that had grown to mind-numbing complexity. The machine at Citigroup, which supplied Cioffi with much of his product, and other underwriters as well, had now mastered the art of the CDO inside the CDO, known as the CDO squared. This odd concoction was supposed to boost returns because it doubled down on the high-yielding subprime-related loans. But on the way down, losses were compounded because inside the CDO squared were as many as 100 mortgage bonds made up of other mortgage bonds made up of actual mortgages, all of which were falling in value, a *Time* magazine analysis showed.

On Wall Street, complexity isn't something to be avoided—it allows smooth-talking salesmen to obscure simple concepts like risk and losses. Its unclear how much of this complexity Cioffi explained to his investors or even weighed himself as he managed his two hedge funds to strong returns through the first part of 2006.

He was weighing it now as the prices of mortgage debt began to falter and his hedge fund became a proxy for the mortgage market much like the ABX Index. Cioffi and Bear Stearns were slowly becoming emblematic of this, the latest era of Wall Street excess, as news spilled out, slowly at first but eventually with greater intensity, that Bear Stearns, the expert in mortgage bonds, had created two hedge funds that had badly misjudged the market and were bleeding money.

Bear and Cioffi were now being bombarded by calls, some of them from major brokerage clients, looking to withdraw their funds. Cioffi, one of the most even-tempered people at a firm of wild personalities, went into a panic. The ultimate bullish CDO buyer, he did a 180—he shorted the mortgage market, seeking to capitalize on the market's decline, betting that the ABX Index would go down.

But oddly, it recovered in late February and through March. In the words of the young Larry Fink, Cioffi was being "whip chained"—he was losing no matter what he did. Cioffi then reversed his position in April and May, going long on the mortgage market, but he lost again as trading conditions for CDOs weakened.

When most people think of bank runs, the image is that of the Great Depression and people waiting in long lines to withdraw their deposits from a bank, depleting it of its capital and forcing it to close. The modern-day version, particularly when it comes to rich hedge fund investors, is a bit different, but its outcome is the same. Cioffi was now being bombarded with requests to withdraw funds. Some clients and money managers pulled their money out, while others could be reasoned with. "You will make a bad situation worse," Cioffi told the investors.

His point was pretty simple: the mortgage market will come back—it always has. But investors would lose even more if they just yanked their money out at the depressed prices. Through May, every day was a battle to figure out how to keep the fund, and his career, from imploding. Cioffi knew that at a brutal meritocracy like Bear Stearns, his career was on the line. In e-mails to Matthew Tannin, his co-manager, he waxed philosophic, telling Tannin that the troubles they were facing paled in comparison to those of the kids fighting in Iraq. In other e-mails, he questioned his strategy and whether the market would ever return. To investors, the best bond salesman on Wall Street did what he did best: he sold. This time he sold the idea that a market prone to violent upheavals would again turn positive. He was holding back the deluge, but just barely.

Wall Street, of course, had benefited greatly from housing expansion through its origination and underwriting and increasingly through investing in the bonds that were derived from mortgages. The question became: if Cioffi's funds are taking a beating, can Bear Stearns be far behind?

"No way" was the answer Cioffi's ultimate boss, Jimmy Cayne, gave the press as the condition of the funds continued to spiral out of control with every tick downward of the ABX Index. Even so, speculation swirled that Bear's own books were infected with Ralph Cioffi–like bonds. And for obvious reasons: Bear was one of the largest underwriters of mortgage debt. Cayne's position was pretty simple: we don't invest in them, we underwrite them and sell them. His press department was more emphatic; BSAM didn't have the same risk controls used by the firm's trading desk, so there was no way Bear Stearns itself was exposed to the toxic stuff that was crushing the funds. Dick Fuld, the CEO of Lehman Brothers, gave a similar answer: "We're not in the warehousing business." Citigroup and Merrill began circulating the same spin.

Whatever the CEOs told the press, by mid-2007 Wall Street was heavily and dangerously invested in mortgage-backed bonds, much of it directly vulnerable to subprime defaults and a deteriorating market that had grown beyond anything Lew Ranieri or Larry Fink could imagine.

With the hedge funds imploding, Cayne finally had something far more serious than contributions to a Democratic presidential candidate to hold over Spector's head. Bear itself did cut back on carrying CDOs, but the firm's balance sheet was stuffed with other forms of mortgage debt nearly as toxic. Bear was saddled with tens of billions of dollars in mortgage-related debt, which Spector had told top executives like Cayne was "hedged," according to people with knowledge of the matter. What they were hedged with, Cayne didn't know and, at least until the hedge funds started to crumble, didn't care.

But now he did. Cayne suddenly realized just how little he knew about vast portions of the company, namely its bond department and its asset management unit, which he had left to Spector to run. He didn't even know exactly how much of Bear's own money was in the Cioffi funds—first an investment of $20 million and then an additional $25 million.

There had been talk privately among some board members at various times over the previous year about firing Spector as some directors came around to Cayne's assessment of him. The move to oust Spector, though discussed, had never been carried out because nobody could come up with a suitable replacement. But both Cayne and his board knew the time would eventually come when they could push Spector out, and they couldn't wait for the day to arrive.

Spector, of course, would say just the opposite: that he wasn't made part of the team, that Cayne had cut him out of key decisions and basically ran the firm without his input. For that reason, Spector didn't consult with Cayne when he placed an additional $25 million of the firm's money into the hedge funds as they were running out of cash. Cayne later exploded, calling the move a "stunt," but Spector had had complete authority to do it.

Either way, the blowup of the funds underscored the severe dysfunction at Bear. Looking back, executives now concede that at some point, with the firm rolling in so much profit, with cheap financing everywhere, and with the bond markets rallying, the risk management committee meetings increasingly became perfunctory affairs; fifteen minutes of chitchat led by the aging Ace Greenberg, and then they were back to work trading. Cayne himself often bragged that he hadn't gone to the risk management committee meetings in years because they were "like a report card" and he didn't "want to hear how some guy had a position in Colombian bonds and another had Ecuadorian."

Add all of that together and throw in Bear's leverage of more than 30 to 1, sometimes as high as 40 to 1, that had grown its balance sheet to $400 billion worth of every type of security imaginable, including a large position in mortgage-backed securities, and you have a recipe for disaster.

And the disaster was just beginning. Cayne might not have known the first thing about risk or mortgage bonds, but he did understand the business of Wall Street, and as the cries from investors grew louder that either they had been misled by Cioffi or they wanted their money back or both, he grew increasingly concerned. At one point, Spector recommended to the firm's risk committee that Bear start crafting settlements with investors, paying an estimated $200 million to get the issue resolved and out of the press; Bear's stock price was getting killed—it had now fallen to $100 from its high of above $170 just a few months earlier, which represented billions in lost market value. "We have really unhappy clients," Spector said. "Why not settle this and tamp down the anger because once we fight them in court, even if we win they won't be our clients anymore."

But Cayne, who was opposed to settling just about any case that Bear had faced during his tenure, wasn't in a giving mood, particularly since the request now came from Spector. "What do these fucking people want?" Cayne demanded of the increasingly despondent Cioffi one afternoon as investors' skittishness reached new heights. "You had two bad quarters after three great years." He was right, of course. Cayne had a soft spot for Cioffi, and it had nothing to do with his laid-back personality.

Cioffi had made Bear a lot of money, and at bottom, that's what counted. There was also a fairness issue, as Cayne said at the time: who were these bozos trying to fool that they hadn't known the risks involved in Cioffi's style of investing? No matter how much financial advisers claimed they had been fooled into thinking that Cioffi had been running a supersafe government bond fund, there was no way he could have produced those returns without taking on at least some leverage and risk, which anyone with a basic economics class under his belt could have figured out.

But even with Cayne's support, Cioffi's professional life was in turmoil as the threats of mass redemptions grew by the day. At one point, he decided to lower the gates: there would be no more redemptions. Such a move is always controversial, and this one was even more so, given the high profile Cioffi had attained. Investors described it as a desperation play, and creditors of the fund, such as Merrill Lynch, which had lent Cioffi money and received collateral in return, were getting nervous now that the news of a relatively small hedge fund imploding had become a proxy in the media for the state of the entire $1-trillion-a-year mortgage-backed bond market.

Cioffi couldn't deny any of this—the funds *were* in trouble—but he kept telling investors he remained confident about the mortgage market and his ability to pick winners and losers among the securities the funds had purchased. The only thing the fund needed was time for the market to recover.

In reality, Cioffi didn't know what to think; though the mortgage market had seen its ups and downs in the past, he knew this time was different. It wasn't just investors wanting their money back; the fund's creditors were getting skittish themselves, chief among them Merrill Lynch. Cioffi had financed the fund's purchases of CDOs and other bonds with money borrowed from several big banks, including Merrill, Citigroup, and Barclays. In return for their cash, the banks had received collateral in the form of CDOs; if the funds ran into trouble, as they were now doing, and the creditors were worried that they would not get their investment back, as they were now doing, they could seize the CDOs and sell them to recoup their investment.

Merrill was the first firm to start telling Cioffi it was getting ready to "seize" its collateral. The move wasn't as simple for Merrill as just getting its money back. To be repaid, Merrill would need to take its collateral in the fund, in this case CDOs, and then find a buyer for the securities, no easy task as the market for CDOs began to dry up. More than that, a sale would create what's known on Wall Street as a "mark," or a market price. Accounting rules force the big firms to mark the value of their holdings up or down based on their prevailing price. For now, the firms had some leeway when it came to marking their holdings—they could use an internal model to come up with a price.

Merrill didn't realize it at the time, but either way it was screwed. If it kept its collateral in the fund, it would have to believe that Ralph Cioffi had some magic formula to turn its performance around. Not even Cioffi believed that was the case. But if it took the collateral and sold it in the market, its losses might well be compounded; in a matter of months, the "mark-to-model" pricing system would be changed to a more strict "mark-to-market" system, in which holdings were priced according to the last trade available. Merrill would clearly be establishing "a mark," or price, on the bonds, and that price would likely reflect the deterioration of mortgage debt over the past year, the same debt that was held on the books of Bear, Lehman Brothers, Morgan Stanley, just about every firm on the Street, including Merrill itself, which held between $40 billion and $50 billion in CDOs on its own balance sheet, many of them the same quality as those held by Cioffi.

As O'Neal and his team pondered what to do with their Bear Stearns hedge fund exposure, Merrill's risk managers assured Stan O'Neal that the firm's exposure to the broader mortgage market was secured and hedged with insurance. In other words, Merrill believed it could continue to hold its CDO inventory and other mortgage debt at full value, or "par," even after they seized and sold the collateral.

But there had already been warning signs that the marks at which Wall Street valued its mortgage debt—something close to 100 cents on the dollar or at full value—were way too optimistic. The bond insurers that covered Merrill's CDOs were on shaky ground; meanwhile, the ABX Index was hardly showing strength. In early May, Goldman had sent Cioffi a market rate for CDOs that was wildly below where he had been marking its inventory. Like the rest of Wall Street, Cioffi was still marking his portfolio at full value. Goldman had come back with a mark in the range of 50 cents on the dollar, about half of what Wall Street had been valuing CDOs at, and the move had forced Bear to restate its April losses in both funds, with the biggest impact on its Enhanced Leverage Fund, the riskier of the two, whose April losses grew from 6 percent to 19 percent.

Meanwhile, Jimmy Cayne, having largely ignored the despised Warren Spector's recommendations to cover investors' losses in the funds, was constantly on the phone to his in-house attorneys. The board launched an internal investigation to see if there was in fact any fraud in the way the funds had been sold. At bottom, Cayne was comforted by the fact that every investor had signed an agreement with the firm that it understood the risks involved in buying something as convoluted as the High-Grade Structured Credit Strategies Fund.

But that comfort wouldn't last long.

At the UBS Global Financial Services Conference in May 2007, the talk wasn't just about the Bear Stearns hedge funds or the simmering subprime crisis but about a slide show presentation given that afternoon by Dow Kim, the head of capital markets for Merrill Lynch, and the firm's investment banking chief, Greg Fleming.

Merrill had never been stronger, they assured the crowd; 2006 had been a record year in revenues, and the first three months of 2007 had beaten the same period in 2006 by 43 percent!

Merrill was running on all cylinders: investment banking, global trading—and, more than anything else, risk taking.

Kim, who ran the trading desk, was particularly proud of the way Merrill had made so much money taking risks, bragging about the expansion of the firm's "risk-taking capabilities while maintaining our client-centric approach." Trading in derivatives was a big part of the Merrill success story. Kim said that activity was "driving our ROE higher," to the 30 percent territory that O'Neal lusted for.

Fleming explained that Merrill had risen to become the third best firm in accumulating investment banking fees, rivaling Goldman Sachs, the perennial winner. He said that the firm under O'Neal had been particularly good at "connecting the dots," meaning cross-selling investment banking deals through trades and other services to Merrill's clients. Merrill was doing this, he said, all over the world. It had deep ties to companies and their CEOs and CFOs in China, India, Russia, Japan, and Germany, "and the list goes on."

When they were almost finished, both men were exhausted—with so much to say and brag about, who wouldn't be? Then they came to the last slide, which was different from the charts and graphs that had dominated the presentation.

It was a cartoon entitled "How Stan O'Neal Transformed Merrill Lynch." O'Neal wasn't at the presentation—he normally hated such events and was lousy at giving speeches. But there he was in the flesh—or almost flesh, a cartoon of him standing in his trademark black suit between two men in their underwear, the one on his left overweight and balding, with a resemblance to Dave Komansky. That was Merrill "before" O'Neal's arrival and makeover. The other, trim and steroid buff, with a tag that read "Best Investment Bank" across his wide chest, stood for the Merrill after O'Neal had made his mark.

The crowd, many of them investors and competitors, erupted in laughter, not only at the fat guy's resemblance to Komansky but also at the cartoon of O'Neal. He was smiling, and most of them knew O'Neal almost never smiled.

Fleming was quick to point out that it wasn't just he and Kim who had made that assessment; *Euromoney* magazine had named the firm "Global Investment Bank of the Year 2006."

"Dow and I have had fun with this slide since it appeared in *Euromoney* last summer," Fleming later said with a laugh. "Everybody disavows knowing the guy on the left, and, depending upon who's got the podium, we take credit for the guy on the right."

O'Neal was taking the most credit for the man on the right, and who could blame him? Even enemies couldn't argue with his success.

Merrill, of course, wasn't walking away from the mortgage market or its investment in CDOs even as the first warning signs of the market's eventual implosion began to spread. O'Neal's marching orders were the same: take risks, use leverage, and make money.

And so Merrill's holdings of CDOs continued to grow. Merrill packaged $32 billion of the stuff during the first half of 2007, most of which couldn't be sold and so was carried on the firm's books. With the concerns over Bear's hedge funds, investors weren't exactly in the buying mood, and it was unclear when they would return. Bond chief Osman Semerci, however, told O'Neal he believed the investors would return, according to

O'Neal's recollection, by Labor Day at the latest, which would allow Merrill to sell off its inventory at favorable prices.

In any event, the bonds were hedged, Semerci said; he assured Merrill's senior management that the CDOs were triple A, and they had insurance that they believed would limit losses. Even without AIG underwriting CDO insurance, Merrill was heroically and successfully scrambling to find other types of insurance, from the so-called monolines such as MBIA and from the smaller company ACA.

But there were no heroes in this story.

"Sell everything, we're too long!" said Dale Lattanzio, one of Semerci's lieutenants, according to an account in the *Wall Street Journal* that described how the firm's bond department was desperately trying to reduce the firm's exposure. According to the *Journal*, the day-to-day task of selling out or hedging the firm's mortgage bond was known by various accounts as the "mitigation strategy" or the "risk transformation strategy" to reduce and eliminate the possibility of the firm's CDO holdings imploding, and according to Lattanzio and Semerci, it was producing great results.

According to an internal Merrill Lynch document labeled "confidential," in June 2007 Lattanzio gave a presentation to the finance committee of Merrill's board in which he provided additional details about this alleged success story.

Merrill's board had been remade by O'Neal over the years in his bid to consolidate power, remove pockets of dissent, and approve his business model of increasing risk.

But the topics discussed at the finance committee meeting also illustrated just how little the board knew about the business model O'Neal had embraced. Lattanzio began the presentation with a chart labeled "What Is a Collateralized Debt Obligation?" He went on to describe Merrill Lynch's "market leadership in [this] high growth market." Through the first half of the year, Merrill had underwritten $34.2 billion in CDOs; Citi was a close second, with $30.1 billion. Lattanzio described how the mortgage market had evolved since earlier in the decade, spurred by low interest rates, "house price appreciation," and "increasing subprime origination" in mortgages, and how subprime mortgages were experiencing "increased delinquencies and losses" that were having a negative impact on the bonds tied to those mortgages, namely the CDOs.

Yet, he explained, Merrill had dodged a bullet. In November 2006, a few months after he and Semerci had taken over the bond department, the firm had had "substantial risk in the warehouse" that it had reduced through a "risk transformation strategy." By converting the raw mortgages in the warehouse into CDOs, now on the firm's books, Merrill had created less risk, not more. The bonds could be sold off at a profit as soon as the market recovered.

That was, of course, the plan. Lattanzio then discussed how the catalyst of the market's current turmoil, the implosion of the Bear Stearns hedge funds, had gone down and how Merrill had "moved quickly and decisively" to seize its collateral in those funds (what he didn't say was that that seizure had actually set off the chain reaction that was causing the market's troubles).

He then turned again to the question of risk: what was Merrill's exposure, and how much could the firm lose if the market continued to unwind? Much of the firm's risk was concentrated "in lower risk super-senior" CDOs, he said. One slide he showed the board stated, "Multiple risk management options enable effective strategies for stable or volatile markets."

Lattanzio's assessment: all the actions taken by the firm's risk management team would lead to a modest loss of just $73 million based on the firm's models, as long as the firm continued its strategy of "discipline in financing of illiquid assets."

Sometime in the spring of 2007, O'Neal and Kim gave a similar presentation to Merrill's board, which was obviously concerned about Merrill's exposure to the subprime market. Their message was just as upbeat: Merrill had been busily reducing its exposure to mortgage debt all year, and what was left could be managed through insurance and other hedges.

Then something odd happened.

"What do you mean, you're quitting?" That's what the incredulous Stan O'Neal shouted when, in mid-May 2007, Kim told him that he planned to leave Merrill and start a new hedge fund. He wanted O'Neal's help, of course—could the firm make a significant investment? O'Neal was too shocked to answer that question; instead he tried to keep Kim from leaving and demanded a face-to-face meeting.

O'Neal had in his head an organizational chart about whom in his inner circle he could do without and whom he couldn't. Men such as the investment banking chief, Greg Fleming, he considered "legacies," holdovers he needed to make the trains run on time and appease the old guard as he remade Merrill into a leaner, meaner fighting machine. But there were a handful of people he thought he couldn't live without. One, of course, was Fakahany, and the other was Kim, a savvy trader who made a lot of money and carried out O'Neal's orders regarding risk and trading to a fault, or so O'Neal thought.

They met right away at a restaurant in Manhattan and O'Neal dispensed with the pleasantries. He had once described Kim as "inscrutable," because no one could figure out what he was really thinking and sometimes what he was saying. During a meeting where Kim was weighing whether to buy the NYSE floor trading or "specialist" business from Bear Stearns, he repeated himself three times before a representative of Bear figured out that he was asking "Do specialists make money?" and he could answer Kim with a simple "yes."

Tonight, at least in O'Neal's mind, was no different. O'Neal couldn't get a straight answer out of Kim as to why he wanted to leave and why now. For months Kim had been telling people inside Merrill, O'Neal included, that he wanted to leave to run a hedge fund, but each time O'Neal had convinced him to stay. O'Neal thought he had put the matter to rest by awarding Kim a 2006 bonus of $30 million. How much more money was he going to make running his own hedge fund?

"Can you give me until the end of the year?" O'Neal asked, according to a person familiar with the conversation. "Are you sure this is something you want to do now?"

That's when Kim put his head down, stared at his plate of food and, after what seemed like an eternity to O'Neal, raised his head and said, "Yes."

O'Neal left the meeting thinking there was no way he was going to be able to keep Kim at the firm, so he began to weigh Plan B.

The next day, Merrill issued a press release announcing that Kim "has informed the company that he plans to leave Merrill Lynch by the end of the year in order to establish a multi-strategy, private investment firm. The company said it expects to be an initial client of Mr. Kim's new firm and that Mr. Kim also would serve as an advisor to senior management for a transition period."

Losing such a senior guy in charge of such a vital part of the firm's business left a huge void and would have to be addressed in some way. O'Neal immediately named Fakahany and Fleming co-presidents.

The two people who couldn't have been bigger rivals were now running the nation's largest brokerage firm, and O'Neal couldn't have cared less. He believed Fakahany was CEO material, and he needed Fleming around for his deal-making abilities and ties to the old guard. Moreover, he believed that Merrill was on the right track. He told people it might take another decade or more to transform it into a top-notch company, something along the lines of General Electric when it had been run by Jack Welch.

"It took Jack twenty years to make GE a successful company," O'Neal said at the time, and he was moving as fast as he could to beat Welch's time frame with the help of people like Fakahany and Osman Semerci.

As for the simmering subprime crisis, the talk among analysts, hedge fund traders, and short sellers was that it was only a matter of time before the big Wall Street firms disclosed their holdings and their losses in a market that continued to deteriorate as the new accounting rules, including mark-to-market pricing, got closer and closer to becoming law. Every firm would be taking the hit as the new mark-to-market rule kicked in, and that included Merrill, the largest underwriter of CDOs, which had some of the most toxic of the depressed mortgage debt on its books.

The Wall Street firms fired back at what they considered rumormongering by short sellers and analysts such as Richard Bove of Punk Ziegel & Co., who they said were working for the short sellers. "It's bullshit," Jimmy Cayne remarked at one point. Dick Fuld assured reporters that Lehman wasn't a warehouse but a clearinghouse of mortgage debt. Jeff Edwards, Merrill's CFO, attempted to dispel the mortgage worries by repeating what Lattanzio had told the board; he called whatever was on the books "manageable" and certainly not something investors should be worried about, at least not yet.

By the end of June 2007, Cioffi's hedge funds were still alive, albeit barely, though that was about to change. Merrill was no longer threatening to seize the collateral on the $1.5 billion it had lent Cioffi, it was in the process of doing so, effectively killing the hedge funds once and for all.

Merrill might have thought it would get out of the funds without a blemish given all the hedges and insurance on its CDO holdings, but that would soon change as well. O'Neal and his executive team appeared oblivious to the time bomb they were about to detonate as they sold into a largely frozen market.

But not the traders at Bear Stearns.

"We couldn't believe it," said one Bear Stearns executive. "If the mark came back low, they would be repricing their own balance sheet," because Merrill and Citigroup, the two largest underwriters of CDOs, had been warehousing billions of dollars' worth of the securities on their balance sheets over the past year.

Traders at Bear didn't know how much Merrill had on hand, but they knew it was a lot and they didn't buy the notion that the CDOs were hedged. But one of the things the Bear Stearns traders knew, which apparently the risk managers at Merrill didn't or failed to remember, was that there is no such thing as a complete hedge in a rapidly declining market, and with AIG out of the business, Bear doubted that Merrill could buy enough insurance to cover its losses as the bond markets continued to fall. Just how reckless Merrill's decision was can be seen in the way the bond market was behaving in June 2007. Unlike stocks, bonds aren't priced on a centralized market; for the most part they're traded and priced between individual dealers.

The prices of bonds are often stated through their return or yields and how they compare to the yields of the safest bonds available, U.S. Treasury bonds. If the yield differential between Treasuries and other bonds widens, that means investors are buying up Treasuries (yields go down when prices rise), and that's exactly what was happening.

By June the ABX Index was gaining more and more attention as it fell even further. More dramatic was the flight to quality of investors selling corporate bonds and snapping up supersafe Treasuries. In June 2007, the difference between the three-month LIBOR rate and the 3-month Treasury bill rate, which had stayed below 0.5 percent most of the year, suddenly shot up to more than 0.8 percent as investors flocked to the safety of 3-month Treasury securities.

Other measurements showed similar results. Even the spread between Triple-A rated corporate debt and 10-year Treasuries widened to 0.72 percent from 0.58 percent, showing that investors had come to the conclusion, as fear spread, that the risk of buying any bond other than Treasuries was a risk possibly not worth taking.

Merrill, which boasted the top bond-underwriting house on Wall Street, seemed to ignore the market's deterioration and the losses that it might produce by selling into it as Stan O'Neal gave the order to take the collateral and sell the CDOs.

"Merrill's being a hard-ass," a Bear Stearns public relations spokesman said as Merrill did indeed seize $400 million of its collateral and began the process of finding buyers for the bonds.

Cioffi wasn't concerned about Merrill's balance sheet or, for that matter, Bear's balance sheet. He worried what the pulling of the collateral would mean for his fund as the controversy surrounding the funds continued to swirl. Cioffi, for example, planned to off-load some of the fund's investments and place them in a new company, which would have issued shares of stock. Cioffi, in effect, would have transferred the risk from the CDOs now in the fund to public shareholders. Following press reports about the move and the possibility that public shareholders, not allegedly sophisticated hedge fund–eligible clients, could now own pieces of Cioffi's fund, Bear pulled the offering. Cioffi then responded with a plan he thought would buy him and the fund additional time for the markets to return. Bear itself would provide a line of credit to the fund (Bear would ultimately put $1.5 billion into the hedge funds to provide liquidity); the rest of the creditors would hold off for a year and not seize collateral and sell it into the market.

The move, Cioffi believed, would buy the market time to recover. He let his creditors know he had the deal sold to Bear's management, including his buddy Warren Spector. Merrill delayed its collateral sale, and Cayne, who now received regular briefings on the funds' condition, was jubilant. "The bottom line, this could be great for everyone involved," he said.

But it wasn't. Merrill ultimately went through with the bidding process, and its traders began to digest what it means to be a hard-ass in a soft market. They thought the bonds would sell close to par value, or at worst at 99 cents on the dollar. But when the bidding began, the first price Merrill got back was significantly lower, close to 65 cents.

Panic had yet to spread through Merrill—Semerci and his team reminded the firm's top executives about the hedges and the fact that the bonds were triple A rated—but something close to panic also began to percolate at Bear, particularly on the mortgage bond desk. when traders heard the news of the lowball bids, not from Merrill itself, as was usual, but from secondary sources in the market. The problem for Cioffi's portfolio was obvious as word of the trades filtered back; the funds were being exposed as holding truly toxic material. If the CDOs in question had to be priced at 65 cents on the dollar, they were trading at junk bond levels with or without their triple-A ratings, and Cioffi's hedge funds were finished.

As the head of the Bear Stearns bond department and asset management division, Warren Spector was informed of Merrill's disastrous attempt to sell its collateral, and he immediately began making calls to Merrill's trading desk to find out what had gone wrong. He believed Merrill was being not just dumb because it held on its books the same junk as

Cioffi did, and now had to take losses when it wrote the prices of those securities down, but also overly secretive. Bear had a right to know the marks, Spector believed, for no other reason than to price its own holdings appropriately.

O'Neal was alerted that Spector was on the warpath. O'Neal couldn't be bothered with speaking to Cayne's second in command (he and Cayne were chums, after all), so he ordered one of his underlings to handle the issue. Investment banking chief Greg Fleming was brought into the discussion, not because he's an expert at mortgage debt (he wasn't) but because he could hold his own in a verbal shoot-out with Spector, which this was likely to be.

Every Wall Streeter in existence has some of the same DNA; they're driven, type A personalities who above all else love to make money and believe at their core that their life can be summed up in their net worth. But to find two people on Wall Street less alike than Spector and Fleming would be an amazing feat. Let's start with their appearance. Spector wore thick glasses, and his even thicker black hair was parted neatly on the side, while Fleming seemed like a fortyish version of Richie Cunningham, from the sitcom *Happy Days*; he was freckled, and his hair never seemed to be combed.

Spector was brooding and intellectual; even friends say they couldn't remember the last time he laughed at a joke. Fleming had become one of the best investment bankers on Wall Street because he was equally adept at strategy and schmoozing. It's why Larry Fink had liked doing business with Fleming so much; he knew Fleming was always a great dinner companion, and the two often ended the workday at Fink's favorite restaurant, Sistina, downing an expensive bottle of wine (occasionally costing as much as $1,000) and laughing about the day's events.

Spector certainly wasn't in a laughing mood when Fleming returned his call and started to explain why Merrill had seized the collateral of CDOs—management was worried about the hedge funds pending collapse, pure and simple. Spector cut him off. "Do you understand how the process works?" he asked in a manner Fleming later described as dripping with condescension. Spector made it clear that Merrill owed Bear a call to tell them at what price the CDOs were being sold so it could mark its own book properly and value the Cioffi hedge funds, but Fleming wasn't budging, offering what Spector believed were bland excuses. "He was speaking Chinese," Spector would later say.

Actually, the reason Fleming was equivocating was that Merrill might have seized the collateral, but it was choosing not to sell the debt at such depressed prices as the bids it was receiving. It was holding them, trying to figure out what to do next.

So was Cioffi. Merrill's seizure and failed sale spread through the market like a cancer as the ABX Index continued its decline. The entire market now knew that Merrill couldn't sell a CDO for anything but a depressed price. And it wasn't just CDOs; mortgage bonds were not moving as defaults were surging not just among subprime borrowers but among less risky ones as well. Housing prices were falling, and bank foreclosures were on the rise.

For Cioffi, it was pretty clear that his hedge funds were doomed. It was one thing to stop investors' redemptions, but he couldn't stop the creditors from demanding their collateral, and that's what they were doing as fast as they could; and just as Merrill had, everyone was now finding the mortgage market unforgiving as bids began to reflect the reality that the housing bubble had popped.

At Bear's most senior levels, the finger-pointing was everywhere, starting with Merrill. Cayne, who had used to enjoy Stan O'Neal's company, now blamed him for putting the final nail in the coffin of the hedge funds. His legal staff, which had initially said that liability over the hedge fund losses was isolated inside BSAM, now had to defend a broader problem. There would be investor lawsuits not just against Cioffi and his number two, Matt Tannin, but against the firm over the losses and Cioffi's claims about not having invested in the subprime market. The SEC too began nosing around to determine whether Cioffi had properly disclosed the risks of his investment strategy and how much upper management was involved in the decision-making process.

If Cayne had hated Cioffi's boss, Warren Spector, before, he hated him even more now for the feeding frenzy of negative news about the firm and about Cayne himself.

News stories had him spending much of his time on the golf course; CNBC reported that his country club had investigated whether Cayne had cheated by inflating his golf scores. The *New York Times* uncovered that Richard Marin, Cioffi's nominal boss, had a blog in which he opined about working through the crisis and reviewed his favorite movies, prompting Cayne to break his silence. Although Cayne loved to hold court with a few selected reporters, he was still among the most reticent CEOs on Wall

Street. He shunned major interviews and almost never appeared on television even when the firm had record earnings. Now he was opening up to the *Times*, and it was clear why he had stayed away from on-the-record interviews: he was lousy at them.

He all but admitted that Bear as a firm had screwed up. His strategy to keep the hedge funds separate from the firm itself became a joke the moment investors read this sentence: "When you walk around with a reputation for being the most rigorous risk analyzer, assessor, controller and this is trashed, well, you have got to feel bad. This is personal."

A round this time, Cayne called a meeting of the firm's senior executives to discuss the hedge funds. He asked at one point how much money the firm had in the funds. The answer came back: $45 million. Cayne loved the fact that BSAM was a separate unit from Bear, and thus he believed his exposure was limited to the shit that was hitting the fan. But the $45 million exposure surprised him. He had thought it was just $20 million. Spector soon admitted he had authorized the additional loan.

"Without telling anyone in this room?" Cayne snapped. Spector, obviously embarrassed, said he had "fucked up." The room went silent. Cayne just sat with his cigar in his hand, staring at Spector to let the moment set in.

The general consensus inside Bear Stearns, particularly among some board members, was that Spector was increasingly less motivated for his job—one of the biggest on Wall Street since he had controlled so much of the firm's business units—and now it had finally caught up to him. Still, the focus wasn't on Spector but on Cayne.

In a business built on trust and credibility, Bear was losing both quickly.

In late July, Bear made it official and finally pulled the plug on the hedge funds. Marin was ousted, and Cioffi was put on a kind of modified leave; everyone knew he was a dead man walking (not even Spector spoke to him now), even though he was supposed to be helping wind down the portfolio—sell what could be sold—and then leave the company. That's the way things work at Bear; you screw up, and you're out, and Cioffi had done more than screw up.

Cioffi's problems were only beginning, as were Bear's and soon the rest of Wall Street's. Jacob Zamansky, a prominent investor attorney, had been getting calls from the funds' investors, who said they had been misled by Cioffi into thinking they had bought a fund that despite the standard disclaimer related to risk was actually hedged against the type of wild swings that were now occurring in the market. Even worse, those investors recounted to Zamansky the meetings Cioffi had had with their financial advisers, in which, they contended, Cioffi had explained his investment strategy as decidedly low risk.

Zamansky, who had made his name bashing, in court documents and also in the press and on CNBC, scandal-tarred Internet analysts like Jack Grubman and Henry Blodget years earlier during the dot-com bubble (information from Zamansky cases had been used by Eliot Spitzer in his investigation of the Wall Street research analysts), now set his sites on Cioffi and Bear. His suit against Cioffi and Bear included many of the same elements as his prior work allegations of self-dealing and conflicts of interest, not to mention all those meetings in which Cioffi had allegedly assured investors about the low-risk nature of his fund.

At Bear, Cayne referred to Zamansky as an "ambulance chaser" who would lose in court. But Zamansky's lawsuit was taken more seriously by regulators. Just as Spitzer had turned to Zamansky for his investigation into fraudulent research, the SEC and the Justice Department, both looking for a poster child for the latest era of Wall Street greed, turned to Zamansky as well. Just after filing his civil case against Cioffi and Bear, Zamansky received a call from the U.S. Attorney's Office for the Eastern District of New York to give a presentation about what his clients were saying about their experiences in the Bear Stearns hedge funds. Within a few weeks, prosecutors began taking depositions from those clients.

Nearly a year later, both Cioffi and Tannin were indicted after prosecutors uncovered e-mails showing that they had both questioned the subprime market's resilience at the same time as they were telling investors the market was about to make a comeback. The criminal case was remarkably similar to Zamansky's civil lawsuit.

Cioffi was an easy target, given the fund's prominent role in what was now the complete unraveling of the mortgage market, as prices of mortgage bonds fell and the bond market began to "lock," meaning it was getting increasingly difficult for just about every player except the U.S. Treasury to sell bonds. But in reality Cioffi and his fund were merely symptoms of the mortgage bubble, not one of its root causes. More than anything he had done, the absurd rationale fueling the mortgage market had to do with lax regulation, the expansion of the welfare state into the mortgage business, greed, and, of course, the blindness and venality of the big rating agencies.

The rating agencies that had fed the bubble with their omnipresent triple As on mortgage debt were now making up for lost time. The combination of the constant news surrounding the Bear hedge funds and the fall of the ABX Index prodded the rating agencies to do what they should have done years earlier, and they began to downgrade mortgage-backed securities in droves. According to the trade journal *Mortgage Banking*, in early July Moody's placed 184 types of CDOs on negative review. Standard & Poor's followed suit, as did Fitch. The message was clear: the credit quality of the mortgage market had been greatly inflated, and now the great unraveling had begun.

The downside hit Wall Street hard. Firms on the Street began an investigation into how much of the debt they held on their books—and began to reveal the size and scope of the problem to their management and boards of directors. In one of the great ironies of the era, CEOs such as Jimmy Cayne and Stan O'Neal at Merrill had been paid historically high bonuses in 2006 because of gains largely from the selling, trading, packaging, and carrying of mortgage-related debt, yet as they delved deeper into their mortgage holdings they realized that they had no idea of how the business really worked and how much of the stuff they were now carrying on their books.

In mind-numbing detail they now got their education; one Merrill Lynch document was used to explain the structure of a CDO. It depicted a complex tangle of boxes with a maze of arrows pointing in various directions to show how cash from mortgage payments by home owners ultimately flowed to the bottom line. Others showed how CDO squareds, or "synthetic CDOs," were created through complex arrangements between counterparties to create a CDO without even having to buy mortgages. Cayne, like many of his fellow CEOs, couldn't stand the detail. He just wanted to know one thing: "Are these fucking things hedged?" he asked Spector during one executive committee meeting. Spector said they were, and he later gave the same response to the firm's risk and audit committee.

Cayne and even Stan O'Neal might not have been able to grasp all the nuances of the CDO squared, but they understood the nuances of a market that was now, in the words of the risk managers, "locked": meaning nobody was selling mortgage bonds, particularly CDOs.

Bear was particularly vulnerable, and not just because so much of the firm's profits flowed from a market that was for all intents and purposes dead. Its problems were also reputational. Cayne now found himself spending most of his day (when he wasn't golfing or playing bridge) trying to separate the reputational damage of the hedge funds' implosion from further impacting the bigger firm. Wealthy individual investors who held brokerage accounts, such as the Wall Street PR impresario Robert Dilenschneider, took their money out of the firm because they had lost faith in its management, feeling that any firm that could allow its hedge funds to blow up like this couldn't be run by people who knew what they were doing. Large institutional investors began questioning their salesmen at Bear how the firm that was supposed to be so smart when it came to mortgage debt could have acted so recklessly and stupidly.

Bear was now Wall Street's laughingstock; its stock price was sinking fast, and investors were now demanding tougher terms when lending the firm money in the overnight repo market. Merrill Lynch produced a report for its board of directors entitled "Bear Stearns Asset Management: What Went Wrong." Among its many conclusions about the Cioffi hedge funds was that "the situation grew quickly out of control as a result of the leverage being carried" and that "Bear Stearns . . . naively assumed Wall Street creditors would hold off on margin calls especially once the big investment banks were able to discount fears of systemic risk."

That last statement was odd because it underscored how blind risk managers were to the broader consequences now sweeping the Street— most notably Merrill itself—as a result of the hedge funds' collapse. By the middle of the summer the possibility of systemic risk, or the risk that a single occurrence such as the Bear Stearns hedge funds' collapse would spread losses throughout Wall Street, was real and evident. The secret was out; Wall Street titans that had assured investors that they merely sold mortgage bonds to others were now starting to fess up that they weren't doing anything all that much different from what Ralph Cioffi had done.

've been at this for twenty-two years. It's about as bad as I have seen it in the fixed-income market during that period of time," Bear's CFO, Samuel Molinaro, said on August 7, 2007, during a hastily arranged conference call with Wall Street analysts who cover the company, which, despite his dour assessment, was designed to calm fears that the hedge funds' problem might engulf the entire firm. It didn't.

It was a sweltering August morning. Cayne briefly led the discussion, assuring everyone that the firm had, during his four decades on Wall Street, survived crises in the past and would again. It didn't help that after his initial remarks and after saying that he and Molinaro "are here with others to try and address the concerns in the investor community," he abruptly left the meeting, which added to the perception of disarray.

The call lasted about an hour; much was said and discussed, but it was Molinaro's remarks, sparked by a question from the noted banking analyst Michael Mayo, that sparked a feeding frenzy in the markets. Seconds after he made the statement, wire service stories and CNBC headlines began featuring Molinaro's honest and scary assessment of what was now a financial crisis every bit as large as what Wall Street had faced in 1998. Also making news was Cayne's odd behavior; after making a brief statement during the call, he had suddenly disappeared without even saying good-bye, forcing Molinaro to nervously answer a question directed at his boss.

Bear's public relations staff ("Bear" and "public relations" by now was an oxymoron) later said Cayne had to make an important client call. Investors took it as another sign of disarray and weakness. That day, the Dow lost nearly 250 points; the ABX Index continued to fall and the spread between Treasuries and other bonds widened significantly, meaning investors were dumping their holdings of debt, particularly mortgage debt, and diving into Treasuries. Financial stocks were decimated, but no stock price fell farther than Bear's, which was so strongly tied to the fate of the bond market; in one day it plunged nearly 8 percent.

Bear had built up credibility among analysts for its string of record earnings; now it was losing credibility almost by the day. The firm had yet to completely disclose its holdings of mortgage debt or the losses from its portfolio. But a general consensus was building among investors that before long Bear would have to come clean. Bear had been among the top underwriters of mortgage debt even as investors' appetite started to wane, so it had to have gobs of the stuff on its books.

The analyst Dick Bove began to pounce most prominently, stating emphatically that the firm was holding massive amounts of some of the same toxic waste that had blown up the hedge funds on its balance sheet. Cayne again denied this in an interview with me, but then Standard & Poor's placed the firm's long-term debt on credit watch; the rating agency didn't come out and say that Bear's books were fraught with risk, but it indicated that at the very least the controversy over the firm's hedge funds could have long-term implications.

"Reputation damage" may sound like a minor issue, but it wasn't to Molinaro, who as Bear's CFO was the person in charge of securing financing. Because of Bear's reliance on short-term debt to fund its operations, reputation meant everything; if Bear's counterparties on repos or banks that had provided lines of credit were to lose confidence in it, they would simply stop lending it money. It had happened to Lehman Brothers during the LTCM debacle, and it was now happening to Bear Stearns. Bear financed its balance sheet, now at $400 billion, through a variety of short-term funding vehicles, but most prominently through repos, in which Bear lent securities to firms, mainly banks and other investors in return for cash.

Repos were the most cost-effective way the firm could finance its operations because it could lend out its large inventory of bonds, including the tens of billions of dollars' worth of risky mortgage debt it held, for cash. But now, as the subprime crisis spread, the banks were getting picky—they were balking at taking mortgage debt as collateral, thus squeezing Bear to find funds somewhere else.

The growing precariousness of Bear's financial condition and its waning credibility among investors did little to convince Cayne to think twice about his next move: the firing of Warren Spector. Cayne was livid that Spector was at the same bridge tournament as he was as the bond markets were imploding, the hedge funds began their final descent, and, most important, Bear's funding sources showed the first signs of drying up. With that in mind, Cayne decided to call a board meeting and end Spector's career at Bear once and for all.

Of course, Cayne had been itching to do the deed for years, and the simmering financial crisis gave him an excuse. It wasn't just Spector's mismanagement of the hedge funds, the $25 million investment he had made in the funds without alerting Cayne or the board, or even that he had been playing bridge while Rome was burning. Cayne told the board he couldn't care less about Spector's knowledge of the mortgage business; he simply couldn't work with Spector anymore, and the best way to save Bear was to kill the source of the problem, which he believed was Warren Spector.

Alan Schwartz would become his heir apparent. It didn't matter that Schwartz was a banker who had never handled risk as Bear's traders had.

He had put in his time. He was going to run the place, Cayne said, when he stepped down, whenever that day came.

Cayne told Spector his decision face-to-face. Cayne later said he thought Spector was going to cry when he broke the news (Cayne had for years questioned Spector's masculinity and had more recently began to refer to him as "Little Lord Fauntleroy").

Spector has told people that he remembers the meeting somewhat differently: he was more "pissed off" and shocked than anything else. After the initial surprise of being fired, for the first time in his adult life, he said, he regained his composure and began taking notes about the conversation. He asked Cayne if he was sure he was making a wise move; he was, after all, one of the experts in the part of the market that had landed the firm in big trouble. Cayne said he was—he was more than sure, as a matter of fact. "I can't work with you anymore," Cayne said, adding "someone has to go." In any event, Spector had no say in the matter. The board had agreed that his time at the firm was up.

After the firing Cayne felt as if a lead weight had been removed from his chest. In an interview just hours after the decision was made official, Cayne said that he was planning a trip to China's CITIC Group, the country's largest investment bank. He said he felt great, he said, even though he didn't look it; his face was showing visible signs of aging and, having dropped something like 30 pounds, he suddenly looked older than his seventy-three years. He attributed his sudden weight loss to a new diet and exercise regimen, but friends say the cause was stress. At a recent golf outing he had nearly passed out and couldn't finish the course, something that had never happened before.

But the firing of Spector appeared, at least for the moment, to energize Cayne. "You don't realize the damage this guy has caused," he said, adding that Spector's support among the firm's rank and file had fallen to zero, and he thought investors felt the same. It was a gross overstatement. Spector may have been arrogant to a fault, and lately he might have spent more and more time playing bridge and hanging out on Martha's Vineyard than in the office, but the general consensus was that one of the smartest guys in the mortgage market had been fired at a time when the firm needed his expertise the most.

That realization hit home in mid-August, when Cayne heard new reports from Bear's creditors. The firm's share price continued to show signs of strain, but that was just one of the problems. Bear's creditors were now closing their lines of credit. It wasn't the occasional hard-ass demanding additional collateral for repo trades. Both JPMorgan Chase and Citigroup, the firm's two biggest banks, were now balking at extending credit to Bear.

Sam Molinaro's assessment was as simple as it was frightening: Bear was facing a funding crisis of such magnitude that it was threatening the firm's very existence.

For Jimmy Cayne to save Bear, it would turn out, would be a lot more difficult than getting rid of Warren Spector or going into a joint venture with the Chinese, as he had planned.

Now Bear was reeling as, in addition to the looming funding crisis, the Justice Department, investigating the collapse of the hedge funds and the SEC, began delving into the firm's finances, namely its holdings of mortgage debt, which senior officials at the firm had said was hedged and manageable. The Sunday following Warren Spector's firing, SEC officials held a series of meetings in Bear Stearns' forty-second-floor conference room, where the head of each division talked about the health of his unit. Senior Bear management dubbed it the "proctology exam," with six SEC officers poring over everything on Bear's books. The funding was the last topic to be covered. Paul Friedman, who was in charge of repo, along with the rest of the finance team, spent nearly three hours going over all of Bear's credit facilities—its repo lines, its short-term funding obligations, its cash and cash equivalents on the books—stressing that Bear's liquidity was fine. By the end, the SEC guys were staring at Friedman with glazed eyes, but Bear had passed the test.

Friedman actually felt sorry for them. "I think they might have understood about twenty-five percent of that," he said to a colleague after the presentation.

No one on Wall Street was feeling sorry for Bear because Bear's problem was now *their* problem. They now had their first true brush with panic; a credit crunch had developed in which lending was starting to dry up, and the ABX Index fell to the 50s, a decline of 50 percent in less than a year, as mortgage traders continued to bet against the subprime market, where loans tied to risky borrowers continued to fall into default.

More tellingly, signs of credit turmoil began to emerge in the more widely followed corners of the credit market. The yield gap between triple-A-rated corporate debt and Treasury bonds blew out, as people increased their buying of the supersafe Treasuries and shied away from corporate bonds, junk bonds, and especially mortgage-backed securities.

The bond market's pricing woes had implications for the economy. Such a widening of yields meant that corporations—even the best and most solvent corporations in the world—had to pay much more to borrow money. Corporations with ratings below triple A faced even worse conditions and much higher borrowing costs, and for those even lower on the credit scale, the squeeze was really on.

Wall Street's business model of heavy borrowing was being exposed in all its imperfections as it became more difficult for Wall Street firms to raise money from their usual sources.

Nearly every firm on the Street had warehoused now-illiquid mortgage securities, and that included the big banks that had dived into the mortgage underwriting business once the walls of Glass-Steagall came crashing down and that were also Bear's biggest repo lenders. Tom Maheras's bond department at Citigroup had warehoused around $50 billion of CDOs, though the exact amount was still unknown to the outside world and would be for some time.

Compounding the problem for Citigroup, as well as the other banks, was their off-balance-sheet exposure to the mortgage market through SIVs. These off-balance-sheet entities were perfectly legal under the bizarre accounting rules of the banking system, and they warehoused billions of dollars in higher-yielding mortgage securities, often the really risky ones. The benefit to the banks: by keeping those assets off their balance sheets, the banks didn't have to keep any regulatory amount of capital against them.

But now, with the prices of the securities in those vehicles falling, many of the banks were forced to mark them down and put them back onto their balance sheets. The move soon forced them to post huge amounts of capital against them, but it also tempered their willingness to take in more of those same troubled assets in exchange for cash, which was fast becoming a scarce resource.

And that meant increasingly less access to the all-important repo market for Bear because the banks didn't want any more of its mortgage debt on their books. In mid-August, just after Spector's firing, Bear's management and compensation committee, plus members of Bear's executive committee—essentially the dozen most powerful people at the firm—held an emergency meeting in the firm's executive conference room.

This, of course, was a meeting Cayne couldn't skip for a bridge match. The executives plotted how to repair Bear's fallen reputation following the hedge fund collapse and now the firing of Warren Spector.

Some people in the meeting said they were concerned that Spector's firing had fueled rumors of a full-on crisis within Bear, which could eventually jeopardize its funding, particularly in the short-term repo market. After all, who would want to extend credit to a firm in chaos? "We need to get our bank lines lined up," Overlander, the co-head of fixed income, said, "just in case the market gets even more skittish."

A ll I hear is bad stuff about Chuck Prince," Cayne said one afternoon before paying a call to the Citigroup CEO. "I bet he's not such a bad guy."

Cayne was trying to think positive before his meeting with Prince, and later with JPMorgan Chase's Jamie Dimon, Bear's other big creditor. Jimmy Cayne wasn't used to wooing people like Prince and Dimon. After all, just a couple years earlier Dimon had been wooing him when he'd tried to buy Bear Stearns. Cayne barely knew Prince, but the word on the Street was that he was a lightweight who'd landed in a heavyweight job because he had bailed Weill out of trouble a few years back.

But with the survival of Bear on the line, Cayne was now willing to kiss up to anyone, including men he had once considered his inferiors.

Hat in hand, Cayne asked both banks to open up additional "long-term" lines of credit to the firm so it wouldn't be so exposed to the fickle repo market for funding.

Cayne still liked Dimon and thought their meeting went well. The Citigroup meeting made Cayne sick. The meeting had been set up by Citigroup board member Dick Parsons, the former head of Time Warner, and Bear's Alan Schwartz, who had a close relationship with Parsons. It didn't matter.

Cayne made it clear that he wasn't looking for an emergency loan, just a line of credit and "just in case" things got really bad. The people from Citigroup made no promises, but by the end of the meeting, Cayne thought he had won them over as he had Dimon.

Instead, he got the same answer from both, but not the one he was looking for: no way. Both banks were demanding onerous terms; they wouldn't accept mortgage securities in exchange for collateral. They didn't say so outright, but the terms they were demanding were those you demand from a company you think might not make it.

When Cayne heard the news, he felt as if he had been sucker-punched.

He would later make the same pitch to the China Development Bank, including a 10 percent stake in Bear Stearns, and get the same rejection.

"It's Armageddon out there," Cayne told his board member Vincent Tese, who responded, "Based on what I hear, Citigroup is probably in worse shape than we are." "You're probably right," Cayne answered.

And he was. One of the things that the implosion exposed was the myth that the guys at Bear Stearns—from Cayne to Greenberg to Spector—were expert risk takers, particularly when it came to bond market risk. It also exposed the myth that every firm perpetuated in its spin to the media: that its holdings of risky and now-imploding mortgage debt were modest and somehow hedged against losses.

The Street suddenly began yet another period of mass self-examination, as it had in the late 1980s, 1994, and 1998, the last time the bond markets had crumbled and the Wall Street business model had shown its flaws.

Like a partygoer after a long night on the town, Wall Street felt the hangover begin. The chilling reality was setting in: lower profits and losses from their holdings of now-illiquid bonds.

The question was how long it would last. In the past, postbubble hangovers had been relatively short, largely because the size of the problem had been containable, and Greenspan's solution to every financial crisis of the past three decades—slashing interest rates—had worked like a charm.

But this time was different. First, the Fed was slow to react to the crisis, so slow, in fact, that it elicited a wild outburst from the CNBC commentator Jim Cramer, who said the new Fed chairman, Ben Bernanke, who had replaced Greenspan in 2006, was "asleep" and "has no idea how bad it is out there." But the problem was bigger than Bernanke—and more precisely his ability to pump "liquidity" or money into the financial system through low interest rates—because the entire regulatory apparatus had broken down and Wall Street had taken full advantage of it.

The SEC, which was supposed to be monitoring the capital levels at the investment banks, was clueless to the now-emerging fact that the buildup of risky assets was so huge that just about every firm on the Street needed billions of dollars more in capital as a cushion for losses if the crisis continued. The Fed had missed the same buildup at the banks, failing to appreciate that when Citi had been allowed to combine investment banking with commercial banking it was really mixing mortgage bond trading with savings deposits and any losses would be shouldered by average people.

In the past, massive amounts of liquidity had repaired the system as borrowing became cheaper, and firms had been able to trade their way back to profitability. But this time was different because the stakes were many times bigger—trillions of dollars in debt were held by banks and securities firms, leverage was at extremely high levels, and the types of bonds held on the books were some of the most volatile ever created. Long-Term Capital Management's losses had been staggering at $4.6 billion. This time Bear Stearns alone was sitting on a powder keg, its 30-to-1 leverage leading to a balance sheet of trades and investments of $400 billion, including billions of dollars' worth of mortgage debt that was losing value seemingly by the minute.

Because of the size of the leverage, Wall Street's holdings of increasingly illiquid mortgage debt were so large that almost no amount of Fed easing—even as Bernanke got the hint and began lowering rates—could restore the mortgage market and with it the banking system to health anytime soon. As a result, in the late summer and early fall of 2007, the mortgage market didn't just slow as it had during past contractions; it shut down. Even Merrill, the leading underwriter of CDOs, had basically stopped underwriting by the second half of the year.

More than that, lending activity to everyone didn't just get tougher; it was starting to grind to a halt. Banks began to hoard cash in anticipation of losses. One of the consequences of the repeal of Glass-Steagall was that banks had transformed themselves into risky casinos that were now facing huge losses because they had warehoused these money-losing bonds on their balance sheets as Citigroup had done.

It was free and easy money, and Wall Street had made it for so long that it had forgotten that there is no such thing as free money. With that a new term was introduced into the daily conversation about the economy: "the credit crunch," a widespread panic among lenders that there was no such thing as a borrower with good credit.

Average investors, myopically focused on the stock market, were now getting a broader education in the bond markets and just how important lending was to the economy's well-being. The problem that had started on the balance sheet of Wall Street began to spread to Main Street. The only question was how long would it last.

Even so, regulators still very much looked at the smoldering crisis as a passing storm. Treasury Secretary Hank Paulson and his counterparts at the Federal Reserve, including Chairman Bernanke, had been meeting with Wall Street executives since the Bear Stearns hedge funds had imploded. Jimmy Cayne had complained to the Fed and Treasury about the situation at Bear and in the markets. "No one is lending to anyone," he said.

Paulson had known Cayne for years and was well acquainted with Cayne's propensity for glib remarks (his frequent references to Goldman Sachs as "Goldman Sucks" had made their way back to Paulson's office before he left the firm for Treasury), so he didn't consider the Bear Stearns CEO to be an expert on the health of the financial system. Moreover, both Paulson and Bernanke still seemed unconvinced that the credit crisis was anything more than a much-needed correction to teach Wall Street a valuable lesson in risk management. It was their belief that the overall economy, even the banking system (except for a few bad apples) was still sound. Private equity firm Blackstone had just become a public company, and other private equity firms were considering the same. It was a vote of confidence in the markets and the financial system in general.

Then in August 2007, top Fed officials, including Bernanke, were at their annual conference in Jackson Hole, Wyoming, where key policy makers and market players join Fed officials to think big thoughts about the economy. This conference was remarkable for how little was discussed about the Armageddon that was being described by Cayne, felt in particular by Bear Stearns but now starting to infect every firm on the Street. That is, until Bernanke, huddling with two of his key advisers, Fed Governor Kevin Warsh and Vice Chairman Don Kohn, had received a report about the credit markets. It could be summed up in one word, he said: "Ugly."

Bernanke, his advisers now concede, had no idea just how ugly the markets would get. Maybe that's because the general consensus not just among regulators but among many of their contacts on Wall Street (apart from Cayne, of course) was that as ugly as things were getting, they would get better. According to the *New York Times*, in mid-2007 at Citigroup, Tom Maheras gave nearly the identical response that Spector had previously given Cayne and Semerci had just given the Merrill Lynch board, assuring Chuck Prince that Citigroup's mortgage holdings were not becoming toxic—they were holding their value, a tribute to the risk management abilities of his team.

It's unclear if Prince or the Citigroup board asked for an exact inventory of the holdings that the bond group had generated, but a few others were doing so. Mike Mayo, the veteran bank analyst, had directed his team to begin scouring the Street to determine the banks' and securities firms' exposure to subprime debt, which was slowly losing value, if you believed the default statistics and the ABX Index. Mayo was surprised that during the second quarter Citigroup showed surprisingly strong results—its profit rose 18 percent over the same quarter the prior year—but most surprising was that its revenue growth, even in fixed income, exploded. Other banks were showing similar results.

Mayo was in the same position as Jamie Dimon had been earlier—unable through the publicly available information to determine how much of the increasingly toxic mortgage debt Citigroup had been holding because of the opaque, and in some cases nonexistent, disclosures.

The only way to obtain a full accounting of the potential problem was to ask the management of the banks and Wall Street firms to voluntarily provide a list of their holdings. The answers he received from Citigroup and the rest of Wall Street were astonishing, he thought at the time. He was being told that the firms either had no exposure or were hedged and as a result the holdings weren't a problem.

He was also being told that the man in charge of the Citigroup bond department could one day be running the entire enterprise. By all accounts Chuck Prince didn't necessarily like Tom Maheras as a person, but he respected Maheras's knowledge of the bond markets, and, more important, so did Robert Rubin, who considered Maheras's team the best in the bond business. Inside the firm, Maheras had now become the odds-on favorite to replace Prince. Maheras's purchase of the company's stock the year before signaled his intention to remain with the firm and fight for the CEO spot.

One of the oddities that pervaded Wall Street at the time was how valuable information that should be shared broadly and conclusively with shareholders was locked down, warehoused among a few people. At Merrill, a firm with more than 50,000 employees, the mortgage department had about 50 people underwriting and selling the bonds; at Citigroup, a firm with roughly 375,000 people, about 100 people worked in that department, among them Maheras and his team.

By now the old Salomon Brothers–Smith Barney rivalry had been replaced by a rivalry between the sales staff and the traders. The salesmen complained that the firm spent so much time trading and figuring out what bonds to hold on its books that it forgot about helping clients make money in the markets by selling bonds.

They also complained that Maheras spent too much trading for himself. In addition to his personal trading, Maheras was trading a pool of company money worth about $500 million. It was called the Fixed Income Management Account, and it was essentially Citigroup's proprietary trading portfolio, where stocks and bonds could be traded and bets made. Maheras, given his long and up until now strong record of trading, would seem like an obvious candidate to have fairly wide discretion over which trades were made from this portfolio and which ones weren't. There was just one problem, according to people inside the firm: Maheras was now the head not just of the bond department but of all of capital markets. That meant he oversaw a vast empire that included not only sales and trading but also the origination of bond issues, giving him an immense amount of market information that people outside the firm couldn't get.

There was never any proof that Maheras had done anything wrong, but his dual role of overseeing the trading while heading all the other functions raised eyebrows, particularly among some of the company's salesmen, who were concerned primarily that Citigroup's 180-degree turn in the Sandy Weill–Jamie Dimon era had fundamentally altered the firm's mandate.

Trading was the path to success, just as it had been at Salomon Brothers years earlier, and serving clients—that steady but low-profit business—was a thing of the past at Citigroup. Indeed, all three of Citigroup's top salesmen quit between 2004 and 2007; in exit interviews, Maheras promised them he would change the culture, but he didn't. People who worked with Maheras say there was no incentive for him to change; it was the short-term trading culture that helped him earn $30 million in 2006, a salary larger than that of most CEOs in the country, and it was trading that put him just inches away from becoming Citigroup's next CEO.

By late summer 2007 and despite the profits generated by the bond department and elsewhere, Citigroup was still considered a lumbering giant. Its stock price had been stalled at around $45 to $50 for the past five years. Investors began to call for Chuck Prince's head. CNBC even started laying odds on how much longer he would last, even without knowing what was brewing on the firm's balance sheet. Jim Cramer bluntly predicted that Prince would be fired because there was no reason he should be in the job in the first place other than as payback for having saved Sandy Weill from being charged by Eliot Spitzer in the Grubman research scandal.

With Maheras's assurance that all was fine, Prince decided to create some positive buzz for the firm and himself, granting an interview with the *Financial Times* in which he tried to calm investors' fears about the mounting losses from subprime debt and whether Citigroup was cutting back on providing credit to clients. Quite the contrary, according to Prince; there was so much liquidity in the market that the subprime mess couldn't possibly cause a retrenchment, and Citigroup was right there providing liquidity to anyone who needed it.

In his words, "When the music stops, in terms of liquidity, things will be complicated. But as long as the music is playing, you've got to get up and dance. We're still dancing."

Maybe Prince forgot his meeting with Cayne. In any event, the interview did not have the desired effect. That day, telephone lines between angry Citigroup shareholders lit up across America. "That's the dumbest thing I ever heard a CEO say," said one large investor. Citigroup's stock price had languished for the past five years, so the last thing investors wanted to hear was Prince comparing his tenure to a "dance."

Meanwhile, the dance was about to end. In the executive suite at Citigroup, word began to spread about the firm's holdings, not just straight CDOs but all the risky debt that had made its way to the firm's off-balance-sheet holdings of mortgage bonds in SIVs. They were possibly the biggest ticking time bomb if the mortgage market didn't return to some type of normalcy sometime soon because they were hidden from public disclosure. Investors had no idea that Citigroup would have to take losses on those investments because the accounting laws allowed banks to keep them hidden, until market losses occurred.

It was, of course, perverse logic because the whole object of financial disclosure is supposed to be to enlighten the investing public to excessive risk. In the case of the SIVs, Citigroup and other banks would have to begin disclosure when excessive risk began taking its toll and the banks started to write down the losses.

Prince and Citigroup's board were assured again that the holdings were fine; this wasn't Bear Stearns, a sleazy trading shop. This was Citigroup, with its massive balance sheet, hundreds of billions of dollars in customer deposits, and offices all over the world, with Tom Maheras, the best in the business, guiding Citi through the credit crisis.

Maheras's assurances notwithstanding, Prince was starting to get nervous. Perhaps it was that Dave Bushnell, the chief risk officer, and Maheras were so close. Or maybe it was what he was reading about the spreading pain in the mortgage market and the credit crunch. Then again it could have been the fact that after years of unbending optimism, the rating agencies were making up for lost time and downgrading pools of mortgages. Or the speculation in the market that Citigroup's rival in the CDO market, Merrill Lynch, was going to report large losses tied to mortgage debt.

Prince was always a careful lawyer, and he ordered a massive review of Citigroup's holdings of mortgage debt. The review came as Citigroup's stock price began to fall. A megafund created by the big banks to bail out the SIVs held by Citigroup, Bank of America, and JPMorgan Chase failed to materialize, primarily because JPMorgan CEO Jamie Dimon saw no reason to help his old firm and now bitter rival. Citigroup's write-downs of bad debt began to grow from insignificant levels to those that show up on financial statements and prompt news stories.

Then came Meredith Whitney, an analyst for the small brokerage firm Oppenheimer & Co. Whitney had been making a name for herself warning that Citigroup was a firm in disarray—too large, too undercapitalized, and too costly—to survive in its current shape. What Citigroup refused to admit was that it wasn't even close to creating the financial supermarket that Weill had envisioned; its systems weren't set up for consumers to tap into all the banking services offered by the firm; many overseas operations barely spoke to the home office about cross-selling products; and investment bankers and commercial bankers rarely spoke with one another. Whitney described all of these problems, often in acerbic prose.

Prince and Citi's supporters initially laughed and brushed Whitney off as a bit player looking to make a name for herself by taking on the biggest bank in the business. But with each passing week, Whitney's predictions seemed more and more dead-on; Citigroup's costs continued to grow, and now it no longer had bond-trading and underwriting income to paper over its high-priced plan to create the world's largest financial supermarket.

Though Whitney didn't say so, at least not yet, Citigroup's risk taking was setting the stage for even bigger problems in the months ahead.

ear Stearns' problem was more immediate: securing its funding and restoring its reputation. Alan Schwartz, now Cayne's heir apparent, went to work. He was a great investment banker who knew how to woo clients, but none of his schmoozing worked on Citigroup as he prodded and pushed the bank for large long-term lines of credit and other funding that Cayne had gone begging for earlier. He called up Citigroup's CFO, Gary Crittenden, and blamed Citigroup for creating the mess at Bear by selling the mortgage securities to Cioffi that had blown up the hedge funds and destroyed confidence in the firm.

Crittenden held his ground; Bear wasn't getting money primarily because Citigroup didn't want the collateral Bear was offering in exchange for the funds—speculative mortgage-related bonds. "I don't understand this," Schwartz shot back. "You guys underwrote and sold these bonds to Ralph, and now you're balking at taking them?" Crittenden said there was nothing he could do.

Cioffi was, of course, one of Citigroup's biggest customers for the CDOs that were now the bane of the capital markets. But now Citigroup was chock full of its own inventory of CDOs and there were no Ralph Cioffis to sell them to.

The deteriorating conditions had Bear's management scrambling to line up other sources of funding as a cash crisis loomed. One major issue: the possibility that its hedge fund clients with large cash balances in their prime brokerage accounts might desert Bear and yank out billions of dollars in cash, out of fear that the firm could become insolvent.

For years, Bear, like the rest of the Street, largely ignored the risk associated with their reliance on prime brokerage accounts for business purposes. No longer. Cayne worked the phones to get assurances from Bear's top prime brokerage clients that they would remain at the firm. For now, Cayne's efforts were working. James Simons, a former math professor who ran the massive Renaissance Technologies hedge fund, stated that he would stay with Bear, as did others.

Meanwhile, Friedman and Overlander met with executives at foreign banks to secure repo lines and other lines of credit. The broker Kurt Butenhoff's relationship with the billionaire commodities trader Joseph Lewis proved useful as Lewis took a 7 percent stake in the company, while Cayne's trip to China paid off; Bear and China's CITIC Group reached "an agreement in principle" for a deal in which Bear would get $1 billion from the bank (great for its balance sheet) and CITIC would get an investment in the form of a convertible bond from Bear.

The crisis had been averted, though just barely. In late August and through September, the credit markets didn't exactly recover, but at least they stopped deteriorating, and regulators breathed a sigh of relief. Bear's stock price also recovered somewhat, and so did Cayne.

J immy's fine, he just has a little cold" was the word from a Bear Stearns spokesman when CNBC inquired about the strange disappearance of the firm's CEO in the midst of the firm's burgeoning troubles. A couple of days earlier, Cayne had nearly fainted on the golf course, and now, despite the assurances from Bear's PR staff, Cayne wasn't fine—in fact, he was in the hospital, nearly dead. A urinary tract infection had developed into something far worse; as the infection spread through his body, it was developing into a condition known as sepsis, which often results in death. The doctors gave him just a 50 percent chance of surviving.

Pat Cayne was at her husband's side when he went into shock and was taken to Columbia Presbyterian Hospital in Manhattan. Pat herself had survived a bout with cancer, become a devout Jew, earned a PhD, and now, in her sixties, looked much younger than her age. In many ways, they had a storybook marriage: a beautiful daughter, a great son-in-law they both adored, and wonderful grandchildren. They had more money than they had ever dreamed of when they'd first met, when Cayne was a part-time municipal bond broker while playing bridge the rest of the time.

Through it all, Pat Cayne had been the strength behind her husband; she had supported him and prodded him to take chances and move through the corporate ranks. She had advised him in delicate corporate issues, and she had certainly helped him navigate the treacherous politics at Bear. She also put up with his antics, which she summed up with a standard warning she gave the wives of senior executives at Bear during gatherings: "Keep your eyes open and pay attention, because the world is looking for a wealthy guy."

Now she was watching her husband, who had battled long and hard to win his job, nearly lose his life fighting to save it.

But as so often at the bridge table, Cayne won this hand and survived, vowing to return to Bear as soon as possible. As he recovered, he worked out of his hospital room. "I'd give anything to take a normal piss," he said at the time, sounding like Hyman Roth from *The Godfather Part II* as he groused about having to take the drug Flomax so he could urinate.

But the pressure on the financial system continued to mount. In late September, Bear would post sharply lower earnings for the third quarter as it wrote down losses in its mortgage holdings. The analyst Dick Bove had been right; investments in the now-illiquid mortgage-backed bonds weren't confined to the hedge funds, as Bear's PR staff wanted everyone to believe; the firm held substantial amounts of the illiquid debt, and because of the new mark-to-market rules it had to account for the deteriorating price by taking losses.

And it wasn't just Bear Stearns that had bitten off too much of the mortgage market.

16. THE DANCING STOPS

I f you ask me, Merrill is in worse shape than we are." That was Jimmy Cayne's assessment in September, the same month Bear was writing down risky assets and taking losses because of its mortgage holdings. While recovering from his near-death experience, Cayne was looking to take the heat off Bear as the market's whipping boy, and he began telling friends, associates, and reporters (including me) that Merrill was next in line for public excoriation.

How did Cayne know this? He and his team watched the league tables like everyone else. Merrill had continued to underwrite CDOs even as investors' demand ground to a near halt. They had to be holding billions of dollars' worth of the stuff just ready to explode when the company's risk officers got the guts to come clean as Bear was doing.

What made the possibility that Merrill's books were conceivably more screwed up than Bear's so unthinkable was that for most of the year Merrill had been telling investors it was flying above the raging mortgage storm; its investment in First Franklin, the large subprime mortgage originator, gave it a more comprehensive view of which parts of the market to avoid, plus its investments were hedged.

"I think many of the reports that I've seen have exaggerated and misunderstood the nature of the business and how it's managed," Stan O'Neal told Dow Jones Newswires in late April 2007 after a speech at a conference hosted by the Wharton School of Business. "It's not consistent with what I would assess the state of the business to be." Jeff Edwards, the CFO, was more emphatic, describing Merrill's exposure to the subprime market as "limited, contained, and appropriately marked."

What gave O'Neal, Edwards, and above all, the firm's chief risk officer, Ahmass Fakahany, comfort was that the "risk transformation strategy" enacted by Semerci and Lattanzio was supposed to have reduced the firm's exposure to the mortgage market even as the firm increased this exposure through the purchase of First Franklin.

But the fears over Merrill's exposure to toxic mortgage debt were far from exaggerated. Merrill's risk managers appeared wrong at just about every turn; they had stated that the firm's exposure was limited and manageable, when in fact it was becoming less manageable because (as we'll see in a bit) the counterparties that had sold them the insurance and hedges they had bought to limit losses were about to blow up.

Moreover, they believed the market would return and allow the firm to clean house, sell off the CDOs and other mortgage bonds that were sitting like a lead weight on the balance sheet, but all the indicators pointed to just the opposite occurring. The bad news in the mortgage market kept on coming day after day, with stories about home foreclosures mounting and the ABX Index continuing to fall. One story that underscored the depressed state of the mortgage market was the travails of Countrywide Financial, the largest originator of mortgages. Ironically, it was a Merrill analyst who predicted that Countrywide might have to file for bankruptcy if mortgage default rates continued. It sent the markets into a tailspin and Countrywide's CEO, Angelo Mozilo, who had built the company into a colossus through his advocacy of subprime lending, into a tizzy.

Mozilo, great salesman that he was, couldn't spin the obvious reality: Countrywide had begun handing out subprime loans nearly at the top of the market, and those loans were now imploding and so was Countrywide. With its stock price falling and funding lines beginning to disappear, it was Mozilo who now needed a loan. He sought out and received a line of credit from Bank of America that staved off a full-blown crisis, albeit only temporarily. Bank of America would later buy the entire firm in an even more distressed state.

But not before Countrywide, the subprime king, had concluded that the market it had built its business model on was radioactive. In the late summer of 2007, the company announced it would no longer make subprime loans and would lay off twelve thousand employees; the move sent shivers through Merrill Lynch. In one internal document, Merrill regarded it as a "key event" in the growing "dislocation" of the mortgage market because it underscored the severity of the problem with the loans that were at the heart of Wall Street's holding of illiquid bonds.

Angelo Mozilo, who had proselytized about the need to lower lending standards, ignore credit scores, and do God's work by lending to people who held low credit ratings, now declared the business to be dead.

"It's over," Mozilo told the trade publication *Mortgage Strategy*.

The bigger question was whether Wall Street—which had used subprime mortgages to make billions of dollars and now faced losing billions—was over as well.

In the fall of 2007, even as the bond market continued to contract, the gravity of the situation had yet to bleed into the stock market. Stocks, as measured by the Dow Jones Industrial Average, remained at lofty levels, as the Dow reached its all-time high of 14,093 in October 2007 despite Wall Street's mortgage woes.

Apparently, average investors really did believe Merrill's claim that its losses were limited and contained and the worst of the crisis was over.

It wasn't, of course. What O'Neal would soon find out was that not only was the firm's projection that the mortgage market would return wrong but that its CDO pricing models were just as flawed. The models, by relying on a convoluted risk measurement system known internally as the "DV01 Metric," had never taken into account the possibility of a large decline in housing prices.

DV01 measured the possible decline in CDO prices only in small increments (DV01 meant that the bonds would move in a deviation of one basis point, just $\frac{1}{100}$ of 1 percent) and had never taken into account a continued decline in the mortgage market, including the hemorrhaging of CDO prices that ramped up in late summer and early fall of 2007, which would push prices down further. People who know Stan O'Neal say that one of his many peculiarities is that he's at his best when he's challenged. During his career at Merrill, colleagues say he periodically dropped out of sight when he was bored or felt that no one cared what he thought. O'Neal would later concede to just one such incident after he was named president of the firm and continued to weigh whether he really wanted the top job. But Paul Critchlow, a longtime and highly respected Merrill Lynch communications chief, was forced to track O'Neal down on several occasions both as president and later as CEO because O'Neal had gone into what he would describe as a "funk."

It was the type of dark mood where he refused to answer telephone calls or respond to e-mails, as if, after surviving so much and achieving so much, he was looking back on his life and questioning whether it was all worth it.

As Critchlow recalls, O'Neal emerged from the funk only when presented with a challenge—such as his fight to be named Komansky's successor or his fight to push out Komansky faster than the former CEO wanted to go. That's when O'Neal was at his best and most focused.

O'Neal, after his funk, which resulted in his playing endless hours of golf, much of it alone, was put to the test again in the late summer and early fall of 2007. O'Neal had been telling those around him that he was fairly confident in the advice of his risk managers that Merrill's mortgage-bond investments, including its billions of dollars in CDOs, had the appropriate hedges and insurance to keep their triple-A ratings, so even if the subprime market fell, Merrill wouldn't have to write down losses; the bond insurance would keep the securities at full value. Semerci believed there was little to worry about because he and his team had massively deleveraged the firm's mortgage holdings. He provided the numbers to O'Neal showing the declining exposure; over the year they had sold off somewhere close to $30 billion of mortgage holdings. O'Neal several times congratulated Semerci for a job well done.

But the job was tougher than anyone realized, and O'Neal was discovering that one of the downsides of losing people like Jeff Kronthal was that their knowledge cannot be replaced. All told, the handover from Kronthal to Semerci had resulted in the departures of ten bond executives with a combined two hundred years of experience, one senior executive at the firm estimated.

With experience, of course, comes knowledge, and what Jeff Kronthal would have known, according to people close to the situation, was that deleveraging the way Merrill was doing it—getting warehoused mortgage bonds off its books and creating CDOs—was actually creating more problems. When asked to describe a CDO, traders liken it to a stew of many different bonds, their main ingredients being mortgage bonds consisting of subprime loans plus other asset-backed securities. The problem was that by packaging all those ingredients together, Merrill had compounded the problem, creating a security, the CDO, that was more toxic than the sum of its parts. As the markets continued to fall, traders at Merrill came to a terrifying conclusion: they could have sold the CDO ingredients such as asset-backed securities at much higher prices than they could the entire CDOs. In a sense, Merrill had created more toxicity in its effort to create less.

It's difficult to know when O'Neal began to think that Merrill was now facing a crisis from its holdings because the people around him appeared so clueless to the coming storm, Fakahany chief among them. The accountant, who was good at numbers, had left the heavy lifting involving the bond book to Dow Kim, who had now left the firm. Later Fakahany conceded that he was shocked to learn that the firm's CDO holdings had grown so much, even if they were hedged and covered by insurance, as Semerci and Lattanzio assured him. His excuse: he was spending too much time dealing with the company's international business, which also came under his watch.

One possible tipping point for O'Neal was a telephone call he received from Treasury Secretary Hank Paulson during the summer, one of the days the stock market momentarily cratered amid fears of the spreading credit crisis. Paulson, along with Timothy Geithner, the head of the New York Fed, had been calling on all the heads of the big firms to see how they were handling the subprime storm. O'Neal vividly recalls the day he received the call from Paulson because the stock market was getting crushed; there was panic selling, and that day the European Central Bank injected $131 billion of cash into the European banking system, the most it had ever done in a single day. It was a desperation move to restore confidence to the European banking system; just as Bear was being squeezed out of the repo market, there was little if any interbank lending going on in Europe.

O'Neal suddenly realized that whatever was happening was larger than just a couple of crappy hedge funds going down; he checked with his trading desk. The only mortgage bonds now being traded were those of Fannie Mae and Freddie Mac because of their implicit government guarantees. Thank God for those triple-A ratings, O'Neal thought to himself as he pondered Merrill's own holdings.

Then he received the call from Paulson. The call was unusual because of its timing on the heels of the ECB action and the deteriorating markets and because Paulson, to O'Neal, seemed genuinely scared. "How do you feel about things?" Paulson asked. O'Neal, who wasn't someone who telegraphed his true feelings, replied, "I feel okay" and then discussed what he knew about the firm's holdings of mortgage bonds.

O'Neal's version of the events may be accurate, or it may simply be spin intended to deflect criticism over the mounting troubles Merrill would soon face. But at least a dozen former Merrill executives (many of whom, it's important to note, despise O'Neal both personally and professionally) believe he was unaware of the dangerously high level of toxic mortgage debt Merrill was holding, which may be worse than knowing about the exposure all along.

As the first tremors of the crisis Merrill was about to face started making its way around the firm, the finger-pointing began—not just at O'Neal but at his senior team as well: Dow Kim, the head of capital markets, and Ahmass Fakahany, who in addition to being O'Neal's eyes and ears was in charge of risk, as well as Semerci and Lattanzio, who had been handed the job to clean up (and by some accounts, bungled the effort).

But only O'Neal can be blamed for the singular vision of turning Merrill Lynch into a casino. "We were told time and again by O'Neal that we had to be like Bear and Lehman in subprime, like Goldman in risk trading, and like Citi when it came to the CDO market, and that's what we did," said one former senior executive.

Being like Bear and Lehman in subprime wasn't where Merrill wanted to be by the fall of 2007. Merrill's models may have protected the firm from everything short of a once-in-a-lifetime calamity, but that's exactly what was happening as the prices of mortgage debt began to fully reflect the dismal state of the housing market. Housing prices were falling, in unison and all over the country.

The mortgage-bond markets were getting worse, not better, despite what O'Neal's team had predicted. All you had to do is look at the trading of the 3-month Treasury bill, the market's ultimate antidote to fear because it's the safest bond in the world. Its price was now spiking dramatically, sending its yield crashing from 5 percent in July to around 3 percent. Traders were continuing to sell mortgage debt and were running so scared that they were buying not just 30-year Treasury bonds but those that matured in three months.

Despite the lack of disclosure that glossed over Wall Street's holdings of toxic assets, senior managers at most of the firms were now concluding that the entire banking system would be facing huge losses unless the credit market somehow changed course. Insolvency was clearly on the table for some firms (the smallest, like Bear) because they had borrowed so much money and had so little cash on hand to cover these losses, and because they couldn't borrow their way out of this situation. They would have to raise capital (and thus dilute current shareholders' holdings) from places where there was still capital, namely from overseas investors like sovereign wealth funds; merge; or sell themselves at prices well below their bubble highs.

This was the conclusion that Stan O'Neal was slowly coming to. By the early fall of 2007, O'Neal was bitter and felt betrayed by his staff. He recalled their assurances: in April he had been told that the losses from the mortgage bonds were manageable—close to $500 million would have to be written off and subtracted from earnings because of the deleveraging campaign that had begun earlier in the year. By August that number had grown to $1 billon, and now he was told it was growing even higher as the insurance contracts began to fail, the so-called hedges weren't working, and Merrill's traders couldn't unload anything on their books because there were no buyers. O'Neal ordered a review of the firm's holdings of risky mortgage paper and how much it was worth or—to be more precise, how much money Merrill was losing. He selected a senior risk manager named John Breit for the task.

Breit's assessment wasn't pretty. He told O'Neal that the real exposure of the firm was much larger than the rosy scenario he and the board had been supplied; an internal Merrill Lynch document showed the firm was carrying around $46 billion of CDOs that were now among the most toxic of the toxic bonds in the mortgage bond market. Even worse, the firm's hedges were faulty. They included insurance by ACA Capital, which was heading into insolvency, and some marginal coverage by MBIA, a big-league bond insurer, which had agreed to cover the losses in event of default. But the market was now questioning MBIA's ability to meet its obligations.

O'Neal began to focus on the size of the problem; he didn't know, or claimed not to have known, the magnitude of the CDOs on the firm's books. "How can there be that there's so much of this shit?" he demanded.

Breit calmly described part of the problem was in Merrill's CDO pricing model, DV01, which didn't take into account the possibility of massive declines in housing prices; it had calculated that the downward deviation in CDO prices would be small and manageable and had thus failed to pick up what happened in the mortgage market after the Bear Stearns hedge fund fiasco got worse, particularly in August, when CDOs took an unprecedented nosedive.

In other words, as his risk managers were saying everything was fine, the CDOs held on Merrill's books were actually imploding.

O'Neal couldn't believe what he was hearing. Breit's analysis revealed several things, but especially just how interconnected Wall Street had become—and just how badly his risk managers had screwed up. The fortunes of Merrill, Citigroup, Lehman, and to a degree Morgan Stanley and even Goldman Sachs had been tied to a singular event: the crashing of the Bear Stearns hedge fund, which revealed the flaws in their business model, built on a wild bet on a market that was now crashing.

Since the hedge fund blowup, newspapers, CNBC, and other television outlets had been chronicling mismanagement at the top of Bear Stearns; overnight, it seemed, Bear had been transformed from the toughest, smartest firm in the bond market, led by the wily Jimmy Cayne, to Wall Street's buffoon. The reality, however, was that Merrill's management wasn't much better. That's when it dawned on O'Neal that Merrill was in trouble—big trouble. The firm's lifeblood, of course, like Bear's, was short-term funding—borrowing through repo arrangements, short-term debt, and bank loans. What sane creditor would lend Merrill money knowing it had so much illiquid debt on its balance sheet?

O'Neal kept his composure, but his stomach was churning. He asked if Merrill could now change its systems to immediately produce an accurate snapshot of the totality of the losses, which, if everything O'Neal was now hearing was right, were increasing by the day.

"I don't know," Breit replied calmly.

When Breit left his office, O'Neal began to assess responsibility. "I could have put an ax through his forehead," he would later say about Breit for his calm assessment of the firm's financial Armageddon and because O'Neal believed Breit was supposed to be managing risk. But in some ways, Breit was nothing more than the messenger of bad news. Others were more directly to blame, including O'Neal himself, for the overall strategy of the firm, and his good friend Ahmass Fakahany, the executive directly in charge of risk. He believed Osman Semerci hadn't been in the job long enough to stop the financial tsunami caused by the firm's dalliance with mortgage bonds, which had begun before he took over.

He couldn't say the same for Dow Kim, who, O'Neal believed, more than any other single person in the firm, had known, or should have known, about the buildup of risk that Merrill had undertaken and done more about it. It was around that time that O'Neal received a call from Kim, who by this point had left the firm and was launching his hedge fund. He wanted O'Neal to invest some Merrill money in it, as would be customary for any longtime employee who left on good terms. But based on what O'Neal was hearing and feeling, those terms had changed.

O'Neal had been openly accusing Kim of being responsible for the buildup in CDOs and other risky bonds. It's unclear whether O'Neal made this point directly to Kim during the meeting (Kim, through a spokesman, says he didn't), but one thing he did make clear was that Merrill wouldn't invest in his hedge fund, which was later disbanded. O'Neal's world was turned upside down. He had trusted the wrong people, who had trusted the wrong safeguards: the ratings agencies and their triple As on CDOs that had never been and could never be as safe as Treasury bonds; and the bond insurers, which could never have had enough capital to guarantee trillions of dollars of debt. The more O'Neal investigated and the more he spoke to people who understood the mortgage market, the more he realized the enormity of the problem: there were no buyers of this junk because everyone was trying to sell, having made the same stupid assumptions.

That's when it dawned on Stan O'Neal that Merrill Lynch was in deep trouble. And so was he.

There was once a time when the idea of selling the mighty Merrill Lynch to anyone, much less a bank, would have been considered sacrilege. This was after all, the firm of Charlie Merrill, the "thundering herd" with its vast network of brokers and a massive international presence that under Stan O'Neal had become among the most profitable on Wall Street.

Even for a person like O'Neal, who hated the old Merrill culture and wanted to change it any way he could, selling was a bitter pill, but that's exactly what he was prepared to do now. O'Neal saw the mortgage market deterioration through the prism of his LTCM experience and believed that with the holdings of so many illiquid CDOs, Merrill would need to team up with a bank to survive; and he knew just the bank to talk to.

Bank of America was run by Kenneth Lewis, who secretly lusted for Merrill despite having professed during the fall of 2007 that he wasn't interested in buying an investment bank. (His exact words: "I never say never, but I've had all the fun I can stand in investment banking at the moment.")

Lewis was an empire builder—he had worked at Bank of America his entire career, starting as a credit analyst in 1969 and working through a series of acquisitions as his mentor, CEO Hugh McColl, built the tiny North Carolina National Bank into NationsBank, which dominated banking in the Southeast and Texas.

But McColl wanted more. His signature deal was the purchase of the San Francisco–based Bank of America in 1994, creating what would become the largest bank in the country.

McColl may have changed the name of the combined entity for marketing reasons to reflect Bank of America's better name recognition, but he and his minions made no bones that they were running the show. And the show had to do with growing Bank of America into a powerhouse that rivaled Citigroup, JPMorgan, and the rest of the banks around the world. In effect, he wanted to move the world's banking capital from New York City to Charlotte, North Carolina.

As did the man who followed him; Lewis took over the company created by McColl in 2001 as his handpicked successor. Despite his attempt at investment banking, Lewis believed that Merrill's brokerage branches, which under O'Neal seemed like an afterthought, were really the power behind the firm. Providing advice was a business that might not be as sexy as running a carry trade with mortgage debt, but it was the future of the financial business, Lewis believed. All those brokers combined held $3 trillion of customers' assets. Those assets needed services that Bank of America could provide. Some of the customers were CEOs, meaning they needed investment banking advice. The possibilities of such a combination were endless.

According to people with knowledge of the talks, O'Neal didn't approach Lewis through normal channels; he never went to the entire Merrill board for the green light. Nor did he take with him investment bankers to start crunching the numbers on what a deal would take to complete. Instead, these people say, he met with Lewis privately and came away from the conversations with a preliminary bid of $90 a share. Merrill was trading at $50, and, knowing what he knew about the firm's balance sheet, he wanted to do the deal immediately.

But the board members didn't. They had been apprised of the firm's deteriorating condition, though the final numbers were not yet in. O'Neal told the board that he was worried about the firm's funding—how it could dry up as it nearly had during the LTCM crisis, which, he believed, was happening again now, only bigger and more broadly. Merrill did take steps to reduce its reliance on repos, but the markets were getting nasty—the credit crunch resumed during the fall of 2007, and borrowing became increasingly difficult even for Merrill, one of the largest securities firms on Wall Street. The secret was out: size doesn't necessarily equate with soundness. Merrill's balance sheet had grown from around $500 billion when O'Neal was named CEO to close to $1 trillion, a direct result of the firm's increased appetite for risk. Merrill was holding not just hard-to-price CDOs on its books but other positions that had soured during the crisis.

In other words, Merrill needed a well-capitalized partner. Bank of America was as good as any.

O'Neal sent out feelers about a possible deal to various board members, including, he says, to the lead board member, Alberto Cribiore (Cribiore says he has "no knowledge" of such a conversation), and the response came back the same: no way.

First, they weren't happy with O'Neal winging the talks by himself. In addition, such a deal would be messy—they would have to lay off thousands of employees. Merrill's brand name would be wiped off the face of the earth by a bank in Charlotte, North Carolina. Maybe most of all, Merrill would be selling out cheaply at $90 a share.

O'Neal was astounded by their arrogance. "None of these guys have skin in the game," he later remarked. What he meant by that was that the board as a whole didn't hold much Merrill stock, so their interests weren't aligned with the company's.

Except for a brief five-month hiatus, Robert McCann was a Merrill Lynch lifer, having begun there in the 1980s. In his twenty-plus years at the firm, he had been in key positions in sales and trading; he was part of the team that had worked on the bailout of LTCM. He was now the head of global wealth management, Merrill's thundering herd of brokers. Despite his status and lofty title, he had never made it into O'Neal's inner circle. It was a source of consternation for McCann, who had designs on higher office, possibly even the CEO spot.

In O'Neal's mind, however, McCann represented the past—the dark years when Merrill Lynch had eschewed risk, instead selling stocks through its brokers, while the firm was run by a club of Irish-American drunks: marginally smart men and a few women who had gotten their jobs through the patronage of "Mother Merrill" and who could barely contain their racism as an outsider like O'Neal rose through the ranks to become the first African American to run a major Wall Street firm.

McCann was hardly a racist; in fact, he had been wooed by O'Neal to return to Merrill in 2003, when O'Neal needed someone with a clue about how to run the brokerage business. Now O'Neal was wooing McCann to stay at Merrill even longer. That's because McCann wanted out; his internal rival, Greg Fleming, another Merrill lifer, had become co-president of the firm. Though McCann was on the executive committee, he knew how O'Neal worked; information and decision making were confined to a few trusted people: Fakahany, Kim when he was still at the firm, and Semerci. Now Fleming had found his way into the inner circle, and McCann figured O'Neal would have room for only one of the longtime Merrill executives.

So he was ready to quit, and he called O'Neal to alert him. O'Neal asked him to come up to his office. McCann had always marveled at O'Neal's office; aside from a few pieces of African art, it was modern and sleek, like the firm he was trying to build, a stark contrast to the wood-paneled, down-to-earth style of his predecessor Dave Komansky, who had imported an occasional painting from Japan and Asian artifacts to show that he actually did some traveling overseas.

Being in O'Neal's office could be a terrifying affair, and McCann, like most people at Merrill, viewed O'Neal with a mixture of fear and dread. He was a demanding boss; he rarely smiled during meetings, and he demanded quick answers. That's when he was around, which was something McCann had noticed O'Neal doing recently with greater frequency. It had caused McCann to quip, "What's up with Stan? He's usually gone for three months in the summer golfing; now he only took a three-week vacation."

McCann noticed that Fleming and Fakahany were around the office more as well.

O'Neal now disclosed why he was spending less time golfing as he tried to keep McCann from leaving. "Bob, we have big problems here, and we need people like you to stay." O'Neal said that losing a senior executive at this time would send a signal to the markets that Merrill wasn't just losing money, it was imploding. McCann was blown away; it was the most O'Neal had disclosed to him in years.

Despite the recent public comments about its minor exposure to risky mortgage debt, O'Neal said, just the opposite was true: the firm was sitting on $45 billion in CDOs that were tanking and taking the firm's balance sheet with it.

He added ominously, "I might not survive."

O'Neal knew by raising the possibility that he might be ousted, McCann would never leave. That's because several years earlier, McCann had once confided in O'Neal over dinner that his grand ambition was to run Merrill Lynch. "Somebody has to, so why not me?" McCann told O'Neal.

Still, McCann was in a state of shock at O'Neal's frank assessment of not just his job prospects but also the condition of the firm. McCann had been with Merrill for so long that he thought he had good intelligence, and the word he had received basically correlated with the public statements: the CDO exposure was contained at $5 billion. Now he discovered from the ultimate source that the exposure was many, many times larger.

O'Neal also said he was going to make some changes—Semerci and his lieutenant, Dale Lattanzio, were out. He laid the blame not on them but on the now-departed Dow Kim, who had been in charge of the bond department at the time. McCann didn't really buy the argument because he knew how much O'Neal loved Semerci, and in the brutal corporate setting that was Merrill Lynch, Dow Kim hadn't been calling all the shots when it came to risk.

But he wasn't about to argue with O'Neal, not about this. If O'Neal was on the ropes, that meant the entire management team responsible for the losses was in trouble as well. And he was nowhere near this crap. In short, McCann saw the possibility of running Merrill—if, of course, the firm survived.

"Okay, Stan, I'll stay," McCann responded. "But I'm not staying because of you. I'm staying because I've been with Merrill for so long."

And because, at least in his mind, he might soon be running the show.

Osman Semerci never saw it coming, according to people close to him. A week before, he had received an e-mail from O'Neal about what a good job he had been doing in risk management, which at least back in July had appeared to be true when he claimed to have reduced Merrill's exposure to risky CDOs dramatically by taking down Merrill's risk significantly during the year he had been in charge.

But now he was being questioned, mostly by Fakahany, about just how he had reduced the risk and how he had disclosed Merrill's exposure to the deteriorating mortgage market to the board as the hedges and insurance he had crafted seemed totally inadequate to handle the mounting problems. To say Fakahany had been getting nervous about Semerci's risk management skills is an understatement. Fakahany always thought Semerci was one of the most organized people at the firm; a vision of Semerci carrying a notebook with his to-do list was etched in Fakahany's memory, which is why he had supported him for his current job. But now he believed Semerci and his number two, Dale Lattanzio, were in over their heads—"overwhelmed," was the way he described it at the time—with the financial crisis that had spread from the mortgage desk he was in charge of and threatened the future of the entire firm.

Semerci's being overwhelmed posed many problems for Merrill and of course for Fakahany, who was ultimately in charge of risk, and Fakahany told O'Neal that he wanted Semerci and Lattanzio out. "Are you sure?" O'Neal asked, still believing that Semerci had inherited rather than created the problem. Fakahany said he was for many reasons, including the facts that he couldn't get what he believed was a straight answer from Semerci about the firm's mortgage holdings and that the trading desk was now in open revolt against management. Something had to be done fast.

Fakahany's decision to fire Semerci and Lattanzio was carried out in vintage Stan O'Neal style. They were called to Fakahany's office and given the news. Both were shocked, even more because they were asked to leave the building immediately.

Semerci might not have been the best risk manager over the past year, but he was in many ways a bit player in a bigger scheme to ramp up risk taking on the orders of senior management, to be more like Goldman in using the firm's balance sheet on trades and more like Citigroup in underwriting CDOs. In that sense, he had been in the wrong place at the wrong time, and now he was out of a job in a corporate shakeup that was designed to appease angry investors when they heard what was happening at Merrill.

McCann recalls receiving a call right after Semerci got the news. He had known Semerci as a very good salesman who had simply been elevated above his skill set by O'Neal and Fakahany. Still, he and Semerci had remained friends, even as O'Neal elevated the salesman to a job that was senior to the one McCann held, at least in O'Neal's bizarre pecking order.

Semerci thanked McCann for having been a friend for many years and then added, "It was a wild ride."

The statement took McCann by surprise. "Wild ride?" he thought to himself. The firm might not survive, and he's summing it all up as a wild ride? When the conversation ended, McCann called O'Neal and relayed the conversation to him, adding "That's the problem with Wall Street these days; too many kids think it's just a wild ride."

O'Neal didn't disagree.

The firm was prepared to announce to the markets that when it released its earnings report for the third quarter in late October, there would be losses—big losses. The company was holding many more billions of dollars' worth of risky bonds than it had originally believed, and the losses were far from manageable; it would have to write down $5.5 billion in losses due to its investments in mortgage-backed bonds.

It was a surprise announcement because Merrill had for so long publicly denied that it had anything but contained exposure to the market.

The announcement would create a public outcry over its past disclosures, and Merrill was preparing for an SEC inquiry into whether the disclosures made earlier had been fraudulent. In short, the outside world would now know that Merrill and O'Neal were in deep trouble.

Stan O'Neal was never a public presence at Merrill; he was an executive who delivered his messages through subordinates, mainly Fakahany, because he was uncomfortable in public settings and also because he didn't like to deal with mundane tasks such as communicating directly with the troops. The mortgage crisis forced him to come out of his shell, if for no other reason than self-preservation. He issued a statement explaining that "the outlook for fourth-quarter revenues remains difficult to predict" but that the firm continued "to see evidence of strong long-term growth trends in each of our global businesses."

Later, in a video message to employees, he admitted that the losses had been partially his fault. "I missed it," he said, according to the *Wall Street Journal*.

At Merrill, a collective "No shit!" could be heard from its headquarters in lower Manhattan, accompanied by anger from investors and employees, many of whom were paid in stock that was now getting crushed.

During the O'Neal years Merrill had never been a happy place; now it broke out into civil war. Brokers and traders who lined up with Bob McCann, the brokerage chief, attacked Greg Fleming, the head of investment banking and now president of the firm, lumping him in with the incompetent management. Fleming's loyalists in the investment banking department attacked Fakahany, who was allegedly in charge of risk management. One thing that everyone could agree on was O'Neal's culpability. He was now universally reviled by traders, investment bankers, and brokers. If there were ever a truce at Merrill, it would be to celebrate O'Neal's beheading.

The problem for O'Neal was that as he had run the place like a dictator for so long, the only friends he had were the people who were responsible for the mess. Mother Merrill, the culture he had vowed to eliminate, and former top executives who cherished this culture despised O'Neal and everything he stood for. A few years back, several former Merrill executives had even approached Jamie Dimon to ask JP Morgan Chase to make some type of hostile bid for Merrill, which they believed was heading for trouble under O'Neal's autocratic leadership. But getting rid of O'Neal then was the ultimate long shot given the firm's surging profits.

Now it wasn't. Former executives, such as a past brokerage chief, John "Launny" Steffens, the man whom O'Neal had replaced years before, began pressing colleagues still in the firm to lead a revolt against O'Neal. Steffens was aided by former CEO Dan Tully, the man most responsible for the Mother Merrill culture and whom O'Neal had once labeled a "racist," according to McCann. Tully also wasn't that fond of O'Neal. In addition to destroying the culture, Tully blamed O'Neal for having pushed out his son, who had had a promising career at the firm and been regarded as a savvy investor in his own right.

Tully might have been out of the firm for about a decade, but he still had contacts with numerous top producers, mostly in the brokerage network, which he used to foment revolt with Steffens. Their goal: to oust O'Neal in a proxy vote where they would present to shareholders the absurdity of his First Franklin purchase and what they believed was possible fraud by assuring the markets that the firm's risk was contained when it obviously wasn't. Their first choice was to have McCann replace O'Neal, but they would have settled for anything other than the status quo.

Chuck Prince might have been mocked by the rank and file at Citigroup as "One Buck Chuck" for the fact that the firm's stock price had barely moved higher than $1 since he had taken over amid the biggest boom in the market's history, but he had never been hated like Stan O'Neal. Employees who met Prince said that he was a gentleman whose biggest mistake had been taking a job above his skill set. No such thing would ever be said of O'Neal.

But he did have something in common with the Merrill CEO. Like O'Neal, Prince had thought his mortgage bond risks were contained. At least that was the word he had been receiving from his risk managers, led by Tom Maheras, whose risk models for weighing the prices of mortgage debt, particularly CDOs, according to an account in the *New York Times,* bore a striking resemblance to those at Merrill in that they hadn't taken into account the possibility of the rapid and massive decline in home values that was now in full swing across the nation.

Merrill's mea culpa for its recently disclosed losses now forced Citigroup to reexamine its earlier assumptions. According to the *Times,* Prince dispatched a team of risk managers separate from the Maheras group. It shouldn't have taken much for Prince to figure out that by now the fixed-income department at Citigroup was a mess due mainly to the fact that Maheras had a built-in advantage when it came to risk taking—his chief risk manager, Dave Bushnell, was a close friend, and his other close friend, Randy Barker, the co-head of all of fixed income, had consistently leaned on Bushnell to approve increasingly complex trades.

The review came back, and it had disaster written all over it. The firm's total holdings of subprime-related bonds were massive, nearing $80 billion, and that wasn't counting the off-balance-sheet exposure in the form of SIVs. For now, Citigroup would have to write down losses of $5.9 billion, and its profits would decline more than 50 percent when it announced earnings in late October.

People who know Prince said the size of the losses was a shock; other heads needed to roll before his did. He restructured the firm's investment bank, which meant that the long and controversial reign of Tommy Maheras and his team was over.

The press accounts were pretty glowing, given Maheras's track record. Several newspapers said that traders cheered when he left the floor for the last time. He was, after all, the last and, according to some, best of the breed of traders that had been spawned by Salomon Brothers, and over the years he had helped remake the stodgy, client-driven firm Sandy Weill had set out to create into a risk-taking dynamo.

That Maheras still had many friends inside Citigroup, and even outside in the form of Weill himself, is a testament to his pleasant manners and people skills. Even among people who sparred with him over his risk-taking policy, Maheras was considered a great guy to have a drink with. Others would say it was hardly Maheras's fault that Citigroup, under Prince and with Robert Rubin's prodding, had ramped up its bond and risk-taking operations to pay for the costly implementation of its strategy to create the world's first one-stop-shopping financial services supermarket.

At some point risk taking has a price, and that price would have been paid, if not by Maheras, then by someone else.

But that's just one side of the story. Many others, including past and present Citigroup executives who had watched their retirement savings of Citigroup stock get squeezed, cheered Maheras's departure, because to them Maheras represented the worst of the trading culture. Under his reign, they believed, Citigroup had lost its focus—it had cared more about making market bets than it had about serving clients. They point to what they describe as Maheras's obsession with trading in the Fixed Income Management Account and his own personal account.

More than anything, Maheras's detractors would say, his penchant for risk taking and leverage had spread through the firm like a cancer; he was no different from John Meriwether, his old Salomon Brothers colleague: a gambler, one with a world-class college degree but a gambler nonetheless. Only Maheras had been gambling with bigger stakes: the balance sheet of one of the world's largest banks.

It's unclear if the policy makers who advocated the repeal of Glass-Steagall ever considered the ramifications of mixing the likes of Maheras with the savings and checking accounts of Middle America, but in time they would.

Prince, meanwhile, was scrambling to remake Citigroup's management ranks and save his job. He named Vikram Pandit, a former longtime Morgan Stanley executive, who had started a hedge fund named Old Lane Partners that Citigroup had purchased for $800 million, head of the firm's investment bank, thus his number two. It didn't matter that the performance of the fund would turn out to be abysmal (it would eventually be shut down); Pandit had been considered a top talent during his years at Morgan Stanley and a candidate to run it some day. The thinking inside Citigroup, including on the board, where Robert Rubin pushed for the deal, was that the bank was paying $800 million not just for the hedge fund but for Pandit, and he became known derisively inside the firm as the "$800 million man."

Chuck Prince was being derided too for not seeing that Citigroup should have been broken up into more manageable pieces and sold because of the firm's stagnant stock price and mostly because of the continued disclosures about its exposure to rapidly deteriorating mortgage debt.

In addition to its CDO holdings, Citigroup also held around $60 billion in SIVs; when Treasury Secretary Hank Paulson saw the full weight of the exposure, he proposed a super-SIV bailout; a plan in which all the big banks would pool their capital to buy the SIVs' risky mortgage assets in an attempt to prevent massive write-downs of losses in the banking industry. There was just one problem: such bailouts work only when everyone is sharing the pain equally, as in the case of LTCM, when all the big firms had had a stake in preventing the meltdown of John Meriwether's hedge fund because of their nearly equal exposure to his soured trades.

In this case, the bailout would have helped one bank in particular, Citigroup. Paulson and the New York Fed president, Tim Geithner, believed Citigroup was worth saving, given its size—a balance sheet of trades and investments worth close to $3 trillion. They worried about the "systemic risk" of Citigroup's faltering. But Jamie Dimon, the CEO of JPMorgan Chase, wasn't about to throw Chuck Prince a lifeline. Within a few weeks, the super-SIV bailout fizzled, as did Chuck Prince.

It was only fitting that Prince and Stan O'Neal suffered their indignities at the same time, as both had charted similar courses through the mortgage meltdown, first assuring investors that their firms had more than adequate risk controls and were hedged against large losses, then slowly admitting to increasingly larger losses as they admitted to inadequate risk controls and to firing key lieutenants in charge of those controls, before finally finding themselves on the chopping block.

By now Prince, ever the lawyer, was fighting for survival through negotiations with board members and the firm's largest shareholder, Saudi Prince Alwaleed bin Talal, whom he traveled to meet in Saudi Arabia. Jamie Dimon had always thought it was a "joke" how Citigroup senior management, beginning with Weill, had bestowed so much prominence on "The Prince." Alwaleed was, after all, just one shareholder, yet he had nearly final say on major management appointments and the firm's continued embrace of the financial supermarket concept, which clearly wasn't working.

Through the last several years of chaos and controversy, the prince had supported Chuck Prince as the right man to lead Citigroup and carry on Weill's vision, but he was now having second thoughts. Weill himself had traveled to see him to discuss the company and its future, and Weill had been making no secret of how he thought Chuck Prince was running the company into the ground, even though the firm had begun its descent on Weill's watch. (He had also made no secret of his desire to return to the company, which the board would never accept.)

In any event, without the Saudi royal on his side, Chuck Prince's days at Citigroup were numbered.

O'Neal, meanwhile, was looking for some grand gesture to save his job, and that came in the form of a Hail Mary pass involving a merger with Wachovia, Bank of America's crosstown rival.

According to people close to O'Neal at the time, his desperation began to grow when he discovered that his initial write-down announcement was far lower than the reality; in early October, Merrill had predicted losses from soured mortgage debt at $5 billion. His risk managers now said that the losses would be closer to $8 billion.

O'Neal alerted the board to the deteriorating situation, according to people with knowledge of the matter. He even, at one point, offered to resign, which the board rejected, those people say. What he didn't tell the board was what his next move would be. According to people close to O'Neal at the time, he secretly contacted the CEO of Wachovia, G. Kennedy "Ken" Thompson, about a possible merger. Thompson said he was interested, and O'Neal believed he had found a way to save the firm, and possibly himself.

O'Neal was thinking like the CFO he once had been; funding and liquidity were the keys to Merrill's survival. Since every Wall Street firm had borrowed as much as thirty times more than it had in available capital, the key was making sure their funding sources remained open. But there is no assurance when the only thing a firm can offer a creditor is confidence. With all the write-downs and losses looming on the horizon, O'Neal didn't see any reason why a creditor would lend Merrill money. Teaming up with a bank was the only solution; banks can borrow, but, more than that, they have deposits, so in a pinch they can finance themselves with deposits. They also have the ability, through something called the Fed discount window, to borrow from the government when worse comes to worst.

It was such a no-brainer argument that O'Neal once again didn't tell the board he had opened negotiations on a potential merger. He did, however, tell Greg Fleming, whose expertise was in investment banking and who had worked closely with the people at Wachovia, including Thompson, on deals. Fleming launched the initial stages of what would be a merger of one of the biggest banks in the country with the nation's largest brokerage firm.

O'Neal was prepared to alert the board about his private talks with Wachovia at its next scheduled meeting in late October, just days before he was going to tell the markets and release third-quarter results showing that the bank's mortgage market losses had been bigger than anticipated; in fact, they were of historic proportions.

He kept the circle of people who knew about the Wachovia deal tight; Fleming and his team and no one else. Pete Kelly, who had told senior Merrill executives that he nearly ripped O'Neal's head off when O'Neal had fired Jeff Kronthal a year earlier, had buried the hatchet and was back at work. O'Neal didn't bring him into the deal talks, but Kelly was at the board meeting to present a plan that he believed would enable Merrill to recover from its mortgage market losses and possibly remain independent.

Kelly went to the meeting thinking that the plan that O'Neal would present was his: how to work out of the balance sheet mess. A presentation to the board laid out the problem facing the firm in great detail, and it was in stark contrast to the rosy scenario presented in July. According to an internal Merrill Lynch document, the board was given the same grim assessment O'Neal had received from Breit, including a chronology of the key events involving Merrill's CDO exposure going back more than a year to the summer of 2006, as a backdrop to why the firm was facing its greatest crisis since its founding many years earlier.

When Semerci and Lattanzio replaced the Kronthal team, Merrill's "CDO inventory continue[d] to ramp up through year end," the document stated, from $7.2 billion, when Semerci had taken over, to $17.3 billion in about six months. Merrill had held more than $30 billion in CDOs and the securities needed to create those CDOs, just as the market began to deteriorate, the timeline continued. Even so, in the first part of the new year, Merrill had continued to underwrite CDOs, "over $10 billion in deals." The document described a firm in denial that its bet on CDOs—underwriting them and then holding them to earn the carried interest—was destructive and wrong.

Bizarrely, the document pointed out, the firm's management had originally concluded that there was still a market for CDOs, and there was an "expectation" that the market would come back to its former vibrant self. But if Merrill wanted to sell its buildup in bonds, it would have to take a loss. As a result, "significant senior and mezzanine tranches are taken into inventory," thus explaining the various pieces of the CDOs now held on the firm's books and accumulating losses, "senior" being the highest-quality part of the CDO and "mezzanine" among the lowest.

In reality, there was no differentiation when it came to CDOs; the market in late 2007 had labeled them all toxic waste. Among the key events leading to Merrill's current near meltdown, the document cited the departure of Dow Kim, the former head of capital markets, whom O'Neal personally held responsible for the mess, and of course the implosion of the Bear Stearns hedge funds. It was at that point that "liquidity dissipate[d]" and the market became "impaired."

"Risk management" deserved most of the blame for the disaster hitting Merrill, according to the presentation. It didn't blame Osman Semerci and Stan Lattanzio by name, but it did suggest that both had seriously misread the deteriorating condition of the market and through poor hedging strategies had made a bad situation considerably worse.

In the coming days, when Merrill made these disclosures to investors and the markets, the finger-pointing would be intense, and Semerci and Lattanzio would get much of the blame—as they deserved. But the problem for Merrill was bigger than two men who had traded bonds. It stemmed in part from a board of directors that had largely rubber-stamped O'Neal's mandate to take more risk, his decisions to remove those who were risk averse like Jeff Kronthal, and, maybe most of all, the perverse compensation incentives that had led to the massive risk taking. At bottom, compensation at Merrill, and for that matter the rest of Wall Street, was driven by how much money a trader could make in a day or week, rather than by how much he might lose in the future.

Compensation wasn't something either O'Neal or Kelly addressed in their presentation to the board, but it should have been. At Merrill those perverse incentives had been magnified by O'Neal's management structure, with Fakahany running risk management from his perch as COO. He was O'Neal's eyes and ears at the firm, and O'Neal paid him well for it. In 2006, he had earned $30 million. But he was also in charge of risk at a time when his incentive wasn't simply to manage risk but to expand it so that he and O'Neal (who had been paid $50 million in 2006) could make more money.

As the presentation pointed out, the firm had taken risk to new and more dangerous levels; risk management had "fail(ed) to note the growth of CDO inventory" and overrelied on the DV01 metric, "which is an insufficient metric to evaluate the true nature of the risk." Maybe the most appalling aspect of the whole sordid mess was that the "growth of position and aggregate exposure [was] not highlighted to senior management."

As the presentation to the board continued, Kelly noticed something strange about O'Neal's body language. It was as if he were just going through the motions because he had something else of greater importance to discuss.

Kelly laid out how the firm could recover—it had already taken steps to minimize the impact on the balance sheet and set up funding sources just in case creditors got antsy and cut off short-term credit. He was then asked to leave the room, after which O'Neal unveiled the Wachovia negotiations. The reaction by the board members was anything but positive. Their concerns were similar to those expressed earlier when O'Neal had unveiled the possibility of a merger with Bank of America, only with heightened urgency because now the stakes were bigger: Merrill was getting ready to announce an $8 billion write-down.

O'Neal's strength with the board had always been that he seemed firmly in control of any given situation; in flirting with Wachovia—a bank with billions of dollars in deposits but also billions of dollars in exposure to decaying mortgages—he now appeared desperate and flailing. Most of all, boards hate being surprised, and this one just blew them away.

The board rejected the merger out of hand, leaving open the question of when it would reject O'Neal—at least for now. As bad as things were, board members told O'Neal, they were in agreement that Merrill would recover. Board member Alberto Cribiore thought Merrill would now be selling out cheaply because the environment was so shitty. The general consensus was that the Fed would keep lowering interest rates, the bond market would bounce back, and all those CDOs on the firm's books would by some miracle start showing profits.

O'Neal was astonished. After all, he had access to the same data the board had, and, based on what he knew about the CDO market, the chances of prices rising anytime soon seemed remote at best.

When O'Neal announced the third-quarter results the following Wednesday, he replaced Ralph Cioffi as the new face of the mortgage meltdown. Rating agencies that had once praised O'Neal's management style now criticized his risk management. Merrill posted a record $2.25 billion loss after factoring in $8.4 billion in write-downs.

O'Neal's conference call with analysts was tense; according to the *Wall Street Journal*, he conceded that "some mistakes were made," though by whom he didn't say. The paper concluded that O'Neal's job didn't appear to be in jeopardy, as he blamed the losses on an "unprecedented liquidity squeeze." Investors didn't exactly celebrate the prospect of more O'Neal; the firm's share price fell nearly 6 percent, and its bond ratings were downgraded.

The management style O'Neal had brought to Merrill Lynch had backfired; with losses mounting, he had no one except a small contingent of friends inside Merrill, now reduced mainly to Ahmass Fakahany, to turn to. The Mother Merrill network he so loathed was now exacting its revenge as well, as former executives worked behind the scenes to spread the word about O'Neal's disastrous run as CEO.

O'Neal loved objective analysis in the form of raw data; he had used that raw data to oust Komansky as he proved empirically that Merrill was falling behind the rest of Wall Street by any measure of profitability. Now the numbers were working against him. His grand plan to make Merrill competitive with Goldman Sachs in trading, and with other firms in the high-margin business involving mortgage-backed securities, was, according to the data, a huge flop.

In 2006, Merrill had more than tripled its CDO issuance to $44 billion from when it had first launched its grand push into risk. That year, Merrill had earned $700 million in fees and O'Neal himself had earned one of the largest bonuses on Wall Street, around $50 million. But now Merrill was bleeding billions of dollars and its franchise was in tatters; simple math showed that the risks involved had never been worth the short-term rewards, and neither O'Neal nor his chief risk managers had to return a dime of their bonus money based on results that had suddenly become illusory.

Despite the losses, despite O'Neal's own assurances about limited losses that turned out to be wrong, and, maybe most of all, despite the hatred now directed at O'Neal from just about every quarter, he still might have survived as CEO except for what happened next: a leak to the *New York Times* about the Wachovia deal and the board's reaction to it.

O'Neal first heard about the pending story as he was sitting in the Upper East Side restaurant Sistina having dinner with Larry Fink, the CEO of BlackRock and a person widely considered to be a possible replacement for O'Neal if he should be ousted. He received a BlackBerry message that the *Times* had the story; he later discovered that Fleming had been called as well and the story was going to run in the morning papers.

"I got something important to attend to," he told Fink, who O'Neal recalls had responded ominously, "Trouble with the press?" O'Neal didn't answer, and, knowing how secretive O'Neal was, Fink didn't inquire further. O'Neal quickly left the table to plot damage control, though that last comment stuck in his mind all night.

Media leaks are troubling for corporations because management loses control of the situation; it becomes reactive rather than proactive. Now O'Neal and the board were reacting to charges that O'Neal had at the very least violated normal board protocol by freelancing a huge possible merger. At most, he now made Merrill look desperate to paper over its massive exposure to risky mortgage debt. O'Neal had heard that a powerful contingent of directors was already plotting his forced exit, so he beat them to the punch, alerting the board that day of his intention to step down. They accepted.

And with that the reign of "Stan bin Laden" was over. It would take a few days before it was made official and the board announced that it had begun a search for a new CEO. Fink, given BlackRock's relationship with Merrill, was already on the short list, and he went so far as to tell friends that the board was duty-bound to offer the job to him first. But board members had other ideas and opened the search up to others, including most notably John Thain, the former co-president of Goldman Sachs and the current CEO of the New York Stock Exchange.

Alberto Cribiore, one of Merrill's lead board members, believed Thain had done wonders with the NYSE, converting it into a public company, modernizing its trading systems, and instilling a Goldman-like culture there. He believed Merrill would do well to be more like Goldman, and he had a point. Goldman was the only firm not caught up in the subprime crisis so far because it had done just the opposite of Merrill and most of the rest of the Street; while Merrill and Bear and Lehman and Citigroup had doubled down and carried massive amounts of real estate–related debt on its books (Merrill even bought a mortgage originator when the market was about to tank), Goldman did the opposite, and, like the hedge funds run by John Paulson and Stan Druckenmiller, it shorted the mortgage bond market.

Unlike the hedge funds, Goldman's short wasn't a market call where it made a large concentrated bet, but rather an ass-covering hedge, approved by CFO David Viniar in late 2006 to soften the losses that would come from its own exposure to the sector.

Goldman's short was now huge news in the market, making its way into a long *Wall Street Journal* story that profiled the traders who had executed the short and the role that Viniar, now considered one of the best

CFOs on Wall Street, had played in the move. More than anything else, the story underscored Goldman's prowess in the business of risk or, as one scribe at the *New York Times* described it, how the firm had reduced risk taking to an "art form."

Goldman, of course, was much more than a trading house. The firm was powerful beyond its modest size (25,000 employees compared to more than 300,000 at Citigroup) largely because Goldman Sachs executives are littered throughout the upper reaches of government, including the Treasury secretaries in both the Clinton and Bush administrations. With so much political power concentrated in one firm, the rest of the Street had always believed Goldman had the system rigged: it could sway policy to favor its business model of risk taking. There is, of course, more than a little truth to the notion that economic policy over the past three decades centered on ways the banks and the big firms can increase risk by reducing regulations, and Goldman has pushed for the lessening of those rules to enhance its bottom line.

But Goldman wasn't the only beneficiary. Its former chairman and later Treasury secretary, Robert Rubin, pushed for the demise of Glass-Steagall before he could join the firm that most benefited from the combination of commercial and investment banking, Citigroup. Goldman now had to compete with a firm that because of its sheer size could offer clients more services than any other in the business.

What Citigroup, or for that matter any other firm on the Street, couldn't offer was Goldman's degree of expertise. Larry Fink refers to it as "intellectual capital," and as much as Fink never trusted Goldman (Fink, like many money managers, believes Goldman often takes advantage of its clients' trading positions), he respected the intellectual capital its bankers and traders brought to the financial business. And it was that intellectual capital that in late 2006 and early 2007 allowed the senior executives at Goldman to appreciate the notion that risk can lead to losses as well as gains and see the possibility of the coming storm when the rest of Wall Street saw nothing more than good times ahead.

One firm that obviously didn't appreciate risk as art form was Merrill Lynch, which at the behest of O'Neal had spent the past four years trying to copy Goldman in every way and failing in just about every way to emulate the way it managed risk. That didn't stop former Goldman Sachs CEO and current Treasury Secretary Hank Paulson from calling O'Neal not long after the news of his pending resignation hit the wires. Paulson declined to be interviewed for this book, but people who were dealing with him during this time say his mood toward the end of the year was still pretty upbeat, as if the worst of the crisis had passed; despite Merrill's losses and strong evidence that the credit crisis was getting worse, not better.

Paulson's call to O'Neal was somewhat upbeat as well; Paulson, according to his recollection, told him that he should be proud of his achievements at Merrill—his moves to change the firm's culture, modernize it, and make it more competitive. Paulson didn't see the O'Neal years as a seminal event in the firm's demise, as others had.

That made O'Neal feel better, but remarks from others were not so kind.

He had heard that the news of his imminent dismissal had been warmly greeted by the rank and file and that many of the people he considered the "old guard" had been preparing to move forward with a proxy fight when the news hit the wires; former CEOs such as Dave Komansky and Dan Tully made no secret that they despised O'Neal for how he had changed the culture of the firm. Both were former brokers who didn't know the first thing about pricing a CDO squared. But they could read the newspapers, and Merrill was a patient on life support. Its culture had been destroyed, and so had its balance sheet. Saddled with billions of dollars in debt, the firm needed immediate attention—it would have to be downsized, remodeled, and remade to survive.

"The whole thing makes me sick," Komansky said at the time.

It was around this time that brokerage chief Bob McCann recalls receiving a message on his cell phone from O'Neal. McCann quickly returned the call. O'Neal dispensed with the pleasantries and deadpanned, "Bob, soon it will be announced that I'm stepping down" as CEO of Merrill Lynch.

Then came a long pause, as if O'Neal were expecting some kind words from McCann about how great their time together had been, the kind of stuff usual among Wall Street executives, who are accustomed to saying nice things to people they hate. But McCann had had enough of O'Neal and simply said, "Thanks for the call" before he hung up.

McCann went back upstairs to his bedroom, where his wife was reading the morning papers, which contained numerous stories about Merrill's worsening financial health and predictions of O'Neal's demise.

"Who was that?" she asked.

"It was Stan," he said. "He's resigning from Merrill."

"Good," she said as she continued reading. "He ruined our company."
McCann couldn't have agreed more.

The thundering herd agreed as well. In the brokerage sales force, the place where the old Merrill culture maintained its presence most prominently, brokers could be seen cheering and high-fiving in the hallways. Sparking even more euphoria was the likelihood of the removal of the rest of the O'Neal team. The chief risk officer, Ahmass Fakahany, would leave the firm after a transition, sometime in the beginning of 2008. Jeff Edwards would step down as CFO and go back to a job he was actually good at: investment banking.

Having been so much out of the O'Neal inner circle, Bob McCann stayed on as head of the brokerage division. He was, of course, a throwback to the Komansky-Tully era, having been with the company in various capacities for twenty years, which now wasn't such a bad thing. Even better, he had a legion of supporters, both past and present Mother Merrill types, who were actively floating his name as the next CEO of Merrill Lynch.

Like McCann, Greg Fleming had designs on the top job at Merrill, but he couldn't succeed O'Neal because he was, after all, tied to the Wachovia deal attempt. He remained at the firm, praying that the job wouldn't go to McCann, whom he loathed (the feeling was mutual).

Fleming knew he might not be loved by the brokers—McCann's sales force blamed him for being part of the O'Neal team, even though he had little to do with the firm's mortgage market business—but he was needed at a firm that might soon need a partner. Merrill wasn't out of the woods, not by a long shot.

Merrill now had to rebuild its capital base, which had been shrunk by the losses stemming from its bet on mortgage bonds, which the firm still had on its balance sheet in the tens of billions of dollars. Given the direction of the markets, Merrill, like Citigroup, Bear Stearns, and the rest of Wall Street, couldn't just grow back to health. Wall Street would have to sell out—to foreign investors by selling pieces of itself or to commercial banks that had bigger balance sheets (under accounting rules, banks also had more leeway than brokers in marking down the prices of mortgages).

Fleming had the best contacts in the investment banking business; he knew every CEO at just about every major bank in the business. With more write-offs being predicted for Merrill and the rest of Wall Street, Fleming's skills would be in more demand than ever before.

O'Neal, meanwhile, didn't want to go down without a fight. For days he tried to figure out who had leaked the story to the *New York Times*, which he believed had sealed his fate. In the end, he came to believe that most likely the leak had to be from Fink, who had the most to gain from his departure. O'Neal believed Fink had by now almost anointed himself the new CEO in the media, and because he had contacts at the *Times* it wouldn't take much to have planted the story.

There was something else that made O'Neal point to Fink. During their dinner, O'Neal also recalled that Fink had referred to all the money O'Neal had made as CEO as if to remind O'Neal he would be financially secure during difficult times. Then O'Neal recalled the quip he says Fink made just before O'Neal left the restaurant, when Fink asked him if he was having "trouble with the press."

When confronted with O'Neal's version of their last supper, Fink says he made no such remark. "I had dinner with O'Neal that night because no one else would," Fink says. "I felt sorry for him."

Maybe so, but over the days and weeks ahead, O'Neal would develop an enemies list of all sorts of people he says were responsible for his ultimate demise. It would include Fink, Fleming who also might have leaked the story (Fleming was the person who messaged O'Neal while he was eating dinner with Fink), McCann who could have gone to the *Times* himself because he so desperately wanted to be CEO, and then maybe someone on the board who was leading the charge to have him removed. That would have been any number of board members who had turned on him, including Armando Codina, the board member he believed had become most vocal for his ouster.

"Fuck you" is what the old Stan O'Neal would have liked to say to Codina and all the others, but he didn't. The fight was finally out of him.

huck Prince didn't leave after a fight either. During the late summer and early fall of 2007, the talk on the Street wasn't if Chuck Prince would resign, it was when. The losses that had seemed to take the CEO by surprise were one of the final nails in the coffin; he spent the last year at Citigroup on a much-touted cost-cutting drive, but as Meredith Whitney, now the most influential analyst on Wall Street because of her early warnings about Citi, pointed out, the company's vast and far-flung operations were still operating with no coherency, and costs were still out of control.

Prince seemed resigned to resigning as well. People who know Prince say that during his last tumultuous days at the firm, he seemed almost serene, as if a great weight had been removed from his chest.

The latest bit of bad news had come in the form of bigger losses from the firm's mortgage holdings; they had now grown to $11 billion, and by most accounts more would be coming down the road. Citigroup's CFO, Gary Crittenden, broke the news to Prince that the mortgage-related losses were now bigger than first expected—bigger than anyone had ever imagined. That the firm's risk management was bungled was an understatement; it appeared that as the maelstrom had picked up steam, Prince had still been getting reports that everything was fine.

The stunning new disclosures underscored just how little Prince and Citigroup's board knew about or understood the wild gamble of the past five years executed by Tom Maheras's team. The moment Prince heard about the new numbers, he was devastated. The next moment, he concluded he would have to resign.

In an interview with *Fortune*, Robert Rubin admitted that, like Prince, he had no understanding of a central tenet of Citigroup's exposure to the CDO market; it was called a "liquidity put." In its competition with Merrill to win honors as the largest underwriter of CDOs, Citigroup had added an incentive to entice investors to buy their CDO deals.

Citigroup told investors they could sell, or "put," their CDOs back to Citi at full price, 100 cents on the dollar, if the value of these bonds declined. When the markets were running fine, putting back bonds was rare, almost nonexistent. But now that the CDO had imploded, the puts came in droves, as much as $25 billion of them, adding to the firm's massive exposure to risky debt.

It's unclear at what point Prince found out about the put feature (it was a footnote in the company's financial statements), but its existence, which was first disclosed to investors in early November 2007, underscored that the dysfunction in risk management at Citigroup was every bit as severe as it was at Bear Stearns and Merrill. Each of the boards of these firms appeared shocked at the magnitude of the bets. Were they incompetent? According to traders who had given the board briefings, the answer is yes.

"These guys wouldn't know a CDO from a PowerBar," one said, "and they didn't want to learn."

Rubin asserts almost proudly that he had "no operating responsibility" at Citigroup, meaning he had no role in the management of risk even as a board member, even after having earned millions of dollars a year for his role as chairman of the executive committee, whatever that means. He says he relied on the expertise of the bond department, which was "supposed to be the best in the business," a statement that provides a bit more insight into how Citigroup went so wrong. One of the problems that board members like Rubin faced was that their briefings on risk were usually given by the risk takers themselves. They were being assured that gambling was minimal and hedged by the gamblers.

Prince may have been an ineffective CEO, but he worked hard, maybe harder than any other Wall Street CEO, staying up long hours and constantly checking his e-mails. According to his critics at Citigroup and those on the Street, One Buck Chuck would never be found on the golf course or at a bridge tournament, particularly when business was on the line. Given the size and scope of the problem he had inherited and the business model the Citigroup board wanted him to maintain, could anyone else have done better?

Maybe that's why Robert Rubin, who had advised Prince for so long, went public to make sure the press knew that Chuck Prince had left on his own accord; he hadn't been fired or pushed out, he had chosen to leave. Like most messages delivered during this bizarre era, it was mostly spin. On the Citigroup board, the general feeling was that Prince had simply made the job of his removal much easier by resigning.

For both Merrill and Citigroup the departure of their respective and much-maligned CEOs was a chance for a new beginning, except for the fact that the old management was still very much in charge. The boards of directors at both firms remained intact, still calling the shots and now attempting to repair the damage they were responsible for inflicting on shareholders as much as their departing CEOs.

Most disconcerting for many investors was Citigroup's elevation of Robert Rubin to interim chairman (the interim CEO job went to Sir Win Bischoff, the head of Citi's European operations). Rubin held considerable clout on the board and with management, but his standing with investors was slipping. Mike Mayo, the prominent securities analyst, was one of the first to point out that Rubin had earned around $100 million in compensation during his time with the company without any real job other than to give advice (which turned out to be wrong), from risk taking to his support of Weill's failed universal banking model. William Smith, an investor who led a yearlong campaign to dump Prince and spin off the various Citigroup units he and many other investors believed could never work as a coherent whole, pointed out in stinging commentaries on CNBC that Prince had never really downsized the firm, even after announcing cuts to its workforce of more than 370,000. He had called Rubin Chuck Prince's "enabler" and compared the former Treasury secretary to a benevolent but misguided prime minister of a socialist country who had worked behind the scenes in Washington to pass the legislation that had made Citigroup legal, then refused to own up to his mistake, earning $15 million a year for keeping the structure together, when all the evidence pointed to the model's failure.

"This guy couldn't care less about shareholders," Smith said. "He's looking to maintain the welfare state that is Citigroup."

Early one night in the fall of 2007, Morgan Stanley CEO John Mack sat at San Pietro laughing it up over dinner with his bond chief, Zoe Cruz. And why shouldn't they be laughing? Mack had looked like a genius after he had put Cruz, an aggressive commodities trader, in charge of the bond group over the objections of senior executives at the firm and she had delivered amazing results. He had also looked like a genius when he had changed the firm's business model, making the firm more competitive and aggressive in trading, more like its main rival, Goldman Sachs.

That is, until the latter part of 2007, just a few days after their dinner, when the bill for the risk Cruz had been taking came due: trading in mortgage-backed debt had cost the firm $7.9 billion; the firm was now holding billions of dollars' worth of now-decimated mortgage-related debt.

The net result: Mack ousted Cruz, and Morgan Stanley was set to report its largest quarterly loss since it had gone public.

Mack was facing the biggest crisis of his two-year tenure as CEO. The firm's balance sheet was impaired in ways that seemed to defy Morgan's image as among the best risk managers on Wall Street. He had two choices. One was to merge with another company, which would be messy, with layoffs, not to mention the fact that he might lose his job as CEO. The other option was more palatable: he could sell a chunk of the firm to one of a number of foreign countries looking to make investments in U.S. financial firms. In the end he sold a $5 billion chunk of Morgan Stanley to China Investment Corporation to plug the gaping financial hole.

The Wall Street financial crisis was about to reach unprecedented heights. Wall Street wasn't reeling just from losses stemming from its bets on mortgage bonds; the fact was that it wasn't making much money anywhere. Doing mergers and acquisitions advisory work, underwriting stocks and bonds, selling IPOs, and selling stocks to small investors were not just low-paying businesses, they were drying up and showed little signs of reviving anytime in the near future.

So, faced with write-downs on mortgage debt that would continue into the new year and a decaying business model, America's big financial firms scrambled for capital and, in doing so, sold out, literally, to capital-rich foreign countries that operated massive investment funds and wanted nothing as much as to own a piece of something as uniquely American as the Wall Street financial system, particularly if they could buy it cheaply.

For the first time ever, Wall Street needed to sell out to foreign interests in order to survive.

It didn't matter that the big firms, such as Morgan Stanley, which traced its roots to the old House of Morgan, were selling themselves to countries whose national interests, to put it mildly, don't always reflect the goals and principles of America. Wall Street needed money and didn't really care where the money came from. Merrill received billions of dollars in new capital from Singapore's Temasek Holdings and then billions more from funds in the Middle East and Asia. Citigroup followed suit; after it sold a 4.9 percent stake in the bank to the Abu Dhabi Investment Authority (ADIA), the prospect of further write-downs caused it to hit the sovereign wealth cash machine once again in December as the bank received another $7.5 billion from the fund.

By selling out to foreign interests, Wall Street was selling out its shareholders by diluting the value of their holdings through the issuance of new shares. But bigger problems loomed. Citigroup, Bank of America, and other firms had consistently rebuffed analysts such as Mike Mayo when he had asked for an accounting of their mortgage bond holdings. Now Mayo no longer had to ask for these positions. New disclosure rules now forced the big firms to disclose their assets and break them up into "Level 1," "Level 2," and "Level 3," the last being securities that were hard to value and for which there was no active market. This was the category in which the mortgage-backed debt was being classified, and now it was there for the entire world to see. To be sure, the disclosures were still difficult to wade through but not so difficult for short sellers to understand and begin making their bets. Short sellers make money by betting that stocks will fall. And the most aggressive among them—traders such as Phil Falcone, John Paulson, and Stan Druckenmiller—had made billions betting against the housing industry and the firms that had benefited from the housing bust.

The new accounting rules only confirmed their suspicions about housing and the Wall Street firms that had fed off its inflation for so long. As the disclosures pointed out, every Wall Street firm had loaded up on Level 3 assets, even Goldman Sachs, which had shorted the subprime market in 2007, a bet that had more than made up for the losses on other securities it had held—at least for now. The bet had been so successful at Goldman that its CEO, Lloyd Blankfein, was one of the few Wall Street CEOs to be paid a bonus—$67 million—but it obscured the time bomb that existed at Goldman as well: Merrill had off-loaded some $13 billion of CDOs to Goldman that were insured by AIG.

In effect, Goldman's future was now tied to AIG's. Its former CEO Hank Greenberg had been complaining for months about the insurer's poor earnings, and analysts now began questioning its exposure to subprime debt through the credit default swaps it had issued. AIG stopped issuing swaps on the most toxic mortgage debt in early 2006, but many analysts feared the damage had already been done.

In October 2007 Joseph Cassano, the head of AIG's financial products group, assured investors that the company was fine, its exposure was minimal and manageable, and the market would have to sink much further from its already beaten-down level for there to be any discernable impact. The people at Goldman Sachs were hoping and praying he was right.

The short sellers were now betting heavily that he was wrong—and their winnings were starting to pile up. Stan Druckenmiller, for one, was ecstatic. He had begun shorting subprime loan–related investments as far back as 2005 after he received briefings, ironically from analysts at Bear and Lehman, that the housing market was about to tank. Druckenmiller took big losses in 2006 when subprime was all the rage, one of his worst years ever. But he was now both making up for those losses and completely baffled why the CEOs of the Wall Street firms hadn't listened to the warnings of some of their own best people as he had. "Maybe they were too busy listening to the cash register going *cha-ching*," Druckenmiller would later say. Another person ringing the cash register on Wall Street was John Paulson, who had also been shorting CDOs for a while, and now he was looking more broadly not just at the securities but at the firms holding them and writing down their losses. The short sellers were putting together two scary pieces of data: the Street was loaded with illiquid debt, and its business model, based on a high degree of borrowing, had put it at the mercy of its creditors.

The bet was that Wall Street was about to come to an end.

Jesse Eisinger, a reporter for *Portfolio* magazine, didn't come to that exact conclusion, but he came pretty close. He wrote an article titled "Wall Street Requiem" in which his conclusion was that Wall Street was so saddled with Level 3 toxic junk that it would need more capital to survive.

High on Eisinger's list was Lehman Brothers, which had escaped much of the turmoil of the past year and ended 2007 with a strong stock price and its management team in place. But Eisinger concluded that the firm's holdings of Level 3 assets had soared, and in the third quarter of 2007, "Its total leverage was more than 30 times capital, up from 25.8 times a year ago." His conclusion: Lehman couldn't survive as an independent company.

"Bear is in the crosshairs now," Eisinger wrote about the growing fear that the firm couldn't survive without a suitor, "but it may soon have company there."

That company would be Lehman, short sellers began to conclude.

W e don't warehouse this stuff," Dick Fuld said in an interview in 2007, just before the mortgage meltdown started, "we sell; we don't keep this stuff on our books."

Fuld was feeling pretty cocky in early 2007, as Lehman's stock price pushed to new heights, and he wasn't feeling so bad now, as the year came to an end. Several Wall Street CEOs, including John Mack and Jimmy Cayne, hadn't been given bonuses this year, but Fuld had: a 4 percent raise from 2006, with a stock bonus of $35 million.

It was, according to many analysts, money well spent, because Lehman, for all its leverage and ties to the mortgage market, had somehow escaped the carnage. November 2007 was generally regarded as the worst month of the year for the credit markets, yet Lehman continued to make money, earning nearly $1 billion in profits during the fourth quarter alone—down from the previous year's record results but amazingly positive given all that had happened that year.

Many analysts were celebrating Fuld and his risk team—President and

COO Joe Gregory and the new CFO, Erin Callan—as brilliant market strategists. But some savvy investors were taking a deeper look into Lehman's earnings (pumped up by onetime gains) and, thanks to the new accounting rules, the disclosure of its ties to risky real estate, particularly some massive commercial real estate properties.

One of the first to identify the problems with Lehman was the short seller David Einhorn. In his *Portfolio* piece Eisinger predicted that Lehman wouldn't survive as an independent company—that it would have to be sold sometime soon, possibly in the new year, to a bank with more capital and less exposure to those hard-to-price Level 3 assets. But Einhorn went a step further. His conclusion was that Lehman was likely to become insolvent if what he was seeing in this analysis was accurate.

Einhorn, the head of the hedge fund Greenlight Capital, had taken his lumps in the subprime market with his investment in New Century, the now-bankrupt subprime lender. But that had been a mistake in a career of other brilliant calls, and in late 2007 and early 2008 Einhorn believed he had made the best call of his career by betting that Lehman Brothers was in deep trouble.

In December he began to build his short position in Lehman quietly and without press attention (that would come later). Several factors went into his thinking, but most of all he took a completely different view from the common perception on Wall Street that Lehman was run by the smartest people in the business. They weren't that smart, he felt, just gamblers who had gotten lucky.

Einhorn read Lehman's annual reports dating back to 1998; his conclusion was that Fuld and his team had finagled Lehman's books in 1998 during the LTCM crisis, disguising the true nature of the losses until the market improved. And he should know. He remembers being on vacation at the time and consumed with bond-related losses at Greenlight Capital, his hedge fund. Looking back, he couldn't understand how Lehman had largely avoided such a brutal month when his shop was posting massive losses on what he believed were many of the same trades. So he looked on some press reports and public filings and determined that Lehman must have been massaging its losses into gains.

He believed it was doing the same again. The tip-off: in the middle of the subprime crisis in late 2007, nearly every firm had been deleveraging, but Lehman had been increasing its leverage and, he believed, not accounting for losses as they spread from residential real estate to commercial holdings. What astounded Einhorn was that Lehman had seemed to get a free pass in the markets in late 2007 despite taking a huge stake in a commercial real estate deal known as the Archstone-Smith Trust, a complicated land transaction, at a time when property values were plummeting across the country.

If you believed the dire predictions about the economy, as Einhorn did, commercial real estate bought with leverage would deflate as residential real estate had, and so would Lehman. As far as Einhorn was concerned, it was only a matter of time.

B y the end of 2007 Wall Street was clearly changing. The financial crisis had swept top CEOs out of office. The stock market was finally digesting the news of the mortgage crisis, not to mention the new disclosure rules that forced banks to tell investors the quality of their balance sheet assets. With that, stocks, particularly those of banks and brokerage firms, began a steady decline and the Dow descended from its high of 14,000 in early October.

The credit crisis continued thanks to the simple fact that having so much capital tied up in risky mortgage debt meant the banks had to scale back their lending to businesses and individuals, and in late 2007, the nation's economy was beginning to fall into recession.

Wall Street's business model was bruised and beaten too, and Bear Stearns was among the most battered of the lot. Amid all this, Jimmy Cayne continued to assure investors that the company was solid and strong. He couldn't wait for the joint venture with the Chinese bank to start in the new year. The firm's top investors were on his side, and as for Bear's board of directors, his best friend, Vince Tese, was the lead board member. Enough said.

Still, with each passing day of November and December, Cayne's once-mighty hold on his job seemed to weaken. Cayne may have believed he was running Bear Stearns as if it were late 2006, when the stock had been trading at about $170 and Bear had been scoring billions in the mortgage market. But it was 2007; Bear's business model was weakened by the fact that it had for so long relied on a mortgage market that had disappeared from the face of Wall Street and might not return, at least not soon. In the past, Jamie Dimon would have given his left arm to buy Bear; now he wouldn't even give it a line of credit. He wasn't alone. Cayne made a bid to sell a 10 percent stake in the company in exchange for a line of credit from the China Development Bank and was rejected.

Bear's stock was barely holding at $80 a share and was trending lower, and it was increasingly the target of short sellers, who were engaging in a new maneuver that put even more pressure on the firm's stock price: they bought credit default swaps and shorted Bear's stock. When the prices of credit default swaps began to rise, word spread about Bear's problems—after all, why would investors be buying credit protection if they thought Bear would surprise everyone with good results? The speculation was repeated in the press, and shares of Bear declined. As they fell, the short sellers reaped huge profits.

The irony of Bear being hammered by short sellers is that the firm had been the shorts' best friend for so long. It had lent money and stock through its clearing operations to facilitate short sales that had been the ire of many small companies that complained about the excesses of this trading strategy. But now Jimmy Cayne began sounding a lot like Patrick Byrne, the controversial CEO of the company Overstock.com, who had led a years-long campaign on the evils of short selling, particularly naked short selling. In a regular short sale, a trader borrows stock, sells it immediately, and repays the borrower at some later date, hopefully when the shares decline in value and the trader will make a profit.

In a naked short sale, the trader doesn't borrow the shares to create a short position; the practice is technically illegal, although regulators generally don't crack down on naked shorting because the sentiment is that short sellers serve a vital purpose of keeping the market informed about stock promoters. Even though regulators never charged the firm, Byrne believed that Bear Stearns was one of the major instigators of naked shorting by allowing customers to short stocks without securing the borrowed shares of companies such as Overstock.com.

Cayne bragged that when he had tossed Warren Spector earlier in the year, most of the firm had cheered. Spector hadn't been loved, and now, as Cayne came to realize, neither was he. Alan Schwartz was itching to take Cayne's place, and he had powerful support; many of the top executives had come to the conclusion that Cayne should relinquish his post. That conclusion would grow because of some unflattering press Cayne was about to receive.

The same week O'Neal and Prince succumbed, the *Wall Street Journal* ran a long story that questioned Cayne's management of the firm during the crisis; it brought up his frequent bridge trips and golf outings while the firm was falling victim to the subprime mess and shares of Bear fell to their lowest levels in years.

One part of the story disclosed to the public for the first time that Cayne smoked pot after bridge games and in what it called "more private settings," although it didn't mention the early allegations of his office pot smoking that so many people at Bear recalled and that Cayne still denied had ever taken place, at least on Bear's premises. Pot smoking might sound like a harmless recreation, but Bear's board of directors took it seriously enough to ask him if the account in the *Wall Street Journal* was accurate. Cayne said it wasn't, though it's unclear whether the board asked him when he'd last smoked a joint.

The vision of a seventy-three-year-old CEO smoking pot became the talk of the gossip sheets; the *New York Post* ran a clever illustration of Cayne with love beads and long hair, which infuriated him. The feisty online blogger Bess Levin of DealBreaker did so many satirical posts about Cayne's alleged pot use that Bear blocked the Web site from its computer servers.

Cayne continued to deny the pot-smoking allegations, telling some people he hadn't smoked in years, while explaining to others that he had once been diagnosed with multiple sclerosis and an occasional toke relieved him of the discomfort. For the most part, he was angered and hurt by what he believed was a private matter that had nothing to do with his work as CEO. But he couldn't help but crack a smile for a few moments, just after the story first broke, when he told me that he had never before had so many requests to play golf since the *Journal*'s pot story had come out.

While Cayne was handling the pot controversy, he was hit with another personal matter, which involved Ace Greenberg, the legendary trader and Cayne's predecessor as CEO. Greenberg had just turned eighty when a woman about half his age who worked at Bear in sales made the following claims to Cayne: she said Greenberg had touched her in a sexually provocative way and at one point had kissed her, going so far as to stick his tongue down her throat.

The charges—which Cayne didn't seek to prove—would normally have made him laugh, given his wild sense of humor. But he wasn't laughing when the woman made demands: $10 million, or she would go public with the charges. Cayne at first told the woman it was an outrageous sum of money she was asking for; she claimed she had witnesses. Cayne discussed the situation with several top executives.

Bear was in a bind; if it fought the charges, she would certainly file a suit. Having seen the way the tabloids were playing stories about Bear lately, Cayne decided the best thing to do was settle. People with knowledge of the matter say he wasn't about to pay $10 million. So he offered $2 million. The woman accepted, and Greenberg was saved the humiliation of seeing his photo in the *Post* with a headline like "Ace in the Hole."

However, Cayne's humiliation continued. Bear's board—the same board that he had once referred to as "my board"—was now openly questioning whether he should remain as CEO. The joint Justice Department/SEC investigation of the hedge fund collapse was moving up the ladder and looking at not just Cioffi but other senior executives at the firm as well. The biggest danger to Bear was Cioffi, who was now out of the firm; if he were indicted, he might point the finger at others. Cioffi was already making noises that if he had done anything wrong, Bear's senior management, particularly in the asset management ranks and compliance staff, had been there with him, approving his every move.

But by far the biggest problem facing Cayne was the potential for future losses on Bear's mortgage bonds; according to Eisinger in *Portfolio*, the firm had $18 billion in Level 3 assets, or 135 percent of its market value.

"Jimmy, why don't you just write them off and we take our hits and move on." That was the advice given to Cayne by one of his closest associates, the broker Kurt Butenhoff. What made Butenhoff so powerful inside Bear and influential with Cayne was his gold-plated clientele, which included the man who was now Bear's biggest investor, the commodities billionaire Joe Lewis. While Cayne owned 5 percent of Bear, Lewis now owned 7 percent, and his stake was shrinking with every negative story about the firm.

Still Cayne wasn't about to take the losses, not yet anyway. There were no buyers. He also spoke about a possible "bounceback," a Wall Street term meaning that even as the firm wrote down losses on its mortgage holdings, it would eventually write up gains when the market returned.

In November Bear told analysts that it would write down $1.2 billion in mortgage bond losses. CFO Sam Molinaro said he believed the credit markets were firming up and Bear had seen the worst of the crisis.

It didn't matter that the Oppenheimer analyst Meredith Whitney said just the opposite; Bear's stock rose, and the money management firm Legg Mason upped its stake.

Then, in December, Cayne joined the growing list of CEOs who had overpromised and underdelivered. The firm reported $854 million loss for the fourth quarter in 2007. It was the first loss the firm had taken since becoming a public company in 1985, and, according to the *New York Post*, it was four times as big as expected because of a massive devaluation of mortgage bonds—a $1.9 billion write-down of losses, or $700 million more than the firm had thought it would have to take just the month before.

Cayne had always said that Bear didn't hold on to its mortgage debt like other firms when defending Bear against analysts such as Dick Bove, who had repeatedly warned that the firm was essentially no different from Merrill or Citigroup in this category. Now Bove was looking pretty smart. The same couldn't be said for Cayne.

Cayne knew his job was on the line. In late December 2007 the board began formal discussions on finding a replacement. The likely candidate was Schwartz, but Cayne still refused to accept that he was about to join O'Neal and Prince as high-level casualties of the spreading mortgage crisis.

"If I'm going down," he told me at the time, "I'm going down like a samurai!"

17. THE END BEGINS

As 2007 came to a close, Wall Street collectively breathed a tentative sigh of relief. The Armageddon predicted by Jimmy Cayne hadn't occurred, at least not yet. Credit still wasn't flowing freely to business and consumers as banks imposed stricter lending terms, largely because securitization—the engine of the previous lending bubble—was virtually dead. After the second half of 2007, not one CDO was priced and sold; regular mortgage-backed deals as well as asset-backed deals ground to a halt.

But the stock market managed to regain its footing as the Dow Jones Industrial Average hovered around 13,000, reflecting a general optimism that also continued to be felt by policy makers in Washington, including the two who mattered most, Ben Bernanke and Henry Paulson, that the worst of the credit crisis was probably over. Yield spreads began to tighten a bit, and the ABX Index showed some signs of life.

Wall Street workers saw their bonuses fall—several CEOs agreed not to take a bonus for 2007—but not to an extreme degree; according to the Associated Press, the average bonus on Wall Street in 2007 was around $180,000, just 4.7 percent lower than in 2006, one of the best bonus years in the history of the financial services business.

It seemed that with a little luck Wall Street had dodged the bullet once again. Despite tens of billions of dollars in losses tied to write-downs of mortgage debt, the firms were sticking to their business models, which stressed risk taking in bonds and leverage; they were acting as if the massive losses experienced by nearly every firm in 2007 were the aberration their computer models said they were. The prevailing wisdom was that a return to an era when Wall Street had worked simply for fees would be the end of Wall Street. Without risk, without leverage, the Street couldn't produce the returns it needed to satisfy investors, who demanded healthy returns. A Morgan Stanley spokeswoman, speaking for CEO John Mack, summed up Wall Street's attitude this way: "John just believes you can't return to the past."

Morgan Stanley wasn't alone, of course. Firms such as Goldman Sachs were pressing securities regulators to maintain the status quo of a business model that relied heavily on short-term borrowing, and it was working. In early 2008 Secretary Paulson, Fed Chairman Ben Bernanke, and New York Fed President Tim Geithner held occasional gut checks on the condition of the banking system, and the answers were the same as before: they were in broad agreement that once the dust settled (as they believed it would soon), the mortgage debt held on the balance sheets of the banks could be sold at less distressed prices or simply marked up to reflect the market's resurgence, Wall Street would once again return to its old self.

Part of the reason for their optimism—and the optimism on Wall Street in general—was the changing of the guard at the top of the firms in the most trouble.

"It's a beautiful day and we have a new CEO, and I feel pretty good" was the assessment of Kurt Butenhoff, one of the top brokers at Bear Stearns, to the news that after a year of losses, controversy, and allegations of everything from pot smoking to cheating on his golf scores, not to mention his brush with death, Jimmy Cayne had decided to quit Bear before the board forced him out.

Cayne maintained that the decision was his own; there was no pressure or palace revolt that forced his hand. But there was. Bear's board was getting numerous calls from investors inquiring whether the pot-smoking allegations were true—even as Cayne denied them publicly—and what the plans were to get the firm back on track.

One such call came from Bruce Sherman of Private Capital Management. By early 2008 Joe Lewis had surpassed Cayne as the largest shareholder in Bear Sterns, but Sherman's stake was large as well, just under 5 percent. For months Cayne had bragged that Sherman was on his side. No longer. "I think it's time for Jimmy to go," he told the board. He was joined by a growing chorus of people inside the firm, including many of Cayne's former allies and breakfast partners, who wanted him out as well.

Cayne considered Alan Schwartz one of his allies, but Schwartz told various board members that as much as Spector could be blamed for the implosion of the hedge funds or the firm's holdings of mortgage debt, he had heard from many senior executives who still believed the wrong guy had been fired, that Cayne had lost his touch and was simply too old, too incapable, and too tainted by the actions of the past year to bring the firm back to health.

Even Cayne would agree with part of that assessment; by early 2008 he was spent, drained. Over the past year he had tried everything he could think of short of selling Bear at a fire-sale price to try to repair the damage done to the firm, and none of it had worked. The Chinese joint venture still wasn't operational. Potential deals had fallen through; "Who would want them?" quipped Larry Fink when he heard that Bear was holding exploratory talks with the hedge fund Fortress Investment Group. Investors were still jittery about the firm's future; its share price fell to its lowest level in four years, around $70, making people yearn for the days it had traded at $100, much less its early 2007 high of $180.

Employees just wanted the free fall to end, for the controversy and the endless debates about the company's future (and the peccadilloes of their CEO) portrayed almost every day on CNBC and in the press to go away.

Board members, including Cayne's friend and most loyal ally, Vincent Tese, now conceded that "Jimmy had to go," but in a way that preserved a modicum of respect. Cayne would give up the top job and remain as non-executive chairman, which meant that despite the lofty title he would have no operational duties. In other words, he would be free to play as much bridge as he wanted without coming into the office, and from nearly the moment the switch was made, he did exactly that, while the new management figured out how to repair a franchise built largely on the backs of mortgage bonds that was now facing a future where its main profit center was at best mortally wounded and at worst extinct.

Given the new direction the firm was taking (no one knew exactly what that direction was, other than that it would involve less risk taking and less mortgage bond trading and underwriting), it appeared logical that Cayne's replacement wasn't another executive in the bond department but instead a capable investment banker.

Alan Schwartz was one of the most decent men on Wall Street and at fifty-seven about fifteen years younger and healthier than Cayne. As former Time Warner CEO Richard Parsons put it, Schwartz was a "smooth operator."

Kurt Butenhoff, for one, was elated at the transition, and his reaction reflected the prevailing opinion inside Bear; Schwartz was relatively young, well liked across the firm, considered able and effective, and above all, not Jimmy Cayne. But Schwartz was also someone whose experience left him completely unprepared for what lay ahead. For all his savvy and connections with CEOs in corporate America, Schwartz had never played a leading role in crafting Bear's business model.

Yet here he was assuring the world that Bear was on the cusp of a massive revival, despite the tens of billions of dollars in near-worthless mortgage debt on its books and a bond market that was hardly back to normal. "I think people realize that we have righted our ship and our businesses are off to a strong year," he told *Institutional Investor*. Schwartz told the magazine that Bear was "going back to playing offense" and returning to its old self as the toughest trader on the Street. In interviews, he sidestepped questions about selling Bear, probably because he knew he couldn't sell the firm for anything close to the $180 a share that many top executives used to calculate their wealth, even as the share price fell to $70 and lower.

In meetings with senior executives and regulators, Schwartz downplayed the need to raise new capital despite the massive amounts of money being raised by other firms. Paulson, who had made a habit of conversing with the Street's CEOs since the Bear Stearns hedge fund disaster, considered Bear the problem child of Wall Street. Neither he nor Bernanke had yet hit the panic button, demanding that the firm find a buyer. But Paulson did begin badgering Schwartz about a potential strategic alliance, such as the one with the Chinese, that would open up business and a line of capital, which was one of his concerns.

Bear had $18 billion in capital on hand but a balance sheet many times that—close to $400 billion. Bear relied, maybe more than any other firm, on short-term borrowing in the repo market to fund the trades and positions of its balance sheet, meaning its business was largely dependent on investors' confidence in Bear's ability to repay those loans, not exactly the best position to be in considering everything that had happened in 2007.

Yet when Schwartz was asked by regulators in early 2008, and even prodded internally by some senior executives, why he wasn't raising capital through a stock sale or from outside sources, as Merrill and much of the rest of the Street had just done to bolster their balance sheets, Schwartz appeared more concerned about Bear's stock price, which would be further diluted by an equity offering. "Our share price is down too much," he said.

"We'll do it when our share price returns." As one senior official at the Fed said, "Alan was very worried about dilution and was loath to do an equity raise until the value of the stock was higher."

With that in mind, Schwartz knew his first order of business was to stop the bleeding at Bear—end the losses and the bad press. Then, as much as he didn't want to admit it yet, given Bear's culture of independence, to sell the firm in an improved market. Whether he could achieve any of the above was still an open question.

In late 2007, while Citigroup faced the biggest crisis in its history, management had come up with a new reengineering plan. It wasn't focused on how to wash away the billions of dollars in toxic debt on its balance sheet or the SIVs that it had hidden from investors until losses had forced Citi to take them back. No, Citi, according to a memo issued by Bob Druskin, one of Chuck Prince's top lieutenants, had bigger and loftier goals to "embed a continuous improvement philosophy and capability throughout Citi, which will become a permanent and important part of how we manage the company."

The two-page memo said management would be "instilling ownership in the businesses and regions . . . appointing re-engineering owners within each sector and function . . . developing a non-bureaucratic approach to [management]," and "allocating some of our savings to fund investment opportunities."

But Druskin and Prince himself were now gone from Citigroup as Robert Rubin chose the former Morgan Stanley executive Vikram Pandit to restore the megabank to its former glory. Like Schwartz at Bear, Pandit seemed a logical choice for the job for many reasons, one being that Citigroup had already invested so much in him.

By 2008, the fact that Pandit's Old Lane hedge fund was itself quickly faltering and on the cusp of closing down didn't stop Rubin from pushing for Pandit's appointment as CEO of the nation's biggest and most dysfunctional bank. Several board members felt that Pandit should get the job for no other reason than to show they had gotten *something* for their $800 million.

What wasn't logical was the board's unyielding support of Citigroup's "financial supermarket" model. The board received a report from McKinsey, the large consulting firm, that recommended that the firm be broken up, that it sell off units, including the brokerage business, and that it return to a much simpler model, namely its core commercial bank. In truth, the board didn't need a consulting company to tell it this—many of Citigroup's top investors and analysts had been demanding it based on Citigroup's stock price, which had gone nowhere for nearly a decade. Put simply, the Citigroup experiment had failed on a number of levels, the McKinsey study pointed out. It had brought together risk taking in the investment bank with a business that demanded strict risk management, namely, holding customers' deposits. So as Citigroup's losses had mounted because of its bad bets on CDOs, the entire bank and its billions of dollars in deposits had been put into jeopardy.

But more than that, the experiment had failed because it simply hadn't made enough money.

Pandit's first order of business was to travel the world to find new capital to support the bank's mortgage bond–battered balance sheet. His second was to announce modest job cuts in Citigroup's workforce, which at 370,000 employees was the size of a midsized American city. His third was to let the world know that Citigroup would not be broken up despite the widespread concern of investors, analysts, and even some of his senior management that the company was just too big to function efficiently.

It's unclear if Pandit came to that conclusion independently or was influenced by the twelve men who had given him the job. Another candidate for the job, New York Stock Exchange CEO John Thain, told the board that he would break up Citigroup, which immediately hurt his chances. The Citigroup board was dead set against what seemed so logical to many investors, namely, breaking up the bank into more manageable pieces, and much of that opposition came from Rubin, the most influential single executive and board member at Citigroup, who despite his lack of responsibility had so much power he could perpetuate the failing status quo.

Pandit, a former business school professor with a PhD in finance, was the means for Rubin to maintain the status quo, and Pandit had no problem telling Rubin and the board what they wanted to hear, namely, that the "supermarket" model, for all its complexities and expenses, could be saved. In meetings with analysts and investors, Pandit reminded the naysayers of the firm's glory days before the subprime crisis. when it had generated $25 billion net each year; the business areas in which the firm dominated; and how Citigroup was the only true global financial services empire (even if parts of that empire had no idea what they were doing).

It was exactly what Rubin wanted to hear, even if Citigroup's glory days had never been that glorified—$25 billion disappears pretty fast when subsumed by a cost structure as massive as Citi's—and even if Pandit, from nearly the moment he was named CEO, seemed incapable of convincingly articulating his case to investors. His first interview with CNBC, in which he was supposed to lay out his vision for Citigroup, was a disaster.

The anchor, Erin Burnett, prodded him with questions he should have been well prepared for: would Citigroup survive in its current form, and would the firm experience additional write-downs? Throughout, Pandit had a deer-in-the-headlights look, and he sounded even worse, as he refused to give clear-cut answers to just about all the questions he was posed. Investors watched in horror and continued to unload their shares even as Pandit promised a new beginning, including a bottom-up review of Citigroup's business in which "everything was on the table," which presumably meant he might sell off some nonessential parts of the Citigroup empire to pay for the firm's heavy costs, although no one really took him at his word.

After all, what was there to review? That was the question being asked by investors already well versed in the firm's problems, its bloated cost structure, and its massive investments in toxic assets.

By early 2008 Citigroup's share price had fallen by nearly 60 percent from late 2007 through the first two months of the year as analysts, most notably Meredith Whitney, ignored Pandit's optimism and predicted further losses on the firm's bad investments and costly infrastructure, combined with its aimless business model.

The concept of cross-selling was now so fully exposed as a sham that not even analysts who had supported Citigroup in the past could figure out how the bank was going to make money. Certainly not Whitney, who began to predict that in addition to raising capital, Pandit would have to cut the once-sacred dividend to make ends meet.

In the face of this critique, Citigroup's board did what one might expect: nothing. Unless, of course, you consider decisive action to be ordering Pandit to immediately begin taking public speaking classes to learn how to reshape his message of optimism about the future. Rubin, meanwhile, assured anyone who would listen that Pandit was the right man for the job. "Vikram has earned a reputation as one of the most respected leaders in the financial services industry," he told *Investment Dealers' Digest*. "The combination of his deep executive experience and long history as a strategic thinker makes him the outstanding choice to be Citi's CEO."

If Rubin had been completely honest, he would have also pointed out that Pandit's biggest attribute might have been his stated desire to save Citigroup's business model, which Rubin, one of its creators, held near and dear to his heart. As a result the Citigroup board passed up some of the most qualified candidates in the industry, who made it no secret that they would break up the mess known as Citigroup if they were CEO.

In fact, two of those candidates were competing to replace Stan O'Neal at Merrill Lynch.

They have to go through a process, give them a break." That was Larry Fink's assessment following his meetings with Merrill's board. By now Fink was telling everyone—almost literally—that the job was his if he wanted it, as he surely did. How could he not make the cut?

In fact, Fink had told so many people that his near-certain appointment had leaked to the press. On paper, of course, he was the obvious candidate. He had been one of the developers of the mortgage-backed security, he had built one of the nation's most successful money management firms in BlackRock, in which Merrill owned a 49 percent stake, and he had seen the subprime meltdown early and even managed to profit from it.

But more than that, he had spent a career redeeming himself of his past sins. The 1980s swashbuckler who had nearly blown up First Boston by trading in complex mortgage bonds had spent the past two decades trying to understand the complexities of risk—and making a good deal of money doing so. At BlackRock, he had created a unit of high-tech systems known as BlackRock Solutions, which had managed distressed portfolios for companies such as GE in the mid-1990s and many others along the way. He was expecting similar assignments for BlackRock as Wall Street crawled out from under the current mess, which Fink believed was bigger than 1986, 1994, and 1998 combined, even if the Street hadn't yet accepted that reality.

Having spent so much time studying his own near downfall from taking risk and then profiting from a business that cleaned up the risk-taking follies of others, Fink understood risk and its limitations better than almost anyone else on Wall Street. Fink had matured in the way that Tom

Maheras or Jimmy Cayne never had; like gamblers who never lost, they had bet right for so long that they had never thought they could bet wrong, until, of course, they did. Getting burned as Fink had, at such an early stage of his career, had tempered a man who could roll the dice with the best of them.

John Thain had had no such losing experience during his long career on Wall Street; in fact, his career can be summed up as holding one large winning hand after another, from graduating from Harvard Business School to holding various positions at Goldman Sachs, where he had never seemed to bet wrong. Though he had never actually traded mortgage bonds, he had become a protégé of the bond-trading honcho Jon Corzine, who would eventually run the firm and make Thain the manager of its mortgage bond department. After Goldman suffered losses tied to the LTCM debacle, Thain joined the investment banking chief, Hank Paulson, and pushed out Corzine as CEO. It was an act of both betrayal and survival. Instead of being ousted as a Corzine loyalist, he was named to the ruling triumvirate of Goldman. When the tide turned again and trading profits shifted the power back to the risk takers, led by Lloyd Blankfein, it was Paulson's turn to leave.

Even before Paulson resigned in May 2006 to become secretary of the Treasury under George W. Bush, Thain seemed to have sensed his career at the firm might be coming to an end and jumped ship to the New York Stock Exchange still reeling from the controversy over CEO Dick Grasso's pay package, which ultimately forced the NYSE chief to resign in disgrace.

When offered the NYSE job, Thain didn't think twice, and in an efficient, clinical fashion he began to remake the 118-year-old stock exchange, ditching the way the NYSE had done business for centuries—matching buyers and sellers of stock through human floor brokers and specialists— for a computerized trading system. He bought an electronic trading platform named Archipelago (majority-owned, it should be noted, by his recent employer Goldman Sachs) to make trading more efficient. Thain became known on the NYSE trading floor as "I, Robot" for his almost sterile, emotionless approach to his job, as if he had a calculator in his brain that weighed facts while ignoring the human consequences. He once told a floor broker who complained that the new system would land him on the unemployment line that McDonald's was hiring. That day, traders feasted on McDonald's hamburgers on the floor in a display of defiance, and Thain barely cracked a smile.

Thain saw Merrill as his next big trade. While Fink met with the board and asked to review Merrill's mortgage debt holdings, Thain systematically laid out his plan to rebuild Merrill: he didn't have to sell the firm, as O'Neal had advocated. Instead, he would write down all the losses as soon as possible and return Merrill to profitability after the first quarter of 2008. He wanted $50 million to do the job. He never asked to study the firm's mortgage holdings.

Thain's single-mindedness impressed the Merrill board. Fink told the board his move would be complicated because he owned most of Black-Rock—Merrill couldn't be seen favoring BlackRock over other customers by selling its funds through the firm's brokerage sales force. So Merrill would either have to cash Fink out of his stock, valued at hundreds of millions of dollars, or find a way to insulate his holdings from conflicts of interest.

Moreover, repairing Merrill would take time and would be complicated by the firm's holdings of billions of dollars in CDOs and other mortgage debt. Fink and his risk management team at BlackRock would have to come in and study the holdings before he could even take the job. Even so, several board members told Fink the job was his and asked what more it would take to get him in the CEO's office. The board also hired an outside search firm, whose lead executive told Fink that based on his conversations with the board, Fink was the board's lead candidate.

But one of the leading members of Merrill's board, Alberto Cribiore, a Milan-born hedge fund manager and one of the directors who Fink says all but offered the him job, was now ready to appoint Thain. Cribiore argued that Thain came with a clean slate, not to mention a winning record. More than that, Thain agreed with the general sentiment of the board that Merrill could survive on its own as the markets improved. Fink, on the other hand, had had that messy experience at First Boston in the 1980s, even if he had redeemed himself thereafter. And he was hardly an optimist; Fink wasn't as cocksure about the bond market's growth to recovery and Merrill's easy path to profits as was Thain, which might mean he would sell Merrill as O'Neal had advocated.

Ultimately, the board agreed with Cribiore's assessment and the job went to Thain.

One of the many tragedies of the era of excess was the failings of the gatekeepers. The housing crisis was able to spread as it did because the regulators, bond raters, risk modelers, and, to some extent, the business media didn't understand the perils of risk and leverage. Wall Street's government watchdogs, the Federal Reserve, the SEC, and the Treasury Department, were even now clueless about the potential for massive losses and a large systemwide meltdown of the financial system even as nearly every major firm on the Street held tens of billions of dollars' worth of illiquid mortgage debt.

But a special place in the pantheon of irresponsibility belongs to the boards of directors who looked away from the risk taking; believed the best-case scenarios presented by management when they should have been preparing for the worst; and, worst of all, left in charge management that didn't have a clue how to manage leverage and risk.

Cribiore, like the rest of Merrill's board, had followed Stan O'Neal's management philosophy at Merrill until it was obvious that O'Neal had nearly destroyed the company, and now the board was about to double down on its mistake.

The Wall Street conventional wisdom, and that of some regulators like Paulson, was that Bear's balance sheet was the most problematic when it came to its holdings of illiquid securities. But Merrill's wasn't much better; it had a better brand, and its capital surpassed Bear's by billions of dollars because of money flowing in from sovereign wealth funds, but its balance sheet was stuffed with debt that was conceivably more toxic than Bear's: Merrill held tens of billions of dollars in CDOs, for which there was no bid, unless you consider pennies on the dollar to be a bid.

Which was why picking the most competent risk manager on Wall Street as its next CEO would mean life or death for Merrill Lynch.

John Thain wasn't incompetent in the risk management department or as a Wall Street executive—in fact, he is highly competent but in a very narrow way. As smart as he is, as accomplished a manager as he had become, he had missed the defining event that had made Larry Fink ask to study Merrill's mortgage holdings before he would take the job. That defining event was, of course, the misery of losing money and nearly losing a career. It had made Fink cautious and plodding, at least when it came to risk; the lack of it made Thain utterly confident, even in matters he had never dealt with during his illustrious career.

Alberto, you have to tell Larry, and if you don't I will!" screamed Greg Fleming, Merrill Lynch's president and investment banking chief, as board member Alberto Cribiore, who was leading the CEO search, broke the news to him that Thain, not Fink, was getting the post. Cribiore made sure that Fleming was one of the first to find out because he and Fink were close friends (the board wanted Fleming to remain at Merrill) and Fleming was the banker who had executed the Merrill-BlackRock deal, which Cribiore believed was more important than ever before if Merrill were going to dig out of the hole Stan O'Neal had dug.

That's because BlackRock was printing money, and Cribiore wanted to make sure Merrill kept getting its share of the winnings. If BlackRock was so important, Fleming asked, why piss off Fink even more by waiting until after the closing bell to break the news to him, as Cribiore was planning? The likelihood of a leak was high. If the story hit the press first, Fink would feel worse than slighted; he would feel betrayed since he had gone to the BlackRock board to alert them that he had been all but offered the job a couple of weeks earlier. Cribiore said it was best to wait—Thain hadn't signed his contract yet, so the appointment wasn't official.

It didn't matter; hours before the planned announcement, a report on the DealBreaker blog by the reporter John Carney said that the board was eyeing Thain as the CEO. Then the *New York Post* published on its Web site that it was official: the job had gone to John Thain.

Fink was everything you might expect: saddened, hurt, and, above all, pissed off. He even claims that the executive recruiter hired by Merrill to help the board choose its next CEO told him that Cribiore had given the headhunter the green light to begin official contract negotiations. But it was all a ruse, he now believed, and as Fleming had feared, Fink discovered that the job had gone to Thain when he watched a headline scroll across his computer screen flashing Thain's name as the new Merrill CEO.

"Those fuckers" was all he said.

Merrill's board didn't care that it had just slighted one of the most able CEOs on Wall Street. That's because for all Fink's knowledge of mortgage bonds and risk, the market loved the Thain appointment; shares of Merrill rebounded in the weeks following the announcement, even if inside the management ranks, survivors such as Fleming were convinced that the wrong guy had gotten the job. The rank and file was impressed as well. "Thain's the right guy," said a senior trader, who pointed out Thain's successes at the NYSE. "No one knows this stuff better."

During Thain's first meetings with his senior staff, he told them much of what he had told the board—that Merrill would take all its losses up front and then watch the business grow. Thain's confidence even surprised Fakahany, who in one of his last acts as a Merrill employee (before turning to running his Manhattan restaurants full-time) suggested to people inside the firm that Merrill should sell much of its CDO holdings, which he believed could fetch sixty cents on the dollar, to clean up its balance sheet, as well as its stake in the financial information company Bloomberg LP, which Fakahany said at the time could bring in $8 billion. Fakahany has said that Thain rejected the offer because he believed the CDOs, now marked down, would be marked up.

Thain, through a spokesman, denies that Fakahany ever made such a suggestion to him, but one thing Thain can't deny was his confidence that despite all the troubles of the past year, the tens of billions of dollars of illiquid debt on its books, Merrill was poised for a massive, and rapid, comeback. How Thain came to that conclusion is difficult to unravel. Part of it stemmed from his view that the credit markets would improve. He also believed that the mortgage debt wasn't as impaired—it had mortgages that weren't in default—as the market pricing suggested.

But most of it had to do with the lofty confidence John Thain had in himself, in his abilities to manage and win.

Soon rumors swirled that Thain would bring in former buddies from Goldman; top executives at Merrill had expected as much, but what they didn't quite understand was his optimism given the size and depth of the problem. He had already begun telling the press that Merrill would make a swift and certain return to profitability. The firm, he said, wasn't about to contract massively, despite some modest layoffs; in fact, he was about to enact a growth strategy.

Larry Fink listened to Thain's optimism and thought it was Thain, not Jimmy Cayne, who had been smoking dope. Merrill was still holding some $40 billion to $50 billion in CDOs, and the credit markets weren't even close to coming back to the point where anyone could be confident that the securities could be sold at anything but a steep loss, which could make Merrill insolvent. In early 2008 the growing likelihood was that the opposite was about to occur: the mortgage market was more likely to get worse than better. But Thain didn't want to hear any wavering; in his mind losses and possible failure didn't compute.

One of Thain's first orders of business wasn't merely to reassure the markets and rebuild Merrill but also to build himself a new office. As Thain discovered when he took over, Stan O'Neal may have loved the life of a top CEO, which gave him access to any golf course in the world, but he wasn't much for offices. For him, a desk and some simple but modern furniture and pieces of African art would do. "If the whole thing cost $100,000, I would be surprised," said one longtime Merrill executive.

Thain wanted something different, grander. What he wanted, he didn't immediately say to anyone at Merrill. In fact, he left the redecoration to his wife, Carmen, who reached out to the famed interior designer Michael Smith to do the makeover.

Smith, often with Carmen Thain at his side, spent tens of thousands of dollars on rugs, furniture, and a wastepaper basket. Even for a guy like Fakahany, who loved the expensive wines and floral decorations in the executive suites, the extravagance was mind-blowing as he witnessed some of the redecorating, which included knocking down walls and installing chandeliers. The official bill showed dozens of expenditures on items more befitting a monarch than the CEO of a company suffering massive losses. Thain spent $87,000 on a rug and another $44,000 for an area rug. Another $25,000 was spent on a mahogany pedestal table, $68,000 on a nineteenth-century credenza. A sofa ran $15,000; four pairs of curtains totaled $28,000; a pair of guest chairs, $87,000; a George IV desk cost $18,000; six wall sconces went for $2,700; six chairs in his private dining room for $37,000; a mirror in his private dining room for $5,000, while a chandelier was priced at $13,000; fabric for a Roman shade cost $11,000; a custom coffee table, $16,000; Regency chairs, $24,000. He spent $5,000 on forty yards of fabric for wall panels and $1,400 on a parchment paper waste can. Senior executives who saw the final bill were most baffled by one line item in particular, $35,000 for a "commode on legs," because they were unsure if it was an expensive toilet or just an expensive dresser.

At Goldman, Thain would have had to pay for his office out of his own pocket. But not at Merrill; documents also show that Thain signed off on the purchases personally and that they came out of the firm's budget. He himself didn't spend a dime on either the office or the labor costs, which combined totaled more than $1.2 million. The spending spree was largely a secret inside Merrill, but it enraged the few senior executives who witnessed its size and extravagance.

Thain, of course, had inherited not just a drab office but also a dismal fourth quarter from Stan O'Neal, something he made clear in all his discussions with investors and analysts as he vowed to change the direction of Merrill immediately and return it to profitability. Though he was aloof and secretive with much of his management team and senior executives were just waiting for him to begin naming old Goldman pals to key positions, at least for the moment investors liked what he was saying. He went out and found sovereign wealth funds to invest billions in Merrill Lynch, diluting shareholders' holdings but providing a massive cash and capital cushion to withstand just about any additional losses. That didn't mean there would *be* additional losses, Thain said. Merrill wouldn't need to raise more capital, he assured investors; in fact, he said, it was on the cusp of profitability.

"John Boy has absolutely no idea what he is talking about" was Larry Fink's assessment one afternoon in early 2008 while having lunch at San Pietro. Fink may have been embarrassed by Merrill's board, but, given BlackRock's profitability, he was still Wall Street royalty. Today he sat along the famed "chairman's row" of tables at the popular luncheon spot with other Wall Street luminaries such as Jack Welch, Richard Grasso, John Mack, and Sandy Weill. Fink was reacting to the press accounts of Thain's optimism, and he could barely contain his anger. Thain was adopting a more optimistic view of the future that seemed to reflect the views of the economic policy makers who were keeping the closest eye on the financial crisis, his old boss at Goldman and now Treasury Secretary Hank Paulson, who believed the subprime crisis had largely been contained, and Tim Geithner, the president of the New York Fed, who shared that view.

Fink liked and respected both Geithner and Paulson; in fact, he was in touch with them almost daily about his outlook on the credit markets. Fink was decidedly less optimistic than they were, however; the Wall Street business model, he told them and others, wasn't broken but possibly beyond repair. Leverage helps firms make money on the way up, but it has a crushing effect on the way down, and with mortgage defaults continuing, the damage had yet to run its course.

"I don't see how Wall Street comes back," Fink warned friends including John Mack, the CEO of Morgan Stanley. Though his press department tried to spin the massive losses of the past year as an isolated event, Mack knew better. He had been around long enough (more than thirty years) to know that isolated events are rare, particularly since Morgan held its own illiquid assets, bonds, derivative trades, and the like, on a balance sheet of close to $1 trillion. Morgan wasn't in Bear territory—it had good management, lots of capital, and the confidence of its creditors—but, given the shape of the credit markets, it wasn't out of the woods yet, and Mack told associates he was concerned about the future.

That's why Fink liked Mack so much—he was a realist—and that's why Fink couldn't stand Thain and started calling him "John Boy." In Fink's view, Thain, at fifty-three, was merely two years younger than he was but a relative child when it came to understanding how the bond markets and Wall Street worked. He also hated Thain, now a BlackRock board member (an outgrowth of the agreement between the two firms). The two rarely spoke, and when they did, the conversation was often strained.

Part of the problem was personality—Fink was an extrovert and Thain extremely guarded—but they had also both vied for the same job, and only one had won. Whatever the reason, Thain began to freeze Fink out from information about Merrill, even details of decisions that affected Black-Rock. Fink complained he got more information about Merrill from retired executives, including the former CEO Dave Komansky, who as a Black-Rock board member began serving as a go-between.

The deep freeze imposed by Thain extended to members of his executive team. Greg Fleming may have been the president of Merrill, but Thain rarely consulted with him on major issues, except when he needed to raise money, one of Fleming's expertises. Even O'Neal had used to plug in top executives on calls with regulators such as Hank Paulson so they could get a feel for what policy makers were thinking. Thain handled all such calls privately with the door of his new $1.2 million office closed.

Thain's circle of advisers was small and exclusive. It included Nelson Chai, who was now the CFO of Merrill Lynch despite just a short stint as CFO of the NYSE, and Margaret Tutwiler, a former undersecretary of state and chief spokeswoman for former U.S. Secretary of State James Baker, whose knowledge of Wall Street was based for the most part on her equally short stint as Thain's chief media handler at the NYSE. Thain also intended to expand the circle to include a couple of his cronies from Goldman, long-time investment banker Peter Kraus, and trading desk veteran Thomas Montag.

Fink could understand why Thain kept him out of the loop, but he couldn't understand how he could freeze out Fleming, who in addition to being head of investment banking was also still the president of Merrill.

"You mean he talks to Margaret Tutwiler more than you?" he fumed. Fleming said that was the case, and he wasn't sure how long he could put up with such bullshit.

Fink told Fleming he had a standing offer to work at BlackRock, which Fleming said he would sleep on; meanwhile, Fink told anyone who would listen that he believed that despite all the positive spin Thain was getting from the press, he would in time ruin Merrill Lynch.

D uring the first months of 2008, it seemed as if Wall Street looked for a silver lining in every piece of bad news that came its way. In late February AIG made a shocking announcement: it had lost $5.3 billion, the largest loss in its history, the direct result of a massive write-down in losses stemming from its insurance of CDOs. That insurance, of course, was known as the credit default swap. Just a few months earlier, Joseph Cassano, who ran the financial products unit of AIG, had assured the world that he couldn't see any scenario in which the business would show losses; now, for having made that statement, Cassano was out of a job.

Cassano soon became a whipping boy for all the ills that faced AIG; the firm was reeling, and CEO Martin Sullivan was on the hot seat as well, as former chief Hank Greenberg stepped up his campaign for boardroom change and Sullivan's ouster. By now, it was pretty clear that Spitzer's decision to oust Greenberg back in 2005 might have been among the most costly actions taken by any regulator in the past decade. The numbers don't lie: with Greenberg gone, risk taking had soared, and the implications for the firm would be increasingly painful in the weeks and months ahead. Put simply, new management just didn't have the ability to rein in the cowboys.

Though Cassano deserves his share of the blame for misdirecting the market, he was only reporting what was the conventional wisdom in late 2007: that the markets were in fact recovering, the credit crunch somehow abating, with a "light seen at the end of the tunnel," even if all the evidence showed otherwise.

By the end of 2007 the banks had written off $100 billion in losses tied to bonds that invested in subprime mortgages. The theory was to take big losses up front and make up for them later in the year, when business conditions improved. John Thain reiterated his belief that Merrill was poised for a comeback, as did Vikram Pandit at Citigroup. The Lehman Brothers board was so confident about the future that it awarded CEO Dick Fuld $22.1 million in compensation for the previous year. But Fink understood the vagaries of accounting better than the regulators did and knew that the write-downs didn't represent the true market value of the subprime debt; only a sale could determine that, and Merrill, like the rest of Wall Street, wasn't selling because no one was buying except for people like himself; BlackRock was in the market to snap up mortgage bonds at prices much lower than Merrill and its brethren were willing to sell.

"There's a difference between writing this stuff down and selling it," he repeated every time he heard about the worst being over.

But was anyone listening? John Thain certainly wasn't, perhaps because the words of wisdom came from a man he had begun to despise. Or maybe Fink was right that "John Boy" Thain just didn't have the chops to run a big Wall Street firm and didn't understand that with rising subprime defaults and bigger losses at AIG, it was becoming increasingly clear that the condition of Merrill, like the rest of Wall Street, was far worse than anyone imagined.

The reality that Thain and the rest of Wall Street was missing started to emerge in February 2008, around the time Thain told senior executives at his firm that Merrill would begin returning to profitability, when the once-obscure business of insuring bonds suddenly became the new face of the credit crisis. Bond insurance was yet another once-stable business made unstable by a wild gamble during the mortgage bond bubble. Bond insurers had once been cash cows; they had made stable and fat profits guaranteeing the principal and interest of municipal bonds, which rarely defaulted. For a fee, municipalities could transfer their lower rating for the triple-A rating of the bond insurance companies.

Some investors believed the bond insurance business was nothing more than a scam because of the low risk associated with municipal debt, which rarely fell into default. But the municipalities kept purchasing bond insurance because they could save more money on selling bonds with lower interest rates by upgrading their rating to triple A than they had to pay in insurance premiums. The system had worked well for years, and everyone seemed happy, except, it turned out, the bond-rating agencies that bestowed the triple As on the insurance companies.

They wanted "diversification," meaning they wanted the insurers not to be tied to the zigs and zags of just the municipal bond market if they were to keep their AAA ratings. In diversifying, the bond insurers discovered the mortgage bond market.

It would be too easy to blame the insurers' push into mortgage-backed bonds, especially the risky CDO market, solely on the bond raters; in fact, the insurers saw mortgage-backed bonds as a riskless gold mine, since the bonds were the ultimate in diversification and produced massive fees for all the players involved in bringing them to market.

The two biggest bond insurance companies, Ambac and MBIA, were public companies that had to face the same investor demands for relentless growth that had prodded Stan O'Neal and the rest of Wall Street into the mortgage market, and with that, the big bond insurers became major players in the mortgage bond market. They got a boost when AIG left the CDO insurance business in late 2005; during 2006 and into the next year, almost no mortgage bond issue went to market without the insurers playing some role.

And like O'Neal, the insurers were now suffering for it. Already one bond insurer, the undercapitalized ACA, had gone belly-up. Its implosion had been one of the reasons that Merrill had been forced to take such large losses on its CDO holdings in 2007; billions of dollars' worth of Merrill's CDOs had been insured by ACA.

But Merrill also had billions more insured by Ambac and MBIA, widely regarded as blue-chip outfits, except when Meredith Whitney, the analyst who had been among the first to raise questions about Citigroup, began to put both companies to the test. A year earlier Whitney had been an unknown, working for a second-tier firm and fighting to get her research aired on CNBC and elsewhere, where producers viewed her as little more than a publicity hound looking to make a name for herself by using her good looks, blond hair, and strident views on the banking system's problems, particularly her skepticism about Citigroup, to propel her to stardom. What they failed to see was the analytics behind her research; she showed that Citigroup's costs were running out of control, while its revenues were flat. Her view: Citigroup would have to unload $100 billion of assets to make up for losses and lower revenues. That had been her assessment in late 2007, and it would turn out to be as accurate as one can get on Wall Street.

Whitney was applying the same analytics to the bond insurance business. The problem, Whitney believed, was bigger than ACA's implosion or the possibility of big losses at MBIA and Ambac. Bond insurers were one of the last lines of defense to prevent the complete implosion of the mortgage-backed securities business and maybe the entire financial system. Banks such as Merrill and Citigroup had sought insurance as a way to keep risky assets on their books at full value—essentially buying the triple-A ratings of the insurers and placing them on their mortgage-backed securities. This trade occurred at a time when the CDO market had few buyers, meaning that without the insurance, the CDOs were worth, on a market basis, close to zero.

But Whitney believed the triple As assigned by the rating agencies were about as safe as unexploded land mines, and she became one of the first analysts to raise questions about the bond insurers' ability to cover the mortgage debt they had insured. If Ambac and MBIA couldn't meet those demands, further huge write-downs and losses weren't a mere possibility for the banking system, they were all but guaranteed, no matter how many times John Thain assured the world that Merrill was on the road to recovery.

Just as Whitney's report on the banks dispelled the false optimism in the fall of 2007 that the worst of the credit crunch was over, her report on the bond insurers underscored the fragility of the financial system going into the new year. Whitney and the rest of Wall Street had once believed that Ambac and MBIA were different from ACA, which had lost its triple-A rating and been liquidated at the end of 2007. Her prevailing assumption, and that of the rest of the market, was that they were in the Citigroup category, so to speak: too big to fail. Because of their reputation in the market—they were among the oldest and most profitable insurers in the country, and they guaranteed trillions of dollars in bonds, everything from CDOs to municipal securities issued by states and cities—regulators would never allow them to go down, which would truly give the phrase "systemic risk" a new name.

No longer. Whitney was now saying the insurers were indeed candidates for failure because their capital seemed inadequate to cover the growing problem of mortgage debt and because regulators probably didn't have the stomach to cover all of their likely losses.

If they failed—or at the very least lost their triple-A ratings—the ramifications for the market were huge. That would place an additional $40 billion to $70 billion of mortgage debt held by the banking system in jeopardy of further write-downs and losses.

Under her microscope the dirty secret of bond insurance was exposed. The rating agencies had made diversification into insuring mortgage securities a condition for the bond insurers to keep their triple-A ratings, but the price of diversification had been huge, as Whitney's analysis pointed out. The market, meanwhile, was betting that, given all the risky mortgage debt the insurers were now covering, the bond insurers were on the verge of collapse.

Into February and March 2008, insured bonds, both mortgage-backed securities and municipal bonds, began to trade as if they had no insurance whatsoever. Shares of both MBIA and Ambac, meanwhile, traded as if the companies were about to go out of business.

The rating agencies, embarrassed about missing the subprime crisis, tried to get ahead of the coming fallout. Fitch Investors was the first to move, downgrading Ambac and placing MBIA on its suspect list. The other agencies announced that they were considering similar moves. The fate of the bond insurers became a proxy for the market, and for the economy as a whole, as the prospect of losses on tens of billions of dollars of mortgage debt grew by the day.

Y ou people created this mess . . . and the headline on this is going to be: 'How Wall Street Ate Main Street.'"

That's what New York State's insurance superintendent, Eric Dinallo, was telling some thirty executives of the big Wall Street firms and banks who had assembled for a meeting in his Manhattan headquarters. Dinallo was well known to the Wall Street community. A former prosecutor for New York Attorney General Eliot Spitzer, Dinallo had led an investigation a few years earlier that had forced many of the people in the room to pay $1.2 billion to settle charges that they had misled investors into buying worthless technology stocks with fraudulent stock research.

He had become insurance superintendant about a year later, after Spitzer was elected governor, and now he thrust himself into the bond insurance mess by employing some of the same strong-arm tactics his boss had used when attacking his Wall Street targets. Dinallo now wanted the big firms to put up billions of dollars to bail out the bond insurers. He said it wasn't just their patriotic duty but also in their best interest—once the insurers went down, the losses on their books would pile up and the markets would bleed.

Dinallo's entry into the dilemma facing the bond insurers was somewhat self-serving since, as insurance superintendent, he had oversight responsibility for the bond insurance industry that now had blown up on his watch. More than that, Dinallo was largely oblivious to the massive risk taking of the giant insurer AIG, another company now under his jurisdiction. AIG lost its triple-A rating and significantly ramped up its insurance of risky mortgage bonds as soon as Hank Greenberg was ousted by his boss, Spitzer, then the attorney general and now the governor of New York.

During the meeting, the attendees, mostly general counsels from the big banks, were polite, even gracious; but when they left, several called the Treasury and the Fed, their primary regulator, and reported that Dinallo was overstepping his authority.

But Dinallo had touched a nerve. His plan attracted serious media attention—after all, he was the first public official to actually try to do something concrete to stem the further breakdown of the financial system unleashed by the mortgage crisis. Both Paulson and Geithner continued their regular briefings with the heads of the big firms, and with some senior officials, about the markets and access to credit.

The general consensus among regulators was the same as it had been in August 2007, when Bernanke had remarked that the bond markets were getting "ugly": whatever issues the financial system faced regarding losses, they would be overcome in time. Put simply, the regulators believed that they had a handle on the problems facing the Street and the financial system, nothing that couldn't be handled with lower interest rates and easy money supplied by the Fed, as it does during every crisis. Dinallo, meanwhile, believed something bigger was needed, at least in his corner of the market, the insurance industry, which he wanted to save as the firms' financials continued to deteriorate. In doing so, he was taking aim not just at Wall Street but at the short sellers who were betting that Meredith Whitney was right and the insurers were toast.

One short seller, William Ackman, became a favorite target; for years, Ackman had been sounding the alarm bells on the insurers, particularly MBIA. At one point he had issued a plan to "save" the bond insurers, which the insurers themselves rejected as a backdoor attempt to kill their companies. Maybe it was, but his plan forced even greater scrutiny of the insurers' financial condition, and the short positions continued to mount as investors rejected the insurers' assurances that they had strong capital levels. At one point, Dinallo even raised the possibility that he might be forced to impose a little-known state law that prevents investors from spreading rumors about insurance companies.

The irony of the Dinallo-Ackman battle is that they agreed on the central issue: the bond insurers were in trouble. Only Dinallo wanted to save the business (the last thing he or his boss, Eliot Spitzer, needed was a major industry blowing up on their watch), while Ackman wanted it wiped out. With that, Ackman's plan to save the bond insurers and Dinallo's response became one of the most closely watched stories in the markets through the first three months of the year. The markets swooned and recovered with every disclosure about Dinallo's rescue attempt or Ackman's next analysis of the insurers' dismal prospects. By late February, the "bailouts" began to emerge. Though they were nothing more than plans to recapitalize the companies in the difficult market, with Dinallo prodding the banks to underwrite the stock even if investors weren't eager buyers, they had the desired, albeit short-term effect: MBIA and Ambac managed to survive with their triple-A ratings intact for the time being and the markets recovered a bit, as did the share prices of Ambac and MBIA.

Ackman, who had become a household name on Wall Street for his analysis of the insurers and his fight with Dinallo, had lost the battle, but he wasn't going away quietly, and neither were the short sellers, who quietly began to pick bigger targets, namely, the Wall Street firms themselves.

To be sure, Wall Street bristled over Dinallo's strong-arm tactics, comparing him to his autocratic boss, the hated former sheriff of Wall Street and now governor, Eliot Spitzer. His attacks against Ackman at times even seemed juvenile, since Ackman didn't have to spread rumors; he had been issuing negative research on the insurers for years, so his position was well known. Dinallo's strong-arming was off-putting and dangerous.

Dinallo may have been worried about his career, as all politicians do, in trying to save the bond insurance business, but he saw the future, and it didn't look bright. Wall Street had in fact created a mess that threatened to destroy Main Street. The mortgage crisis was now in one of its lulls as the big banks weighed their holdings, took the best scenarios in their pricing projections, and in the process forgot about the $100 billion in losses they had taken and the tens of billions more they were going to take unless there was a miracle.

That miracle, of course, would be that housing prices would recover; that subprime borrowers would somehow outperform their credit histories and get well-paying jobs so they could afford the inflated mortgage payments that kept coming due; and that all of this would filter into the mortgage securities held by the big banks and change market psychology so prices would recover to their 2006 levels and the firms could begin writing up prices, rather than writing them down.

But there would be no miracle, not for Wall Street and certainly not for Main Street.

Mr. Estronza, my name is Steve Davis, and I'm here to investigate your mortgage."

There were few New York City cops tougher than Steve Davis, a former detective and retired captain who had had twenty years on the job before retiring to become a private investigator. Davis had seen it all during the 1970s and '80s, the worst years to be a cop in New York City—murders, rapes, muggings, and an occasional Mob hit or two. Nothing seemed to faze Davis, who after seeing the worst of human behavior had developed an almost clinical approach to his work. That's what made him such an effective private investigator: he could compartmentalize his emotions.

But what he was doing now made him yearn for the time he had been dodging bullets in some of the seediest parts of New York City. He had recently built a reasonably successful business conducting security for large Wall Street firms and insurance companies. One of the growth areas in 2007 and 2008 was investigating mortgage fraud. Davis's job on behalf of the insurance companies seemed pretty simple: interview the borrowers who defaulted on their mortgages, find out if there was fraud of any kind, and report back. The reality was more complicated because Steve Davis was now face-to-face with the human toll of the mortgage crisis, which went beyond fired CEOs and traders who lost billions betting on collateralized debt obligations.

Part of that human toll included Jose Estronza, a polite man in his midforties, who let Davis into his home in the working-class town of Corum, New York. Davis had seen many crime scenes in his time, certainly his fair share of blood and brains spread across dark alleys and along sidewalks. It had left him hardened and somewhat impervious to his emotions. Yet the scene in this home for some reason hit hard.

Maybe it was because Jose Estronza was so polite; he immediately offered Davis a cup of coffee. Or maybe it was the home itself. Davis's previous mortgage fraud cases had been largely investigating people living beyond their means. They fit the stereotype of the subprime fraudster: a family borrowing big bucks to live in a McMansion with wide-screen televisions, chandeliers, and expensive artwork paid for through deceit by the mortgage broker, the borrower, or a combination of both.

In contrast, Jose Estronza seemed to be an honest, hardworking guy. He was dressed in hospital scrubs. His McMansion was no more than a modest two-bedroom home with peeling paint and crab grass on the lawn. He had purchased the house for $329,000 a couple years earlier with a subprime loan. What was striking to Davis was that it now had almost no furniture. Estronza's sixteen-year-old daughter sat on the floor watching a smallish color television set perched on a box. The only other item of note in the house was a table littered with what looked like bills and collection notices. It was a scene of economic devastation, and it made Davis sick.

"I would offer you a seat, but as you can see, I don't have one," Estronza remarked, sounding embarrassed.

"No problem, let's go in the other room so we can talk," Davis said, looking to save Estronza the indignity of having to admit in front of his daughter that he was a deadbeat. "By the way, where's all the furniture?" Davis inquired. Estronza said he'd had to sell it to make his mortgage payments, which had ballooned just a few months after he took out the loan.

"Mr. Estronza, that's what I want to talk about. It says here on your mortgage application that you make about $70,000 a year and that you work for a consulting firm in Corum." Estronza was bewildered and said he'd been working as an orderly in a hospital for twenty years, earning about $30,000 a year. Davis then asked about another item on the loan application. In Davis's file there was something that purported to be a Citibank account statement with Estronza's name on it. It said he had $30,000 in savings.

"Mr. Davis, I have never had a Citibank account," Estronza said. "I have an account with Banco Popular, and I don't think I have ever had more than $1,400 in it." Davis showed him the documents with his signatures, the bank accounts with his name on it. "Sir, all I can tell you is that I have never seen that stuff before."

It didn't take Davis long to realize that the Estronza case went beyond the simple irrational exuberance of a mortgage broker looking to tweak Estronza's FICO score or embellish his salary. This case smelled of outright fraud, which was confirmed by what he learned next.

Davis asked Estronza how he had come upon the house in the first place. Estronza said he had been living with his daughter and girlfriend in an apartment in Brooklyn. He'd always wanted to own a home in the suburbs, a chance to raise his family among trees. He'd seen an ad in a newspaper for "affordable housing, no money down." He'd called the number and had spoken to a representative of a local contractor who was building homes on Long Island starting at $329,000. Estronza told the sales rep he couldn't afford that because he made only $30,000 a year.

The sales rep said it wouldn't be a problem. There were people willing to lend money to people like himself—decent, hardworking people. There were ways to make mortgage payments affordable. Estronza said he was told to trust the experts, which he did.

Within a few days, he was in front of a loan representative, a mortgage broker, and an attorney who handled the details of the closing, and after signing a few papers, Jose Estronza had his own home.

This is when the scam became crystal clear in Davis's mind; the builder had had it all rigged, he believed. The loan officer and the alleged attorney both worked for the contractor. They had forged the documents that had been submitted to the mortgage company, including phony bank statements and W-2 forms.

Their fees had been embedded in the loan costs. In an era of low interest rates, easy money, and securitization, the bank had never checked on whether the documentation was real—and why should it have? The loan was not its problem because it would be sold, to Bear Stearns, Citigroup, or some other bank that would package it and many others into mortgage-backed bonds.

In many ways the banks' taking such huge losses on those securities was a form of rough justice; they had been the ones that had demanded "product" for their bonds, thus spurring the lower lending standards that had led to people receiving loans they could never repay. And the banks hadn't cared, at least at the time, about the consequences. Estronza said he had never received a follow-up call from the mortgage company that had made the loan. He had merely signed where he was told to by "the lawyer" supplied to him by the mortgage broker.

Davis's previous case had been a twenty-five-year-old woman who had defaulted on mortgages used to purchase seven homes in New Jersey. She had relied on her brother, himself a mortgage broker, to make the numbers work. He'd had no problem chasing down that woman. She'd been looking to dupe the system and known what she was getting into.

By comparison, Jose Estronza was an innocent. Maybe he should have known better, asked how much his mortgage payments would go up, known he couldn't stretch a small salary to meet those payments. More likely, he had been prodded and pressured into thinking he could afford a $329,000 home on a salary of $30,000 a year.

"Mr. Estronza, thanks for your time," Davis said, concluding his work. But before he left, he saw Estronza's daughter watching television, and again he started to feel ill. That's when he asked Estronza, "So what are you going to do now?"

"Mr. Davis," Estronza replied, "this is more than terrible. I not only lost my home, but I lost everything I had trying to pay for this." Estronza said he was going to stay with friends and maybe family until he got back on his feet. He didn't say what would happen to his daughter.

Davis left the house saddened. He went to a local bar and downed a couple of vodkas, and that's when the old Steve Davis, the analytical detective, returned. Davis knew enough about the business to know that the original loan granted to Estronza was no longer held by the mortgage originator; it had been sold to a bank and was now embedded in a mortgage bond.

That bond was likely to be held by a bank or a Wall Street firm that was looking for his client, the insurance company, to make good on its promise to cover the cost of the loan in the event of default. But the insurance company would pay only in the absence of fraud, and Davis hadn't seen a case so far that didn't include fraud of one form or another.

What that meant for all those mortgage bonds held by the banks, the possibility of further losses piling up on their balance sheets, didn't really concern Davis, at least not now. That's because as much as he drank he still couldn't get out of his head the scene of that house without furniture, the girl seated on the floor watching a television perched on a box while her father worried whether his family would have a roof over its head.

By March 2008 government policy makers had taken several modest but significant steps to unfreeze the credit markets, from accepting just about any collateral a bank could pony up for overnight borrowing to the Fed's cutting interest rates to unprecedented levels. Even Larry Fink was starting to feel a little better about the overall health of the economy and the market's ability to recover soon.

At an investor conference in Mumbai, he remarked that the subprime crisis wasn't over; indeed, there would be additional losses by banks around the globe. But he pointed out that despite all the alarm in the United States over the banking system's woes, the U.S. economy had weathered the crisis pretty well. The economy was still showing positive growth, and "if there is at all any recession," Fink added, "I don't think it will be a deep one. It will be very shallow, very narrow."

The stock market, as measured by the Dow Jones Industrial Average, isn't the best barometer of economic health, but it does provide a pretty good gauge of investors' sentiment about the future since so much of the nation's wealth is tied up in stocks. It was hovering at around 12,000, not bad considering everything that had happened over the past year and what was continuing to percolate on the banks' balance sheets.

Professional investors had spent the past six months with access to the best information that had ever been publicly available about the big banks and securities firms. FASB 157, the accounting rule that had forced the firms to price their investments to what the market said they were worth, and the now better disclosure of their assets in terms of so-called Tier 1, Tier 2, and, the most risky, Tier 3 capital.

Hedge funds and other investors looking to make bets on the future of the banking system were taking sides on the banking system's health; for every Joe Lewis betting billions on the future of Bear Stearns looking bright or Temasek Holdings betting $4.4 billion on Merrill's future under John Thain, there were nameless others, mostly hedge funds, lining up with massive short positions on the banks after poring through the new disclosures and coming to the conclusion that for all the CEOs' happy talk, the banks were in no different a position than they had been in late 2007; they still held billions of dollars' worth of risky mortgage debt that no one wanted to buy except at prices that would put the major securities firms in jeopardy of survival.

How could two sides be interpreting information so differently? Lewis, of course, is a commodities trader with a gambler's instinct. In some ways he was just riding out his preexisting bet on Bear (he would later remark, "What's a billion dollars? I still have eight left."). It didn't seem like such a gamble at the time; Merrill Lynch's veteran securities analyst Guy Moszkowski had been predicting for some time that the firm would be sold to a bank. After all, sovereign wealth funds such as the Singapore-based Temasek were thinking strategically—the United States was selling chunks of its crown jewel banking system relatively cheaply, and now might be the chance to establish a financial beachhead there.

The hedge funds, meanwhile, saw the glass as being half empty. They were looking at the numbers, the massive debt and leverage, and the feckless management. In addition to the massive holdings of mortgage debt, the Wall Street firms and banks clung to a business model that relied on others through heavy borrowing. Most of the big firms continued to remain highly leveraged, even if they began to reduce their short-term borrowing somewhat at the end of 2007.

Bill Heinzerling was, at least for the moment, out of the business, working on creating a broker-dealer and investment advisory firm. He was also part owner of a popular restaurant in downtown Manhattan, The Harrison, just a block away from the Citigroup office where he had once worked. Life was good, primarily because he had left Citigroup; in 2005 he had walked into Tom Maheras's office and said that after twenty years with the firm, he had had enough. The business had changed, the company had changed and not for the better, and he wanted to do something different. Maheras agreed that the time was right for him to leave, but not with his assessment of the firm.

Based on the events of the past year, Heinzerling's gut, which had told him Citi was a sinking ship, was looking more and more accurate. Even so, Heinzerling didn't think his old firm, no matter how dysfunctional the management or how low the stock price fell, was in imminent danger of collapse. "They're a bank and too big to fail," he said.

But Bear wasn't.

"I've been trading this stuff for years. You can smell it," one trader told me around this time. "Things aren't getting better over there. They have all this stuff on their balance sheet. They can't sell the place, and no one knows how to manage it. They're going out of business, mark my words."

By early March 2008, that trader's speculation was becoming universally accepted not just by academics like Nouriel Roubini but by hedge fund managers. The hedge fund manager and prominent short seller James Chanos had seen this point of view become more accepted in the hedge fund business since the demise of Bear's hedge funds, and now it had become common knowledge, to the point that his colleagues were actively talking about which firm was so loaded with bad debt that it couldn't survive. The name on most people's lips: Bear Stearns.

Chanos's firm is called Kynikos Associates. Kynikos is Greek for "cynic," which is a good way to describe short sellers in general and Chanos in particular. Chanos ran a hedge fund that had been shorting stocks since the mid-1980s. He had built one of the most successful businesses among those that specialized in betting on companies' failure and then profiting from it. He was well known on Wall Street, despised by some but respected for his analysis and savvy. The investors in his fund loved him, of course, because he was very good at what he did.

He hadn't shot to stardom (or infamy, depending on your point of view) until his most prescient call became public. In 2000 Chanos had warned investors that the energy giant Enron was a house of cards that had misstated its earnings and just about everything else regarding its financial health. It was a gutsy call even for Chanos; Enron's stock was trading at $90 a share at the time, with revenues of $100 billion. Its chairman, Kenneth Lay, was a friend of President Bush, and its CEO, Jeffrey Skilling, was considered one of the best in corporate America.

But Chanos stuck to his conclusions: Enron, he said, wasn't as much an energy firm that drilled for oil as it was an intermediary like a big investment bank that took positions in far-flung energy markets trading oil, gas, and various forms of electricity. It was a complex business model, and the markets in 2001 were hurting as a result of the dot-com bust and the 9/11 terrorist attacks. Chanos began to ramp up his short position as Enron began to disclose more and more of its problems; by the end of 2001, it was clear that Chanos hadn't overstated Enron's predicament. Enron had hidden its losses in a massive shell game of off-balance-sheet partnerships. The stock price fell to a mere $1, and the firm filed for bankruptcy in what was then the largest accounting fraud in U.S. history.

Chanos didn't think the big banks were guilty of fraud as much as they were guilty of clinging to a business model that left them in a precarious position in times of trouble. He understood how the firms funded themselves with short-term repo financing that could dry up at a moment's notice. Moreover, he had around $600 million with Bear Stearns in a prime brokerage account, which would disappear if Bear went bankrupt, as some of his brethren in the short-selling business were now predicting.

Chanos wasn't in the Bear-going-bust camp just yet, but he was aware that the firm faced significant problems. The first sign of trouble he could recall came in late 2007. He and Cayne were having dinner at San Pietro. Cayne was coming under increasing pressure at Bear. Chanos thought Cayne would press him for money, possibly to keep more funds in his prime brokerage account. But he didn't want to talk business, just golf and bridge. Cayne seemed distracted and sounded as if he were hiding something, Chanos thought at the time.

A few weeks later, Chanos attended a wedding for a friend and ran into a high-ranking Bear executive there. He asked what was happening at the firm with all the focus on Cayne's pot smoking and the daily scrutiny in the press. Chanos ribbed him that the firm might go under, like Enron, so maybe he should start shorting Bear.

"Don't laugh," the executive said. "In a year from now, we're not going to be around."

Chanos dropped the subject, but now the guy in charge of his "back office" operation, Alan Best, was revisiting it. Chanos had been shorting the financial sector stocks since 2006, though he says he didn't short Bear.

Best's job was to keep tabs on all Kynikos's administrative needs, including the money at its various prime brokers. He joined with five other partners to tell Chanos that even if he didn't short Bear, the growing consensus was that the firm might not survive and as a result Chanos might want to safeguard the firm's cash. The write-downs and losses of 2007 had been just the beginning, investors were beginning to conclude. Mortgage debt wasn't trading now because banks weren't looking to take any additional losses by selling their toxic waste. So it was just sitting there on their books.

But when the next "mark" came from a trade, the system would have to adjust and write down the additional massive losses. All the banks from Citigroup to Morgan and even Lehman and Goldman, which had barely been touched by the crisis (like Fuld, Goldman CEO Lloyd Blankfein had earned a massive bonus for keeping the firm relatively clean in 2007), would eventually fall prey because they too held all the same debt, used massive leverage to support their operations, and would now have to write down their losses, or so the theory went. It was what Jesse Eisinger had laid out in the November 2007 issue of *Portfolio* magazine: eventually, Wall Street would be toast.

But Bear would get burned immediately. All the controversy surrounding the firm made it the symbol of the crisis, and it was hard to argue with the facts. Its relatively small size made it more difficult for it to weather massive losses, and its leverage was huge. Most alarming was the fact that the hedge funds that supplied the firm with so much cash could also expedite its downfall by removing money from their prime brokerage accounts. Sitting through all that made Chanos—who, after all, had made his fortune off of doom and gloom—queasy. "So what does this mean?" he asked.

What Best had begun to hear is what could best be described as the beginning of a run on the bank. The hedge funds seemed to be moving first. Like Chanos, they held money at Bear through their prime brokerage accounts and were moving the funds into "custody," meaning they were off limits for Bear to fund its operations.

Given the firm's tenuous funding position, Bear needed every funding source it could muster. With hedge funds pulling out money, others would stop doing business with Bear as well. Repos could dry up, and issuing short-term commercial paper would be difficult if the crisis of confidence picked up steam, as Best was suggesting it would, and then Bear would be in trouble.

Chanos hadn't seen Cayne since their dinner, and now he knew why. In late February Chanos had hosted a short sellers conference in Miami and several conversations focused on what would happen if an investment bank did declare bankruptcy. Could the funds get access to the money in their prime brokerage account if a bank was facing court-ordered liquidation? No one could answer the question. Some had said it was up to the bankruptcy court. Everyone had concluded, why take the risk?

After digesting what Best had told him and thinking back to that conference, Chanos made the same decision about his relationship with Bear: why take the risk? During the second week of March he ordered Best to safeguard the firm's $600 million in prime brokerage funds and place it in custody. He didn't short the stock—he wouldn't; Chanos might be a hated short seller, but he was a client of Bear's and, being hated, he knew that shorting the stock while pulling out his funds in this environment would be a recipe for a call from the SEC.

He wasn't alone.

H ey Jimmy, it's John, we need to talk."

Like Jim Chanos, John Angelo was a longtime Bear Stearns client. Angelo and Jimmy Cayne were close friends and frequent golf partners, and they belonged to the same country club in Deal, New Jersey, where they both had homes. He had seen the toll the past year had taken on his friend. When Bear had run into trouble in the summer of 2007, he had been among the first to notice Cayne's poor health; Cayne had nearly passed out during a golf game, and Angelo had prodded him to seek medical help. Cayne, of course, had survived, but Bear kept getting worse.

And the worst was yet to come. Cayne no longer had an operating role at Bear Stearns. He was a nonexecutive chairman, and in early March he was preparing to take advantage of not being responsible for the daily grind at the firm and play in a bridge tournament, just as dozens of firms, including Chanos's and now Angelo's, were pulling money from Bear's prime brokerage account.

When Angelo broke the news to Cayne, he took it surprisingly well. He didn't argue or fight but accepted Angelo's decision as a matter of business. "I have to protect my firm," Angelo explained. The rumors about Bear's impaired financial condition, its holdings of illiquid debt, and its shaky finances had yet to make it into the press, but they were floating around every trading desk.

What Angelo didn't tell Cayne was that he wasn't just working on rumors; he knew the firm had many problems, including exposure to distressed debt, which happened to be Angelo's specialty. All the mortgage bonds on Bear's books might have been diversified and hedged, as Bear was telling investors, but they were worth crap on the market.

Angelo had been around Wall Street long enough to know how fast funding could dry up for a firm when the markets turn ugly. He had seen what had happened to Lehman Brothers in 1998. As a matter of caution, he said, he had now taken his money out of Bear. He told Cayne he would return once the crisis, or the rumors of the crisis, subsided.

John Angelo's decision, coupled with Chanos's earlier move, should have sounded alarm bells in the executive suites at Bear, since they were both longtime clients and friends of the former CEO, but by all accounts they didn't.

Cayne left for his bridge tournament. Alan Schwartz had been running the firm for only six weeks, though he still hadn't done much to convince investors that Bear was turning things around.

Schwartz was simply too overwhelmed to realize the significance of what was happening. He was dealing with angry investors, a brand name that had been beaten and bruised, and a need to attract talent if he were ever to grow the firm's business enough to entice a buyer. Bear's holdings of toxic debt would have to be unloaded systematically if the firm were to survive, so he needed to hire a capital markets chief. Schwartz was so desperate that at one point he held extensive discussions with Tom Maheras but never extended him an offer.

He also had to do the things that CEOs need to do, that is, schmooze with clients and investors. Sometime after the first week in March, he left for Palm Beach to host Bear's annual media and telecommunications conference at the posh Breakers Hotel, where the speakers included some of Bear's biggest clients in the media business, such as News Corporation's Rupert Murdoch and Viacom's Sumner Redstone.

If Schwartz was overwhelmed, the regulators were oblivious. For all their conference calls with Wall Street executives and those seeking market insight from Larry Fink, the economic brain trust of Paulson, Bernanke, and Geithner had yet to act in any significant way even as the rumblings of Bear's implosion started to pick up steam in early March. The SEC chief, Christopher Cox, a former Republican congressman from Orange County, California, seemed the most clueless. Under the new regulatory agenda, the SEC had frontline responsibility to make sure the big investment banks had adequate capital to withstand financial storms. But the SEC under Cox gave Bear a clean bill of health as the bank run continued through the first week of March and beyond.

What's striking about Cox's inaction, and for that matter Paulson's, Bernanke's, and Geithner's, is that Bear's actual unraveling was a slow-moving event, beginning in earnest the prior year as it faced all the difficulties Alan Best had laid out, including massively high leverage that made access to borrowing—or, in Wall Street parlance, "liquidity"—the key to its survival. Paulson, Geithner, and Cox had time to react, yet they failed to do so, beyond engaging in a few conference calls with Alan Schwartz and Jimmy Cayne.

Which is why short sellers make so much money.

What made Brad Hintz a particularly effective stock analyst was that he had actually worked in the business he covered; he had served as Lehman Brothers' CFO in the 1990s. Investors trusted his insight because he knew how the sausage was made. Hintz wasn't the first analyst to raise concerns about how the mortgage meltdown would spread losses through the banking system; that distinction goes to Meredith

Whitney. But he was increasingly raising alarms about the banks, and now he was raising a specific alarm about Bear Stearns.

On Monday, March 10, Hintz recommended that investors treat bank and brokerage stocks like the plague; he had a "market perform" rating on Bear, his old firm Lehman, and Goldman Sachs, meaning that his official stance was neutral, but his warning was anything but. "We believe there are several more quarters of write-downs to come as the financial leveraging that had benefited the group and the overall financial market during 2004 to the first half of 2007 continues to unravel," he wrote in a research note that morning.

Bear, the weakest of the brokerage stocks, began a free fall; the share price dipped precipitously at the market open, and other brokerage stocks followed. Hintz's comments were a blast of reality to a market that had been spun for months by Wall Street CEOs about the brightness of the future; he was saying just the opposite, and holding those stocks was like holding lit dynamite. A second blast came from Moody's Investors Service, the big rating agency that had been asleep through much of the subprime crisis. Moody's had finally awakened to the fact that Bear Stearns in particular had exposed itself to mortgage bond deals packed with alt-A mortgages, the less risky but still toxic cousins of subprime mortgages. Moody's downgraded fifteen Bear deals, many of which had produced debt that was still being held on Bear's books.

Bob Pisani, a veteran CNBC reporter who worked the floor of the stock exchange during the day, noticed the wild activity. Pisani, a smart, amiable reporter, considered it one of his main jobs to debunk rumors that spread through the floor and influenced trading.

The rumor he heard now—that Bear was facing a funding crisis, meaning that some of its creditors were backing away from their loan commitments while others were demanding large levels of collateral on trades—seemed impossible to debunk. There was no overt evidence that Bear was imploding other than the market and the persistent rumors among traders that the firm was in deep trouble, yet Chairman Jimmy Cayne and the new CEO, Alan Schwartz, weren't speaking to reporters.

Ace Greenberg broke the silence. Even in semiretirement, Greenberg carried considerable clout at the firm. He still traded on the floor. He ran the risk management meetings, and people sought out his advice. Cayne had become the face of Bear in recent years (something Schwartz was desperately trying to change; he was sick of seeing Cayne depicted in the *New*

York Post wearing beads and braided hair and surrounded by marijuana plants), but in the eyes of the public Bear would always be considered the firm that Ace built.

And Ace was and always would be a guy who marched to this own drumbeat. It's what made him a media darling; the press loved his mannerisms, from the magic tricks he performed on the trading desk to the fact that he answered his own telephone calls. Cayne saw the dark side of Greenberg's personality; it's why he had never doubted the sexual harassment story.

As crazy as Cayne seemed, Greenberg could match him in being off the wall. It was, after all, Greenberg who had once donated $1 million to a hospital so homeless men could enjoy sex by having access to free Viagra. He had made a splash of it, making the announcement in an article in the *New York Times* without alerting Cayne, who first heard it when he picked up the paper that morning and nearly hit the ceiling.

"Are you fucking kidding?" Cayne screamed at Greenberg after reading the story. "A million bucks so homeless men can jerk off? How does this make the firm look? How does this make *me* look?" Cayne snarled before slamming down the telephone.

Greenberg's habit of opening his mouth when the firm would rather he keep it shut continued. On Monday just after noon, CNBC's Michelle Caruso-Cabrera reported that Greenberg had said the rumor about a liquidity crisis was "totally ridiculous." Maybe so, but by talking about the situation, he was sending a signal to the market that management was concerned enough about the speculation that it felt the need to shoot it down. In other words, by dignifying the rumors, Greenberg suggested they might be true.

Bear's stock continued to fall. Inside Bear's offices, CFO Sam Molinaro was sweating. First, he didn't like the fact that Greenberg was talking on behalf of the firm. Second, he couldn't figure out why the market had turned so negative against the firm so quickly. To him, it seemed the situation wasn't any worse than it had been all year.

He spoke to Schwartz, who was still in Palm Beach at the media conference. Schwartz saw no reason to return, although before the end of the day, he did issue a statement that Bear's "balance sheet, liquidity, and capital remain strong." But that couldn't stop the carnage. Bear's stock price fell 11 percent that day, to $62.30. Its market value was $8.5 billion, which, according to a report on the Dow Jones wire service, made it not just the smallest firm on Wall Street but not much bigger than the motorcycle company Harley-Davidson.

The lone bit of comic relief amid all the tumult in the market was the parallel implosion of the former sheriff of Wall Street, now the governor of New York State, Eliot Spitzer. Just as the trading in Bear picked up steam that day, news reports said the governor was involved in yet another scandal; he had been implicated in a prostitution ring. It was yet another blow for the man who had targeted Wall Street abuse and parlayed that into a meteoric political career. In recent months he had become the focus of an investigation by his successor as attorney general, former HUD Secretary Andrew Cuomo, over allegations that Spitzer had used state troopers to spy on his chief political opponent, New York State Senate leader Joseph Bruno.

If the so-called Troopergate had taken Spitzer's reputation down several notches, news of his tryst with a twenty-two-year-old prostitute, Ashley Alexandra Dupré, was devastating. Lurid stories about their liaison filled the press for days. Spitzer had recently appeared in Washington to give testimony to Congress on the Ambac bond insurer situation, but, as the *New York Post* reported, the previous night he had had an encounter with Dupré in his Washington hotel room. Another *Post* story revealed that he liked to have sex with his knee-length black socks on; they were described as an "identifying characteristic" of "Client No. 9."

Within a few days of the first story Spitzer was forced to resign, to the tremendous glee on Wall Street among those who'd been the target of his aggressive, no-holds-barred tactics. Bear took a bit longer to meet its end.

There were, of course, twists and turns along the way. Sam Molinaro on Tuesday attempted to calm the waters by telling CNBC that everything was fine, liquidity was strong, and he had no idea where the rumors of the firm's funding problems had come from. He chalked the commotion up to rumors started by short sellers, nothing more.

In retrospect, Molinaro appeared totally unaware (or unwilling to admit) that what was happening at Bear wasn't rumor but reality: an avalanche of orders from prime brokerage customers who were now yanking their money from the firm on top of the growing skittishness of creditors about dealing with a firm loaded with so much toxic debt.

At first Schwartz didn't want to speak about the chatter about the firm because he thought it would give credence to the speculation of a possible funding crisis. On Tuesday, around the time Molinaro spoke off camera to

CNBC, Schwartz decided that he had to do something bigger and grander: he would grant an interview with CNBC to set the record straight once and for all

The members of Bear's management and compensation committee had just concluded a meeting when David Faber broke into CNBC's coverage with an exclusive interview with Schwartz. "Turn it up, turn it up!" someone hollered. On the screen in the sixth-floor conference room, live from Palm Beach, was the embattled boss of Bear. Schwartz had considered leaving Florida and conducting the interview from headquarters. But he was between a rock and a hard place: if he left, it might incite more fear that the company was in trouble. But if he stayed in Palm Beach, it might seem as if management was asleep at the wheel. Either way, there was no winning.

The interview didn't get off to a good start. Faber began with a stunning revelation: that some Wall Street firms would not agree to be a counterparty to Bear Stearns on a particular transaction. Schwartz initially appeared composed. He denied it was the case; there had been problems, he said, "administratively," with processing some trades, but, he added, "we're not being made aware of anybody not taking our credit as a counterparty."

Faber's next question was more specific, namely, that not just any firm but Goldman Sachs "would not accept the counterparty risk of Bear Stearns. You're saying you're not aware that would be the case?"

The Goldman question was important in that the firm by now had the reputation of having been the first to understand the severity of the subprime crisis and had shorted the market. If Goldman was smelling trouble at Bear, maybe the "rumors" were more than just rumors; investors, counterparties, the entire market would head for the exits. All of which made Schwartz's reply critical to the firm's survival.

His answer, however, did not instill confidence. "I'm not aware, and you know, on a specific trade from one counterparty to another," he said, "and where you're a third party, ah, we have direct dealings with all these institutions, and we have active markets going with each one, and our counterparty risk has not been a problem."

It was a rambling discourse, and to at least some people even inside Bear, it seemed to suggest that the market, and Goldman in particular, had concluded that it was best not to deal with Bear.

"Why isn't he explaining this properly?" one of the executives in the conference room demanded. "He's missing the point of the question."

As it turned out, Goldman would indeed not agree to be the counterparty to Bear on the trade in question. But there was a more complicated reason than simply a lack of creditworthiness. When the market panic surrounding Bear had begun to intensify earlier in the week, many trading counterparties that had entered into swap agreements with Bear had become nervous that they could not rely on the firm to meet its end of the bargain, so they had sold their contracts to other brokers, namely, Goldman Sachs, Credit Suisse, and Deutsche Bank, which agreed to take on the risk because they believed Bear was in better shape than the market realized. Eventually those brokers reached their self-imposed limits for exposure to a single counterparty, in this case Bear.

So when Goldman said it would not accept Bear as a counterparty, what it was *really* saying was that it had reached its credit limit with a single entity, rather than that it feared Bear wasn't a creditworthy counterparty. (Goldman said that it eventually accepted the Bear counterparty risk.)

Making the interview even more difficult for Schwartz was something totally out of his control; the anchor, Erin Burnett, suddenly broke in with the news that Spitzer would resign as a result of his sex scandal. It made a bad situation even worse; Schwartz seemed stunned by the bizarre news and never regained his composure about the single biggest deal he had ever had to sell: that it was safe to do business with Bear, despite all speculation to the contrary.

"Jesus Christ, he can't get his point out!" another executive yelled as Schwartz explained the company's party line, first articulated by Molinaro the day before and now underscored by the CEO: the speculation about the firm's liquidity problems, about counterparties such as Goldman not wanting to do business with anything that had to do with Bear, was nothing but "rumors"; the firm had $18 billion in cash on hand, and "when speculation starts in the market with a lot of emotion in it and people are concerned about volatility, then people will sell first and ask questions later . . . our liquidity position has not changed at all. Our balance sheet has not weakened at all."

That's where things got tricky for Schwartz; what many hedge fund clients knew was that the firm's balance sheet, though not yet in tatters, was beginning to become impaired. Hedge funds, such as Chanos's, John Angelo's, and now many others, were protecting their cash balances and pulling money out of their prime brokerage accounts with Bear. The withdrawals weren't lethal yet, but it was a stretch to suggest that everything was fine, and the word spread among short sellers and hedge funds that Schwartz was so desperate that he was willing to say anything to save the firm.

The longtime Bear broker Kurt Butenhoff didn't know about the prime brokerage run; he was more focused on the look on Schwartz's face as he answered Faber's question. "We're in trouble," Butenhoff thought.

A day before Schwartz's broadcast, shares of Bear had rebounded, up 6 percent on the day, the result of a combination of Molinaro's off-camera comment about the firm's strong liquidity position and a comment by SEC Commissioner Christopher Cox that Bear had passed the SEC's capital tests. Now the share price was falling sharply. In his uneven performance Schwartz seemed to confirm the market's worst fears, that Goldman and others wouldn't deal with the firm and the problems were as bad as the recent sell-off of Bear stock suggested. Not only did the stock price dive significantly; if there hadn't been a liquidity crisis before Schwartz's broadcast, one had developed now.

The walls were closing in; hedge funds were now leaving the Bear prime brokerage business in droves. Bear was forced to dip into the $18 billion in cash Schwartz and Molinaro had bragged about as a cushion against losses. Yet an even bigger problem was brewing elsewhere: hours after the Schwartz interview, smaller brokerage firms that cleared their trades through Bear also yanked their cash. That had also been a place where Bear could borrow to fund its other customers. Now it was gone. Even worse was Bear's standing in the repo market; it had turned into an untouchable. Banks were starting to close down repo lines.

That $18 billion in cash was disappearing quickly.

The typical Bear Stearns executive was paid a chunk of his or her salary in stock (top executives received nearly all their compensation in stock), and the past year had been a dismal one. The share price had fallen from $170 to $60, and after Schwartz's CNBC interview it headed lower. The people at Bear were now angry at management, at the markets, and most of all at the media. Highest on their list was CNBC. Many of the bankers and traders came to believe that the media coverage was driving Bear into the ground and that the same management that had loaded up on risky mortgage debt, built a business model that was narrowly focused on a risky part of the market, failed to raise capital at key times, and allowed a bond salesman to run a major hedge fund were not mostly to blame.

"Look at these clowns. It's like tabloid TV!" screamed one Bear Stearns broker as CNBC covered Bear's problems like the major networks cover a presidential campaign. The broker, a woman who had worked at the firm for two decades, might have hated the nonstop coverage, but like many of her colleagues she was glued to the television, and like many of her colleagues she believed CNBC and the rest of the media were "rooting for us to fail."

"Are they considering what they are doing to this firm? Have they thought about the margin clerks? The back-office people? We're not all millionaires."

Her sentiments were being channeled through to management. At bottom, Schwartz, Molinaro, and all the top players at the firm now believed that the shorts, through their sources in the media, had the firm in the corner—hedge funds were spreading false rumors that were driving down the share price, causing investors to buy credit default swaps on Bear's debt that added to the uncertainty. Inside Bear there was no recognition of management's mistakes dating back to years before, even months before, when Cayne should have sold out to Jamie Dimon at JPMorgan Chase at a reasonable price, or if and when Schwartz should have raised capital while he was assuring the markets that all was fine when it obviously wasn't.

By Thursday, the run on Bear became even more deadly. It had now extended to the broker dealers—not just prime brokerage clients but now traders were either demanding additional collateral on repos or refusing to repo with Bear at all. Retail clients were getting nervous as well. Despite CNBC commentator Jim Cramer's wild pronouncement that their cash would be safe in reaction to a question from a viewer during his show *Mad Money* ("No, no, no! Bear Stearns is fine," he said. "Do not take your money out. . . . If anything, it will be taken over."), this money kept fleeing the firm. Bear was discovering that $18 billion in cash dissipates pretty quickly when you run a firm with a balance sheet approaching $400 billion and no one will lend you money. Only a few days earlier, Schwartz and his CFO had been on CNBC telling the markets that Bear was fine. By Thursday, he and his team were in panic mode.

Schwartz called in the firm's banker and lawyer, Gary Parr at Lazard Frères and Rodgin Cohen at Sullivan & Cromwell. They made contingency plans: Parr, to possibly sell all or part of the firm; Cohen, to pressure the Fed to open up a loan program to investment banks immediately. Tim Geithner at the Fed seemed cool to the idea, but he did tell Cohen that if the problems at Bear were so bad, it was Schwartz who should be on the telephone with him, not his attorney.

Schwartz waited. People who know him say he was still looking for a Hail Mary pass of some sort, something to buy the firm some time, help it live through Friday, and then he would have the weekend to find a buyer.

Jim Chanos didn't know Alan Schwartz well; in fact the two had met briefly during a breakfast a few years back and had never spoken again until now.

On Thursday night, Chanos and a couple of friends were on their way to the Post House steakhouse on Manhattan's Upper East Side for an early dinner. The markets had just closed, and Bear stock had taken yet another pounding. The next day he was planning to travel to Europe to meet with investors. That's when he received a call from a "272" number, the prefix of all the telephones at Bear Stearns.

On the line at 6:15 New York time was Alan Schwartz. As Chanos recalls, Schwartz asked him if he could go on CNBC and make a comment about Bear that might calm the markets. Chanos told Schwartz he wouldn't do such a thing; based on everything he was hearing, Bear was on its last legs. Schwartz, according to Chanos's recollection, countered that it was all a bunch of rumormongering from the short sellers and Bear was fundamentally fine. Chanos didn't bother to argue, but he didn't change his position either.

Chanos left for Europe early the next morning, only to discover when he landed how right he had been all along.

Despite his loyalty as a longtime Bear client, he must have regretted missing out on the short sale of his lifetime. Later that night, despite making assurances to Chanos that Bear's financial condition was strong, its balance sheet healthy, Schwartz was facing oblivion.

Schwartz had now convened an emergency meeting in the conference room of CFO Sam Molinaro's sixth-floor office. Sitting with him was the brain trust at Bear: CFO Sam Molinaro, Robert Upton, the firm's treasurer; Paul Friedman, who ran repo; Jeffrey Mayer and Craig Overlander, the co-heads of fixed income; Tom Morano, who headed mortgages; Steven

Mayer, the head of equities; and Wendy de Monchaux, who headed the principal strategies group.

Upton was running the meeting, going through the numbers and trying to figure out just how much cash was left in the firm's coffers. When he finally announced the tally, silence descended on the room. Staring back at him, scribbled on a pad in front of him was the number two with a capital B next to it. It was $2 billion. That was all the cash Bear had left. Just three days before, the firm's cash balance had stood at $18 billion. But in the course of just a few days, a firm that had survived a Great Depression and world wars, other recessions and difficult times, that had always seemed to win in difficult markets, as it had done in 1986, 1994, and 1998, was done, killed by a mad dash for cash.

Alan Schwartz looked up from the numbers in front of him and said very calmly, "We have to call the Fed."

Schwartz's tranquillity wouldn't last. Bear needed money fast. Schwartz's bankers had called everyone, including Jamie Dimon, about a loan or a possible deal to buy the firm. Dimon had been interested in Bear before, particularly in its prime brokerage and clearing business and certain parts of the firm's trading operations, such as its commodities trading business, but he hadn't been able to afford Cayne's sky-high price.

Obviously he could buy the firm much more cheaply now. But Dimon didn't operate that way; by March 2008, it was clear to everyone from top regulators to his peers on the Street that Jamie Dimon's moment had come. In JPMorgan he had built the strongest of the nation's—maybe the world's—financial institutions. He had done it by not chasing fads or gimmicks such as bonds tied to the subprime market. He had made some bad loans, to be sure—JPMorgan Chase's books were filled with credit card receivables that could blow up if the country's economy fell into recession—but he had built a strong bank based on the simple notion that excessive risk taking almost always leads to trouble and that the only way to manage risk was to manage the risk takers, which he did in his own way. JPMorgan hadn't chased subprime loans as Citigroup and Merrill had because he was a natural skeptic.

Now he needed to be convinced that Bear, not the 2004 or 2005 version but the 2008 version, was something he should buy while it traded at $60 a share and lower. Even so, he wasn't rushing into anything. He understood the concept of systemic risk as well as any regulator. The financial crisis had reached a new level—a firm, and a fairly large one at that, was on the brink of bankruptcy. His first order of business was to make sure JPMorgan survived the market turbulence once Bear bit the dust.

Dimon told Schwartz he wasn't willing to make any type of commitment, at least not yet.

Systemic risk was staring Geithner, Paulson, and now Bernanke in the face. Bear had run out of its $18 billion or was pretty close to doing so as soon as the markets opened Friday morning. If Bear fell, it would have to unwind a massive balance sheet, trades around the world, and billions of dollars that were stuck in prime brokerage accounts. Even so, their natural inclination was to let Bear fail and file for bankruptcy liquidation. They recalled Jimmy Cayne's intransigence during the LTCM bailout negotiations. Geithner had been a mere kid back then, working at the Fed—Cayne recalled that he had looked like a clerk taking notes—and now he was one of the three men who had the fate of his firm in their hands. They also recalled Alan Schwartz's intransigence, his refusal to raise capital that would have now come in handy (Schwartz disagrees, saying that no amount of capital would have saved the firm).

Bernanke had outlined a doctrine that had been in place for a year; it called for a piecemeal approach to dealing with the financial crisis: lowering interest rates, which he did dramatically, and giving banks greater latitude in borrowing from the Fed's discount window. It didn't call for bailouts of troubled firms, particularly those such as Bear Stearns that bet big and lost. That would create what was known in regulatory circles as "moral hazard," or the tendency of financial players to act recklessly because they knew the government would bail them out of their bad bets.

In other words, why would Lehman Brothers, Merrill Lynch, and others clean up their act if they could be bailed out with the taxpayers' dime like Bear Stearns by simply blaming everything on evil short sellers spreading rumors through the media?

In war, there may be no atheists in foxholes, and on Wall Street there are no pure capitalists when systemic risk rears its ugly head. Though Adam Smith might be turning in his grave at the prospect of moral hazard, Geithner, Paulson, and Bernanke were now questioning their initial optimism. The "correction," the "credit crunch," and all of the other euphemisms used to describe the state of the financial system over the past year had grown into something more devious with the implosion of Bear Stearns. What it was exactly, they didn't know, but it was something certainly bigger than the market's becoming "ugly" as Bernanke had been told in Jackson Hole late the year before. Bear was a clear marker that the crisis was indeed a crisis that wasn't going away and that, if not dealt with completely and quickly, it could mushroom into something bigger.

Bernanke, the student of economic panics, understood all too well what the onslaught of "systemic risk" would mean: a resulting crash or an implosion of the entire financial system as counterparties sold anything of value to make up for losses tied to Bear Stearns's demise.

Bernanke, Paulson, and Geithner worked through the night, occasionally speaking to Schwartz, who had his own team weighing the dwindling set of options available. Conspicuously absent from the meat of the discussion—i.e., whether there would be a government bailout of Bear Stearns—was the SEC and its chairman, Chris Cox. The SEC, at least on paper, was supposed to be monitoring the firm's capital position, which it had deemed satisfactory—until it wasn't anymore—which had left a lot of people on Wall Street openly questioning Cox's abilities.

At Bear, executives had had several conversations with SEC staffers. "They all had this deer-in-the-headlights look, as if they didn't know what was happening and were afraid to do anything on their own," said one Bear Stearns executive with direct knowledge of the matter. It's unclear if the big three policy makers felt the same. But there was a practical reason why Cox took a backseat: the SEC didn't have money. In other words, only the Fed and Treasury could bail out Bear. And that was what they were now weighing—whether or not to give Bear enough money to help the firm survive another day, prevent a market meltdown, and get the firm sold as soon as possible.

Before the market opened Friday, Schwartz had his answer: the Fed would provide a credit facility through which JPMorgan Chase would serve as a conduit to make good on any trades Bear had. In parsing the details, it initially seemed that the Fed was giving Bear not just a lifeline to survive the next twenty-four hours but, it appeared, twenty-eight days to complete a deal of some kind. Having JPMorgan there was the government's way of telling the market, Bear included, that it wanted the firm sold, preferably to the big bank, which was the healthiest of the major financial institutions. Dimon agreed to put the firm's name on the deal, and when the markets opened, the desired effect occurred: Bear's share price jumped. It seemed the firm would be saved—by the Fed, JPMorgan, or both.

Inside Bear, traders and bankers cheered. But as the morning wore on, it became clear that Bear didn't have the lifeline; a more careful reading of the press releases put out by JPMorgan and Bear itself revealed that Dimon had merely agreed to explore options about doing a deal with Bear. As the day wore on, Bear continued to lose customers. Traders would not accept collateral from the firm, and for good reason: a couple days earlier Schwartz had been saying everything was fine and been trying to get Jim Chanos to say the same thing on CNBC. Now he was telling the world that Bear needed a bailout from the Fed.

After the brief rally, Bear Stearns stock went into free fall. Paulson and Geithner came to the conclusion, even if Schwartz hadn't yet, that the firm needed to be sold before the end of the weekend. They didn't make an announcement to that effect, but they didn't have to. Standard & Poor's chimed in by cutting Bear's bond rating three notches to near junk, a BBB rating. Schwartz, the media investment banker, might not have understood the significance of the move, but Larry Fink, the expert in debt, certainly did. "Okay, Bear is finished; they are done," he said. "Those jerks at S&P just put a stake through the firm's heart."

The rating agencies, of course, had lost their credibility during the past year for having missed the subprime crisis, but one of the oddities of the market is that their judgments were still considered gospel. By cutting the rating so deeply now, after all that had happened, S&P was signaling that it didn't think Bear would survive. "No one is going to do business with Bear," Fink said. His prediction: JPMorgan was going to work over the weekend and then decide Sunday whether to buy Bear at a fire-sale price.

"Why does it have to be them?" I asked.

"Who else is available?" Fink countered with a laugh. "They're the only game in town."

Back at Bear's headquarters, after the initial elation about the Fed bailout, reality set in. No one wanted to trade with Bear, Fed backstop or not.

It would be sold, and hopefully Schwartz would get a good price; shares of Bear had fallen that day by more than 60 percent; they were now trading at about $30.

Employees were more than shocked—they appeared nearly comatose. It was as if the entire place had been lobotomized—they were just staring at their computer screens in silence. Ace Greenberg was one of those who couldn't believe what was happening. Even amid the fury of the week, Greenberg came into the office every day to trade and invest for a coterie of clients. That day, when the broker Kurt Butenhoff asked, "Hey Ace, how's it going?" Greenberg just shook his head mutely, shocked that the culture that he more than anyone else had epitomized was now clearly doomed.

Butenhoff then turned to a more immediate problem. What worried him and others at the firm was what he knew about Jamie Dimon. The JPMorgan Chase CEO hadn't survived this long because he did favors for competitors. Dimon was a ruthless deal maker, a skill he had learned from Sandy Weill. If Dimon was the only bidder, what would stop him from playing chicken with the Fed? Shares of Bear settled at $30, and Sam Molinaro was telling analysts the firm had a "book value"—the total net value of its assets—of $80 a share. But no one thought it was worth $80, and Dimon would be stupid to be the only bidder and pay $30. And Jamie Dimon was anything but stupid.

That afternoon, bankers at JPMorgan developed a number of scenarios for Dimon to consider. One was a partial acquisition of Bear; another was a full acquisition. Bear was relatively free of subprime exposure, but it still had a little more than $30 billion in mortgage bonds, many of them tied to the less risky but still unsalable alt-A mortgages, and who knew how they were valued? Bear was looking as if it were going to report a small profit for the first quarter of 2008, which made JPMorgan think that Bear was being very generous in its valuations of the securities on its books.

The final scenario was the doomsday one, conducted to determine what would happen to Wall Street if JPMorgan just walked away from the Bear deal. Bear, of course, would go, but so too could every other major investment bank. Already shares of the big Wall Street firms had tanked in unison with Bear's on the common belief that Bear's problems were the financial industry's problems as well. Citigroup was a commercial bank, and the government would do everything in its power to prop up an institution with $3 trillion of deposits. But in short order, the systemic risk would kick in. Given the leverage used by Lehman and Merrill, they would implode as well. That would leave Morgan Stanley and Goldman Sachs, which had better balance sheets than the rest of the Street. But they had more than dabbled in risky debt, and according to the analysis their days could be numbered as well.

What made this scenario so enticing for JPMorgan was all the business it would then pick up; it would be the last man standing, and when Wall Street recovered, as it eventually would, the bank would clean up as never before.

It's unclear how seriously Dimon had considered this, the "nuclear option," as it were, but his negotiations over the weekend suggest he had it in the back of his mind. Bear Stearns executives knew he wasn't going to pay $80 or even $30 a share. They were now settling on some figure north of $8 and south of $15. It seemed only fair, they reckoned; the Bear Stearns building alone was worth $1 billion, and it was a hell of a lot better than Dimon's digs on Park Avenue. Initially, JPMorgan seemed to agree. But then Dimon came back with a no-go. It was based on a fundamental analysis, he said; he couldn't price the $30 billion in mortgage debt—there were not many bidders, and the last thing he could do was put his own firm in jeopardy.

Fink was right: JPMorgan was the only game in town, and Dimon knew it, as did Geithner, Paulson, and now finally Bernanke. J.C. Flowers & Co., the large private equity fund and investment house, had been setting up shop inside Bear as well as poring over the books and trying to come up with a deal of some kind. But it had none of the things regulators were now looking for in a savior for Bear. How would the markets react to a J.C. Flowers deal to buy Bear? No one wanted to know.

So the government upped its ante; the "Bernanke doctrine," which until now had consisted of a policy of lowering interest rates and standing as the lender of last resort to repair troubled banks, would be stretched again. If Dimon didn't want to assume the $30 billion in risky debt on Bear's balance sheet, maybe the Fed would? The government's backstop would cover everything but the first $1 billion in losses. Dimon might go for that, depending on the price. A series of conversations took place among Geithner, Paulson, and Dimon, who was now interested in buying the firm for around $8 a share, a price he believed he could sell to his board. But Paulson, the old Goldman Sachs banker who had priced so many deals in his career, gave Dimon the green light to bid even lower. The last thing the government wanted to do was reward excess, and Bear Stearns had been a model of excess over the years. Paulson also knew the government's machinations would be controversial. Average people who paid their bills on time and had watched Wall Street get rich in recent years as real wages stagnated wouldn't have much use for arguments about systemic risk. They would see any deal at anything other than a nominal price as a bailout, helping people like the alleged pot-smoking chairman hold on to their billions. With that in mind, he said, Dimon should bid as low as he wanted, and $2 a share would be a nice round number.

The $2 bid shocked just about everyone who saw it, from the reporters who believed $8 to $15 was still on the table to the Bear Stearns board, its financial advisers, its attorneys, and now Cayne, who made it back from his bridge tournament for the final deliberations late Sunday afternoon, just before the deal was made public. Cayne's absence had caused yet another stir on Friday; after all, he was still chairman and he was playing bridge while the company was imploding. (One CEO began referring to Cayne as the "Eliot Spitzer of Wall Street" for philandering in the form of playing bridge while the company teetered.)

Yet as the Bear Stearns board and senior management were weighing the $2 bid, Cayne was advocating the same stance he had successfully taken ten years earlier during the LTCM deliberations: just say no. He may have been egged on by Frederic Salerno, a longtime board member. Salerno wasn't in the best of moods. The firm's implosion had begun as he and his wife vacationed together with fellow board member Vincent Tese and his wife in Palm Beach. Earlier in the week, they had been told by Schwartz that everything was fine, only to be disturbed during dinner Thursday night with a phone call that the firm was coming unglued.

Now Salerno snapped, "What happens if we just say no?" Cayne agreed, stating emphatically that the firm should consider Chapter 11 bankruptcy because its shareholders could get more in a bankruptcy than this deal would provide.

Cayne, of course, had much to lose. His vast fortune, once valued at $1.6 billion, had been hammered as Bear's shares fell. At $2 per share, his holdings of Bear stock (friends say he had periodically taken money out of the company and had other sources of wealth) would be reduced significantly. According to Salerno's recollection of the meeting, he and Cayne and several others initially voted no to the deal. The thinking was pretty simple: fuck Jamie Dimon, fuck Paulson, and fuck everyone else. There's not that much difference between $2 and zero.

Bear's attorney Rodgin Cohen said there actually was. Cohen joined the law firm Sullivan & Cromwell in 1970, and he'd been involved in one way or another in just about every major bank-related issue, from failure to merger, ever since. Aside from being the best bank attorney in the country, Cohen was a straight shooter; there would be no spin from him, no emotion, just facts. And here's how he laid them out: the shareholders would have to approve the sale to JPMorgan, which left time for a miracle; possibly a white knight or something else would come along and force JPMorgan to sweeten the bid. Bankruptcy, on the other hand, would mean that Bear would turn over its fate to the courts; the market would react violently to a bankruptcy filing; its share price would fall to almost nothing. That was something that caught Cayne's attention, given how much of his money was wrapped up in the firm and the strain his shrinking net worth was having on his marriage. If nothing else, Pat Cayne was a sensible financial planner, and over the past year she had been harping at her husband to "diversify" his net worth, which he had done grudgingly but not enough to her liking, according to people who know them. Pat had gotten used to the lifestyle of a billionaire, and rightfully so, since she had been there every step of the way, helping her husband climb the corporate ladder, and now many of the fruits of that labor were vanishing.

Cayne's net worth wasn't something the group either cared about or considered, but the prospect of shareholders and employees who had sunk their lifesavings into firm stock getting nothing as the markets digested what bankruptcy meant was weighing heavily on them all. They all hated the deal; they were ripping mad at Dimon, who board member Vincent Tese thought was being "cute," even if he had been given the okay for the $2 bid by Paulson. But no one wanted to go through the uncertainty of a bankruptcy, Cohen urged.

Tese had been undecided about how to vote, but after hearing Cohen's clear analysis he said that approving the deal was their only chance to get a better offer, or at least any offer. Think logically, he told Cayne. We have $2 now, and in bankruptcy there's a good chance we'd have nothing. Tese later said that in the back of his mind he thought the shareholders would vote down the deal and they might be back to the $8 to $15 level.

And with that a deal that every member of the Bear Stearns board hated was approved.

From his sparsely furnished rented office in Tenafly, New Jersey, Ralph Cioffi, the man who many believed had started Bear's downward spiral, watched in horror as the events unfolded at the firm he had worked at for more than twenty years. One of the little factoids revealed in Bear Stearns' yearlong slide, and the losses that had spread to other firms holding illiquid mortgage bonds, was the importance of the Cioffi hedge funds to the functioning of the mortgage market, particularly the most risky of the debt, the CDOs, just before the crash.

Bear's board had concluded during one of its many examinations of the now-notorious subprime hedge fund that Cioffi was possibly the most aggressive buyer of CDOs in the market; he had bought CDOs in late 2006 and early 2007, when others wouldn't touch the stuff, and when he couldn't buy any more in early 2007, the market had shut down and the prices had imploded. In many ways, Cioffi had been the engine that kept the CDO market running just before its complete collapse by being one of the few large buyers of the securities and propping up their prices. So when he was out of the market of CDOs and finished, so were Citigroup and Merrill Lynch, and to a degree so was Bear Stearns as the meltdown spread to other mortgage debt.

Many of Cioffi's old colleagues considered him a pariah and held him personally responsible for the firm's problems. Warren Spector, who had been his champion for many years, hadn't even returned Cioffi's telephone call when Spector had been axed by Cayne a year earlier. Cioffi, meanwhile, had dealt with the guilt that his actions had put the place he loved into jeopardy by doing what he had once done best: trading. He had been day-trading for himself for about six months, the only job he could reasonably hold until the Justice Department made its move and decided to indict him.

But for the past week Cioffi hadn't been able to trade because he had been glued to the reports on CNBC of the troubles of his onetime employer. Then there was that horrible interview his old friend Alan Schwartz had given to CNBC. What struck Cioffi was how bad Schwartz had looked. He hadn't seen him for about a year, but the images of his old friend from Palm Beach stood in stark contrast to the powerful, confident man he'd known. Schwartz looked pale. Gone was his full, confident voice, replaced instead by a meek whimper when answering questions.

Cioffi thought that Schwartz looked beaten because he was, and so was Bear.

On Saturday night, as the Federal Reserve, the Treasury, and his old colleagues continued to hammer out a deal with JPMorgan, Cioffi and his wife shared a steak dinner with Craig Overlander and his wife, Nancy, at Café Gray in New York. Overlander, Bear's co-head of fixed income, held huge amounts of Bear Stearns stock. Likewise, Cioffi had held on to much of his Bear Stearns stock since he had left (he would later say out of loyalty to his old firm, but also out of necessity). One of the issues being investigated by the feds was that he had moved some of his own money out of his hedge funds before they had imploded; the last thing he needed was the Justice Department charging him with selling his Bear stock as the firm itself went down.

That's why, for much of the night, the couples discussed "the number," or the share price JPMorgan might pay.

"It could be anywhere between five and fifteen dollars a share," Overlander said. At least that's what he was hearing from his sources who had been present at the negotiations. "There's no way we would get anything close to Friday's close," he added, referring to Bear's closing price of about $30 a share. Cioffi agreed.

Overlander didn't realize how right he was.

Cioffi was watching television that Sunday night when his wife, Phyllis, started crying. He asked what was happening. She had just gotten off the phone with Nancy Overlander, who had given her the news that was now being broadcast to the world: Bear had just been sold for a paltry $2 a share.

That's when it hit Cioffi: he wasn't just out of work and possibly to be indicted. He was nearly broke as well.

He wasn't alone. If Bear's fourteen thousand employees had been shell-shocked on Friday, by Monday, when news broke of the deal, they were seething. The urban legend that a group of hedge funds had conspired to spread rumors, influencing journalists to misreport Bear's real financial condition, was accepted as gospel by large numbers of Bear employees and even senior executives (Cayne, Schwartz, and various board members like Salerno at the top of the list). In bars and restaurants over the next few weeks, they would drink and curse short sellers and journalists (including me) in the same breath.

The attorney Jake Zamansky thought he could tap into the anger and make some more money. Smelling another case to add to the one he had involving the Cioffi hedge funds, he arranged meetings with various Bear investors to sue JPMorgan over the losses that resulted from the low-ball offer. Aside from Cayne, no single investor had suffered bigger losses than Joe Lewis, and he was seething as well, vowing to help convince shareholders to reject the deal. But he wasn't mad enough to sue JPMorgan. Lewis had lost $1 billion on the offer, but he still had around $8 billion left in the bank, and, as he confidently told one associate, he didn't have to waste time with a lawsuit because "I can make $1 billion in a year."

Too bad that wasn't the case for most Bear employees, who had much of their retirement savings in the company stock. Adding to their anger was a decision by the Fed, just minutes after the final paperwork of the sale was finished, to allow Wall Street securities firms, for the first time ever, to borrow directly from the government.

It was the most aggressive direct move by the Fed to deal with the crisis, one clearly designed to head off any chance of systemic risk spreading to Lehman in particular, the next smallest and most highly leveraged of the investment banks, but also the rest of the firms as the short sellers took aim at the next victims.

But it was taken as the final "fuck you" to the firm that obviously had no friends. After the $2 deal was struck, Salerno was in his car, heading to the airport to return to Palm Beach, when he heard the news about the Fed's move on the radio. "Those bastards," Salerno muttered. The Fed's discount window had been reserved for commercial banks, but for months Bear management had been lobbying the government to open it up to investment banks like Bear as the crisis spread. To be sure, the new "lending facility," as it was known in Fed-speak, wasn't allowing the full access enjoyed by commercial banks, and to borrow from it was a black mark—a bank that turned to the window signaled to regulators that it was in trouble. Bear's management would have accepted the stigma to survive. Bear would have used an emergency loan to buy time and find a better offer than $2 a share—except that now the deal was already signed.

What upset Bear's management team so much was that they felt that through all the negotiations, Geithner, Paulson, and the rest of the government team working on the rescue package had been, as Salerno put it, "stand-up guys," in that they wanted to deal fairly with Bear. They would later discover that those standup guys had wanted Bear dead and buried. It was, after all, Paulson who had come up with the idea of the $2 bid, even if Dimon quickly seized on it.

What Salerno, Schwartz, and just about every decision maker at the firm failed to realize was that, given the firm's history—its outsider status, its notoriety for being run by a pot-smoking former CEO, and its long, tortuous road to its current predicament—the regulatory community had had enough of Bear Stearns. Letting it fail was considered the most equitable arrangement, maybe not for Bear shareholders but for the rest of the financial system. Paulson believed a lesson needed to be learned, and what better firm to make an example of than the Animal House of Wall Street?

Bear's share price being beaten down to $2, and the Fed's announcement about the discount window, weren't the final indignities for the once-proud firm. The Bear Stearns crew felt used and abused; Dimon had screwed them, they felt, because he could get away with killing a wounded animal. But his lowball offer was now seen as a further sign of his greatness. He was now fêted at his favorite restaurants like a conquering hero. CNBC dubbed him "The King of Wall Street."

It was a huge turnaround for Dimon; only ten years earlier, after being ousted by Sandy Weill at Citigroup, his phone had stopped ringing, and when it had, he wished he hadn't answered, like the time Jimmy Cayne had offered him a job as head of Bear's clearing business. Now he was running not just the clearing department but the entire firm and taking over its posh building as well, a key condition of the deal that made it difficult for Bear to walk away from the agreement without relinquishing its prized headquarters at 383 Madison Avenue.

Even Dimon was surprised at how much his stature had grown one night when he walked into the Four Seasons restaurant and ran into Kenneth Langone, the hot-tempered founder of Home Depot. Langone is a guy known to wear his heart on his sleeve—the people he hates, like Eliot Spitzer, are mortal enemies. "I hope his private hell is hotter than anyone else's," Langone said of Spitzer after the governor was busted in the sex scandal.

But he lavishes the people he loves with praise, as he did Dimon, the minute he saw him at the restaurant just after the Bear deal was announced. Langone stood up amid all the stuffed shirts and black power suits at the posh restaurant and yelled, "Hey, Jamie, can you give me a loan?"

Dimon just shook his head and laughed.

Hank Paulson was known as a tireless worker, one who knew how to seize the moment in corporate power plays, a skill that had helped him dispose of Jon Corzine as CEO of Goldman Sachs and later Dick Grasso as CEO of the stock exchange (Grasso would refer to him as a "snake"). One thing he wasn't was a good public salesman. His speaking style is often awkward and at times gruff, and he has a penchant for saying the wrong thing at the wrong time, as when he told Goldman's investment bankers that they might be fired if they didn't bring in more business (he later apologized).

Now he was forced to sell the Bear Stearns deal to a skeptical public. The rank and file at Bear Stearns may have thought the $2 deal wasn't a bailout, but the growing consensus in Middle America was far different. Paulson and his partners at the Fed were being characterized as the bailout brothers and vilified for creating moral hazard by telling Wall Street firms it was okay to take excessive risk because the government would step in and bail them out, even if that bailout amounted to $2 a share.

In a series of interviews, "the snake" looked and sounded like a real one when explaining that the Bear Stearns bailout wasn't really a bailout. "I don't think Bear Stearns shareholders who have been largely wiped out would think that they've been bailed out," he said during one television broadcast.

In other interviews, he tried to position the deal as a game changer and tried in vain to make the notion of "systemic risk" accessible to the average American: by assisting JPMorgan Chase in its takeover of Bear Stearns, he said, the government was basically stopping the spread of a contagion that could make the economy even worse. Credit, he reminded viewers, continued to remain tight as the economy headed for a definite slowdown and possibly a recession. And he was right; after the Bear deal was announced, there was another flight to quality, with investors once again snapping up 3-month Treasury bills and selling just about everything else.

Still, it could have been worse, Paulson argued. A bankruptcy of Bear would have led to losses felt elsewhere—other big banks would have taken major hits because of their exposure to trades from the firm, and thus they would further restrict lending and credit.

But after thirty years of speaking to Wall Street executives, government officials, and bureaucrats, Hank Paulson's communication skills were downright terrible. In Paulson-speak, the government was trying to "minimize the impact of this business—sharp business slowdown, the sharp slowdown in our economy and again to keep our capital markets functioning the way they need to function." Paulson's inability to project a clear, coherent rationale for the Bear Stearns bailout would have long-term implications that policy makers had yet to comprehend as the markets began showing some signs of improvement following the news of the bailout. President Bush proudly declared, "We'll be just fine," when asked about the economy in the context of the Bear deal. The analyst Dick Bove proclaimed that the credit crisis was "over," adding that "actions taken by the Federal Reserve were innovative, dramatic, and, in my view, brilliant because they went right to the problem."

Hank Paulson could only hope so, because returning to the well with another "bailout" for any of the firms sitting on toxic mortgage debt would be a difficult, if not impossible, sale to the public and his boss, the president.

Yet the chance of a Bear Stearns–like event seemed greater than ever. The markets might have liked the deal and its provisions to cover the firm's toxic assets (shares of financial companies rebounded, as did the Dow, in the immediate aftermath), but the overall economy wasn't showing similar signs of strength. Paulson could read the economic statistics as well as anyone, and he now both privately and publicly began to worry about the economy: it was in all likelihood heading for a recession.

It didn't take a financial genius to figure out what the worsening economy meant for all that debt held by the big banks. Residential mortgage defaults continued to grow, and the fear now was that they would soon spread into commercial real estate portfolios. All of which meant that the Wall Street firms and banks that continued to hold those toxic assets would face bigger write-downs in the future, more losses, and possibly another investment bank following Bear's path to oblivion.

At least for the moment, Paulson's hands were tied politically. Public opinion polls showed that average people didn't care much about seemingly esoteric concepts such as systemic risk. What they saw in Bear Stearns was a bailout of bad behavior by people who didn't deserve a bailout: rich Wall Street tycoons who had made tens of millions of dollars rolling the dice while the wages of average Americans stalled.

The public's antipathy to Wall Street was aided and abetted by editorial pages on the left and particularly the right, which had always been skeptical of Paulson's free-market credentials. To conservatives, Paulson, an avowed environmentalist and avid bird-watcher, was a throwback to the old Republicanism that believed big government could be better managed, and they pounced.

Paulson didn't help matters. He brought a Wall Street banker's attitude to a job that demands politics. On Wall Street you're discarded the moment you're deemed unnecessary, and Paulson deemed the conservative Republican congressional leadership, now in the minority, as unnecessary. As a result, he spent most of his time with Democrats, such as his longtime friend House Finance Committee Chairman Barney Frank, House Speaker Nancy Pelosi, and Senator Harry Reid, whom he referred to in conversation as "Nancy" and "Harry."

Paulson's ideological impurity came at a cost. He had now all but alienated House and Senate Republican leaders, who still had enough votes to kill bailout legislation for the entire financial system, an idea that was now increasingly being tossed around by Fed Chairman Ben Bernanke.

It wasn't that Bernanke was in favor of pumping billions of dollars into the financial system just yet, but he was looking for backup. So far, the Fed was doing everything it could with its balance sheet by extending the discount window to brokerage firms and lowering interest rates. Bernanke argued that there was just not enough money at the Fed to do everything if the financial panic grew to even bigger proportions.

Bernanke's academic research—he had been a professor at Princeton University before moving to the Fed—had been on the policy responses to panics, such as the slow response on the part of the federal government to the stock market crash of 1929. He believed that the slow, deliberate approach of the Hoover administration had been at the heart of the financial panic that had led to the Great Depression. Fed policy makers at the time hadn't cut rates fast enough, and the government hadn't spent money fast enough to repair the banking system, which began to shut down.

That's why Bernanke now began to nudge Paulson to do more—to think about getting authority to pump really big money into the system, just in case Armageddon was staring them in the face. Paulson was skeptical that he could go that far, at least not yet, and from a political standpoint he was probably right.

Wall Street executives and fellow regulators who worked with Paulson say the attacks had a profound impact on him. Avoiding systemic risk had been his overriding theme before Bear Stearns, but after Bear's collapse, politics weighed on every decision. Paulson, for instance, had scrambled to take out a segment of a speech his boss, President Bush, was about to give that said the administration wasn't bailing out Wall Street. To average Americans, even if Jimmy Cayne walked away with $1—and according to sources he had stashed away much more than that; Cayne has told people that his net worth is estimated at $600 million—it was way too much.

What happened next didn't help with the bailout perception Paulson was trying to fight against.

In the days and weeks that passed, Bear's trading floors at 383 Madison began to resemble a job center. People came in, but there was no trading. JPMorgan announced that there would be job cuts. A Bear Stearns analyst named Barry Fox would be one of them, and after getting the news, he went home and jumped off his twenty-ninth-story balcony to his death.

Bear Stearns had fostered an "ownership culture" that went far below the management committee to many of the rank and file, who were paid largely in stock and urged to hang on to their holdings. One Bear Stearns executive told me he was effectively broke; he'd thought he had a nest egg of $1 million, but throw in the fact that he had a mortgage on a home in the exclusive town of Greenwich, Connecticut, two kids in private school, and a wife who likes to eat out occasionally, and bankruptcy didn't seem so far-fetched.

Meanwhile, JPMorgan bankers spent days reviewing Bear's books; one thing they discovered was that the mortgage portfolio was in worse shape than they believed Bear Stearns had advertised. The prices of the alt-A securities were much more aggressive than JPMorgan's own analysis had assumed they were. Dimon was livid.

The consensus among JPMorgan's examiners was that when the real "marks" came back, the new, lower prices of the debt would easily eat through the $1 billion JPMorgan had agreed to cover on the bonds as part of the deal. JPMorgan's press officials began telling the media that their $2 "bargain" hadn't been much of a bargain. There would be layoffs, integration costs, and of course those nasty losses in the mortgage book inherited by the firm.

There was more bad news for JPMorgan; in the days following the $2 offer, the market had rallied on the belief that the Fed and the Treasury would do whatever it took to prevent a full-scale bankruptcy of any major Wall Street firm. More than that, Bear Stearns' stock price had risen to over $5 a share as stories about some of the assets Dimon was paying just $2 a share for—everything from a world-class building to a strong clearing and prime brokerage business—were aired on CNBC and appeared in the business media. The market was saying that Bear was worth a lot more, and as a result, JPMorgan might have to raise its bid to convince shareholders to approve the deal.

That's when Bear's attorney Rodgin Cohen, a master at reading the fine print of deal documents, went back to work. He knew that even under the best circumstances deals are never perfect; they have open-ended language that can be interpreted in many different ways, and this was no exception. JPMorgan had believed its exposure to Bear was limited to the $1 billion so-called first loss on the deteriorating mortgage bonds. (The Fed would pick up the rest.) Not so, according to his reading of the deal documents. In fact, JPMorgan apparently had a rather open-ended agreement to "fund" Bear through the year, to make sure it could meet its daily trading operations, even if its shareholders were to vote down the deal, which now looked increasingly likely.

Dimon was livid. For the first part of the week, he was frantically meeting with Bear's star performers, asking them not to jump ship. He was calling his counterparts at Morgan Stanley and Goldman Sachs, pleading with them not to poach good people until he could come up with offers.

Now he was hit with the possibility that in the rush to finish the deal before the markets opened, he and his legal team had screwed up. In fact, it was more than a possibility. The language did indeed leave an open-ended commitment to fund Bear for a year, whether shareholders approved the deal or not. Dimon could, of course, walk away. But Bear would sue for breach of contract, though Cohen wasn't crazy about that possibility; Bear probably wouldn't survive in such an arrangement, he told the board. After all, who would trade with a firm in limbo even if JPMorgan was financing its commitments? What banker or trader would stay with the company?

But the stakes were just as high for Dimon. Not only did he have to fund Bear's operations, but he might lose the company to someone else. A bankruptcy judge could break up the firm and sell the pieces that Dimon coveted, such as Bear's prime brokerage operations, and the building that Dimon now lusted for.

Dimon wanted the funding language out of the deal document. But it was Bear's turn to play hardball. Tese and Salerno, now the two lead board members, delivered a stern message to the new King of Wall Street: they weren't prepared to change any part of the deal except for its price. It was Dimon's move.

After a few days of deliberation, Dimon agreed to a new price he knew shareholders would accept: $10 a share. It wasn't great, but it wasn't $2.

And with that the House That Ace Built, that Jimmy Cayne had remade, and that Alan Schwartz had tried to save was gone. There would be no "Bear Stearns unit of JPMorgan," as is often customary when a larger bank buys a smaller player with a distinct corporate culture, particularly one as distinct as Bear's had been. The company would be totally subsumed into the massive bank. Around half its employees would be laid off. The Bear Stearns logo would be removed from every window and door at the company's old headquarters.

Jimmy Cayne, who had once swaggered around Wall Street as one of its most powerful figures, was now unemployed for the first time in almost forty years. He was also somewhat a pariah on Wall Street, in some quarters blamed for the mismanagement that had led to Bear's downfall, and no longer did politicians and fellow CEOs beg for a few minutes with the man who had built Bear into a powerhouse with a stock price the envy of the world of finance. Still, during the board meeting that finalized the deal, Cayne was as defiant as ever, urging his soon-to-be-former colleagues not to believe what they were reading in the press about the firm, about his management or mismanagement of it. He also cashed in more than $60 million of company stock and promptly left the combined company, rarely to be heard from again. His longtime nemesis, Ace Greenberg, would be heard from: he was offered a job by Dimon and got a contract to write a book about the rise and fall of Bear Stearns, which many believed would focus on the fall of Jimmy Cayne. Greenberg certainly wasn't feeling sorry for Cayne. He later referred to his former colleague as a "pathological liar." Just to make sure Cayne got the message, Greenberg demanded that Cayne pay a commission to JPMorgan on his stock sale.

But as Bear's demise proceeded, larger forces were at work that would unsettle Wall Street in ways not readily apparent in the weeks that followed the spectacle.

The markets seemed to adjust to the fact that there was one less firm in existence; the Dow rose to 13,000 by late April as the Fed kept pumping liquidity into the system, announcing it would lend $200 billion in Treasuries to bond dealers in exchange for toxic mortgage debt; it was a temporary reprieve but one designed to rid the banks' balance sheets of the risky debt until the mortgage markets returned to health.

Fannie Mae and Freddie Mac, the government-backed mortgage companies that had done so much to fuel the housing bubble, were called upon to step up to the plate once again. Federal regulators lowered capital requirements so that both agencies could expand the scope of their activities as a way of bolstering the housing market, which remained in the doldrums.

Amid all this, top Wall Street executives opined that the worst of the financial crisis was over (for all his earlier concern, John Mack called it the "top of the ninth inning"; Lloyd Blankfein, "the third or fourth quarter"); that the credit crunch was loosening its grip on the ability to borrow; and that the firms would soon be making money again. If only that had been the case. Because it clearly wasn't; just a few days after the Bear Stearns implosion, a group of the firm's investment bankers were traveling to Dallas to complete a deal for a longtime client. It would probably be the last such trip for the team, six men in the middle of their career. Each had heard that their future with JPMorgan was in doubt. The firm just didn't need more investment bankers.

They huddled in the back of the jet; they used to fly first class, but now they were flying coach. Many Bear Stearns traders blamed CNBC, including myself, for fanning the flames of the firm's demise. Not those guys. They could read a balance sheet. They conceded that Bear's massive leverage and $30 billion in risky mortgage debt, trading now at just pennies on the dollar, was the final straw that would have left the firm insolvent, even without the poor management.

They also conceded that the media had covered the firm's downward spiral pretty fairly, with one caveat: if Bear couldn't survive with $30 billion in risky debt, how could Merrill or Lehman survive with their holdings? All the firms held virtually the same toxic waste, and, most important, they had "sold their souls to the devil," one of the bankers said, "because we are so leveraged we borrow from our competitors and our enemies, and they can pull the plug on us at any time." The competitors he was talking about were the banks, such as JPMorgan Chase and Citigroup, which, as they competed with Bear for business, Bear had been pleading with them unsuccessfully for credit. The firm's adversaries were, of course, the hedge funds that had sparked the run on Bear by pulling their prime brokerage balances and shorting the firm's stock.

But he said Bear wasn't alone in this predicament. "Mark my words, we aren't the last one going down."

18. "THERE ARE RUMORS THAT YOU GUYS ARE IN TROUBLE"

Dick, there are rumors that you guys are in trouble," Morgan Stanley CEO John Mack told Lehman Brothers CEO Dick Fuld the night the JPMorgan–Bear Stearns deal was announced. Mack prided himself on having the best market intelligence on the Street (Larry Fink was one of his sources). That night he had his executive staff stationed at Morgan Stanley's offices in Manhattan's Times Square to prepare for the consequences of Jamie Dimon walking away from the deal or the possibility that Bear might reject his lowball offer and be forced into liquidation.

That didn't happen, of course; Bear's board relented; the Fed agreed to cover the vast majority of the losses on its toxic assets; and despite the tremors sent through the bond market in the form of widening yield spreads between Treasuries and other bonds, the broad financial calamity that Paulson, Bernanke, and every Wall Street executive had feared—mass selling, demolished asset values, and the markets falling into disarray—had never come.

But the financial crisis wasn't over, far from it. In fact, it was still growing, and it broadened that very night as the target of the crisis shifted from Bear to another Wall Street bond house, Lehman Brothers.

Before becoming CEO, Mack had worked for the hedge fund Pequot Capital, and he had many sources there who had reported back to him the consensus about Lehman, namely that its holdings of mortgage debt were as bad as Bear's and investors should get out while the firm's stock was still trading at decent levels.

The speculation was that Lehman's balance sheet was every bit as toxic as Bear's and possibly worse. Fuld might have sold Wall Street analysts on the notion that he had built a diversified investment bank, with a private client group, a strong investment bank, and investment in private equity. But what drove earnings at Lehman wasn't that much different from what drove them at Bear: investments tied to real estate. Only at Lehman, the firm's mortgage exposure included not just subprime bonds that it had underwritten and held on its books but commercial real estate, which was looking more and more like the next market on the way to oblivion.

Lehman now faced what Bear had faced over the past three weeks: rampant speculation that it was undercapitalized and overexposed to toxic debt from short sellers, who were beginning to pull money from their accounts and refusing to trade with the firm.

Mack asked if the speculation was accurate. Fuld, the Gorilla of Wall Street, answered in his typical tough-guy manner, "It's bullshit." Mack, who had been close to Fuld for years, said he was relieved to hear it and that Morgan would do what it hadn't done for Bear Stearns: his traders would continue to do business with Lehman and ignore the chatter.

In terms of reputation, Dick Fuld was no Jimmy Cayne and Lehman Brothers was no Bear Stearns. Fuld was on the board of the New York Federal Reserve Bank and considered one of the best managers in the business, a man so fully engaged in the business of trading and risk taking that while Cayne was off playing bridge during his firm's meltdown, Fuld was said to have been roaming the trading floor making sure his firm was safe (although traders said such visits were extremely rare). And despite the chatter about its liquidity and holdings of debt, Lehman was considered one of the market's blue-chip firms.

Yet the question of whether Lehman's broader problems were serious would captivate the attention of the markets for the next six months.

W ithin a month, they're done," said Paul Friedman, Bear's former repo chief, to the remaining people on the trading desk as Bear was winding down its operations.

To some, the concept of another major firm going down so soon seemed unbelievable after the gut-wrenching experience at Bear. But Friedman had firsthand knowledge of how quickly and unrelentingly creditors can back away from a firm—any firm. Lehman had borrowed as much as Bear had—$35 for every $1 of capital—to support a balance sheet significantly larger than Bear's, around $600 billion. It also relied on repos for funding, maybe not as much as Bear had but significantly so. It even had a similar exposure to the subprime mortgage market and a ton of commercial real estate it had taken onto its balance sheet and had yet to unload.

"There's no way they can survive," he said. "Who would repo with them? Who would take that risk?"

At least on paper, Friedman was right. There was very little that separated Bear and Lehman except for an intangible, and that intangible was Dick Fuld. From the moment Bear went under, Fuld went to work. He called up everyone on the Street he believed was bad-mouthing the firm, from hedge funds to Wall Street executives, and upbraided them for spreading lies. As he had in 1998, he threatened legal action, including reporting them to the SEC, which had begun a probe into the market "rumors" that had helped expedite Bear's decline—even though the speculation about Bear's toxic assets now appeared to be true.

One of those executives was Lloyd Blankfein of Goldman Sachs. Goldman is the most hated firm on Wall Street, not just because of its reputation for smarts and savvy; it is its ruthlessness in making money that rubs so many competitors the wrong way. The financier Ken Langone once said that bankers on the NYSE board like Hank Paulson, Goldman's former chief (and now the Treasury secretary), had missed so many NYSE board meetings to spend more time at the office ripping off customers. Larry Fink says he became a NYSE board member because he believed that having a centralized marketplace like the exchange blunted Goldman's power in the markets.

Yet as much as it was hated, Lehman joined Merrill and other firms in trying to emulate Goldman's risk-taking business model. In early 2007, senior executives at the firm say, during strategy sessions, people like Fuld's number two, Joe Gregory, would prod his troops to do what Goldman was doing in the markets in order to match Goldman's earnings. There was one key difference: risk management at Lehman was never what it was at Goldman. According to one senior executive at Lehman, "At Lehman when you buy a bond at ninety and it begins trading at eighty-five, people start talking about how much money they can make by increasing leverage and keeping the bond on its books and making money off the carry trade. At Goldman you get a call, and someone says, 'Don't be upset . . . we're not losing confidence in you, but if you don't have a good reason to keep that position, get out at eighty-five.'"

"The leadership at Lehman believed if they could survive 1998, they could survive anything," the same executive said. "Someone would be there to bail the company out."

Fuld epitomized that cocky attitude, particularly when it came to protecting the firm's brand, and he began to lash out at anyone in his way, including Goldman, the firm he believed was just as responsible for the rumors as any short-selling hedge fund.

Fuld told associates he had seen it before: in 1998 he had blamed Goldman for spreading unfounded rumors about the firm's finances and profiting by stealing customers; he had even called former CEO Jon Corzine to protest. Why should this time be any different?

Now Fuld just didn't protest, he threatened. In a testy telephone call to Lloyd Blankfein, Fuld said if he discovered that Goldman was behind the rumors, either to help its hedge fund clients or to support its own short positions, he would see to it that the guilty party paid dearly. Blankfein denied that the firm would engage in anything close to market manipulation and said he would look into the matter.

John Angelo wasn't speculating about Lehman, but he too fell afoul of Fuld. His hedge fund, Angelo, Gordon & Co., with $16 billion in assets, had a prime brokerage account with Lehman; as he heard the rumor spread from Bear to Lehman, he took out his funds, just as he had at Bear.

Jimmy Cayne was obviously disappointed but understood Angelo's decision; Fuld simply exploded. "We have friends, and we take care of our friends!" he snapped at Angelo, a smart, bookish, soft-spoken man, who simply told Fuld what he had told Jimmy Cayne: it was nothing personal, he simply had to protect his firm. Fuld didn't care. "I thought you were a friend," he said before slamming down the phone.

The interesting thing was that Fuld's bullying tactics were working. He bragged publicly about his firm's efforts to reduce risk, its strong capital positions—it raised $6 billion in capital in early April—and the $40 billion it had in cash. The run on the bank that had started after Bear's death died as well.

Fuld was and is an imposing figure on Wall Street. His position on the board of the New York Federal Reserve Bank gave him access to all the top regulators. He was close friends with just about every major CEO on the Street (except now, of course, Lloyd Blankfein). Meanwhile, his demeanor was as tough as ever and his piercing eyes even more menacing when analysts started questioning Lehman's finances or when he felt the traders at Goldman were spreading falsehoods about his company to support their clients in the hedge fund community.

No one doubted Fuld's determination, particularly after his call to Blankfein, nor did they doubt that he had the chops to win. "Fuld is no Jimmy Cayne," the investor Bill Smith told me as he explained why he was buying shares of Lehman after Bear's fall; Fuld hadn't been playing golf or smoking pot when the crisis hit. He had the track record: Lehman had survived the LTCM crisis in 1998 to become one of the most successful of all investment banking franchises. Bear was a one-trick pony, a big bond house. Lehman, so the thinking went, was an international firm that competed with Goldman and Morgan Stanley in top investment banking deals.

And if all else failed, Fuld had something of a secret weapon to seduce both the media and the markets: his newly minted CFO, Erin Callan.

Callan wasn't exactly a number cruncher; in fact, she had spent most of her time on Wall Street as a saleswoman. Though she was a tax lawyer by training, the forty-two-year-old had made her mark inside Lehman as an investment banker and a close associate of Fuld's number two, Lehman President Joe Gregory, who made promoting women a key priority inside the firm.

And Callan was the perfect candidate for promotion. She was smart, worked well with clients, and had movie star looks. Just a few months after being named CFO, she received a fawning profile in the *Wall Street Journal* as she became the point person in the firm's efforts to burnish its image following the Bear collapse. Callan, who is "undeniably fashionable," the *Journal* wrote, "has received sexist e-mails from TV viewers. But she has no plans to dial back her persona."

That persona included high-fiving traders on the desk and receiving a standing ovation, the *Journal* wrote, after deftly answering analysts' questions about why Lehman could never become the next Bear Stearns given all its attributes: its strong capital position, cash on hand, and management expertise. As the article pointed out, shares of Lehman had risen 42 percent after her performance, and she had even earned kudos from no less a skeptic than the superanalyst Meredith Whitney.

Callan had been with Lehman for thirteen years, the last six months as its CFO, yet she seemed undaunted by the challenge of being the chief financial officer of a major firm in the crosshairs of the financial crisis. She also knew what the short sellers and even some analysts were saying: that

Lehman was a big Bear Stearns with a balance sheet crammed with toxic debt. In numerous face-to-face meetings with analysts, she denied, with varying degrees of specificity, the depth and breadth of the problem.

Brad Hintz had been in Callan's position about a decade earlier, before joining Sanford C. Bernstein & Co. as a research analyst covering the brokerage industry. He had been Lehman's CFO. So he knew all the tricks, the spin CFOs put out to analysts to make lemons seem like lemonade. Because of his background, Hintz was pretty familiar with Lehman. Its business was much more diversified than Bear's, with a much bigger investment banking, asset management, and international business.

But like Bear, Lehman was a huge player in the mortgage and real estate business.

"Lehman doesn't just have mortgages and loans on their balance sheet, they have actual buildings," Hintz said, referring to Lehman's holdings in, among other properties, the Archstone-Smith Trust. Lehman had yet to take significant losses on those holdings, which, depending on whom you believed, either was an act of malfeasance (which is what the shorts were saying) or, according to the firm and Callan in particular, simply reflected Lehman's rigorous risk management procedures.

Hintz wasn't sure who to believe, and that's why he was now sitting in Lehman's executive dining room peppering Callan with questions. The Lehman dining room in midtown Manhattan offers a spectacular view of the city. Callan was all smiles as Hintz asked about the success of Lehman's commercial paper sales, which was a good indication of how the market viewed the company's financial condition.

Brokerage firms use commercial paper of varying maturities on a daily basis to fund their operations. Commercial paper is nothing more than a loan that matures in 30, 60, or 90 days. If a firm can issue commercial paper with longer maturities, that means investors have a greater degree of certainty about its ability to stay in business. Conversely, if commercial paper investors get skittish, they'll look to buy shorter maturities and demand higher returns because of the greater likelihood of the firm facing financial problems and being unable to repay the loan.

Callan assured Hintz that not only was Lehman reducing its reliance on commercial paper, it was also extending its maturities of these securities, meaning that, contrary to all the chatter about Lehman following Bear into extinction, many investors were actually betting that Lehman would be around for a long time. She went on to say that the company was having no trouble selling longer-term debt and that the rating agencies were in constant communication with her; the word from the raters was that they were highly unlikely to downgrade Lehman any time soon.

It was the type of straightforward answer Callan had received so much credit for from other analysts; she didn't hedge or equivocate in any way, she answered the question directly. If she turned out to be wrong, if the firm wasn't selling commercial paper with longer maturities, she would have to be completely out of touch with the firm's financial condition. Hintz believed the chance of that appeared remote.

He then asked about Lehman's heavy concentration of real estate assets. Callan was interrupted by Edward Grieb, the firm's investor relations chief.

"You don't need to be that concerned with them," Grieb said. "We don't anticipate too many more write-downs." Grieb went on to explain that Lehman had never marked many of its assets up, so while their value had declined, there was still some wiggle room and value on the books.

Hintz finished his lunch, a walnut-encrusted tilapia, and left the meeting feeling pretty good. He thought Grieb's answer had been a little odd—why not just let the CFO answer a question clearly in her scope of knowledge? Still, if the opposite turned out to be true—that the value of Lehman's commercial real estate wasn't properly reflected in the public filings—he would have had to be spinning wildly or ignorant about basic facts of the company, and Hintz put low odds on that as well. This was, after all, Lehman Brothers. Based on everything he heard, Lehman appeared to be taking steps to deleverage and extend its funding situation. With access to the Fed, a Bear-like blowup seemed remote. Plus, it had more than $600 billion in assets, which in a jam, it could sell, albeit at a loss, to raise cash. Not a bad place to be for a securities firm.

But was it real or simply a spin job, with CFO Erin Callan, now on the job for all of six months, serving as Fuld's accomplice in his latest effort to keep Lehman from imploding as it nearly had in 1998?

The short seller David Einhorn believed it was a spin job—or worse. In April, just as Fuld began his effort to protect Lehman from the shorts, Einhorn announced his short position and began hammering away at Lehman, saying that the firm's claim to have enough capital was patently false; that contrary to the firm's assurances that it had reduced its leverage, it had actually been increasing it.

Put simply, Einhorn thought Lehman's management were liars. By now

Einhorn had fully recovered from his bad bet on New Century; he was cranking out a 25 percent return for his investors in a fund that bet both long and short. One of those shorts was, of course, Lehman Brothers.

In late 2007 Einhorn had listened to conference calls with analysts in which Lehman had seemed less forthcoming than other firms about its finances and leverage. The company had earned nearly $900 million in the fourth quarter of 2007 on revenue of $4.3 billion, but company officials had seemed dodgy during the call. When the analysts had pressed them to break down the change in value of certain liabilities, they had been evasive, declining to give specific answers. When its rival Morgan Stanley had held its conference call weeks later, the contrast could not have been greater. It had provided more detail about its liabilities and what was on its books, and generally appeared to be more forthcoming about its business.

When Einhorn examined Lehman's filing explaining its fourth-quarter results, he became even more suspicious. The company had moved about $9 billion in mortgages from Level 1 to Level 3 assets and didn't appear to have taken a loss. How, Einhorn thought, could a company with $30 billion to $50 billion in mortgage exposure not have more significant write-downs and book losses?

By January 2008, the Lehman PR machine was kicking into high gear. The stock had taken a hit like the rest of Wall Street's, but management told employees it was a buying opportunity at around $50 a share. After all, management pointed out, look what you would have made if you had bought shares during the LTCM crisis. And Fuld had just replaced the old CFO, Christopher O'Meara, with the decidedly more telegenic, Callan. In the words of Lehman President Joe Gregory, Callan had been hired for many reasons but mostly because she was the type of person who could "really explain Lehman's story."

For Einhorn, all the pieces were in place: a firm in his view that stretched the truth in financial documents and a relatively young, financially inexperienced CFO. In January, Lehman announced that it would increase its dividend by 13 percent and would continue the company's massive stock repurchase program.

Callan told analysts that the firm was now reducing its leverage in preparation for a tough year ahead. Lehman, she said, was well positioned to survive and thrive in a difficult market. Investors seemed to agree. After the dividend increase was announced, Lehman stock rose 15 percent over the next few days.

But Einhorn believed it was all a mirage. In fact, he believed that Lehman, in late 2007, had increased its leverage (the huge Archstone-Smith commercial real estate deal raised his suspicions) to dangerous levels, well above 30 to 1, just as other firms were beginning to reduce their borrowing. That leverage had left the firm exposed not just to subprime mortgage debt but to commercial real estate, which was now being slammed as well. Yet Lehman was making money. The popular perception was that Lehman had escaped the worse of the subprime crisis and Fuld was a genius.

Einhorn had done his research. He knew why he had lost so much money when New Century had exploded. Real estate, whether it was commercial or residential, wasn't getting better. Residential defaults in subprime real estate continued to rise, as private investigators such as Steve Davis could attest. And shopping malls in once-hot areas of Nevada and Florida were empty as businesses closed when the economy continued to fall and began heading toward full-fledged recession.

Lehman had exposure to all of the above, and it was saying that it was making money, which only made Einhorn more convinced that it was being propped up by faulty—possibly fraudulent—accounting and, of course, the bullying tactics of Fuld, who had declared short sellers like Einhorn to be Public Enemy Number One.

Einhorn told people he had seen it before; based on his research, he believed that nothing short of mismarking the firm's holdings of risky bonds had allowed Lehman to escape near calamity back in 1998. He also believed that there was good evidence that any CEO who blamed his or her problems on short sellers was "attempting to distract investors from serious problems" as the economy continued to decline. His bet was that the problems plaguing Lehman, whether the firm's leadership was too stupid or too duplicitous to admit it, were worse than any other on Wall Street.

Aiding Einhorn was the fact that in 2008 things were different from in 1998, when there had been no FASB 157 forcing Lehman to mark its holdings of real estate assets at market prices, as there was today. Then there was timing; Callan might have been an elegant vehicle with which to spread the word that the firm was safe and strong, but after what had happened to Bear, with Alan Schwartz assuring the markets the firm was fine just days before it tanked, investors had become more wary of Wall Street spin, particularly about the value of toxic assets.

Einhorn believed that all that was needed was the bright light of publicity, a fight that would illuminate the firm's balance sheet and finances, and mainly the value of all those real estate assets. The only way to do that was to take on not just Fuld but his media darling Erin Callan.

There would, of course, be pitfalls in any public battle with Lehman. Einhorn wasn't the most convincing speaker; his voice is high pitched and nasal; in short, he sounds like a geek, and he was taking on a Wall Street beauty queen and her schoolyard bully of a boss. But Einhorn was tougher than his nerdy image. Maybe he'd gotten it from playing all that poker—he had placed eighteenth in the 2006 World Series of Poker—or maybe it was having survived so long on Wall Street, taking his lumps with New Century and from the SEC when it had investigated his Allied Signal trade.

Whatever it was, he was now prepared to call Lehman's bluff as never before.

We have reduced our leverage from the high thirties to the twenties, and we're not going below twenty-five," Callan stated at the UBS Financial Services Conference in New York City. Callan was attracting huge crowds at conferences and analysts' meetings. Today she was speaking before a standing-room-only crowd at the New York Palace Hotel.

Her statement that she wouldn't reduce leverage below 25 to 1 caught the attention of several consultants who had fought their way into the meeting to get a glimpse of Callan, Wall Street's latest sensation, and began buying shares of the stock when the conference was over. What impressed them aside from her poise and good looks was how she spun the story about Lehman's future. This was a firm, she said, that had done everything right and could weather any storm. Lehman was no Bear Stearns.

Lehman's stock price now seemed to be on the road to recovery, thanks to Callan's public relations campaign. Fuld, meanwhile, had enticed the billionaire investor Warren Buffett to buy a chunk of the firm but then realized he could get a higher price by selling shares publicly, which he did. Outmaneuvering Buffett had made Fuld feel pretty strong, strong enough to keep the pressure on the short sellers, whom he vowed to "kill," Einhorn included, and no one doubted his resolve.

Nor could they doubt Einhorn's. In April, Einhorn gave a speech at

Grant's Spring Investment Conference where he raised all the red flags that had been on his radar: the similarities of Lehman's balance sheet to Bear's; the share buyback and dividend hike; the company's aggressive accounting treatment of Level 3 assets; and its purchase of Archstone-Smith, a commercial real estate partnership that was lying on its books like a lead weight; not to mention the massive leverage that continued to exist and would make the crash in the share price more pronounced once Lehman admitted its losses.

Einhorn had given speeches about Lehman in the past, most notably in November 2007, when he had raised concerns about how Lehman always managed to beat estimates. But the latest speech touched a nerve with Lehman management. Callan was concerned because she was now getting calls from investors asking her for a point-by-point rebuttal of Einhorn's attacks. Lehman President Joe Gregory worried that Einhorn's attacks were gaining traction. Whatever positive momentum the stock had had was gone. It was trading in the mid-20s but likely to go lower if Einhorn wasn't stopped or at least stalled.

Gregory ordered Callan to call Einhorn to set the record straight, and it would turn out to be Erin Callan's Alan Schwartz moment. Einhorn immediately asked her about the firm's accounting treatment of the $9 billion in so-called Level 3 assets, the most risky ones. Specifically, he wanted to know if the firm had marked down those assets enough, because by his calculations, it hadn't.

Callan couldn't provide a rational explanation, Einhorn would later say, and it confirmed Einhorn's opinion about the company. Callan was the CFO of a major Wall Street firm, and she was struggling to answer simple questions about the firm's financial condition. The *Wall Street Journal* said Callan was a straight shooter. Einhorn said he didn't get one straight answer from her during their forty-five-minute talk.

Einhorn made his next move in a venue that was guaranteed the attention not just of the media but of nearly every serious investor in the hedge fund world: the Ira W. Sohn Investment Research Conference, a charity event that also functions as a place where hedge fund managers float their investment ideas and strategies. It was the same place where Einhorn, in 2005, had told the world he was a big investor in New Century, the now-bankrupt subprime lender. It was now the place where Einhorn launched his most vicious attack on Lehman Brothers and Callan.

During his speech, Einhorn all but accused Lehman of fraud, saying it had not properly marked down its massive portfolio of risky real estate assets; in fact, it had mismarked its Level 3 assets. Those, of course, are the most risky of all, and Einhorn said that Lehman's books showed those assets to be increasing in price rather than falling as they should be, given the state of the bond markets.

Einhorn mentioned his call with Callan and how she had failed to live up to her rep as a straight shooter when it came to disclosing losses from the firm's book of toxic real estate assets. "I asked them how they could justify [only] a $200 million write-down on any $6.5 billion pool of CDOs that included $1.6 billion of below investment grade," he said. "Market prices for comparable structured products fell much further during the quarter. Ms. Callan said she understood my point and would have to get back to me. In a follow up e-mail, Ms. Callan declined to provide an explanation for the modest write-down and instead stated that based on current price action, Lehman 'would expect to recognize further losses' in the second quarter. Why wasn't there a bigger mark in the first quarter?"

Einhorn then demanded action from regulators to make Lehman do the right thing and take the losses he believed it was mandated to do by law.

The regulators did nothing, but Callan and other Lehman executives went back on the attack, and the battle now took on a personal tone. Fuld was fuming. The firm drew up a series of talking points designed to address each of Einhorn's points. Among the most prominent was that Einhorn had cherry-picked all the bad material buried in the firm's financial statements and ignored the overwhelmingly good performance.

The talking points list also slammed Einhorn as a publicity hound, as short sellers rarely go public against their targets, though that was beginning to change. Einhorn's friend William Ackman had launched a similar public campaign against the bond insurers with significant success as their stocks fell into penny-stock territory.

Einhorn wanted to replicate Ackman's success, and, as Lehman executives pointed out, he had an added incentive to draw as much publicity to himself as possible because he was now marketing a new book about his work as a short seller and investor advocate.

Lehman was drawing powerful friends to its side of the battle, including the analyst Brad Hintz, who argued in an interview with *BusinessWeek* that the short attacks on Lehman were "overdone." (At one point on CNBC, I argued that Fuld, given his long record of achievement, should be given the benefit of the doubt.) But Einhorn didn't back down. He soon turned the spotlight on Callan; he continued recounting to analysts and reporters that he had had a conversation with Callan in which she seemed to be ill prepared and unsure of the numbers. He stated that the firm would have to raise capital and that the government should step in immediately to make sure Lehman did the right thing for its shareholders.

Einhorn may have been getting rich by his short position; shares of Lehman were now falling to $20 and heading lower as it became clear that the firm, despite all it had said, would probably have to do what Einhorn had been saying all along: raise capital and admit to losses. But Einhorn also believed he was performing a public service because the Wall Street regulatory establishment had been asleep while a major firm was allowed to misstate earnings and trade without meeting capital requirements. He told the *New York Times* that Lehman was being run by "deniers," executives who lived in a dreamworld where, through accounting chicanery, losses become gains.

Callan now found herself on the defensive; she was bombarded by calls from analysts about the firm's mortgage-related assets and the fact that the numbers were not adding up. It was at this point that the reality of the credit crisis finally began to hit home. The summer of 2008 would be much different from 1998—Lehman wouldn't be able to paper over its problem, hide losses in a mark-to-model pricing because of the mark-to-market edict now fully enforced by the SEC and David Einhorn, and with that Lehman would have to take its medicine and begin posting losses, its first since 1994.

If Callan, Fuld, and Joe Gregory had been spinning before, now they were panicking, and so was Hank Paulson. The Treasury secretary had been paying attention to the rumblings between Lehman and Einhorn and was now coming to the conclusion, along with the rest of the market, that Lehman was heading into Bear Stearns territory, and the last thing the markets needed was another Bear Stearns.

His message to Fuld: "You've got to find another partner." For Fuld and the leadership of Lehman it was a bitter pill. They had fought so hard for the firm's independence, and now because of some fucking short seller it was slipping away.

In late May 2008, the Lehman leadership came to the conclusion that the firm not only needed to raise capital but would post a massive second-quarter loss. They took Paulson's advice and began to look frantically for an outside investor, a white knight to bail out the company not just with money but with confidence.

Bear Stearns had been able to push back the inevitable, its complete decline, by convincing the commodities titan Joe Lewis to put chunks of his vast fortune into the firm. Now Lehman was hoping for a similar lifeline; it flew bankers to the Far East to approach the Korea Development Bank for a $5 billion capital infusion. The firm had long ties with the Koreans, who in the past had wanted a closer relationship with Lehman, including one that involves a strategic investment. Not anymore. Maybe they had read Einhorn's reports. When the bankers arrived, the Koreans wanted something closer to a business relationship: sharing investment banking deals but no money.

The next Hail Mary pass went to the Mexican billionaire Carlos Slim. Investment management chief George Herbert Walker (the second cousin of President Bush) called Slim to drum up business for Lehman, which was slowing dramatically amid the crisis. Callan went a step further and tried to convince Slim to buy a stake in Lehman. But Slim read the newspapers, which included Einhorn's research, and said that he wanted no part of Lehman.

The world had suddenly, and relentlessly, turned on Lehman, and Lehman was about to turn on Erin Callan.

"No way, Erin's a star, she isn't leaving, and Joe is staying as well," said an official in Lehman's press department during the first week of June after I was tipped off that a major change was imminent at the top of Lehman that would likely lead to Callan's replacement as CFO and Gregory's replacement as president.

The moves were described to me as the corporate equivalent of hara-kiri; two senior executives taking responsibility for their actions for the good of the organization. It was now clear that the months of assurances by Callan, approved by her mentor, Joe Gregory, were wrong, and Lehman Brothers was preparing to admit that David Einhorn was right—or mostly right. The firm was losing money, and it needed to raise capital.

For several days, Lehman continued to deny that the firm was even weighing whether to replace Gregory or Callan, as meetings took place among senior executives who felt that heads needed to roll, if for nothing else than to make a statement and restore some accountability to a firm that investors were coming to believe had very little. It was also the latest proof of Wall Street's callous approach to the truth—or, some would say, its penchant to spin everything from major management changes to its bad bets on real estate.

Fuld couldn't spin any longer as he announced the gruesome details to the market. Despite all the assurances that its capital position was fine and it wasn't losing money, Lehman had lost nearly $3 billion in the second quarter, largely the result of write-downs from its soured real estate portfolio, and it needed to raise $6 billion in new capital.

Now it wasn't just Einhorn raising issues with the firm. Lehman was downgraded by the rating agencies Moody's and Standard & Poor's, citing the firm's increasingly shaky financial condition. "I am very disappointed in this quarter's results. Notwithstanding the solid underlying performance of our client franchise, we had our first-ever quarterly loss as a public company," said Fuld in a statement.

But it was hard to see the glass as being half full, as Fuld was trying to convince the market it was. His strategy of spin and bluster had utterly failed. Lehman was the new laughingstock of Wall Street. The Gorilla once feared by short sellers and colleagues was now seen as a bullshit artist rather than a tough-guy straight shooter.

Fuld himself was in the crosshairs as well. What made Lehman work— what made Fuld's us-against-them management style work—was the cult of personality surrounding Dick Fuld. Because he had survived so many battles and helped Lehman survive so many brushes with death, Fuld was considered almost superhuman and was revered by his peers and his colleagues. No longer. In a few short weeks of battling the nasal-voiced David Einhorn, Fuld had been reduced to Jimmy Cayne territory.

Part of Fuld's problem was the company's stock price—after rising to more than $70 a share (close to $150 before it split), it was now in free-fall. Even worse was the damage the falling stock price had done to rank-and-file Lehman workers. Bear Stearns' ownership culture had extended to most of the senior executives. At Lehman, even back-office workers and administrative assistants had been prodded to put their retirement savings in stock. Lehman was "one firm," as Fuld put it: everyone bled Lehman green.

But now employees were just bleeding green, literally, as their retirement savings vanished with the firm's stock price, and they blamed Fuld for those losses. Inside Lehman, executives no longer boasted they would run through a wall for Fuld; now they would have preferred to throw him through it.

O ne of the staple lessons of Wall Street is that when the guy at the top gets in trouble, heads roll below. Cayne axed Warren Spector; John Mack pushed out bond chief Zoe Cruz after Morgan Stanley's losses; Chuck Prince whacked Tom Maheras. None of these moves solved the problem of getting toxic waste off a firm's balance sheet and turning losses into profits. They merely told the world that the guy at the top was still in charge, when in reality he was fighting for his job.

It was no different at Lehman. During the week of June 2, when it was increasingly clear that it was Lehman, not Einhorn, who was misleading the market, the uprising against Dick Fuld began, and it began in the firm's investment banking division.

It was the bond-trading department that had propelled Lehman's earnings during the good years and was causing the firm to suffer now, but it was the firm's investment banking department that gave Lehman its cachet as a top-tier firm. Ask any of its competitors, and they would all agree that Lehman had built what was considered a blue-chip investment banking house. Bear Stearns could never snare the really big deals, but Lehman had competed with Goldman and Morgan for the most high-profile banking transactions and mergers and acquisitions advisory work.

And the bankers were furious. Résumés were flooding the Street. It wasn't a great time to look for a job as deal flow began to slow to a trickle, but for a good banker there's always a place to land. And Lehman had plenty of good bankers, which meant the firm was likely to lose its best and brightest.

Unless, of course, there was a change at the top. Hugh "Skip" McGee, the firm's longtime investment banking leader and the man most responsible for making the department one of Wall Street's best, believed the only solution to restore some credibility to the firm was a wholesale change in its senior management.

McGee wasn't alone. Another power center at Lehman was the equity department, run by Herbert "Bart" McDade, who met with McGee, and they agreed that they needed to make the case directly to Fuld that if he wouldn't accept change of some kind, they would demand it, and it might include Fuld himself.

On June 9, Lehman preannounced second-quarter losses of $2.3 billion and said it would have to raise $6 billion in capital. Preannouncements are basically marketing ploys; they are supposed to signal that a firm is trying to release all its bad news early and build confidence in management for the future. This preannouncement did neither, since it was a rebuttal of just about everything the firm had been telling investors since the beginning of the year. It was also the beginning of the end for Erin Callan and Joe Gregory.

That day, during an executive committee meeting, McGee and McDade made their move—they demanded immediate management change, something to hold current management accountable for the firm's dismal state, not to mention the embarrassment of being shown up by a short seller. The change should begin with Gregory's immediate replacement as president. Fuld was indignant. Gregory was of course a friend, but more than that, he said, he and his team were being held to an unfair standard. The second-quarter loss was his first since taking over as CEO in 1994, and look at all he had accomplished in the past.

But the duo, now accompanied by a handful of bankers, said in no uncertain terms that the firm was in turmoil and Fuld had put it there largely by delegating so much responsibility to Gregory. He had no choice; if Gregory didn't leave, there might be a boardroom showdown that could lead to Fuld's firing as well, and the firm couldn't survive such a management upheaval. Fuld relented.

So did Gregory, and ultimately Callan as well. Seeing the writing on the wall, both took lesser positions in the firm (Callan would leave before the summer was over). Fuld of course stayed, but Bart McDade, now described as the firm's new "rock star" by Lehman's PR department, replaced Gregory as Fuld's number two. Ian Lowitt became Lehman's CFO, replacing the old rock star, Erin Callan.

Lehman was hoping for a new beginning. Mike Gelband, who had left because he advocated less risk, returned to work with McDade, who with his new team launched a full-scale review of the firm's real estate holdings; they concluded that the firm was in deep trouble, holding tens of billions of dollars' worth of mortgage debt of the subprime and commercial variety that no one knew how to value. And the magnitude of the possible losses was leaking into the market. The share price of Lehman fell below $20 and stayed there.

By now the power at Lehman had largely shifted away from Fuld and to McDade. He handled most of the day-to-day decisions and at times kept Fuld out of the loop as much as possible; more than that, he ordered a full-scale effort to find a strategic partner and raise capital. Independence and "bleeding Lehman green" didn't matter that much any longer.

Treasury secretary Hank Paulson, was now fully engaged with the "Lehman problem," prodding the new team to do something, anything, to raise capital or find a partner, which in Wall Street–speak means sell out.

Panic set in at Lehman headquarters, where Fuld had made remaining independent his decades-long crusade but which because of the circumstances was no longer an option as the potential list of suitors who had been lined up to buy all or a piece of Lehman just a year earlier, suddenly disappeared. There were talks with the Koreans, Bank of America, HSBC, General Electric, and CITIC. All said no.

It seemed as if the more people looked at Lehman's range of business, the less appealing this diversified investment bank appeared. Yes, Lehman had a good investment bank and a strong equity department. But its overall business model was far from diversified; in fact, it was driven largely and predominantly by mortgage debt, both commercial and residential, two sectors that showed signs of only getting worse. Conversely, Bear had real businesses that were industry leaders, such as its prime brokerage and clearing units, which had made the firm appealing to JPMorgan.

As potential suitors backed away, Lehman began to look at ways to sell assets to raise capital, including its Neuberger Berman asset management unit, once considered a sleepy business of slow but stable revenues but now considered the firm's crown jewel because, unlike other parts of the firm, it was making money.

Yet another Hail Mary pass involved a spin-off of Lehman's bad debt. As part of the deleveraging that Callan had spoken about earlier in the year, Lehman sold a little more than $15 billion of its bad mortgage assets during the second quarter of 2008, which contributed to its massive losses. But it still held something close to $50 billion more in bad assets with no bidders, at least at prices that wouldn't create such huge losses that Lehman would become insolvent.

McDade began the preliminary work to spin off all of the firm's bad assets into a separate company, offering investors a way to profit from underwater real estate bonds when the market recovered.

It was a long shot, of course, but Lehman wasn't in Bear Stearns territory, at least not yet. Unlike Bear, the firm had some limited access to a new Fed lending program. It was still in business, still assuring investors and analysts it would figure out a way to survive with equity infusions from Wall Street superstars like Larry Fink at BlackRock and Hank Greenberg, who was now running an insurance company called C. V. Starr & Co.

But for every Larry Fink and Hank Greenberg, there were many others now joining David Einhorn's crusade against the firm. The firm now made it official: it had lost $3 billion in the second quarter and warned of further problems in the future. Dick Fuld couldn't just blame the problems on those filthy short sellers.

When Lehman announced the second-quarter losses, Fuld said he took full responsibility for the disaster, and in a moment of contrition he reminded the people at his firm who were still angry over Einhorn's relentless assault that it wasn't Einhorn who had lost nearly $3 billion but Lehman.

It was too little and too late.

"Lehman is raising $6 billion that they said they didn't need to replace losses that they said they didn't have," Einhorn told the *New York Post* in what would be one of his last public statements about Lehman Brothers. There would be no more speeches or attacks because the truth about Lehman's problems was out and Einhorn could now sit back and count his money.

19. FREE FALL

One of the oddities of the unfolding crisis was just how little outrage was generated by regulators over the wildly optimistic prognostications made not just by Alan Schwartz but now also by Dick Fuld about the financial condition of their companies, which turned out to be wrong. The Securities and Exchange Commission was now the laughingstock of the regulatory world and was facing justifiable anger over its chairman's statement that had declared Bear Stearns well capitalized one day, only to see it beg the government for a bailout the next. Yet Wall Street's "top cop," as the SEC was known, continued to investigate the short sellers, the very people who had called Bear, Lehman, and the rest of the Street to task about their faulty disclosures, for allegedly spreading rumors that had led to the demise of Bear and now put Lehman on the same path.

In fact, there was barely a peep out of any of the major government officials about the glaring lack of candor (David Einhorn and Jim Chanos would call it outright lying) on the part of Wall Street in coming clean about not just their holdings of risky mortgage debt but how those holdings continued to burn away at their firm's balance sheets. (The SEC would later start an inquiry into whether Wall Street firms had misled investors with upbeat statements that turned out to be false, though as this book goes to press, not a single firm or individual has been charged.) To be fair, Hank Paulson would argue that he had had other, possibly more pressing, matters to attend to; he had been prodding Lehman constantly to find a grand solution to its problems, such as capital from an outside source or an outright sale. The new management, led by Bart McDade, was dialing for dollars every chance they got, albeit with little success.

Paulson and the gang might also have argued that they had been dealing with bigger problems, namely the latest shoe to drop in the financial crisis: in July it became clear that Fannie Mae and Freddie Mac, the government-sponsored enterprises that had done so much to spur the housing boom, were now joining Wall Street as casualties of its bust. Around the time of the Bear Stearns debacle, the prevailing wisdom among some policy makers in Washington was that Fannie and Freddie were going to bail out the economy by regenerating the mortgage market through their guarantees.

Now it became clear to Hank Paulson and Ben Bernanke that the GSEs too needed a bailout to survive. What had started as a noble mission—making home ownership more affordable to the vast and growing middle class—had turned into something far different. Prodded by various politicians, everyone from HUD Secretaries Henry Cisneros and Andrew Cuomo to Representative Barney Frank and Senator Chris Dodd, to insure and guarantee increasingly risky loans to satisfy the political goal of making sure everyone in the country who wanted a home could get one, both Fannie and Freddie were dying.

The GSEs had grown into trillion-dollar entities whose business model was tied to a single business: the rapidly deteriorating housing market, on which they either guaranteed or securitized mortgages in pools sold to investors or simply held them on their own balance sheets and earned the carried income. Since the GSEs were not just government-sponsored enterprises but also public companies bound by the same accounting rules as Merrill Lynch and the rest of Wall Street, they now had to account for the massive balance sheet implosion that the housing meltdown had caused.

The horror show of the housing market could be described in many ways, but now the rating agencies weighed in. Though clearly playing catch-up, the bond-rating agency Standard & Poor's put it pretty succinctly in the summer of 2008, predicting that, based on its analysis, a whopping 85 percent, or $285 billion, of formerly triple-A-rated CDOs issued in 2004, 2005, and 2006 could fall into default.

The irony of a rating agency pointing out the absurdity of its own ratings was lost amid the general despair that continued to afflict Wall Street. Many of the CDOs were held by the likes of Citigroup, Lehman Brothers, Merrill Lynch, and now the GSEs. Based on current default rates for all mortgate-related debt, Freddie Mac was predicted to lose $30 billion and Fannie Mae as much as $45 billion.

The GSEs' supporters would say their foray into subprime had been a relatively recent phenomenon, one condoned by their critics, the Republicans. Both GSEs had exposed themselves to the vast majority of their subprime portfolios (either through direct guarantees or by holding mortgage bonds), from 2005 to 2007, years in which the Bush administration, through its various HUD secretaries, had had the most power to rein in the extravagance.

But like most of the issues surrounding the great financial crisis, the story of the GSEs defies simple explanation. The Greenspan Fed had fueled the mortgage crisis through its laissez-faire approach to derivatives and its easy-money policies earlier in the decade, but it had also urged Congress to scale back the GSEs' risk taking, and Senate Republicans and even members of the Bush administration had tried to do just that, but they had easily been defeated by determined opposition from Democrats and even some Republicans who had received generous campaign contributions from Fannie and Freddie.

As Fannie and Freddie began to die, Wall Street executives recalled with disgust what it had been like to work with and compete with both agencies. Freddie and Fannie had seemed to have the game rigged. They were public companies with shareholders and boards of directors that demanded profits, but with the "implicit" government guarantee on their debt, they could borrow cheaply and in massive amounts to support their operations, which, in order to meet profit expectations, became increasingly risky.

They were suppliers of mortgage debt to the market, meaning they were huge clients for Wall Street underwriters, and buyers of mortgages, meaning they would compete with firms to find mortgages to be packaged into debt. People who work at Bear recall with anger the "scorecard" that Freddie Mac's chief business officer, Patricia Cook, had used to evaluate them by, rating the firm in terms of how well it was working on behalf of the agency, and how Freddie Mac had used this scorecard to pit one firm against another and squeeze everyone on underwriting fees.

In short, the GSEs had had Wall Street coming and going: they had competed with the Street for mortgages from banks for their securitizations and hired Wall Street to underwrite billions of dollars of their debt, making them two of the largest customers of the banking business. But as housing prices exploded, Wall Street had figured out a way to take on the GSEs: each of the big firms purchased mortgage originators, and now the GSEs had to compete with the Street on a near-level playing field.

That's what they had done in the waning years of the mortgage boom, and they had done so by going toe to toe with the Street in the risky yet lucrative market for subprime loans and alt-A mortgages, at the same time fulfilling the wishes of the housing advocates to use the agencies as part of the federal welfare apparatus. It was, after all, Freddie Mac's CEO, Richard Syron, who had been the head of the Federal Reserve Bank of Boston in 1992, during one of the watershed events in the history of the GSEs: the release of the now-debatable study stating that there was rampant racial discrimination in mortgage lending that had prompted regulators and Congress to prod the GSEs into lending to lower-income, riskier borrowers. About a decade later, Syron's company fully embraced the conclusions of the report, and Freddie and Fannie became the biggest players in the market through their guarantees and securitization of subprime and alt-A loans.

And they did so by ignoring warnings, not just from Greenspan, members of the Bush administration, and Congressman Richard Baker but from inside the agency. One Freddie employee who frequently warned against the new risk taking, the deep dive into subprime, the chief risk officer David Andrukonis, was fired after he warned about troubles to come.

Andrukonis wrote a chilling e-mail in 2004 to his supervisors that said many of the loans on which Freddie had placed its guarantees came from "borrowers who would have trouble qualifying for a mortgage if their financial position were adequately disclosed." (In congressional testimony, Syron later said that Andrukonis had been fired for "a variety of reasons, and it wasn't primarily for having views on credit.") Similar warnings were made over at Fannie Mae; an internal document dated June 2005 cited "growing concern about the housing bubble."

But the warnings had gone ignored until now. What analysts had discovered through 2008 was that Countrywide might have issued some of the most toxic of the country's subprime mortgages, but much of it had been sold to the GSEs, which, when they couldn't sell their securitizations to investors, had simply pulled a page from the playbook of Stan O'Neal, Chuck Prince, and Jimmy Cayne and began carrying tens of billions of dollars' worth of their own increasingly subprime, illiquid mortgage debt on their books. It was the ultimate in moral hazard: with friends in Congress, with their "implicit" government guarantee, the leaders of Fannie and Freddie didn't seem to weigh the possible consequences of eating their own rotten cooking.

By the summer of 2008, as it became clear that the federal government would have to bail out the GSEs, Countrywide had already been put out of its misery, having been purchased by Bank of America as it careened toward bankruptcy.

Now the GSEs needed their own buyout—or, to be more precise, bail-out—or they would most certainly need to be liquidated as their portfolio of bonds, trillions of dollars of toxic debt, were now worth virtually nothing. Some Republicans wanted to let the GSEs die and be forgotten. But Paulson saw the systemic risk immediately.

To finance all those guarantees on loans that were packed into mortgage debt, the GSEs had issued trillions of dollars of bonds on their own. Those bonds were treated in the market as equivalent to Treasury bonds, backed up by the full faith and credit of the U.S. government. If they went into default, what would that mean for corporate debt and even U.S. government debt? Would investors continue to believe that "full faith and credit" meant something? Paulson and Bernanke didn't want to find out.

Around this time Bernanke again pushed Paulson to ask for authority not just to bail out the GSEs but, just in case, to give regulators the power to bail out the entire financial system. Bernanke wasn't predicting doomsday—not yet at least. But with each passing day, Fed statistics showed the housing market's continued deterioration. It was infecting the market for commercial real estate, and data showed the economy to be either in recession or dangerously close to it. It was clear to Bernanke that regulators needed broader powers and more money to intervene if conditions got even worse. Paulson agreed, but he didn't think the political apparatus was ready for such a sweeping move as Bernanke was advocating.

He was right. In mid-July he announced that he would ask Congress to bail out both GSEs. It was Bear Stearns all over again, and Paulson was once again attacked by both the left and the right for bailout excess and stupidity and reestablishing moral hazard.

So was Bernanke, who also got an education in the politics of bailouts. He appeared before the Senate Banking Committee to explain the plan and the concept of systemic risk, possibly laying the groundwork for bigger bailout legislation down the road. Then Republican Senator Jim Bunning accused him of spreading socialism. "When I picked up my newspaper yesterday," Bunning said, "I thought I woke up in France. But no, it turned out it was socialism here in the United States of America."

Congress did ultimately agree to a bailout, and the debate about who was most responsible for the GSEs' irresponsibility began (Massachusetts Congressman Barney Frank became a favorite whipping boy for his support of the GSEs' lax lending standards). For all their talk about systemic risk and the need for legislative flexibility to bail out the system, people who know Paulson, as well as Bernanke and Geithner, say actual panic still had yet to set in; they were, of course, worried about Lehman, Fannie and Freddie, and the losses on Wall Street. Bernanke, for his part, wanted broader bailout authority, but such a plan was still very much in the abstract and not fully conceived. Keep in mind that unemployment was still relatively low, even if the economy was moving toward recession, and every now and then a piece of evidence would emerge that the credit crunch was abating, albeit slightly. Lehman Brothers, for all its public troubles, still wasn't cut off in the repo market, at least not yet.

f I was in your seat, Margaret, and I was advising John, I would advise him to stop saying everything is okay when it isn't."

Paul Critchlow had run Merrill Lynch's PR operations for nearly twenty years before he made a career change and utilized his network of contacts in state and local governments to begin work in Merrill's municipal bond department. During that time, he had been through many wars at Merrill, served as a liaison between key Merrill executives and the firm's board of directors, and advised Stan O'Neal on strategy as O'Neal staged the coup that ultimately led to his appointment as CEO.

John Thain didn't know Paul Critchlow, but his PR chief, Margaret Tutwiler, did, and in her soft southern drawl she was now asking Critchlow what he was hearing from his contacts at the firm. Like Lehman, Merrill, after all, had announced layoffs, a massive first-quarter loss, and a likely loss in the second quarter, yet Thain was upbeat. He had spent much of the past year blaming the losses on O'Neal's wild bet on mortgage-backed securities and proclaiming that the firm was on solid footing, even, at one point, predicting that for all that had happened, it might end 2008 in the black.

Critchlow was hearing overwhelming sentiment inside Merrill's investment banking division and elsewhere that Thain should ratchet down the happy talk; that, based on the frozen market for mortgage debt, to keep extolling the firm's financial strength, Thain was either insane or ignorant of the problem. What if he was wrong? The market would lose confidence in the new leadership, in a business where confidence means everything.

"John shouldn't be making those statements," he said. "How can he predict the future?" Critchlow asked. Bottom line: Thain should just keep his mouth shut, he recommended.

Tutwiler said she couldn't tell him to do that; this was, after all, the great John Thain. "He's so smart, and he knows everything," she countered. "How can I tell John Thain anything when he knows *everything?*"

Critchlow was blown away. John Thain was a pretty smart guy. But he couldn't know everything; in fact, Critchlow was pretty sure Thain didn't know the first thing about the burgeoning credit crisis and its impact on Merrill Lynch. Thain was running around assuring the crowd that Merrill was on the verge of profitability in the midst of Wall Street's most difficult period in decades.

That's when Critchlow came to one other disturbing conclusion about his new boss. He believed Thain and Stan O'Neal had at least one major issue in common: they had surrounded themselves with a coterie of yesmen and -women (Margaret Tutwiler at the top of the list) who wouldn't tell their boss when he was screwing up.

Over time, it became difficult to keep track of every statement coming out of Thain's mouth that turned out to be wrong since nearly the moment he had taken over as CEO. In January he told investors, "I don't think we are struggling." In April he said the firm had "plenty of capital going forward and we don't need to come back into the equity market," just days before Merrill raised capital in the form of debt and preferred stock. In May, Thain stated again that the firm had "no present intention of raising any more capital." In July he boldly predicted that the mortgage-backed securities market would soon make a comeback.

It appeared that nothing could cause John Thain to rethink his bright and sunny outlook for the firm despite the growing doom and gloom everywhere else. Thain was so confident that he and Merrill were on the cusp of a major turnaround that he began to hire his old pals from Goldman, bringing them over to help him in the inevitable turnaround. Peter Kraus, a former investment banker and most recently alternative investment chief for Goldman, would run strategy at Merrill. Kraus had left Goldman after a very public blowup of one of Goldman's investments, a hedge fund with links to the housing market, causing one Goldman senior executive to quip, "I feel sorry for Thain if Kraus is planning strategy."

Thain wooed the star bond trader Tom Montag from his old firm as well. He would be the guy who would attempt to limit the losses on Merrill's mortgage debt holdings, and, like Kraus (who was guaranteed a $25 million bonus), he didn't come cheap; Montag received a $39.4 million guarantee for his first year at Merrill.

The reason they wanted so much was Merrill's obvious predicament—obvious, that is, to most investors, analysts, and even employees but not to Thain. By the summer, Paul Critchlow wasn't the only senior executive who wanted Thain to change his tune. A debate ensued on Merrill's board over Thain's public statements. One board member, John Finnegan, the CEO of Chubb, began to question the logic of setting the bar so high, given the uncertainty of the markets. By now there was little in the way of hedging that any firm could do, and the board knew how much mortgage debt the firm was holding: in the tens of billions, including $30 billion of the most risky CDOs.

While that portfolio was hemorrhaging, so was the Wall Street business model. Leverage remained surprisingly high, around 30 to 1, at most of the big firms, Merrill included, as business began to dry up. The magazine *Investment Dealers' Digest*, which tracks Wall Street business activities, explored the gruesome details: IPOs of new companies had been one of Wall Street's biggest moneymakers, but now they had ground to a near halt. During the first six months of the year, all of Wall Street earned just $6 billion in IPO fees, compared to $24 billion during the first six months of 2007. Yet John Thain acted and sounded as if Merrill were about to perform like a dot-com stock from the 1990s.

All of which made John Finnegan uneasy. He began to press his fellow board members about whether their CEO understood what he was doing. "Why is he boxing himself in?" he asked at one board meeting. "How can anyone say they don't need to raise capital?"

Reality came crashing down on Thain as Merrill was preparing to announce its second-quarter results—billions of dollars in losses because of the deteriorating quality of the firm's mortgage-bond holdings. In just one day in late June its shares fell 5.5 percent as analysts began to question Thain's assumptions on losses going forward. One problem was that MBIA, the big bond insurer, had just lost its triple-A rating and Merrill was holding billions of dollars of mortgage debt insured by the company. But there were other problems: Merrill's debt holdings were large and complex, and the CDOs were jammed with so much complex debt that they were almost impossible to value, so in a jam they would have sold for almost nothing.

The confidence Thain had instilled early in his tenure now began to deteriorate with the firm's increasingly toxic mortgage debt. "This guy doesn't know what he's doing," said former Merrill CFO Tom Patrick after returning from a meeting Thain had held with former top executives about the firm's future. Patrick had been an early Thain supporter, believing he had the chops to turn the company around. Now he thought he was as clueless as the guys he had replaced.

"From the meeting it was clear he had no plan," Patrick said.

Thain may have been coming to that conclusion as well, or, at the very least, the conclusion that the current plan needed to be changed, quickly and radically.

The upcoming second-quarter earnings—or, to be more precise, losses—were much higher than expected, around $5 billion. The size of the losses was in direct contradiction to just about everything Thain had said. Merrill's financial condition was getting worse, not better. More than that, Merrill's capital position wasn't strong; it was deteriorating to the point where the firm needed not just a little more capital, but much more. Merrill's stakes in Bloomberg and BlackRock were now on the table as a way to raise capital that would be needed when the losses were accounted for.

As Thain came to these conclusions, he saw his career flash before him: there would be comparisons to Alan Schwartz, Dick Fuld, and Erin Callan. He would lose the confidence of the market, as Bear had done and as Lehman was doing. Like the other firms, Merrill had survived on the fact that it could continue to borrow money from banks and on other investors who believed they were lending money to a company that told the truth.

Thain was becoming unhinged; during a briefing in one of his finely decorated conference rooms that had been part of the $1.2 million office spending spree, people close to the firm said, he completely lost his composure when an aide informed him about the size of the losses. What Thain did isn't exactly clear, but Merrill Lynch had to replace a shattered glass panel that appeared to have been the target of the CEO's anger. (Thain later said he hadn't broken anything, though through a spokesman he said he wouldn't deny that he had lost his temper during the meeting.)

Thain's more immediate problem, however, was Larry Fink, the CEO of BlackRock. Thain had been pretty emphatic in his public comments that the firm didn't need any more equity capital to cover up possible losses. But if statements are read closely, one sees that he left himself some wiggle room to raise capital through asset sales, including the firm's stakes in Bloomberg and BlackRock. With that in mind, he called his team together and told them to start selling some of the profitable businesses Merrill was holding.

Selling Merrill's 20 percent stake in Bloomberg, the financial information and analytics company, was the easy part. Bloomberg was highly profitable, and Thain had one meeting with its majority owner, New York City Mayor Michael Bloomberg, before reaching general terms, then left the heavy lifting to the firm's president, Greg Fleming, to work out the details.

But BlackRock was different; Merrill owned nearly half of the company, which managed $1.7 trillion of assets. BlackRock managed all of Merrill's mutual fund and pension fund holdings, and Merrill brokers peddled BlackRock's funds to their customers.

Meanwhile, Thain's relationship with BlackRock CEO Larry Fink was tenuous at best; they rarely spoke directly and communicated mostly through intermediaries. As bad as the relationship was, Fink believed the Merrill-BlackRock deal was written in stone because it was a moneymaker for both sides at a time when money was scarce. At least until he heard from his people in BlackRock's press department that Thain had said in a conference call with analysts that he was now considering selling his BlackRock stake.

Fink could barely contain his anger, and he didn't want any intermediary to know what he was thinking. "Are you fucking kidding me?" Fink screamed into his telephone at Thain. "I have to read about this shit that affects my company in the press! You're supposed to be my partner! Fuck you!" Fink was pissed not just because he hadn't been given the courtesy of a heads-up but also because news of the sale would spark selling of Black-Rock shares, which then fell to close to $60.

Thain apologized and hung up the phone, as did Fink. Under the terms of the joint venture Fink had the authority to block bidders not to his liking, and he immediately began looking for another partner and found one. Over the years he had built strong ties in the Middle East, and people close to BlackRock say he had tentative approval to sell Merrill's stake to the Kuwait Investment Authority, the sovereign wealth fund of the State of Kuwait.

As news spread of the pending Bloomberg and BlackRock sales, investors began to panic, as one might expect. The fact that Merrill was now looking to unload profitable pieces of its operation to make up for further deterioration of its mortgage holdings only added to their suspicion that Thain might have been fudging the truth.

Shares of Merrill were now heading south both because investors couldn't trust what Thain was saying about Merrill and because he was now selling or looking to sell off the firm's most valuable assets.

"I always said that when they sell Bloomberg and BlackRock they'll also be burning the furniture down there too," Tom Patrick remarked.

They weren't quite burning Thain's expensive new furniture, but Merrill's second-quarter results underscored the seriousness of the situation. Embedded in the $5 billion loss was a $9 billion write-down in the value of its mortgage-related holdings, including its still massive CDO portfolio. The firm sold off Bloomberg for nearly $4.5 billion, but after threats of a downgrade from the credit-rating agencies, Thain thought twice about BlackRock and decided to keep the money management deal in place. The move didn't thrill Fink, who had secured at least one offer based on his being told that Thain was ready to sell his stake.

Thain, meanwhile, was still putting a positive spin on the events, or at least as positive as he could without making a total fool of himself. "This was a difficult and disappointing quarter in terms of the bottom line, but, in spite of this loss, we likely have in our last two quarters more than replaced the capital that we lost," he said during a conference call with analysts.

Once more Oppenheimer analyst Meredith Whitney called Thain's bluff. Harsh comments by Whitney had sent shares of Merrill down more than 6 percent, and during the conference call, she, like everyone else, heard Thain talk about the firm's healthy capital level and its "trillion-dollar balance sheet." It was then that Whitney asked the obvious question: why not just sell all the stuff off, take the hit, and move on? Thain didn't answer, suggesting to some that a sale might produce losses of such magnitude that even the mighty Merrill Lynch might not survive.

The question of why Thain had made all those positive statements would dog him for a long time to come. People who work with him chalk it up to the arrogance that went with working at Goldman Sachs, the '27 Yankees of Wall Street. Some say it was pure spin, but others feel Thain never had a complete grasp of the intractable problem he had inherited.

A few days after he announced the firm's second-quarter earnings, Thain paid a visit to Deutsche Bank to meet with its influential stock analyst Mike Mayo. Mayo was looking for the same candor from Thain that he'd come to expect from others. "Your job is really hard, isn't it?" Mayo asked in one of the first questions to the Merrill CEO.

"I knew it wouldn't be a cakewalk," Thain replied.

"But I have a problem," Mayo snapped. "Why didn't you just clean house the first time, get rid of everything, all your bad loans and write-downs, all your losses at once?"

Thain barely waited for Mayo to finish before answering, "I thought I did."

Wall Street had now racked up about $200 billion in losses from soured subprime debt. As Vikram Pandit, Citigroup's CEO, continued his allegedly unbiased review of the firm's operations, the bank's losses continued to mount and its stock price to flounder even as Pandit himself bragged publicly that since Citi was a bank and had size, scale, and customer deposits it could draw from in times of emergency, it could withstand the hardest punch thrown at it by the financial crisis.

Citigroup itself had written down $40 billion of losses tied to mortgage debt by August 2008, including a second-quarter loss of nearly $7 billion and a likely third-quarter loss as well. Pandit, however, was becoming cocky, people close to the firm say. In May, Pandit announced, "The universal banking model is the right model. Just ask our clients, or for that matter, our regulators. . . . This is a model that . . . will change the game in global financial service." Pandit's view was that regulators liked Citigroup's business model because having a base of trillions of dollars' worth of deposits and unlimited use of the Fed discount window (the securities firms still had just partial access to Fed lending) gave the bank access to capital in times of need. Bear Stearns needed outsiders to fund the leverage to keep the firm liquid. Citigroup needed only itself.

But investors were still saying what they had been saying since late 2006: the model wasn't working. Citigroup's problems were nothing new; its massive cost structure had been one of the drivers of its risk-taking strategy—the money had to come from somewhere. But even in those heady days, Citigroup had never made enough money to justify retaining its business model.

Now its second-quarter losses were larger than those of the first quarter, and analysts were predicting an equally disastrous third quarter. But Pandit kept holding on to the notion that he could make it all work. Maybe it was arrogance created by running the largest financial services firm in the world; maybe it was pressure from the likes of Robert Rubin; or maybe it was that for all Pandit's brainpower, his PhD in finance, he couldn't understand that when all was said and done, the Citigroup "supermarket" simply wasn't working and probably never would.

Whatever it was, Pandit wasn't touching the basic model of Citigroup that the past two CEOs couldn't manage. There would be no sale of the Smith Barney brokerage unit, which could have fetched $16 billion if Citi had sold it in early to mid-2008, before the crisis reached dangerous levels.

There would be no spin-off of the retail banks in Mexico or even of some brokerage business in India. Pandit wanted to keep them all.

Citigroup was proving to be just as dysfunctional under Pandit as it had been under Chuck Prince. Pandit kept talking about cuts in the firm's workforce and expense reductions that never seemed to materialize in any significant way—unless, of course, you consider one initiative started by John Havens, one of Pandit's top lieutenants. Havens, in one memo circulated to employees, believed that one of the world's biggest banks, with a balance sheet of $3 trillion, could save a lot of money by ordering employees to cut back on the use of color copying and personal telephone calls while at work.

The continued spiraling default rates and the continued illiquidity in the mortgage market had only increased the sense of doom among the top executives of the big Wall Street firms. Losses were mounting; their holdings of mortgage debt were unmovable except at huge losses; and the Wall Street business climate was as crummy as it had been in years. In 1998, following the LTCM debacle and Russian debt crisis, Wall Street had at least had dot-com underwriting to make up for the losses taken in the bond markets. Now it had nothing.

Still, among Bernanke, Paulson, and Geithner the ad hoc approach prevailed. At times, the nation's three most influential economic policy makers seemed bewildered by the problems they faced. On one hand, Bernanke urged Paulson to find a way to get enhanced bailout powers; on the other, the Fed chairman gave speeches downplaying the need for direct bailouts for struggling investment banks, à la Bear Stearns. "If no countervailing actions are taken, what would be perceived as an implicit expansion of the safety net could exacerbate the problem of 'too big to fail,' possibly resulting in excessive risk taking and yet greater systemic risk in the future," he said in late August 2008.

Maybe it was Beltway mentality or just plain stubbornness, but for all the problems and press about Lehman's troubles and what was happening at Merrill, there were times in the late summer of 2008 that the big three hung on to the notion that the problem had been somehow contained. Through their rose-colored glasses they saw the following: management at the big Wall Street firms moving in the right direction by raising capital to meet losses when they appeared; Fannie and Freddie being taken over by the federal government and no longer an issue; Lehman's new management finally taking serious steps to find a strategic partner.

The regulatory system and even key congressional leaders seemed to be looking at everything in glass-half-full mode. As Paulson was getting ready to recapitalize Fannie and Freddie, Barney Frank, the powerful chairman of the House Financial Services Committee, declared both to be "well capitalized" and ready to get back to business once again.

Frank, of course, had been one of the enablers of the GSEs' loosened lending standards by virtue of his powerful position in congressional finance matters. Frank may sound like a typical liberal politician, but he had a base of support on Wall Street. Major pieces of deregulation had been approved by his committee and with his support. That's how Frank and Treasury Secretary Paulson had grown so close when the Treasury secretary was running Goldman Sachs.

But Goldman wasn't the only firm with ties to the Massachusetts liberal. Countrywide Financial and its CEO, Angelo Mozilo, were among Frank's many supporters as well. Former staff members still recall one of Frank's last meetings with Mozilo back in mid-2007. Mozilo and Countrywide were under pressure for being among the catalysts of the mortgage meltdown. He went to visit Frank, whose committee was investigating the now-burgeoning crisis.

Frank promised Mozilo no special treatment, at least not of the kind Mozilo had given some in Congress over the years, such as Frank's counterpart on the Senate Banking committee, Democrat Chris Dodd, who had received sweetheart loans from Countrywide, as had executives at Fannie and Freddie. There's no indication that Frank received any such perks from Mozilo, but Mozilo was known to be a supporter of politicians who supported housing.

Whatever the cause, Frank's committee, for all its bluster about Wall Street greed, steered pretty clear of Angelo Mozilo and Countrywide as the financial crisis spread through 2007 and into 2008, even as it became clear that Countrywide's risky subprime mortgages were a major contributor to the losses on Wall Street and now at the GSEs. In fact, the only comment staffers could recall from Frank about Mozilo after his visit involved his bright orange tan.

"What's with that?" Frank asked with a laugh.

On Wall Street, no one spent a lot of time laughing at Merrill Lynch; in addition to its falling stock price and continued write-downs of bad debt, inside Merrill Lynch a culture war was seething. Though Stan O'Neal's crew had been banished, some of the old Mother Merrill gang remained while Thain was slowly but surely bringing into the firm's senior ranks his own people from the New York Stock Exchange and Goldman Sachs.

The Thain people and the Mother Merrill folks agreed on almost nothing. Greg Fleming couldn't stand being in the same room as the "egomaniac" ex-Goldman executive Peter Kraus, and the feeling appeared to be mutual. The former Goldman bond trader Tom Montag was regarded as an arrogant jerk by the longtime Merrill traders he was now trying to manage in his Goldmanesque way.

One thing they all agreed on was that Jeff Kronthal was paying dividends well beyond the modest freelance stipend he had received to come back to Merrill and help fix the firm's mortgage holdings. Kronthal, of course, had been the head of Merrill's bond operations, an expert in mortgage debt, who had been ousted by Stan O'Neal to make room for Osman Semerci as the firm engaged in one of the most lavish risk-taking experiments in its history. When it was over, even O'Neal had had to admit that the firm's troubles were the direct result of actions taken during that time.

Semerci was ousted in late 2007 as the mortgage bond losses mounted. Kronthal was rehired later in the year when Thain realized he needed someone with the knowledge and connections in the mortgage business to finally rid the firm of its albatross, the tens of billions in toxic mortgage debt, still burning a hole in its balance sheet.

For much of the year, Kronthal tried to do just that, but there were no buyers, or at the very least none willing to pay much more than a few pennies on the dollar for Merrill's debt. That began to change in late July, just after the firm released its second-quarter results. Thain might not have known it at the time, but when Meredith Whitney suggested that he simply sell all of his mortgage debt, Kronthal was working out a plan to do just that. Kronthal had enticed John Grayken, the founder of the Lone Star Funds, a distressed-debt company, to take more than $30 billion of Merrill's CDO portfolio. The deal came at a cost; Grayken agreed to pay 22 cents on the dollar for a portfolio valued a month earlier at 32 cents on the dollar, according to the *New York Times*.

The deal was complex; Merrill "financed" 75 percent of the purchase, meaning that if prices fell below a certain benchmark, Merrill would eat the losses. But in one grand gesture, Thain had done what people thought he should have done early on, sold the junk off at once. When the deal was announced, he was ready for all the counterattacks and stressed the simple, though debatable, bottom line: the firm had solved the problem of its exposure to risk mortgage debt once and for all. For the most part, the media agreed.

"With Merrill's net exposure to CDOs now down to $1.6 billion (the rest is hedged with 'highly rated,' non-monoline counterparties)," *The Economist* wrote, "the worst of the pain is surely over—especially since most of the securities that remain date from 2005 and earlier, before underwriting became really sloppy. It has also halved to $7 billion its exposure to leveraged loans. The question now is whether the bank has enough profit-making oomph in other areas to return to full health."

Along with the CDO sale, Thain raised another $8.5 billion in capital; furthermore, he resolved a dispute with XL Capital, a bond insurer that had covered some of Merrill's CDOs with guarantees. XL was allowed to cancel nearly $4 trillion in policies it had written for Merrill after it agreed to pay the firm $500 million. But it was the sale of the toxic assets more than anything else that had Thain euphoric, along with the market, and Merrill shares surged after the announcement.

People at Merrill believed a different CEO might have understood the nature of the CDOs as soon as he became chief executive and sold them for the best price possible (recall Larry Fink's statement to Merrill's board that his team would need to conduct a comprehensive review of the bond holdings before he accepted the CEO spot).

Even so, the deal seemed to rejuvenate Thain. A couple of days later, Thain returned to his usual positive spin, telling CNBC that Merrill would return "shortly back to profitability," and then blaming his predicament on the problem that he had inherited. But in blaming Stan O'Neal for all his problems, he failed to realize a simple truth: the markets have short memories, and Stan O'Neal was long forgotten by investors and creditors. Merrill Lynch was John Thain's baby now, for better or worse.

The summer of 2008 had been a rough one for John Thain, Dick Fuld, and now Ralph Cioffi. In June, Cioffi and his partner, Matthew Tannin, were indicted on a host of federal charges, including securities fraud, over their management of the Bear Stearns hedge funds, whose implosion many in the markets now blamed for having ignited the financial crisis in the first place. Cioffi, for one, wasn't really surprised by the indictment.

His attorneys and Tannin's had prepared them for months about the pending charges and those likely to be filed simultaneously by the Securities and Exchange Commission. The evidence included e-mails from Cioffi and Tannin that suggested that both had believed the market for subprime debt was beyond repair, in contrast to the upbeat public statements they had made to keep investors from pulling out their money. Cioffi was also charged with insider trading for having moved his own money out of one of the hedge funds before its collapse.

The prosecutors attempted to portray both men as setting a new standard for greed and avarice, yet a more dispassionate look at what they were alleged to have done would show that Ralph Cioffi and Matthew Tannin were no different from, say, the former Merrill Lynch Internet analyst Henry Blodget or the Citigroup telecom stock tout Jack Grubman. In one famous incident, Blodget had e-mailed a colleague that a company that had received positive public rating was a "POS," his shorthand for "piece of shit." Grubman had earned a degree of fame for writing in an e-mail that he had upgraded a company that was to the liking of Citigroup CEO Sandy Weill in order to entice Weill to get his kids into an exclusive preschool, the 92nd Street Y in Manhattan, which he did after Weill donated $1 million of Citigroup's money. Yet Blodget and Grubman faced only civil charges, fines, and lifetime bans from the securities industry. As Cioffi and Tannin await trial and a possible two decades in prison, Blodget is now a celebrated financial journalist, while Grubman is earning a comfortable living as a consultant.

The irony was not lost on Cioffi and Tannin, whose attorneys were weighing a legal strategy that would point the finger at Bear. They planned to show that Cioffi's every move had been approved by Bear Stearns' legal and compliance department, which had listened in on conference calls with investors. Prosecutors have said that Cioffi's statements during those calls were part of the fraud.

Jimmy Cayne, now in exile, hoped for something even more daring: Cioffi turning evidence against Warren Spector, whom he believed was responsible for mismanaging not just Cioffi but the entire firm, leading to its downfall. Since the Bear Stearns sale, Cayne had been making rare public appearances at his old haunts, with a few close friends. He seemed to be a different man. One account in the *Times* suggested that he had become religious. Friends say he definitely had mellowed, except when it came to Spector or other enemies, real or perceived.

When the Cioffi-Tannin indictments came down, Cayne made a prediction to a few friends: "Staring at twenty years in jail, these guys are going to turn on Spector, I guarantee it." Cayne didn't say what criminal evidence either had against Spector, but he was certain there was something that they could come up with.

People close to both men say a more likely legal defense might involve not Spector but another high-profile Wall Street executive in the news, Lehman CEO Dick Fuld. After all, how could a Wall Street CEO like Fuld and his team get away with making all those positive statements that had turned out to be wrong, while the investigatory powers of the federal government were brought to bear on a pair of hedge fund managers?

Many investors were asking that same question and simultaneously selling Lehman's stock, which in August had fallen below $20 a share and was still heading south. The summer had been a rough one for Lehman, between the bruising battle with David Einhorn and the constant rumors about the company, including one that had Fuld selling Lehman for a below-market price, which caused another sell-off of the stock.

Though the below-market sale seemed far-fetched—friends of Fuld, such as the former chairman of UBS PaineWebber Joseph Grano, said they knew for a fact it was something Fuld would never do—the reality of the situation wasn't much better. Lehman *could* be sold, but not at the price Fuld wanted—a significant premium above the $20 a share it was trading at. In mid-August, Fuld and McDade were presented with the grim news: Lehman was facing another loss in the third quarter of possibly as much as $2 billion, and its list of options for survival was narrowing. Barring an outright sale, the firm needed to raise capital, and quickly.

There was talk of taking the firm private; the firm first weighed the sale of its Neuberger Berman asset management business and then the whole investment management business, which included Neuberger, the firm's brokers, and its investments in private equity and hedge funds. The stock price continued to sink. The problem for Lehman was one of credibility; if it had any left, it was quickly vanishing. It almost didn't matter how much commercial or residential mortgage exposure Lehman had and whether the market was severely underpricing those securities, as Lehman's executives argued.

The firm, in the eyes of both investors and creditors, couldn't be believed.

Lehman was now paying the price for its bruising battle with Einhorn and subsequent hyped-up statements about the strength of its franchise, and it became an untouchable. In the past, Lehman had been one of the most sought-after public companies; even when its stock was trading at $75 (on a post-split basis), banks around the world wanted a piece of the firm. Even in the summer, the Korea Development Bank, among others, had still been interested, but Fuld wasn't at the price they were willing to pay, close to $25 a share earlier in the summer, around the time of the second-quarter loss report. But now, as the losses seemed more intractable, the Koreans lost interest and backed off.

It seemed that Lehman couldn't do anything right. The firm announced that it was cutting 6 percent of its workforce; such a cut usually leads to a spike in the share price because investors view it as a cost-cutting move, but in Lehman's case, the price fell.

There was still little interest in Bart McDade's plan to sell the firm's toxic assets on the concept that the bad assets would still be bad when the credit crisis abated, partially because investors doubted the crisis would abate soon and also because no one really trusted that Lehman would accurately mark the bad bank's positions. There was also little interest in the sale of the firm's "crown jewel," its investment management business that included Neuberger Berman, maybe the most profitable part of Lehman, because Fuld was eyeing a pie-in-the-sky price of as much as $7 billion.

The mention of a sale of Neuberger was also taken as an omen of worse things to come. Many investors and creditors wondered why Fuld would sell the most profitable part of the firm unless Lehman needed every penny to survive. And that's when the run on the House of Lehman began.

Jamie Dimon and his executive team at JPMorgan Chase had successfully avoided the worst of the Wall Street crisis because they had cut few corners when it came to managing risk. They had coldly and analytically looked at every major trade, every relationship, and determined how much house money to gamble. That was why, despite his personal affection for

Jimmy Cayne, Dimon wouldn't provide a line of credit to Bear in the summer of 2007, nearly sinking Bear several months before it actually sank. It's also why Dimon, knowing he was the only bidder for Bear, knew he could get away, albeit only for a while, with a mere $2-a-share bid for a firm that a little more than a year earlier had traded at $170 a share.

And it's why, as Lehman's problems began to spread, Dimon was quickly cutting his exposure to the firm. In July, JPMorgan Chase demanded additional collateral on short-term loans it had brokered on behalf of clients to Lehman to fund its operations. According to a senior executive at JPMorgan, Lehman provided the collateral in the form of "crappy structured securities" that JPMorgan accepted, although it kept a watchful eye on the firm's situation.

Now, in early September, the markets had turned uglier, credit was drying up, and confidence in Lehman's ability to survive began to wane seriously. JPMorgan was back for more and better collateral, and it wasn't accepting anything that looked like a mortgage bond; the firm told Lehman it needed an additional $5 billion in collateral immediately. Lehman could balk, but that would be financial suicide. No one in the market would lend to a firm that couldn't make a $5 billion collateral call.

But it was money Lehman really couldn't afford; $5 billion here and $5 billion there, and pretty soon, the $40 billion in cash Lehman had boasted about having during the height of the Einhorn battle would disappear pretty quickly, which was what was happening. As it did, Dick Fuld's world was closing in on him; he had once been one of the masters of the Wall Street universe, a member of the board of the New York Federal Reserve Bank, with a close relationship to Wall Street stars such as Larry Fink and Hank Greenberg, and he had had the admiration of many employees. Now he was slowly but surely becoming a pariah. The consensus was building that Fuld's toughness might have brought the firm back from the dead in 1998, but his arrogance since had been leading Lehman back to the death chamber.

It is clear now that Fuld could have sold Lehman earlier, but he didn't because he couldn't take the stigma of selling out at the bottom of the market, even if that meant survival. Perhaps his vivid memories of chafing under the thumb of American Express before he was able to take the firm independent prevented him. And as McDade and his team continued to dissect what had happened, they discovered just how out of touch Fuld had become. They came away believing that he had no idea just how much the firm had deteriorated. More than that, they believed he had relied too much on and delegated too many decisions to the man they believed should be held most responsible for Lehman's problems, his number two, Joe Gregory.

Such a management structure had allowed Fuld to think big thoughts, meet with clients, and proclaim a strategy predicated on making bets on various markets, most recently the real estate market. But it had also prevented him from understanding the true nature of the risk that the firm was taking, leaving it up to Gregory to sort through the massive amounts of commercial and residential mortgage risk the firm had taken on, including the billions of dollars of bad debt through the Archstone-Smith deal in late 2007.

The systemic risk of losses spreading to safer asset classes as investors unloaded everything of value was what had caused the Fed to bail out Long-Term Capital Management in 1998. But the world had changed a lot in ten years. The trading relationships between banks and brokerages were larger and more global. Lehman, for one, held trades and loans from small banks in the Far East and Latin America. As a result, the losses stemming from "systemic risk" would be more significant than ever before.

That's why, as Lehman's troubles began to transform into something more profound in the late summer of 2008, the credit crunch reached a new dimension. By early September, Federal Reserve officials had come to the conclusion that despite the massive reductions in short-term interest rates and all the steps they took to increase market liquidity, the credit crunch had actually gotten worse. The economy was grinding to a halt, and it wasn't just because of American consumers' appetite for debt that they now had to pay back, the catalyst for other slowdowns. Now policy makers were getting their own education on just how much borrowing permeated the economy. Small businesses that needed short-term cash to pay their workers before revenues came in now couldn't get loans because banks were hoarding cash on the fears that Lehman's implosion would mean mass selling of assets. Nearly every Wall Street firm needed to borrow money to stay afloat, and creditors threatened to cut off lines to banks that had appeared to escape the credit crisis, such as Morgan Stanley and even the mighty Goldman Sachs. The vast, interconnected system of loans and borrowing known as credit was shutting down.

The mood inside Lehman was certainly grim. Others were following suit in demanding additional collateral, as JPMorgan had. Lehman's stock price began what traders call a "death spiral," heading toward $10 a share and lower as investors bet that Lehman was about to lose everything.

Dick Fuld was now flailing. On the one hand, he was pressuring the SEC to ramp up its investigation into the short sellers, as if that could change investors' attitudes about the company's financial condition. On the other, he realized he had fewer and fewer friends. Joe Gregory was gone, and instead he had to work with people who blamed him for the firm's current mess.

The long hours and constant criticism had taken its toll; he looked tired and haggard, yet at times he would show flashes of intense anger—at the media, which were repeating what he saw as unfounded rumors; at the short sellers, who he believed were the source of what was malicious gossip; and at Goldman Sachs, which he believed continued to stoke animus against Lehman through its trading counterparties, namely hedge funds that were shorting Lehman's shares.

It was easy pickings for the shorts these days. Less than a week after Bernanke warned the markets that despite what had happened at Bear, investors and the firms themselves shouldn't bet on the government's bailing out Wall Street banks such as Lehman, Paulson finally made his move on Fannie and Freddie, putting the GSEs into something called conservatorship. They would be 80 percent owned and totally run by the federal government.

For Paulson, there was just no other choice. If Fannie and Freddie were allowed to default, the Chinese government, which held not only U.S. Treasury debt but also debt of the GSEs, would have questioned whether the U.S. government might someday default on its Treasury debt. The implications of such reasoning would be huge, since China finances such a large portion of the federal budget deficit. For starters, it would demand much higher yields (that is, far lower prices) to buy U.S. government bonds, but there was no telling if it would still be a buyer of Treasury debt if such a thing occurred. That type of systemic risk could begin to shut the U.S. government out of the credit markets.

Still, the bailout met with stiff resistance. Republicans in Congress, such as the free-market Georgia Representative John Linder, who had supported the initial legislation back in July that had given the government power to step in and fix Fannie and Freddie, now wished he hadn't when he heard from Paulson how much it might cost: $200 billion.

Though some Republicans in Congress weren't appeased, investors seemed to be. With the bailout, the stock market rebounded strongly; Paulson was basking in the unusual glory of being applauded by both the conservative-leaning *Wall Street Journal* editorial page and the left-leaning billionaire investor Warren Buffett, according to an account in the *New Yorker*. It seemed that the Bernanke Doctrine, coupled with the Paulson Plan of ad hoc crisis management, had won the day.

Except it hadn't.

Nearly from the moment the market opened on Tuesday, September 9, 2008, shares of Lehman were in free fall. The rest of the market wasn't faring much better. By the time the day was over, Lehman's stock was down by 45 percent as investors debated the firm's survival. Television camera trucks parked outside Lehman headquarters were greeted by jeers from Lehman executives, who believed that the media coverage of the firm's descent, possibly even more than the tens of billions in mortgage holdings on the firm's books, was responsible for the nightmare that they were living through.

JPMorgan slowly began to tighten the screws on Lehman, which started to tighten the screws on the entire financial market even more as speculation about the move spread from trading desk to trading desk.

Steve Black is the co-head of the JPMorgan investment bank. His relationship with Jamie Dimon dates back to Citigroup, so he knew how his boss thought, and the last thing his boss would want was Lehman going under and owing JPMorgan and its clients money.

Lehman still hadn't paid the $5 billion in cash JPMorgan had demanded the week before. Black called Dick Fuld personally and demanded the money. It wasn't a request, he explained, it was a demand. If the money wasn't paid, Lehman was finished as far as Morgan was concerned. The most creditworthy of all the banks would be shutting off the spigot to Lehman.

Fuld had never really gotten used to begging and pleading with potential buyers and the government to keep the firm afloat, but he now did his best to let Black know he needed a favor: please accept just $3 billion, he pleaded. Black agreed. Fuld thanked Black, but he was livid. He told friends like Larry Fink that Dimon had screwed him and said if Lehman didn't survive, the demise of a once-great firm could be blamed on Jamie Dimon.

But JPMorgan believed it had ample reason to be nervous. That day, Lehman decided to preannounce third-quarter performance results the following morning. Shares of Lehman had fallen dramatically in late August, when it had become clear that Lehman would lose at least $1.8 billion for the quarter. But the actual loss, Fuld and his team discovered, turned out to be much bigger, about double that amount, or $3.9 billion. Most, if not all, of the losses could be attributed to real estate holdings that had been crushed by third-quarter write-downs.

But JPMorgan was concerned about more than just the losses; there was also no plan to raise capital, which executives at the firm pleaded with top officials at Lehman to do immediately. Apparently, Lehman couldn't find anyone willing to invest, not even in a minority stake. The on-again, off-again negotiations with the Koreans resulted in nothing. Instead, JPMorgan had learned that Fuld and McDade planned to dust off the old plan to create a "bad bank" and spin off as much as $30 billion of depressed commercial real estate into a private company. They didn't say who might buy a real estate portfolio filled with illiquid assets, but that was just one problem with the solution Fuld intended to unveil Wednesday morning. Another problem was his intention to sell the investment management division, for which there was still no buyer.

Officials at JPMorgan were livid. They still hadn't received the collateral they had demanded; now they all but ordered Lehman to raise capital immediately. The word from Fuld was simple: this was the best he could do.

The best wasn't good enough for JPMorgan, which continued to demand its collateral, or for the market, which was waiting to unload even more shares of Lehman. Fuld tried to spin the bad news the best he could during his conference call with analysts. But this wasn't "The Gorilla" who had vowed to kill the short sellers just five months earlier.

Though Fuld blamed many of the firm's problems on the "intense public scrutiny . . . which caused us significant distractions among our clients, our counterparties, and also our employees," this was a more subdued Dick Fuld, one who, according to some analysts, seemed at least partially resigned to defeat.

Even in his subdued state, he was still trying to sell the market on Lehman's viability. He said a plan to off-load billions of dollars in real estate debt, coupled with a plan to sell some bad assets to BlackRock, repackage the commercial real estate holdings into a separate company, and sell it to investors, plus raising capital at some point in the future, "will best protect the core client franchise, and create a very clean, liquid balance sheet."

What he didn't say, according to people who know him, was that in his heart he believed that the federal government would at some point step in and save the day.

During the call, the analysts were polite and respectful, almost as they might have been during a wake, though at one point, the normally abrasive Mike Mayo lived up to his reputation, pressing the company's CFO, Ian Lowitt, about whether Lehman needed to raise more capital to survive. Lowitt responded that the firm's "capital position at the moment is strong," even as officials at JPMorgan were pressing Lehman to raise more capital to avoid a meltdown,

Mayo thought Lowitt's answer was a dodge (the phrase "at the moment" stuck in his mind), and he later opined that Fuld's analysis of the value of Lehman's commercial real estate holdings didn't jibe with the fairly large deterioration in the index that follows commercial properties, known as the CMBX.

Mayo said he couldn't recommend the stock now because the plan the firm had unveiled was too uncertain, which was now the consensus on the Street. William Tanona, an analyst for Goldman Sachs, made virtually the same comments. Dick Bove went even further. Amazingly, he kept a "buy" rating on the stock (he believed it was a takeover candidate), but he added the following caveat: "For the past two days I have been trying to write a commentary on Lehman's earnings but have failed. . . . I finally concluded that I was trying to argue that yesterday's announcement of a new strategic direction meant that something had happened at the firm when in fact the only events that have occurred is that the company announced a big loss and cut its dividend. Lehman intends to do nothing. It is increasingly evident that outside intervention in the form of a hostile takeover is a necessity."

By the end of the day, after falling more than 50 percent the past two trading sessions, shares of Lehman fell another 7 percent to $7.25. Over the next two days Lehman would deliver some $8 billion in cash to JPMorgan: $3 billion for recent collateral calls and $5 billion to make up for the mortgage-backed collateral it had previously supplied.

By the end of the week, shares of Lehman were trading in penny-stock territory and the firm was running low on cash. The only rescue plan on the table now involved selling all of Lehman. The scene in front of Lehman's headquarters in lower Manhattan was chaotic; reporters lined the streets, while employees began crafting résumés in the event that management couldn't pull off a miracle.

McDade, Fuld, and their team were now holed up in the "war room," as they called it, desperately calling potential buyers. The firm had now entered talks with both Bank of America and Barclays PLC regarding an outright sale of the firm. The speculation on Wall Street was that the Bernanke-Paulson-Geithner team would somehow facilitate the sale, though it wasn't known exactly how that would work. Both potential buyers wanted some type of federal guarantee on Lehman's billions of dollars' worth of troubled assets, which, considering the political climate, seemed remote.

The best Fuld & Co. could say was that the stock still wasn't trading at zero, at least not yet.

G reg Fleming at Merrill had been following the events at Lehman closely. Fleming knew Lehman was finished not just because of anything Fuld said or the decline in the firm's stock price but also because Lehman was selling the only part of the business that was making money, Neuberger Berman. Fleming had been an investment banker for Neuberger Berman in 2006, when Lehman had bought the outfit, and he knew something else: there was no way Fuld and his team were going to get the $7 billion they were asking, not in this environment.

And he noticed something else: Lehman was going to take Merrill down with it. Despite everything that Thain had done—all the capital he had raised, not to mention the deal with Lone Star to rid the firm of its bad assets—Merrill's shares were falling in tandem with Lehman's, and its creditors were getting nervous.

The market simply wasn't differentiating Merrill from the most impaired firm in the financial business. As far as many investors were concerned, what Fuld had said over the past year about Lehman's finances wasn't much different from what Thain had said about Merrill's. Both had been less than forthcoming, and now they were paying the price; creditors and investors asked why they should take a chance on either now that the financial crisis had reach a new level. Even so, the massive sell-off took Thain by surprise, given all he had done to preserve the bank's balance sheet.

Fleming, meanwhile, went to work. The former investment banker heard about the Bank of America/Lehman talks but knew that Bank of America CEO Kenneth Lewis, though possibly interested in Lehman, still lusted for Merrill Lynch and its fleet of sixteen thousand brokers who handled trillions of dollars in client money. Lewis was building a global corporate and investment bank, but the key to his strategy was wealth management, or selling stocks, bonds, mutual funds, hedge funds, anything, to individual investors. The previous year, as Merrill had been just coming to grips with its bad bet on mortgage debt, Lewis had been willing to pay $90 a share for the firm; now he could get the firm and its brokers with an even cleaner balance sheet for less than half that price.

Not long after the disastrous Lehman conference call, Fleming, without Thain's knowledge, tested the waters with Bank of America. He didn't call the firm directly but instead contacted an attorney, Edward Herlihy, at Wachtell, Lipton, Rosen & Katz, a firm that represented Bank of America on many of its big deals.

Fleming was acting out of a sense of desperation—he knew the financial business and knew just how connected the firms were. Barring some kind of miracle (such as a government-contrived bailout), Lehman was headed for liquidation, and if Lehman went, so would Merrill, the next weakest link in the Wall Street food chain.

Unless, of course, Fleming could find an appropriate suitor such as Bank of America. Herlihy told Fleming he couldn't do anything official, he would need to speak to Ken Lewis first, but he said that, based on everything he knew, Lewis still loved Merrill if Merrill still loved Bank of America. In this climate, Fleming said, how could we not?

Fleming couldn't start official discussions; that would be up to John Thain, who Fleming thought was oblivious to the company's mounting problems, including its standing with its major creditors, which were now circling like vultures. The market now believed that it would be only a few more days than Lehman for Merrill's demise to be complete. Not Thain. According to one senior executive who worked directly with Thain at the time, "This was a guy who really believed that on Monday morning he would still be CEO of an independent Merrill Lynch, when all the signs were that we were toast."

What were the signs? JPMorgan was once again looking to collect, not just from Lehman but from Merrill. Other demands for collateral started flooding the firm. And though Merrill's stock hadn't been beaten down as badly as Lehman's, it was getting killed nonetheless, trading as low as

$16.60 on Friday and closing at just above $17—its lowest level in more than ten years.

Merrill's board was in a state of panic, but Thain seemed cool as a cucumber. There's something to be said for a leader who doesn't panic under pressure. But panic wasn't what the Merrill board was looking for; it was a sense of urgency, and Thain, people on the board believed, didn't have any.

That sentiment was delivered to Thain on Friday during an emergency board conference call. John Finnegan, a board member who had been critical of Thain in the past, was there, and he wanted some answers.

"John, I have a simple question," Finnegan said. "Lehman's going bankrupt this weekend. We're next. Tell me why this will end differently for us."

Thain simply replied, "We are not Lehman," to which Finnegan snapped, "You say that now, but that's not the way the world operates. Why is it going to end differently for us? What's your plan?"

The telephone line went silent for a few seconds before someone else chimed in, asking Thain and his managers to "keep us posted" about developments going forward.

After the initial market euphoria over the Fannie and Freddie bailout, the mood in Washington had changed, particularly among House Republicans, who were in the minority but had enough votes to quash bailout legislation if that was the route Hank Paulson might take.

Paulson had come to Washington with a guarantee from the president that he could run the Treasury the way he saw fit. But he had always had lukewarm support from the conservative wing of the party, given his reputation as an environmentalist and an economic interventionist with an eye toward preserving Wall Street at the expense of the taxpayer, and now that support turned cold.

House Republicans complained openly and loudly that Paulson had a better relationship with Barney Frank than any member of the House Republican leadership, which was, by the end of the week, in open revolt against the Treasury secretary.

Now they had more than ideology to quash Paulson's interventionist instincts. Paulson's bailout of Fannie and Freddie was supposed to prevent systemic risk. It didn't. The almost certain Lehman implosion was proof positive that moral hazard was alive and well, or so they thought. It wasn't just the opinion of many House Republicans and a growing number of members of the conservative media. The notion of systemic risk had never resonated with the general public even two years into the financial crisis, which had spread from Wall Street to the banking system and now into the general economy, which, by all accounts, was headed for one of the most severe recessions in decades.

Polls showed that Americans, by large margins, were fed up with the notion of bailing out Wall Street and its irresponsible business practices. It didn't help the interventionists' case that there were media reports alleging that Fuld had in the past turned down offers to sell Lehman because he was holding out for a higher price and that he had friends in high places, like Paulson, who were part of the insiders' club that put Wall Street's interests above those of taxpayers.

It's unclear exactly how friendly Paulson and Fuld were by late Friday, September 12, when it was clear that, barring extraordinary intervention, Lehman was finished. Paulson declined comment for this book, but he and Fuld had met several days earlier for a dinner in Washington. Paulson pressed Fuld on his progress in raising capital and finding a partner in a testy exchange between the two in which Fuld said he was still at square one and unclear what to do next. After the meeting, officials at the Treasury and the Fed believed Fuld and McDade weren't seriously looking for a private-sector deal but a public-sector one, something along the lines of what had happened with Bear earlier in the year.

One thing is certain: Paulson had lost confidence in Fuld and in Lehman's ability to survive as an independent company. Just a couple of days earlier, during its conference call with analysts, its CFO had announced that the firm's capital position was strong, and now suddenly it wasn't. Paulson made it clear to his senior staff that there would be no bailout of Lehman by the federal government. It would have to sink or swim on its own or with a Wall Street–led bailout.

Meanwhile, Ben Bernanke was losing confidence in himself, according to people who worked with the Fed chairman at the time. Just how rapidly Lehman and Merrill had unwound after the Fannie and Freddie bailouts had caught Paulson, Geithner, and most of all Bernanke by surprise. He was now tired, wary, and incredibly sensitive to the charges that, on the one hand, he had been slow to react to the simmering crisis and, on the other, that the rolling bailouts had bolstered what all regulators try desperately to avoid, Wall Street firms' complacency about taking undue risk, the concept known as moral hazard.

Former Fed Vice Chairman Alan Blinder, now the head of the Economics Department at Princeton University, spoke with Bernanke about a half-dozen times during the year. Bernanke is one of the most even-tempered people in the world; just look at the way he handles all the absurd questions thrown at him during congressional hearings. (During one hearing he was accused of having been the head of Goldman Sachs, and he calmly replied that the only entity he had headed before the Fed was the economics department at Princeton.)

Though Bernanke's disposition never changed, Blinder noticed that his mood had. "There were times where he has seemed almost beaten down, overwhelmed," Blinder recalled. "And who wouldn't be? He's up all night, sometimes sleeping on his couch, barely sleeping, working around the clock. But there are times where he's been very optimistic. Remember, a couple months after Bear, things were improving a bit, and he seemed hopeful. But then September happened and it sort of just starts all over again."

20. THE RESCUE

It was happening all over again Friday night at the New York Fed's austere headquarters on Maiden Lane in Lower Manhattan, where Hank Paulson and Timothy Geithner had called an emergency meeting of the heads of all the major Wall Street firms to discuss the financial crisis's latest victims. The implosion of Lehman Brothers was the stated main topic, though in the back of everyone's mind was the bigger issue of Lehman's demise, sparking a broader panic.

The credit default swaps, one of the measures of risk like the ABX Index that no one had heard much about before the financial crisis, told the entire story: default swaps on Lehman had risen dramatically on Friday, meaning that traders were buying default protection and predicting the firm's liquidation. But not far behind Lehman was Merrill Lynch, and not far behind Merrill now, amazingly, were Goldman Sachs and Morgan Stanley and, most ominously AIG, the large insurer that had sold many of the credit default swaps in the market.

The irony of AIG's worsening financial situation wasn't lost on Geithner or Paulson or on Ben Bernanke back in Washington. This was the company that was supposed to be the last line of defense against further massive write-downs among the many banks that held this protection, and now it too was considered toast for the simple reason that no one believed the firm could cover all the losses in the market, from CDOs to the bonds of the banks that appeared to be heading for the graveyard.

The looming implosion of AIG and the rest of the Street not only took regulators by surprise, they were simply overwhelmed by events. Just a day earlier, Bernanke and Paulson had heard from insurance commissioners, including New York's Eric Dinallo, who had played down fears that AIG—which in addition to having insured all that risky debt had been holding billions of it as well—was going bust. Dinallo had never quite shown a grasp of the depth of AIG's problems (in an interview he said the financial products group fell outside his purview), and now he described AIG as a firm with solid businesses. Outside the credit default swap area, he said, it was a company that made a lot of money selling life insurance policies, and most of all, it was salvageable. He pointed out that there was also interest from JPMorgan in buying the firm.

What a difference twenty-four hours make when systemic risk is happening all around you. The Washington policy makers who had been applying Band-Aids to the festering financial crisis all year now faced so many wounds that they needed to perform regulatory triage, so to speak, so they began with the patients who were on life support: Lehman Brothers and Merrill Lynch.

Bernanke didn't attend the meeting, but Geithner did, as did SEC chairman Chris Cox, who probably shouldn't have based on the reaction of some of the CEOs when they noticed his presence. By now the SEC had lost all credibility as a regulator, mostly with the people the agency was regulating, the heads of the big firms. Dysfunction had been brewing at the SEC for years, of course—it was Arthur Levitt who, since leaving the SEC, had written a book about the sins of Wall Street and remade himself as an investor advocate, who had missed the last great Wall Street rip-off, the dot-com bubble.

But the long-simmering credit crisis, which now exploded into something much bigger, had shown that the dysfunction of the commission had reached new heights. Cox's lack of understanding of the key vices that led Wall Street (the SEC was, of course, the regulator supposedly in charge of monitoring capital levels at the investment banks) had been on full display for two years. He didn't help himself during the meeting. "It was like this guy would say things just for the sake of saying them," John Mack, the CEO of Morgan Stanley, would later remark. For that reason and others, Hank Paulson did much of the talking as about thirty executives of the country's largest and most important banks and securities firms, including Mack, Jamie Dimon of JPMorgan, Lloyd Blankfein of Goldman Sachs, and Vikram Pandit of Citigroup, sat around a large conference table; it reminded at least one CEO of the famous scene in the *Godfather* when the heads of the five families are called to a meeting (ironically, the location shot for the film used the exterior of the New York Fed) to discuss a possible truce, which was pretty close to what Paulson and Geithner wanted: the firms to act not as competitors but as members of the country's financial community and to come up with a private-sector solution to bail out Lehman Brothers.

To some it sounded remarkably similar to the reasons given for bailing out LTCM ten years earlier, only now the stakes were larger. LTCM had had trades and investments worth about $4.6 billion; Lehman was a large securities firm with a balance sheet of $600 billion and trading partners around the globe.

The mood was tense; though the matter wasn't discussed, some executives privately blamed Dimon for instigating the crisis with his collateral calls on Lehman and Merrill, in effect taking advantage of firms that were on the edge of death. (Dimon would later deny the charge, stating that the collateral calls were to protect clients that had lent both firms money.) Noticeably absent was Dick Fuld, the Lehman chief. By now he and Paulson weren't on speaking terms. Paulson blamed Fuld for having dragged his feet on a sale of Lehman and even more for failing to see the gravity of the situation until it was too late and the government had to come to the firm's rescue.

It was a wild gamble on Fuld's part, or so Paulson and other regulators thought. In the world of Dick Fuld, however, his approach made incredible sense: why sell the firm at a depressed price if, when worse came to worst, the government was standing ready with bailout money if it was needed? The moral hazard that the government had created by bailing out first LTCM and then Bear Stearns was now manifesting itself.

There was just one problem with Fuld's calculation: it was unclear if the government, just six months after bailing out Bear, was ready to go through it all over again. Therefore the word from Washington was that Fuld should stay at home, stay in the office, stay anywhere, but stay away from Maiden Lane that weekend as Lehman's fate was decided.

When the attendees took their seats, Paulson wasn't mincing words: Lehman was toast, finished, it wouldn't last the weekend and certainly couldn't open for business Monday morning; but, as he said, the infection known as systemic risk was present and dangerous.

So something needed to be done. Paulson had long since weighed the political risk and come to the conclusion that after Fannie and Freddie it would be difficult to go to Congress and ask for a wad of cash to bail out

Lehman. Bernanke had been telling Paulson that the Fed was tapped out as well. The agency, after all, was guaranteeing some $30 billion of toxic debt from Bear Stearns. It was also extending loan facilities and guarantees to other parts of the financial system, namely, the big Wall Street firms, that had never before come under its purview. In other words, Paulson made it clear, the government was not ready to bail out Lehman, so Wall Street would have to do it. And by saving Lehman Brothers, they would be saving themselves.

The most feasible option was the one Bart McDade had been trying to sell for the past six weeks: a "bad bank" that would be funded by Wall Street to scoop up the firm's bad assets, mostly the approximately $30 billion in commercial real estate holdings that just a few days earlier the firm's CFO had said was accurately mark-to-market priced but now some executives who were getting briefed on the holdings weren't so sure. In any event, the second option Paulson spoke about was bankruptcy and the liquidation of the balance sheet of $600 billion, which no one wanted but the government was ready to condone if a private solution wasn't reached.

Geithner asked every firm to study its exposure to Lehman in the event that liquidation occurred. He broke the group into two to study the two alternatives. What he didn't mention was something that had been bounced around by Bernanke and senior Fed officials. Though Paulson may have told the group that government assistance was off the table, inside the Fed it was still clearly on, as long as someone was willing to buy the firm, even if only its good assets.

"We could have done something if Lehman had a buyer," said one senior Fed official. And at least for now, there were two potential buyers: Bank of America and Barclays PLC, the big British bank.

The tone coming from the regulators was jarring to the CEOs even after one of the worst days in the market that year; the Dow Jones Industrial Average had fallen more than 200 points. Shares of all the big brokerages had tanked alongside Lehman's and Merrill's, and now this. "Hank and these guys were basically saying 'I want you guys to pay fifty billion dollars to save Lehman,'" said one of the CEOs in the room. Paulson was now the Treasury secretary, but he'd used to be one of them, and the last thing he would have been willing to do as CEO of Goldman was bail out another firm.

"So you're asking us to keep a competitor in business?" John Mack snapped at one point. Mack has a reputation for being one of the straightest shooters on the Street. For the past year, he'd kept an amazingly low profile for the reason that he didn't want to call attention to his firm's business prospects and be lumped in with Dick Fuld and John Thain.

"And if we do this, what about other firms that are in trouble?" he added. Merrill wasn't mentioned, but it was the firm on Mack's mind and everyone else's. The other was AIG. Nearly every firm had mortgage debt that had AIG insurance on it. Even Goldman, which so far had been the least affected by the financial crisis, had bought $13 billion of CDOs from Merrill Lynch back in 2005 and would face significant losses if AIG filed for bankruptcy and it had to write down the true value of the CDOs.

Paulson repeated the government's position: there would be no government assistance. His point was not that much different from the one Eric Dinallo had made during the bond insurance crisis earlier in the year (though with all that had happened in the meantime, it felt like an eternity ago): Wall Street had created the mess it was in; it had lost the confidence of the market through excessive risk taking and outright lying about its financial health; and Wall Street, not government, needed to find a solution.

"Stand up for yourselves," Paulson told the group, reminding them of the cost of failing to work together, which came down to the simple fact that "everyone is exposed" to Lehman. With that, the meeting ended with orders to reconvene the following morning at nine.

Greg Fleming, who had spent his entire career merging and selling financial firms, was now trying to do the most important sale of his life. His view was that no simple capital raise or strategic deal, in which Merrill would sell just a portion of itself, would do. Merrill needed to sell itself to a bank, an institution with deposits and protection from the Fed, if it were to survive the coming week. It didn't matter whether Lehman was saved or not. As an independent institution, Merrill was done.

On Friday night and early Saturday morning he had further conversations with Ed Herlihy at Wachtell, Lipton. Ken Lewis was losing interest in Lehman after Paulson announced that he wasn't going to provide any support for a bailout, and now he wanted to speak to John Thain. That was music to Fleming's ears because if Lewis had gotten locked in on Lehman, it would have been nearly impossible to entice him to take a look at Merrill.

First, Fleming needed to entice Thain. Superficially, it would seem like an easy sell: by the end of the week Merrill was facing a daunting number of collateral calls, its stock was imploding, and it was facing a cash drain. Yet Thain needed to be convinced to do a deal. Part of Thain's thinking was that Merrill had sold most of its CDO exposure and was simply feeling the aftershocks of Lehman's demise. He thought if the firm could get through the first couple days of the week, it could survive as an independent company.

Fleming thought that was wishful thinking, and although he didn't tell his boss this, he really believed that Thain's biggest worry was about himself; that in any deal, he would have to give up the CEO position. Ken Lewis was relatively young, just over sixty-one years of age to Thain's fifty-three, and it would be Lewis running the combined shop for a while. Thain would have to go back to playing second fiddle.

By Saturday morning, just before the second Fed meeting, Fleming decided to stop waiting for a call from Thain and instead initiated the call himself. Thain was in his car traveling to the Fed's headquarters in lower Manhattan when he picked up the telephone. Fleming told Thain about Lewis's interest in Merrill, but Thain still wasn't ready to talk about an outright sale. Fleming told him to keep an open mind and impressed upon him the following: the markets were as bad as they had ever been, and Merrill was now in the crosshairs. He would be passing up a grand solution to their problems: by selling Merrill to a bank that had access to trillions of dollars in deposits for support, not to mention the full lending authority of the Federal Reserve, Merrill would be in safe hands. The two agreed to talk when the meeting was over.

Overnight, something significant transpired: Bank of America concluded that without government assistance it couldn't buy Lehman under any circumstances. Ken Lewis, the deal maker who had helped his former boss Hugh McColl build the largest bank in the country, largely through acquisitions, knew, of course, that the Street had been working on a plan to wall off Lehman's bad assets, but he was suspicious of just how clean a bank he would be buying when all was said and done. The reports he was hearing from his bankers were that Lehman's books were a mess. More than that, though, Lewis was suspicious of Lehman, whether there was enough of a "good bank" buried within all the toxic assets to make such a deal worth it.

It was the first indication that the house that Fuld built wasn't so much

a diversified investment bank with bankers, brokers, and money management services (as most investors had been led to believe until recently) but
rather a business that made money primarily through risk, particularly in
the mortgage area. Furthermore, based on what Lewis was hearing from
Herlihy, a bigger prize was looming in the form of Merrill Lynch.

The Saturday-morning meeting picked up with Paulson and Geithner
urging once again for a private-sector solution to Lehman's troubles.
The CEOs broke up into two groups, one to discuss a potential
Lehman bankruptcy, which now seemed very real as news filtered through
the room that Bank of America didn't like what it was seeing, and another
to discuss the private-sector bailout that would cost each firm billions of
dollars to buy Lehman's toxic assets.

The more they negotiated, the more the entire notion of buying Lehman's bad assets seemed difficult, if not impossible, to pull off. Each firm
would have to put up billions of dollars to buy the assets at a time when
money was tight. For the next twelve hours the CEOs of the nation's largest
banks and brokerage houses would be weighing the following: the seemingly absurd notion of bailing out a competitor and giving Bank of America
(which had yet to formally announce it was out of the bidding for Lehman)
or Barclay's a Lehman free of its bad assets versus the prospect of systemic
risk destroying the capitalist system.

During the negotiations, the group heard officially from Paulson that
though Barclays was still interested in Lehman, Bank of America wasn't.
Without government help it wanted nothing to do with Lehman. Its bankers had already flown back to Charlotte.

Around that time people in the meeting saw Paulson and Geithner pull
Thain aside and walk into an adjoining room. John Mack didn't know exactly what was said to Thain, but he had a good idea: Paulson and Geithner
were telling him to prepare for a new, more hostile world come Monday
morning, with Lehman filing for Chapter 11. That would leave Merrill
Lynch the firm most vulnerable to the raging market forces that had pulled
lines of credit from Lehman to the point that it was illiquid. And he needed
to do something strategic, such as selling Merrill.

Mack had what he believed was the best strategy. In his mind, a combination of Merrill Lynch with his Morgan Stanley would be unbeatable in
the new Wall Street business model, which would rely less on risk taking

and more on advisory services, such as brokers selling stocks to small investors. Taken together, the two firms would have more than twenty thousand financial advisers. The cultural barriers would be minimized by the fact that Merrill and Morgan were both Wall Street firms; banks and brokerages never seem to get along. Morgan's brokerage chief, James Gorman, had been Merrill's brokerage chief just a few years earlier

When Thain returned from the conference, Mack approached him and said cryptically, "Maybe we should talk." Thain replied, "That makes sense." They set a time of 8 p.m. at the Upper East Side apartment of Walid Chammah, Morgan's head of investment banking.

What John Mack didn't know as the negotiations surrounding Lehman continued was that just about the same time he and Thain had agreed to meet, the Merrill CEO had also made up his mind to speak to Ken Lewis of Bank of America.

Thain walked outside the Fed building, dialed Lewis's home number in Charlotte, and matter-of-factly suggested that the two firms should do something called a "strategic arrangement." He didn't say exactly what he meant, but Lewis knew it could mean only one of two things: a partial sale of an equity stake in Merrill or a complete takeover.

Lewis wasn't interested in a stake; he wanted the whole thing. He saw Merrill as the crown jewel of Wall Street, the firm that would take Bank of America to new heights. Earlier in the year Bank of America had purchased Countrywide, which, whenever the mortgage business returned, would make the bank the biggest mortgage lender in the country—and the world. It already had thousands of branch offices and billions of dollars in deposits, and now in Merrill it would be getting fifteen thousand brokers, an international banking division, and a wonderful brand. Lewis would be building an empire like no other.

Thain, meanwhile, wanted his own empire and had to be convinced to sell even a portion of the firm, much less the entire franchise.

But those details would be left for later. They agreed to meet in New York in the afternoon; Bank of America owned a corporate apartment in the Time Warner Center. It was a place where the two could meet in private.

With the meeting set, Lewis contacted his deal team; they had just landed at the Charlotte airport in their corporate jet. He ordered them to turn around and get to work on buying Merrill Lynch. He would join them later in the afternoon.

Much has been written about Ken Lewis's lust to buy Merrill Lynch. As the two men sat down to hash out the details of their "strategic arrangement," Lewis didn't seem very lustful, just determined to strike a deal and do it fast. He made it clear that he wasn't looking to make some sort of puny strategic investment in Merrill, he wanted to buy the entire firm. Thain said he didn't want to sell Merrill Lynch, but he was open to Lewis's purchasing a minority stake in the company, where the big bank would own around 9 percent of the firm and Thain would remain as CEO. Lewis repeated that he wanted the whole thing.

It's here that Thain's intentions get murky; was he playing hard to get or simply doing everything he could to hold on to the firm's independence and his job as CEO? Thain isn't the easiest guy to read, after all, but given his disposition late Friday, given everything he had said during the past six months, people who worked with him believe he had yet to grasp Merrill's vulnerability. At the end of the conversation, both men agreed to keep talking.

It was at that point that formal merger discussions began, with or without a commitment from Thain, who would soon have several potential deals going at the same time.

What Greg Fleming didn't know, as he set up a war room in the offices of Wachtell, Lipton in Midtown to work on a Merrill-BofA combination, was that Thain was now weighing not just a deal with Morgan Stanley but, amazingly, a possible deal with his old firm Goldman Sachs.

Peter Kraus, the former Goldman executive who was now head of strategy for Merrill, had come to Thain with a proposal much like what he was looking for from Bank of America: Goldman would make a 9.9 percent investment in Merrill, and Merrill could keep its independence, having raised capital and teamed up with the best firm in the world to soothe the market's fears about its finances. Thain was intrigued by the Goldman arrangement because the possibilities for the future would be enormous, although Goldman's business model was under attack as well, from the market and soon from lawmakers, who would likely demand that the firm cut back its risk profile.

Though it was just a minority stake, eventually a full Goldman-Merrill combination would be likely, and it would be an unbeatable combination, with the best firm in the business of servicing small investors teaming up with the best in the business at handling large ones and corporations.

In the end, Thain might be able to keep his title and have a shot at what he had yearned for his entire career: running Goldman Sachs.

Despite the massive obstacles of creating a "bad bank" for Lehman, by the end of the day Saturday, a draft of a bailout plan had been crafted. For the past twelve hours, Bart McDade and a host of Lehman executives, including the commercial real estate head, Mark Walsh, had been bombarded with questions about what would be going into the bad bank—billions of dollars in bad assets. It was a humbling experience for the Lehman executives; these were, after all, people they competed against, and now they were begging them for money.

"They'd better beg" was the general consensus coming from Wall Street. Lehman's books were a mess was the report that the CEOs of several firms received as their aides crunched the numbers, particularly on the firm's commercial real estate holdings, which had been one of the main drivers of Lehman's growth in recent years but now seemed the most impaired of all its assets. Moreover, a consensus had developed that Lehman wasn't just holding toxic debt but had also inaccurately marked the value of its holdings by a significant amount. Lehman, of course, denied the charge, but the feeling among many of the firms involved in the conversation was that Lehman was attempting to make a really awful situation look only half as bad.

Still, by the end of the day Saturday, it appeared that, at least in principle, the firms had developed the outlines of a plan to buy Lehman's toxic assets, if only out of self-interest; they were, after all, holding the same crap on their books. Back at Lehman headquarters, where Fuld spent the entire day while McDade handled the discussions at the Fed, executives were euphoric.

For all the worries about handing Barclays a "clean" bank and Lehman's possible mismarking of its inventory of toxic bonds, the big firms were still worried about the unknown—the systemic risk that Paulson had warned about on Friday night when he had said they would all be "exposed" to trades and risk if Lehman went belly-up.

Meanwhile, Fleming had created not so much a war room but a war floor at Wachtell; he now had twenty of Merrill's best bankers and analysts with him as they began crunching the numbers on a complete purchase of Merrill Lynch by Bank of America. The option that

Thain seemed to favor, BofA buying only a minority stake in Merrill, barely came up. The group was focused on a full-scale sale of Merrill to the bank. One of the big questions was price; he knew not to push for too much, but he wasn't about to sell for a song either. Merrill's shares had closed around $17 on Friday, but a year ago they had been trading for as much as $90. He knew he couldn't even split the difference. The company would sell for closer to $15 than $90. The question was just how close.

There was another question: would Thain ultimately go for the deal Lewis wanted? Being bought by Bank of America would mean one thing for John Thain: he would be out of a job. He could stay through the transition, but eventually he would be gone. It's just the way things worked, whether on Wall Street or in Charlotte, North Carolina: to the victors go the spoils. But at least he would be able to take credit for having saved Merrill Lynch.

Larry Fink had been planning a trip to Asia for months, and as luck would have it, he was supposed to board his private jet Saturday afternoon and be in Singapore on Sunday morning, the same weekend Wall Street was "bailing out" one of its own. Fink wasn't at the meetings, but he should have been, given his knowledge of the market at the heart of the financial crisis.

In weighing whether he should leave for Asia, Fink thought like an investment banker and tried to read his good friend Greg Fleming. Given Fleming's expertise in financial institutions, he was naturally plugged into what was going on at the Fed with Lehman. Fink didn't know just how plugged in Fleming was to the other big deal now on the table, and Fleming wasn't about to tell him, even though either deal would have a profound impact on Black-Rock; although during one of their many conversations, in a moment of weakness he told Fink he might want to stick around for the weekend.

"For what?" Fink shot back. "I have to think logically about this; if they don't bail out Lehman, it's Armageddon. If they do bail them out, BofA or Barclays will buy them. So I'm going." He knew that Fleming might be playing him, but all day long, Fleming had told Fink he should stick around until the Lehman deal was completed because, this being Wall Street, there was always a chance something could go wrong. Later in the day, Fleming seemed more upbeat. Fink concluded that his good friend would have to have been spinning him for anything to be happening that weekend other than Lehman being saved.

Fink's close friend John Mack was in the middle of the madness as well. He had scheduled a secret meeting with John Thain for Saturday night and didn't alert Fink to the plans for the same reason Fleming didn't: he was afraid Fink would blab it to the world. Ironically, Fink himself had an apartment in the same building as the meeting. The orders were strict: tell Fink nothing; he leaks too much to the press, and the last thing they needed was to have to scramble to find a buyer in the morning's newspapers.

Fink called Fleming one last time before he boarded his plane on Saturday night. Fleming said he was at Merrill headquarters, holding down the fort and awaiting word on Lehman as he took a break from his deliberations with BofA at Wachtell. But he was really working on the deal to sell Merrill to Bank of America, while Thain was preparing to meet with Mack on a possible Morgan Stanley deal.

Fink asked him one more time if he should board the plane. For the first time all day, Fleming told him, "It's your decision to go"; in other words, everything was fine, Lehman would be saved, and Merrill would remain independent and live to fight another day. BlackRock would remain status quo. Fink was now off on a twelve-hour flight to Singapore without contact with Wall Street, while the real action was taking place just feet from his Manhattan apartment.

Mack was so paranoid about Fink changing his mind and staying in the city that he made sure Fink had left town before finalizing the meeting with Thain and his team from Merrill. At the meeting, Thain did a 180. Now he wasn't talking about a minority stake; he wanted a complete merger and said the deal needed to be done immediately.

Like Mack, Thain saw how powerful the combination of Morgan and Merrill would be. The combined firm would have more brokers than any other, two respected investment banks merged into one cohesive unit, and strong managements. Another reason Thain had flipped may have been the succession. The Street had been buzzing that Mack, at sixty-four, would soon be retiring, and that would open the top job to Thain.

"I want to at least announce a deal Sunday night," Mack recalls Thain saying.

Mack was blown away. "John, we haven't even done due diligence," he said. In all Mack's years of dealmaking, he had never heard of a merger announced after such a brief meeting. But Thain was insistent, and now Mack thought he knew why: Lehman was finished, and Merrill was next, and what better way to get the best price possible than to have multiple bidders. Mack believed Morgan wasn't in the danger zone just yet, so after about an hour and a half, the discussions broke off without even an agreement to talk again in the morning.

There's an old saying in U.S. investment banking circles about the Brits: they talk a good game when it comes to deal making, but in the end they always get cold feet. That's exactly what the Fed and the Treasury discovered early Sunday morning—that even with Wall Street nearing an agreement to buy Lehman's bad bank, there was no buyer for the good bank. Barclays informed regulators that it wasn't interested in buying Lehman.

For Paulson, Geithner, and Bernanke and their team it was like a punch in the gut.

The Financial Services Authority (FSA) the lead securities regulator in Britain, was now blocking the deal. It worried whether Barclays itself was strong enough to absorb Lehman, even without its commercial real estate assets, and, more than that, it wanted nothing to do with the U.S. financial system, its leverage, its excesses, and its business model.

As a senior official in the FSA told Paulson, "We don't want to import your cancer."

Lehman, barring a miracle, was finished. Paulson and other officials at the Fed such as Kevin Warsh had been investment bankers; now they needed to complete the most important deal of their career in just a few short hours Sunday because the Asian markets would open later that evening and the systemic risk they feared would begin.

Now everything really was on the table, including a bailout of Lehman by the federal government—if only they could find a buyer for the good assets. Without that the federal government would have been the sole owner of an investment bank, and no one wanted that.

"Trust me, we were calling everyone," said one senior government official directly involved in the calling. "It was like dialing for dollars." The problem was that there were no bidders for the simple reason that Lehman had nothing to offer. The past weekend every firm on the Street had gotten a good look at Lehman's books, everything from its holdings of toxic debt to its franchises, such as its investment bank and asset management division.

What they had discovered was that Lehman wasn't really much of a firm. It was a shop built on a bubble in real estate, much of it the subprime residential kind but also the increasingly toxic commercial variety. And unlike Bear, which had a willing buyer in JPMorgan, no one wanted to touch it. Fuld desperately hit the telephones as well. He called Ken Lewis and couldn't get a return call, even after imploring Lewis's wife to give him the message. She said she would, but Lewis never got back to him. But John Mack would. Over the tempestuous past six months, John Mack had been one of Fuld's closest allies on the Street; in April, when rumors had begun about Lehman's liquidity, he had called Fuld, alerted him to the speculation, and said Morgan Stanley would be there if he needed it. Mack had been there throughout the summer, even shooting down rumors that Lehman might sell itself at a price below where its stock was trading. He was there now, but not the way Fuld needed.

"Is there anything we can do?" Fuld asked. Mack was exhausted and beaten from preparing for a turbulent Monday, so he didn't have much time to chitchat. He simply said, "Sorry, Dick, I don't think so."

Inside Lehman the reality set in: the firm was finished. It would have to file for bankruptcy.

Fleming heard the news about Lehman and thought, "Thank God we have Bank of America." That's when Peter Kraus called, in the middle of working the possible purchase of a 9 percent stake in Merrill by Goldman. Both men were physically exhausted, though Fleming was riding on adrenaline and caffeine from multiple cups of coffee and the monster deal with BofA that seemed at hand. It wasn't $90, but it wasn't close to zero, which is what Lehman was likely to get in the coming days as creditors and others picked apart the carcass of a once-great franchise.

Fleming had never warmed to Kraus, and the feeling was mutual. Both had been bankers who specialized in financial services deals, and they had competed for years. Now they were competing inside Merrill. Kraus's deal would mean that Merrill remained independent; Thain would get to keep his job; the brand would survive. But Fleming believed it was just a temporary fix. Was a 9 percent equity stake enough to satisfy creditors in this market? Was a line of credit from Goldman, with a business model that was coming under pressure like everyone else's, really enough to assuage investors?

Kraus didn't answer those questions; instead he told Fleming that he needed some of his due diligence staff to pursue the deal with Goldman because "we need multiple options."

Fleming, not known for wild outbursts, exploded. "The BofA deal *will* happen!" he shouted. "We're moving forward, and I am not sending people!"

Kraus hung up. Next to call Fleming was Thain.

Thain had met with Lewis earlier in the morning, when he had made the case that Merrill was worth far more than its Friday closing price of $17 a share. Paulson had been pressuring him to announce a deal before the markets opened on Monday, preferably an outright sale of the company. Still, it was clear from his conversation now with Fleming that he wasn't sold on the idea of being bought by the Charlotte-based bank. He began by asking Fleming about the Bank of America negotiations and if he could send some of his people to Merrill's headquarters so they could crunch numbers on the Goldman investment. Fleming went through his concerns about a Goldman deal. He brought up the cultural challenges: how would rank-and-file Merrill executives feel about working for a hated competitor? He questioned whether the money Goldman was prepared to hand over was enough to solve Merrill's real problem, the perception that it was under-capitalized and that it needed to merge with a big bank. He didn't mention the other problem Merrill had, credibility. Like Dick Fuld, Thain had been proved so wrong time and again about the firm's finances that Fleming believed many investors would look favorably at a change in leadership.

Then he spoke about the price: Merrill's shares had ended Friday at $17 and were heading lower. He was eyeing a deal at between $20 and $30 a share.

Moreover, he said, "a deal with Bank of America is in hand. We have got to do it!"

Thain simply answered, "Yes" and hung up.

With Thain's blessing in hand, Fleming's main priority was price and Lewis's was due diligence. Price was pretty easy. Merrill closed on Friday at $17 and wanted a premium, which Lewis agreed to, at $29 a share. Due diligence was a little more dicey. Lewis had hired Christopher Flowers, a former Goldman investment banker who now ran his own private equity firm, to scour Merrill's books.

Deals as complicated as the Merrill-BofA merger take weeks to hash out, involving numerous meetings, presentations, and legal reviews. But this one was nearing conclusion in just thirty-six hours. Still, the due diligence review found no red flags, people at Bank of America claim. Merrill, after all, had unloaded much of its CDO portfolio, and though it had trades and positions in various markets, the stress tests performed by Flowers suggested that the losses would be minimal.

More worrying for Lewis was a provision in the deal terms that would allow Merrill to pay as much as $4.5 billion in bonuses to its employees in 2008. For Lewis it was a shocker. First, Merrill, by all accounts, had had a lousy year, so he asked if they really needed that much money. Lewis, meanwhile, was notoriously tightfisted when it came to compensation, particularly compared to Wall Street standards. He ordered his team to get the provision removed, but Thain refused to budge.

In Thain's mind, he and his team had done great things over the past year, the crowning achievement being this deal to save Merrill. In other words, they *deserved* to get paid. After Lewis was told the bonus issue would probably be a deal killer, he relented, but the acrimony over the issue was so intense that Lewis and Thain were now barely on speaking terms even as they made preparations to announce the deal of the year.

T hose assholes! You got to be fucking kidding!" Larry Fink screamed as he heard the news that not only had Lehman Brothers crumbled, but now, thanks to his good friend, Merrill Lynch would be owned by Bank of America, and that would mean BlackRock would be partly owned by BofA as well. Fink had just landed in Singapore. A few hours earlier, Merrill's board had approved a $29-per-share deal, and Fleming and his team were celebrating.

Not Fink. He had been without e-mail for the flight. But when he landed Sunday night, he received a full report: Lehman was toast and preparing to file for bankruptcy liquidation imminently; Bank of America's board and Merrill's board were set to meet to finalize a $29-per-share deal for the bank to buy Merrill; and the Street was preparing for one hell of a Monday opening.

Fink was furious: at his staff for allowing him to fly off to Asia in the middle of a financial hailstorm; at himself for having thought that the government would ultimately do anything to prevent Lehman from going bankrupt; and at Fleming for not having told him what was going on. Fink began dialing frantically to reach Fleming. When Fleming finally picked up his dear friend's call, he wished he hadn't.

"You fucking asshole, you lied to me!"

"I didn't, Larry," Fleming replied. "My hands were tied."

"Greg, do you know what this means for my firm? We're being bought by a bank. I sold my company to Merrill Lynch, not BofA."

Fink had a point. It had been bad enough when O'Neal had been booted and Fink had had to work for John Thain, someone he despised. Now he had to work for bankers from North Carolina he didn't know and might hate even worse.

"I am so fucking mad!" Fink snapped.

"But this was life and death, Larry," Fleming countered. The firm's future was hanging in the balance, sixty thousand employees, and he couldn't risk a leak to the press that might have killed everything. "I'm sorry, Larry, I just couldn't do it." With that it was Fink's turn to hang up on Fleming.

Back at Lehman, Fuld hadn't left the office as he prepared for the worst possible outcome: a bankruptcy filing. Until the very end both Fuld and senior Lehman executives working on the "bad bank" deal thought that if all else failed the government would come to their rescue, but as they were still without a buyer, both the Treasury and the Fed refused. Much would be made about the government's decision to let Lehman die, as the systemic risk that policy makers always worry about in the abstract began to percolate for real in after-hours trading in overseas markets from nearly the moment the announcement was made.

Lehman's bankruptcy would haunt policy makers as it became clear that the downside of fighting the creation of moral hazard is that in so doing they could add fuel to a financial panic. Not even Lehman short seller David Einhorn believed the government would allow the firm to die and risk the systemic implosion that many had feared and now was spreading. This fear, of course, later caused officials at the Treasury and the Fed to take a firm stand and adopt as a matter of policy the position that "systemically important" institutions would be saved.

But it was too late. Senior Fed officials say Bernanke was convinced now more than ever that despite all the political pressure Paulson had faced, he had dropped the ball to a degree in not seeking even wider bailout powers as he saw the market's immediate reaction to the news of the filing. Paulson, on the other hand, still doubted that he could ever have gotten enough votes in Congress to bail out a firm that clearly didn't deserve a bailout, even if he had pressed the case that by saving Lehman he might be saving the financial system as well.

Bernanke declined to be interviewed for this book, but talks with his aides and his public comments underscore his growing frustration with the ad hoc approach to the crisis he and Paulson used with less and less success as the year went on. According to a lengthy profile of Bernanke by *New Yorker* writer John Cassidy, a clearly frustrated Fed chairman complained that "There was no mechanism, there was no option, there was no set of rules, there was no funding to allow us to address that situation. The Federal Reserve's ability to lend, which was used in the Bear Stearns case, for example, requires that adequate collateral be posted. . . . In this case, that was impossible—there simply wasn't enough collateral to support the lending."

What Bernanke was saying was that without a buyer, the Fed's hands were tied—it could do nothing because its mandate was limited. Lehman's collateral was so distressed, so much more so than Bear's, that the Fed would not have been able to use its balance sheet to prop it up.

It's debatable whether the Fed's balance sheet is as limited as Bernanke says. Bernanke, to some extent, wanted it both ways; on the one hand, his supporters at the Fed tell me he was worried about systemic risk and inclined, if possible, to bail out Lehman if there had been a mechanism to do so; on the other hand, they concede that the Fed chairman was worried about the public perception of bailing out greedy Wall Street bankers and the creation of moral hazard, particularly when it came to Lehman, which would have been a case study in moral hazard if the government gave the firm even a dime.

But one thing everyone agreed on, from Paulson to Bernanke to Geithner and most of the people who worked for them: no one had predicted the size and magnitude of the crash that eventually came. They believed that for better or worse, the ad hoc method of dealing with the broad crisis, one individual firm at a time, had been effective in keeping the panic from spreading globally and that the repercussions of the Lehman collapse were survivable. And they were getting positive feedback from the Street: Monday would be difficult, but all the major firms had a strong grasp on their trading positions with Lehman and believed they would survive whatever "systemic risk" storm was about to come their way.

Late Sunday afternoon and early that evening, discussions about "systemic risk" and "moral hazard" were the last things on the minds of Lehman employees, who could be seen walking into and out of the firm's Forty-ninth Street headquarters with boxes as they cleaned out their offices. The implications of a bankruptcy filing were startling to everyone involved. The firm's equity would be wiped out, and for Lehman employees that would be catastrophic. Bear's $10-per-share forced sale had been tough on the top executives, who were paid much of their salary and bonus in stock and forced to hold it. But Lehman had pushed the ownership culture even further; even secretaries were urged to buy Lehman shares. Now they were holding worthless paper.

Fuld would lose $500 million, though he had cashed out stock recently, so he had at least $137 million on hand before taxes. Other senior executives would be even harder hit, highest on the list being Joe Gregory. He would be forced to sell his helicopter and his home on Long Island and scale back his lifestyle in ways he probably had never thought of when Lehman was riding high.

But even more than money, Fuld and his inner circle had lost their reputations. They felt betrayed by the government as news spread that the Fed would begin to expand the type of securities it accepted for overnight loans, but only *after* Lehman was finished. Aside from the government bureaucrats who had helped doom Lehman, inside the firm the most hated man was Dick Fuld. McDade's supporters quickly tried to distance their man from Fuld, as after the bankruptcy Barclays came back with a bid for nearly everything except the bad assets, at a price of around $1 a share. Not even Jamie Dimon could top that one.

Now the markets were bracing for a great unwinding: Lehman's $500 billion balance sheet would need to be dismantled. Hedge funds that had money in Lehman's prime brokerage department would have their cash tied up in bankruptcy court for who knew how long, particularly in the United Kingdom, where local law made retrieving in a timely manner difficult. Regulators took some comfort from what they were hearing from the Wall Street firms about counterparty risk with Lehman, but officials at the Fed and the Treasury had given little consideration to the fact that billions of dollars in hedge fund money would be tied up indefinitely because of differences in U.S. and U.K. bankruptcy law.

That undoubtedly forced the mass selling of stocks and any other liquid assets on Monday when the markets opened and Lehman began the process of unwinding its trades, credit default swaps, everything, and with that the credit crisis moved into its most dangerous stage yet as a mass selling of assets—stocks, bonds, everything that could be unloaded—spread around the globe. The shares of the big banks took the initial biggest hits, primarily because investors and banks with money, including JPMorgan, stopped lending it out to other banks like Morgan and Goldman to finance their trades. Merrill had found safety in Bank of America, but without access to credit, Goldman and Morgan would certainly follow Lehman into insolvency, and the troubled Citigroup, barring a massive bailout from the government, would be next.

The financial system was in uncharted territory, something not seen in the lifetimes of the men who ran the world's big banks. No longer were policy makers worried about the economy falling into a recession—that was a given. Their bigger worry was preventing a full-scale collapse of the financial system, then the global economy, resulting in a depression greater than anything ever imagined as banks conserve whatever precious capital they have and no one lends, no one trades, and asset values fall to zero or pretty close to it.

The implications of the looming disaster spread from the markets to the presidential campaign not long after Lehman's demise, as the election that had been a near dead heat before the worst part of the crisis now began to turn suddenly and swiftly in favor of the Democrats when, after Lehman's meltdown, Republican presidential candidate John McCain announced he was going to suspend his campaign, even postpone a long-awaited debate, while Barack Obama calmly and coolly stated that he was prepared to do whatever it took to restore the country to financial health.

Obama's cool demeanor and the boost in the polls he received for it said something about the American psyche at the time of the crash, namely, that the country was looking for a leader, someone it could trust to act in a responsible manner to reverse all the irresponsibility that had led the financial system and now the entire economy to its current state.

Not one part of American society had lost more credibility than Wall Street. For the past twenty-five years it had constantly defended and expanded upon its risk taking despite obvious signs that it would lead to the disaster it was currently dealing with. Allegedly responsible leaders of the top firms had hyped and misled to the point that now no one believed the remaining CEOs' assurances that their businesses were fine or that their financial statements suggested the same.

It didn't matter that Wall Street, at least on paper, was still making money (that would change); on Tuesday, Morgan Stanley trumpeted better-than-expected earnings, and even so its shares declined a whopping 15 percent, as did shares of Goldman Sachs, which also said it was still operating in the black. But in the coming days, Morgan Stanley began to follow Lehman in an apparent death spiral, and Goldman wasn't far behind.

In a desperation move, Morgan Stanley CEO John Mack pulled a page from Dick Fuld's playbook and blamed short sellers for spreading unfounded rumors about the firm. But no one cared; shares of Morgan continued to decline because not even one of the real straight shooters of Wall Street could now be believed and trusted.

Without trust, without lenders and investors believing that they could lend a firm a dollar and get that dollar back in a few days, Wall Street, for all intents and purposes, was finished in its current form.

It's highly unlikely that the remaining heads of the big Wall Street firms knew or even contemplated such a thing a few minutes after midnight early Sunday morning, as Lehman officially filed for bankruptcy and began the process of liquidating the company.

But within a few hours, the Asian markets told the whole story: they were falling and fast, signaling that Monday and the week ahead would be every bit as much of a disaster as some had feared but regulators had failed to do enough to prevent.

In the days and weeks that followed Lehman's demise, friends of Dick Fuld weren't worried just about the safety of the system but about his safety as well, as employees angrily cleared out of the firm. There was a report, since denied, that one employee punched Fuld in the eye as he ran on a treadmill at the company gym. What gave the story legs was the undeniable truth that the rank-and-file Lehman employee held him largely responsible for the company's demise. Regulators were circling as well; the SEC was joined by the Justice Department, which opened a criminal probe into the company's explanations about its financial condition during the course of the year. Fuld himself would then be grilled by a congressional panel about Lehman's fall. Fuld told the House Oversight and Government Reform Committee, one of the many bodies looking into the causes of the financial crisis, that Lehman's implosion was "a pain that will stay with me for the rest of my life."

Committee members were hardly sympathetic. "If you haven't discovered your role today, you're the villain so you have to act like a villain," said Representative John Mica, a Republican from Florida.

Amid all this, people still close to Fuld say their old friend had more than lost his old swagger; he looked tired and despondent, his beloved Lehman was toast and his reputation in tatters. Just after Lehman's bankruptcy became official, he would tell his employees in a note, "I know that this has been very painful on all of you, both personally and financially. . . . For this, I feel horrible."

Dunlavy wasn't feeling so good either. He was watching events unfold on CNBC. He hadn't worked on Wall Street for ten years, but the demise of Lehman and the infection it spread in the stock market further decimated his retirement savings, some of it still in Citigroup stock. He was looking to publish a fictionalized account of his life, including his career at Salomon Brothers, which would in some way provide an insight into the culture of risk of which he had been a reluctant participant.

What struck Dunlavy was how much Wall Street had changed since the days he had been at Salomon Brothers—players such as Lew Ranieri, Mike Mortara, and John Meriwether were gone from the scene. The biggest difference was the compensation; he couldn't believe it when he read in the newspaper that little Tommy Maheras had made $30 million in 2006. But he could believe—in fact, he could have predicted—the practices that had earned Maheras $30 million in one year would also be the practices, of course with the approval of his supervisors, that would lead to Citigroup losing more than $30 billion.

Dunlavy had recently heard from his sources at the firm that Maheras was now being vilified by many of the rank and file for Citigroup's continued demise; the big bank had been off the front pages recently, with the news about Lehman and Merrill taking its place. But risk taking was still very much eating away at the firm—it hadn't turned a profit in a year, and the losses were expected to continue. Though its stock traded up, albeit briefly, during the week of the Lehman bankruptcy on the notion that, as a bank with trillions of dollars in deposits, Citigroup was better positioned to withstand what had happened to its Wall Street counterparts, reality soon set in and its stock went into free fall as well.

Dunlavy wasn't a big fan of Maheras, but he didn't think he was the guy responsible—no single person was responsible, and that's what made him a little queasy watching the coverage that night. It hadn't had to happen this way, but the business "had changed and they all got greedy," beginning in the early 1980s and building up ever since. The business had rewarded excessive risk taking with higher and higher bonuses and penalized those who came to Wall Street to make a good living and simply work with clients, rather than betting shareholders' money on increasingly esoteric bonds whose value had no basis in reality.

"They blew themselves up," he said. "They destroyed themselves and everyone else with them."

21. BAILOUT NATION

On Monday September 15, 2008, the Dow dropped 500 points, the largest drop since the first day of trading following the 9/11 terrorist attacks. The credit markets were even harder hit, if possible, as investors sold every bond they could and snapped up Treasuries for protection, as the yield on the safest short-term government debt, the 3-month Treasury bill, fell to nearly zero. Bonds are essentially loans; the disappearance of buyers for anything but Treasury debt meant the lending markets had closed down. Home owners, businessmen, even highly rated companies couldn't borrow money or could borrow only under onerous terms.

Meredith Whitney remembers the start of the week from hell vividly. Most of her stocks were getting hammered; she had deftly predicted as much with her now-famous warnings about the banks' holdings of toxic assets and how they would eat into profits. But she had never thought it would come to this—a true market panic that threatened the stability of the world economy. Just coming to work made her feel ill, nauseous was the way she remembered feeling, and as she received calls from friends and colleagues about the tumult, she made sure she had a wastepaper basket by her desk, just in case she really did get sick.

With the markets teetering, initially John Thain came out like a hero, selling Merrill for an initial price of $29 a share, although during the weeks and months before the deal closed, Merrill's share price would fall farther, thus threatening the deal (more on that later). But not as low as Lehman's, which was now in bankruptcy court while the vultures picked over the firm's carcass. One of those vultures was Barclays; apparently the FSA had no problem with Barclay buying Lehman's investment bank that Monday for about $1 a share, a total cost of $1.75 billion and free of anything that even looked like the financial cancer known as the mortgage-backed security.

Morgan Stanley, it seemed, was heading for $1 a share as well. On Tuesday, it fell 10 percent even after reporting better-than-expected earnings for the third quarter, then fell another 15 percent on Wednesday. John Mack was now in the hot seat, wondering if he should have made a deal with Merrill Lynch. Morgan didn't have the CDO exposure of other firms, something the firm and Mack explained in every conversation with investors and creditors, but it didn't matter. Even Mack's friend and former boss Art Samberg, the CEO of the hedge fund Pequot Capital, had pulled his firm's money from Morgan's prime brokerage accounts, telling Mack that he "had to protect his investors." He wasn't alone; the cost of insuring against Morgan's defaulting on its bonds, as measured by the price of its credit default swaps, soared. It was hard for many people inside Morgan to believe, but the market was betting that Morgan was heading for Lehmanland.

Goldman Sachs was feeling the heat as well, although not yet to the same degree as Morgan. Investors began to unload its shares, and the insurance on its debt, the now-omnipresent CDSs, soared as investors snapped up the credit protection in preparation for Goldman following the rest of the Street into the abyss.

Goldman went on the offensive—an unusual thing for the normally secretive firm—boasting about the strength of its balance sheet and its hedges. But Goldman couldn't spin itself out of the following: Merrill Lynch had sold Goldman $13 billion of CDOs that had been hedged with credit default swap insurance from AIG, which by Tuesday morning was nearing insolvency with the rest of them. The consensus among analysts and traders: if AIG fell, Goldman wouldn't be far behind.

As an insurer, AIG didn't come under the purview of the Fed, and in the view of some Fed officials, the company was "not regulated by anyone of credibility," a thinly veiled shot at the state insurance commissioners who were supposed to monitor insurance companies.

But AIG was "systemically important" to preventing the financial system from further collapse, as both Paulson and Bernanke were quickly discovering. Despite Eric Dinallo's assurances about the strength of AIG's other businesses, the size and scope of the problems in its financial products division, stemming mainly from its sale of credit default swaps, were overwhelming. The Fed had heard of troubles coming out of certain affiliates at the insurer, but it didn't know just how bad things had become until a break in a Fed meeting on Monday when Ben Bernanke and Fed governor Kevin Warsh took a call from Geithner and Paulson, who had dire news about AIG's deteriorating financial status, including its massive losses and the money it was shelling out to cover the insurance it provided on bonds. "This thing is cooked," Geithner said. "We need to throw money at it."

If the market had traded off 500 points on Lehman's collapse, the systemic risk unleashed by an AIG liquidation would be far worse. AIG was a massive enterprise with offices around the globe. It sold life insurance to average Americans, but, more than that, it had been the glue that held the financial system together. AIG insured the bonds of some of the biggest financial institutions, as well as the CDOs held by those entities. The credit default swaps it insured were held by banks around the planet, not to mention one very important bank in the United States, Goldman Sachs.

With Merrill safely in the hands of BofA, Morgan Stanley was now targeted as the bank most in trouble, and creditors were beginning to pull their lines from the firm, but if AIG went under, Goldman wouldn't be far behind, nor would nearly every other bank in the world. Goldman had been telling reporters that its exposure to AIG was limited and hedged. But the market didn't believe it, which was one major reason why, just after Lehman filed for bankruptcy and all eyes were on AIG, Goldman's shares lost a third of their value.

AIG was now threatening to make Lehman's demise look like a cake-walk. Systemic risk was nearing a whole new level. Goldman's balance sheet was larger and far more convoluted than Lehman's, making its possible unwinding scarier than anything regulators had ever imagined, and those of the banks that held much of the debt covered by the default swaps dwarfed Goldman's. Geithner and Paulson were now discussing ways in which they could quickly access funds for an AIG bailout, including tapping the Treasury's Exchange Stabilization Fund (ESF), which was established in the 1930s as a pool of cash the Treasury could touch without having to get congressional approval. But Geithner said there wasn't enough money in that fund. The consequences of AIG's failure would be severe: if it failed, its exposure to other broker dealers could bring the Street to its knees. They had to do something, and quickly.

Bernanke, with help from Wall Street advisers who would benefit from the bailout (one of those advisers, Goldman concedes, was its CEO, Lloyd Blankfein), came to the conclusion that the Fed was now the only entity with a functioning balance sheet large enough to do the deed, so he dusted off an old provision of the Federal Reserve Act and gave AIG the $85 billion it needed to survive in exchange for a 79.9 percent stake in the insurer and onerous interest rates for the money.

The Fed wiped out AIG's shareholders—they would be getting nothing. It threw the proverbial book at the insurer in the hope of preventing a systemic meltdown that would have brought the Street to its knees. The meltdown abated for a time, but it didn't end.

Even with the AIG bailout on Wednesday, shares of Morgan Stanley and to some extent Goldman continued to slide, with Morgan losing prime brokerage clients, always the first step in eventual implosion. The market was obviously picking which of the two to kill first, and it was Morgan Stanley. Why Morgan? After its big loss in late 2007 and Goldman's savvy short sale of the subprime market, creditors, investors, and short sellers simply calculated that Morgan was in worse shape.

But the market was no longer focused just on the type of assets that Morgan held but on its business model: borrowing and leverage that were now deemed too risky to be exposed to. Morgan was on the brink, and the fear was palpable in the executive suite, except for the suite where most of the decisions were made. As the crisis continued to build, John Mack remained a pillar of strength and stability. "The shit was hitting the fan, and there was no cooler customer than Johnny Mack," Larry Fink recalls. Mack told his senior team to remain cool and composed. They were, after all, working at one of the world's premier firms.

But inside his head and in his gut, Mack was frenzied as Morgan's stock continued to tank. He went on the offensive, borrowing a page from Dick Fuld and blaming the short sellers. The short sellers were, of course, doing exactly what they had done all year, first to Bear, then to Lehman, and now to Morgan. They took their short positions, bought loads of credit default swaps, and then watched speculation swirl about how investors were betting that Morgan would implode.

"This is fucking market manipulation," Mack told officials at the SEC. Morgan Stanley and Goldman Sachs were bitter rivals, but now they had something in common. Mack called up Blankfein and tried to form an alliance to expose improper short selling of financial stocks. Blankfein agreed, but he wouldn't go public, unlike Mack, who issued a memo to employees saying Morgan's financial situation was solid but was being attacked by unscrupulous short sellers.

The move was applauded by some, mainly Mack's friends on the Street, but ridiculed elsewhere, particularly by some of Morgan's clients, the hedge funds.

Morgan's balance sheet wasn't as free of toxic debt as it was now claiming to the press. The firm (along with Goldman Sachs) would take a massive loss for the fourth quarter, as it too had to come clean and write down risky assets. It seemed that the short sellers weren't spreading unfounded rumors after all.

The famous short seller James Chanos, for one, removed his prime brokerage money from Morgan, as did other short sellers, as a form of protest against Mack's blanket condemnation, which only worsened Morgan's cash crunch. Mack was now doing everything he could to save Morgan as he sensed the end coming. He opened talks with the China Investment Corporation for a strategic investment and then with Wachovia, the big bank with almost as many deposits as depressed mortgage-backed debt on its books, but nothing would come together.

He called Joshua Bolten, the president's chief of staff and a former Goldman executive, and asked to speak to the president. Mack had been a big Bush supporter in 2004 and had been floated as a possible SEC chief by the president, but more recently, he had switched sides during the presidential campaign, first endorsing Hillary Clinton and later supporting whomever the Democrats were likely to elect.

Bolten said he couldn't let him; he advised Mack to speak to Paulson or Chris Cox, the SEC chairman. Mack had spoken to Paulson enough and was getting nowhere on the short seller issue, and he thought Cox was a lightweight who hadn't a clue about what was happening on Wall Street.

"Josh, seven years ago you asked me to head the SEC, and now I have to show you how to run that agency," Mack snapped before turning the conversation to something more important: Mack had had conversations with Chinese banks—which are essentially arms of the Chinese government—about a significant investment in Morgan Stanley, but there were rules that made it difficult for any foreign entity, much less the Communist Chinese government, to own or hold a significant stake in a U.S. company.

But these were times when old rules needed to be broken. "If you want

Morgan and Goldman to implode, then stop the Chinese from buying a stake," he said. Bolten was noncommittal, and the two agreed to stay in contact.

W hat would it feel like to be on the front lines of the war to end all wars, with superpowers pointing nuclear missiles at each other? That's how Ben Bernanke and his senior team felt on September 17, 2008, as the financial markets seethed with seemingly no end in sight.

A lot had happened since the Lehman bankruptcy just two days earlier; the stock market had fallen far lower than anyone had thought it would; the credit markets had seized up. There had been a near run on money market funds, with investors pulling their money out of what were believed to be ultrasafe investments after the Reserve Primary Fund "broke the buck," which meant its net asset value fell below $1 a share, something that had never before happened to any major money market fund.

AIG's bailout did not help the financial system to stabilize; in fact, it added to the destabilization. It began affecting parts of the market that had nothing to do with mortgage-backed securities, which was the very definition of systemic risk. It also raised issues for Morgan Stanley and Goldman Sachs, the remaining Wall Street securities firms, which started to feel the squeeze of hedge funds pulling their money from their prime brokerage accounts and creditors denying the firms overnight loans. If they went down, what would stop the panic from reaching the big banks like Citigroup, which, though struggling, had yet to reach the Lehman-Bear level?

The fear among regulators, particularly at the Fed, hit a new level. As bad as the markets were, stocks were still trading and credit cards were still being processed. The basic process of extending credit, though constrained, continued. But the shock of another failed bank could put the system into a total meltdown, where just about everything would stop.

It was still theory because it had never happened before, but theory seemed to be moving toward reality in mid-September 2008: a world in which no one would trade, there would be no credit, and the financial system would come to a halt. The values of homes, stocks, and everything else would then fall to zero.

Forget the Great Depression. This would be the Greatest Depression.

Then something else happened, of near-equal importance: the normally unflappable Federal Reserve Chairman Ben Bernanke lost his patience.

Bernanke was sitting in his office with Fed governor Kevin Warsh. They had just dialed Paulson, who was on the speakerphone. Bernanke was tired, concerned, frustrated, and ready to concede that his deliberate approach to the crisis was wrong. The panic he had studied as a economics student and taught as a Princeton professor, often in the abstract, was becoming all too real. The financial crisis was reaching undreamed-of levels, and nothing they had done so far had worked. For months he had asked Paulson, albeit gently, to go to Congress with something bigger and more pronounced. Bernanke, a student of financial panics, knew one thing: panics remain panics unless addressed quickly and decisively by policy. It was now time to act decisively.

During the call, the clearly exasperated Bernanke told Paulson it was essential that the Treasury get involved in stopping the crisis in a major way; in other words, it needed to throw money at the problem and do it fast. "You have to do something," he said. He argued that as a central bank, the Fed had reached the end of its authority—the limit of how much money it could spend. The solution mandated a major expenditure, one that only the Treasury secretary could pull off by lobbying Congress for an appropriation that would match the size of the problem.

"We can't be the only ones out there handing out money," Bernanke told Paulson, aware that the Fed had become the poster child for bailouts. "There's nothing more we can do. You have got to do something. We have to have Treasury more involved," the almost exasperated Bernanke said into the phone.

Paulson, known for his quick tongue, held it for the moment as he digested what the Fed chairman had said.

Friction between Bernanke and Paulson had been building for months. "Ben was convinced Treasury had to do more," said a senior Federal Reserve official. "Bernanke thought Paulson was trying to have it both ways." Paulson was weary of the political backlash of bailouts and left the heavy lifting to the Fed, which had opened the discount window to investment banks and had just bailed out AIG. Bernanke thought much of Paulson's hesitancy was political; he wanted to deflect blame for the creation of moral hazard after Bear Stearns and the bailouts of Fannie Mae and Freddie Mac.

But more than anything, Bernanke resented the fact that the Fed was the only agency taking the hits by bailing out the financial system, while his counterpart at Treasury stayed in the background. "It would be nice if we could get some help here," Bernanke had frequently told Paulson in the months leading up to the call.

Now the critical time had come, and Bernanke said Paulson would have to step forward. Bernanke was calling for an aggressive, multilateral plan to deal with the financial crisis that seemed immune to the measures that had been taken. In the past couple months, the topic of the Treasury's inaction had been discussed openly and frequently, and at times things had gotten testy, with Paulson arguing that it made little sense for him to go to Congress for a grand appropriation of bailout money when the votes weren't there.

"Hank would often interrupt, going on for hours explaining why politically such a plan wouldn't work," one senior Fed official involved in the meetings said.

But this time, after the crises of Lehman, Fannie and Freddie, AIG, and now the near demise of Morgan Stanley and the possible demise of Goldman Sachs, there was no argument. After twenty seconds into Bernanke's pitch, Paulson simply said, "Great idea."

Exactly how much time Paulson put into creating the grand bailout of what was now declared the worse financial crisis since the Great Depression isn't known. It would eventually be called the Toxic Asset Relief Program (TARP), but Paulson knew he had only a few hours to draft legislation before he went to Capitol Hill and began trying to persuade Congress to endorse the plan immediately.

Paulson initially envisioned and planned to sell TARP as a program along the lines of the old Resolution Trust Corporation: it would purge the toxic assets from the bank's balance sheets, buying the CDOs, CMOs, commercial real estate bonds, and debt, holding them until the market improved, and then selling them off.

A key point he made was that Resolution Trust, which had purchased bank loans in the late 1980s and early 1990s, was profitable. The simple fact was that what was toxic now because of a lack of confidence in the underlying collateral—mortgages—might not be toxic in the future. Even if 30 percent of all subprime mortgages were delinquent, that means 70 percent were still current, yet mortgage bonds were trading as if there were 100 percent delinquency.

Investment bankers say the notion of TARP had been bandied about at the Treasury and with Wall Street executives whom Paulson had involved in his various bailouts for some time. Many, like Larry Fink, didn't think it would work. As news of the plan began to leak on Thursday afternoon, Fink dismissed the notion to friends as a Hail Mary pass. "If this is Hank's plan to save Wall Street, so be it," he said during one conversation.

Fink underscored a fatal flaw in the proposal: if the Treasury was buying the toxic debt at market prices and Citigroup was selling it at market prices, Citi would have to mark down huge losses and either be bailed out or become insolvent. The Treasury could buy the bonds above the market price, but that would be a difficult sell to Congress because taxpayers would be financing Wall Street's aberrant risk taking. Fink didn't believe there was enough money in the Treasury to make the plan work.

But Paulson planned to announce the move as soon as possible, going to Congress Thursday afternoon. Fed officials were worried about Paulson's salesmanship. "Hank was not exactly smooth," one said, which, if you have ever heard Paulson speak publicly, you know is an understatement. Compounding the problem, Paulson was hated by the House Republican leadership, now in the minority but still a potent force that could block bailout legislation of this magnitude. They saw him as a lefty masquerading as a fiscal conservative and hated the fact that Bush, a free-market president, had essentially handed over the running of the economy to this man.

One of the great media fallacies is that Wall Street is a bastion of right-wing politics. In fact, over the course of 2007 and 2008, much of the Wall Street leadership, from Lloyd Blankfein to Jamie Dimon to Larry Fink and John Mack openly sided with the Democrats and were now supporting Obama for president.

Paulson, meanwhile, was a Republican from the party's old liberal wing, once known as Rockefeller Republicans after the former governor of New York State and vice president Nelson Rockefeller. But the party had changed much: it was now more southern, more free market, and decidedly less big government. Its agenda was pro-business, with the emphasis on small business.

In many ways the House Republicans viewed Wall Street as a facilitator of the needs of big government, as when it helped the former housing giants Fannie and Freddie; in essence Wall Street had helped Fannie and Freddie sell their debt and finance all the guarantees that made housing, in their view, yet another welfare program. Now they saw Wall Street as being rewarded for this help by benefiting from various government bailout arrangements when it lost money. More than that, the House Republicans just didn't like Paulson. They saw how he'd gotten chummy with Barney Frank and Nancy Pelosi. Nor were they impressed with his record as CEO of Goldman Sachs, the most powerful firm on Wall Street.

On Thursday morning, Paulson had drafted only a simple three-page document asking taxpayers for nearly a trillion dollars to bail out the same Wall Street banks where he had spent his career making a fortune using much of the same risk taking that had led to the calamity he was trying to fix. The plan was intentionally vague and left the Treasury secretary a lot of wiggle room; he would pitch it as an asset-buyback program, something the House Republicans might go for based on the past success of the Resolution Trust Corporation, but Paulson and the officials at the Fed who were now consulting him wanted to give Treasury as much room as possible to change course, possibly making direct capital infusions into the banks if needed.

"Paulson went out and sold it to the public as a bad-asset program, but all along he knew he could do whatever he wanted with it," said one senior Fed official who was involved in the strategy sessions.

"Things were so bad, I was seeing visions of those old 1930s photographs of people standing on breadlines," said one trader that Thursday as the markets opened for yet another dismal day. Then something happened: reports hit that the Treasury was about to unveil some massive fix for the entire financial system, with the government actually buying the toxic assets from the banking system.

The market exploded, rising more than 400 points during the afternoon as reports of the bailout leaked to the press, most notably to CNBC, which described the move as an RTC-like bailout of the government buying the toxic debt from the Street, and Bloomberg News, which ran an interview with Senator Chuck Schumer, who hinted that something grand and bold was in the works. Closely watching the events were senior executives at Morgan and Goldman, and with good reason: the concept of a cleansing of toxic assets, even though both firms said they didn't have them (though their fourth-quarter results would show they did), boosted their share prices more than 10 percent in after-hours trading.

Throughout the day, the Treasury remained mum about its intentions, refusing either to confirm or deny the reports on CNBC and elsewhere, but after the markets closed, as Paulson began to press the flesh on Capitol Hill, the outline of the plan—Paulson had only an outline at this point— became clear.

Paulson's confidence would soon be tested when, at 7 p.m., he met with the Senate leadership. As a group, Senate Republicans were less ideological than their House counterparts.

Judd Gregg, the influential senator from New Hampshire, set the tone when he issued a statement saying, "The decision by Chairman Bernanke and Secretary Paulson to pursue a broader and more comprehensive effort to stabilize our financial system and protect working Americans and Main Street is welcome news. I look forward to reviewing the details of their proposal. I will work to be sure it protects the taxpayers while allowing our economy to move forward with a strong banking and financial sector that can energize economic growth and the creation of jobs." House Republicans were immediately skeptical, even after Paulson, accompanied by Bernanke, tried to make a case that they were voting not for a giveaway but for a deal that could make the government and taxpayers some money, as the old RTC had. Further, he said, there was no choice but to pass the legislation. "We are in days of a total demise of the global financial system," he said.

"This is a [once-in-a-] hundred-years situation. I've never seen anything like it," he added. One congressman, Spencer Bachus, suggested possible direct capital injections into the banks. Paulson didn't want to go there, sensing the political minefield he was entering. Paulson also didn't want to say just how big the plan would be, at least not yet. One congressman asked, "Are we talking $500 billion?"

"I don't want to put a number on it," Paulson responded. "We're asking for enough to get it done."

"If you want something that works, this is it," he continued. "The world has changed." Several members thought it was odd that such a massive plan had come on just three pages of paper. Others were appalled by the lack of specificity in the amount of money that Paulson wanted to spend. It was open-ended.

To the members of Congress in the room, it was hard to believe that the entire economy was about to implode. As bad as things were on Wall Street, the full extent of the financial crisis had yet to hit many of their districts in the South and Midwest. People were still working, still paying their bills, and their credit cards hadn't been turned off. The credit crunch was something real—small businesses were being turned away from loans—but many wondered if a bill authorizing the Treasury to spend what was clearly hundreds of billions of dollars to bail out Wall Street was the answer. After all, wasn't Wall Street supposed to be the epicenter of free markets?

The concept of systemic risk as Paulson and Bernanke knew it was foreign to the House Republicans, but not to president George W. Bush, who now joined the effort to sell the bailout. Bush gave what Bachus describes as "pep talks about how we have to pass" the bill and how it was needed to avoid a global meltdown.

Still, Bachus, like many other Republicans, remained skeptical. "Seems like a huge bailout," Adam Putnam, a Republican congressman from Florida on the House Financial Services Committee, told Bernanke during another of the countless meetings in the days following Paulson's announcement. Putnam believed that the financial crisis, at least right then, was a crisis facing the financial firms because of their bad behavior. He felt the economy wasn't in such bad shape, at least not bad enough for the government to hand Paulson a massive checkbook.

Bernanke had been pressing the flesh on Capitol Hill, trying to get Congress comfortable with what now emerged from Paulson as a three-page, $700 billion spending plan. "Outside the brokers," meaning the Wall Street firms, "earnings aren't so bad," Putnam told Bernanke. "The real economy doesn't seem so bad, right?"

Bernanke's answer to Putnam's question was straightforward. "Trust me," he told the congressman, "it will be."

Bernanke described the situation as pretty bleak even without the bailout; with it, the system could avoid a total meltdown, in which asset values would fall to zero and the banking system would shut down. The Fed chairman's answer may have been straightforwardly apocalyptic, but it was hardly convincing for most House Republicans. The Paulson plan may have been popular on Wall Street, but back home, voters saw it as nothing more than a bailout of Wall Street. Paulson had failed to make the case to the average American that by saving Wall Street, the government was saving Main Street from heading into a severe recession.

On September 30, the House of Representatives defeated the measure by a vote of 228–205. The coalition defeating the measure was an odd compilation of free-market types such as Bachus and liberals who said they couldn't see bailing out Wall Street fat cats. But the real reason was politics. Just before the vote, House Speaker Nancy Pelosi gave a bizarre speech that, instead of calling for unity, bashed the Bush administration and its free-market principles as the cause of the financial collapse.

As he held a copy of Pelosi's remarks, one of the members of the House leadership, Eric Cantor of South Carolina, said, "Right here is the reason I believe why this vote failed . . . frankly [it] struck the tone of partisanship that frankly was inappropriate in this discussion." It was said that Paulson had gotten down on his knees and hugged the House speaker's legs when begging for support, infuriating the Republicans even more.

Others blamed the big-government aspects of the plan. "It will be important that Congress . . . devise an alternative that reflects the American people's commitment to free markets and limited government," said Mike Pence of Indiana. "The American people rejected this bailout, and today, Congress did likewise."

All of which was true to one extent or another. What House Republicans didn't bring up was another, possibly more compelling, reason for the bill's defeat: it was Hank Paulson's plan.

At the Fed, the House's reaction was met with a mixture of amazement, anger, and trepidation. In the wake of the no vote, the market immediately lost 600 points; it finished the day down 777 points, its largest one-day decline in history.

Some weren't so pessimistic. "My initial reaction was the plan would pass," Fed governor Kevin Warsh would later tell people. "I wasn't entirely surprised when it didn't, given the huge political backlash. But the second time it came to vote, despite what some were saying, I knew it would pass. There was no way that these guys wanted to be on the hook if it didn't pass and the whole system crumbled."

Which pretty much summed up what happened a few days later. President Bush held a meeting in the White House with House members to squeeze the Republicans to change their vote. It was a mundane affair, and Bachus got into a tussle with Pelosi over the lack of details in the plan. "We've already discussed this!" Pelosi snapped. Bachus just shook his head and moved on because the country had moved on. The political backlash House Republicans had felt before the vote had turned; they were now being blamed first in the media and then increasingly by their constituents for being obstructionist and not coming up with a viable alternative to what Paulson was proposing.

This time, with a few amendments, the measure passed the House and Senate, after which President Bush signed it into law.

If Paulson and Bernanke thought that would end the financial crisis, they were mistaken. The run on the banks continued in the weeks ahead. Both Goldman and Morgan would have to sell out some more by seeking capital from outside investors once again. The billionaire investor Warren Buffett came to Goldman's rescue with a capital injection in exchange for preferred shares. Meanwhile, the run on Morgan Stanley sent CEO John Mack into a deal-making frenzy; as Morgan's shares were getting hammered, Mack, at the insistence of regulators, held brief discussions with Citigroup's CEO, Vikram Pandit (Citigroup, being a bank, was still holding up relatively well compared to the brokerages, though that would soon change), and Wachovia before settling on a capital infusion from the Japanese bank Mitsubishi UFJ.

The messy details of the Mitsubishi transaction showed just how serious Morgan Stanley's financial status had become. The bank made a $9 billion investment in Morgan, which initially was bigger than Morgan's market value, causing Mack to scramble once again and change the terms of the investment from common shares to a type of preferred shares that prevented the Japanese from owning the House of Morgan.

Even that wasn't enough. The share prices of both firms continued to decline as it became clear that they were losing money on their holdings of toxic assets, and this time it had nothing to do with short sellers' rumor-mongering. So Goldman and Morgan, that same day, did the unthinkable: they applied to become commercial banks, meaning they now had direct access to the complete array of Fed emergency borrowing programs that they hadn't had as investment banks—but they were now under the direct supervision of the Federal Reserve.

The Fed accepted, and just like that, Wall Street was finished. As banks the firms would have to agree to stricter supervision and ditch the risk-taking business model that had made them so much money over the past two decades but that had now nearly brought them down, along with the rest of the financial system. Leverage would have to be brought down to levels not seen in years. Lower—much lower—profits were in store. Thirty percent return on equity would never be seen again; analysts say the firms will be lucky to achieve half that. Lower profits mean lower bonuses; Mack and Blankfein will never again earn more than $50 million for one year's work, nor will their successors for a very long time.

It was a bitter pill for both John Mack and Lloyd Blankfein to swallow. Both had built their careers around risk, and their firms had benefited from it over the years. But with their companies' survival at stake, they had no choice.

Nor did they have much choice about what happened next.

O kay, I'll be there," said the slightly bewildered John Mack after receiving a call on October 14 from Hank Paulson's office demanding his presence at a meeting in Washington at three that afternoon.

He wasn't alone. The heads of the country's major financial institutions were all summoned to Washington to meet with Paulson, who was announcing a major change in the way the TARP program would be conducted. As quickly as he had created the massive purchase of toxic assets, which had yet to be described in any concrete terms since it had been unveiled a month earlier, he was ditching the plan. Instead, Paulson said, the government was going to be making direct investments—tens of billions of dollars—in all the affected financial firms.

Paulson said he had good reason to do so. The credit crisis was getting worse, and the reason was pretty simple: the toxic assets were still toxic. The government plan to buy the assets would take too much time. So it was on to Plan B: The banks would need billions of dollars more in capital immediately to remain solvent, some more than others and some maybe none at all. Either way, he wanted everyone to take the money, first as a precautionary measure but also "so no one is stigmatized." By now Paulson was pretty good at reading the mind of the market. The banks that didn't get the money would be deemed the strongest; those getting money would have a black mark and be ripe for a bank run. The big banks got the most money: Citigroup, Bank of America, and JPMorgan Chase received $25 billion each. People who know Jamie Dimon say he was personally distressed at being lumped in with a firm like Citigroup, but he didn't protest, at least not then. Morgan Stanley and Goldman Sachs would receive $10 billion each. In exchange, the government would receive shares of convertible preferred stock in each of the companies, meaning that at some point the government could own actual pieces of the big firms.

In other words, Wall Street was being sold out piecemeal to the federal government.

The implications were huge: the federal government had taken its first step toward nationalizing the banking system without much of a peep from the alleged capitalists who ran the big banks, except for Wells Fargo chairman Richard Kovacevich, who received $25 billion in TARP funds. Kovacevich later called the plan "asinine" and said that his bank, which had just purchased another problem child of the banking industry, Wachovia, didn't need the money. What people close to Kovacevich said he was worried about more than anything else was the direct involvement of the government in the banking system. Where would the government's role end?

Paulson wasn't ready to answer that question as he handed documents to each CEO in the room to be signed so the deal could be made official. Instead, he made it clear that everyone would have no choice but to take the money for the good of the financial system, and it seemed that finally the heads of the big banks had found something they could agree on.

"Let's not make this into something silly," Ken Lewis said at one point. "Let's sign this and move on."

What they were moving on to, no one really knew, but it wouldn't be the first time Wall Street had rolled the dice.

EPILOGUE

Much more turmoil still lay ahead after the government forced the major banks to take the TARP money in exchange for preferred stock and warrants, making the federal government a shareholder in the nine biggest banks that dominated the financial system. It was in those turbulent days and nights of fall 2008 that the idea for this book took a turn. I wasn't content just to show how the titans of finance and government had abandoned prudence for reckless risk taking or how our elected officials had sold out common sense for a utopian vision of the world that gave us runaway agencies designed to help the poor but made us all poorer in the process; I became determined to understand and answer how and why America's financial system—the envy of the world and arguably the root of much of our power throughout the twentieth century—had come so close to total annihilation. The markets appeared, albeit briefly, to stabilize with the billions pumped into the banks, not to mention AIG and Fannie and Freddie. But as much as I and everyone else hoped the worst had passed, it hadn't, and the events of the end of that year and the beginning of the next would do much to shape those answers and this book, starting with the latest craziness from the Wall Street insane asylum known as Citigroup.

Citigroup received $25 billion in capital from the federal government during the initial round of government handouts, but it wasn't enough. In late November, investors who had given the big bank and its leadership the benefit of the doubt, who had listened to years of promises from management about cutting costs and finally executing the supermarket created by Sandy Weill and now in the hands of Vikram Pandit, finally had enough of the stupidity and began to unload shares of Citigroup as if a bank that epitomized the notion of Too Big to Fail were too big to survive.

It began as most things had during the financial crisis—with a broken promise. Pandit had announced with great fanfare Citigroup's plan to buy the North Carolina banking giant Wachovia, a move he said would put Citigroup on the path to growth once again. Wachovia was emblematic of the financial crisis; it had once been viewed as being among the most pristine banks in the country, with $400 billion in deposits and so many branch offices nationwide that Stan O'Neal had believed it to be the best place for Merrill to find safety back in 2007. But a year later, Wachovia was hurting, mainly from exposure to subprime–lending, though that didn't stop John Mack from briefly flirting with the idea of a Morgan Stanley-Wachovia combo in September 2008 to bolster Morgan's chances of survival during those dark days.

Mack's theory went something like this: Since Wachovia was a commercial bank, Morgan would be part of a company protected by its lead regulator the Federal Reserve, which viewed banks the size of Wachovia as too big to fail. If Wachovia was too big to fail, so too would be Morgan Stanley.

Too bad investors didn't see it Mack's way. Shares of Wachovia went into free fall and depositors began yanking money from their accounts, forcing Mack to turn to another bank, Mitsubishi, for help. Wachovia, meanwhile, turned to Citigroup. Under Citi's plan, approved by the FDIC and other regulators, Citi would pay a paltry $2.16 billion for Wachovia and its bad loans, which had mostly been washed clean with government assistance and guarantees. It was a win-win for everyone: regulators moved forward with their plan to identify problem banks and find solutions to the problems. Wachovia and its relatively new CEO, former Treasury Undersecretary Robert Steel, had gone on CNBC to proclaim that the bank's toxic portfolio was manageable and Wachovia had a "great future as an independent company." Just a few days later, however, Steel joined the ranks of Erin Callan, John Thain, Alan Schwartz, and the rest of the misguided prognosticators and had to concede that the losses were high—much higher, so much higher that the bank with such a great future was facing liquidation. But with the Citigroup deal, for a mere $1 a share, its $40 billion in losses would be covered by the FDIC. Steel, himself a former Goldman Sachs executive, had finally found a partner—the federal government, not Citi—that was truly too big to fail, while Citi would benefit from more than $400 billion in new customer deposits and a strong consumer lending department.

The deal was announced and in the early stages of integration when Pandit was caught napping, literally. Just as teams of bankers from Citi were flying down to Charlotte, North Carolina, and teams from Charlotte were taking tours of their new headquarters in New York, Wells Fargo, so far largely untouched by the crisis and also looking to grow, put forward a deal no one could refuse—except, of course, Pandit. On the day TARP became law, Wells made a surprise offer: it would buy Wachovia for $12.5 billion and without a dime from the federal government.

Pandit received a call from Steel early Friday morning, while he was sleeping, that a Wells deal was in the works. The normally cool and calculating finance PhD went "bat shit," as someone close to him described it. He ordered his legal staff to sue to prevent the deal from closing. He reached out to the federal government for help and found none, for the obvious reason that the Wells bid was a no-brainer from the government's standpoint.

There was general rejoicing in the markets (this was, after all, how the system was supposed to work, healthy banks buying less healthy ones without government help) and maybe even more rejoicing at Wachovia from employees who had just barely escaped being partnered with possibly the worst-run large bank in the history of banking.

Over at Citigroup, Pandit's leadership team voiced outrage, first threatening to sue to prevent the deal (the case has since been dismissed) and then pleading with the federal government, particularly FDIC chief Sheila Bair, to do the right thing and honor the initial agreement. Ned Kelly, one of Pandit's top executives, at one point got into a screaming match with Bair, who wouldn't budge. As she put it, the Wells offer was superior in many ways, price being just one, but, maybe most important, it didn't cost taxpayers a dime.

Citigroup's lawsuit attempting to break up the deal was seen as a final act of desperation by shareholders, who had viewed the Citi-Wachovia deal as the last-ditch hope for Pandit and his team to grow Citigroup out of its problems,

In late November 2008 the share price of Citigroup went into free fall, declining nearly 70 percent in just one week as the implication of the failed bid became all too evident: Citigroup was stuck with a business model that wouldn't work and a mother lode of underwater, unsalable bonds and loans that would continue to produce losses for the foreseeable future.

Citigroup moved into Lehman Brothers territory; its traders were having difficulty borrowing money to finance sales and trading (there was a limit to how much of its customer deposits it could actually borrow from, making the potential for a cash crunch even more acute). Corporations didn't want to work with the firm, and depositors began to bail out as well, pulling money out of savings and checking accounts with everyone fearing that the bank was so messed up, its management so outmatched by the size and scope of the problem ahead, that nothing short of a government take-over—a nearly complete dilution of shareholder equity—could save the bank from extinction.

And that's exactly what happened.

If the concept of too big to fail could be applied to AIG, it didn't take regulators long to figure out that it should be applied to the biggest near failure so far during the financial crisis. After a weekend of meetings in late November, Paulson, then in the waning weeks of his tenure as Trea-sury secretary following the election of Barack Obama, and Bernanke, working with Pandit and his team, hammered out a deal in which the gov-ernment would inject another $20 billion into Citigroup, bringing the total to $45 billion, and agree to assume the vast majority of losses on some $300 billion of bad assets on Citi's books if they began to crater even further.

With that, the U.S. taxpayer was on the road to becoming the largest shareholder in one of the world's biggest, most troubled, and worst-run banks. Citigroup officials, of course, argued that the bank wasn't being na-tionalized at all; the government would receive so-called preferred stock that carried no voting rights, and current management was firmly in control of the situation. But that turned out to be every bit as wrong as had Pandit's promises about making the Citigroup business model work. In the days and weeks to come, government officials reviewed and recommended most major moves made by the bank (including replacement of board members and changes in management), and, as this book goes to press, the next plan to revive the company, by dismantling it. Not long after the bailout, Pandit announced that the Citigroup experiment was over and he would begin the process of selling off vast parts of the company, including its brokerage divi-sion (Morgan Stanley had agreed to buy chunks of the brokerage sales force over a period of years), once the firm's crown jewel but now one of the few assets that could fetch a decent price. The move came as Citigroup was get-ting ready to convert the government's preferred shares to common stock, with all the voting power that goes with stock ownership, demonstrating that the government is not going to be such a passive shareholder after all. Government officials, in particular FDIC head Sheila Bair, have put Pandit on notice to either clean up the mess or clear out of the company.

Shares of Citigroup briefly rebounded following the government's intervention, nearly doubling to $6 before falling again in the weeks and months ahead. It was a bitter pill for many former Citigroup executives who had been paid in stock and now were casting blame for the firm's failures as they watched Citi's market value decline by $270 billiion in less than two years. At the top of that list were company founder Sandy Weill and Robert Rubin, who would soon resign from Citigroup amid growing condemnation over his failure as a board member to have a firm hand in dealing with the bank's mounting troubles and his advocacy of risk taking that had contributed to the bank's near demise. In just a decade, Rubin had gone from the man widely credited with the economic success of the Clinton years, according to some, to one of the chief culprits of the collapse of the country's financial system.

Unlike Rubin, Weill had been long gone from the firm, but he continued to obsess about its failures. People who know him say they believe he had sold much of his stock before the crisis engulfing the firm became acute, allowing him to maintain much of his vast wealth (Weill declined to comment). But money had never been Weill's only motivation; he was an empire builder, even if his concept of that empire was misguided and costly. He now had to watch Vikram Pandit—who was reporting to a phalanx of government regulators worried that the firm would face even more erosion—announce to the world that the Citigroup experiment was over; he intended to break up the empire and return to its roots as a commercial bank.

Pandit might have thrown in the towel, but Weill doesn't hold him primarily responsible for the failure of the Citigroup model to work. People who know Weill say he holds Chuck Prince largely responsible for what occurred, despite, in my opinion, showing little contrition over his own manifold failures as CEO, from his lack of attention to regulations to his elevation of risk takers like Tom Maheras and his team to senior positions to maybe his most important failure of all: the firing of Jamie Dimon. As Citigroup cratered, JPMorgan Chase and Dimon were flourishing; the firm was cranking out profits when the rest of Wall Street was losing money into 2009, and, in so doing, Dimon cemented his standing as the new King of Wall Street, complete with all the powers and perks that go with it, from media attention to the attention of the new Obama administration. Upon winning the White House, the new president began attacking Wall Street greed, particularly as the economy slipped further into recession, even though many of its top players, men like Larry Fink, Lloyd Blankfein, and Warren Spector—men who had been key participants for good and bad in the financial crisis—had all supported him. Even as Obama stepped up his attacks on Wall Street's perks and compensation, Dimon became one of his most trusted unofficial advisers on financial matters

With so much to worry about after Lehman's bankruptcy and the panic selling it touched off, including the continued saga of Citigroup, officials in the Federal Reserve and the Treasury Department could take some comfort that at the very least Merrill Lynch had been saved. The systemic risk of Merrill's folding as Lehman had was far too devastating for either Ben Bernanke or Hank Paulson to imagine. It had a balance sheet twice as large as Lehman's—close to $1 trillion—which meant twice as many trades to unwind and losses to account for. More than that, Merrill was, for all intents and purposes, a big bank, and with that it carried another set of concerns for regulators. Citigroup had close to $1 trillion in customer deposits in checking and savings accounts, but Merrill managed more than $2 trillion in money for small investors through its vast network of retail brokers scattered throughout the country, in small towns and big cities alike.

So Bernanke and Paulson were thanking God that Ken Lewis and Bank of America had the cash and the vision to buy Merrill before the week from hell began. They moved on to other crises, including the final nail in the coffin for Wall Street, when the Fed announced that both Goldman Sachs and Morgan Stanley would become commercial banks with full access to all the ways banks can borrow from the Fed's discount window. After the announcement, neither Goldman nor Morgan promised to open branch offices or hand out toasters and debit cards to their customers. But the designation signaled a new era of deep and possibly painful regulation for the independent investment banks. After all, they had spent the past two decades borrowing to finance their increasingly risky trades under the purview of what was known as "self-regulation," the idea that Wall Street could police itself through its membership in organizations like the New York Stock Exchange, the National Association of Securities Dealers, and the SEC.

The problem was that the self-regulators did very little regulating and the SEC missed too many crises, including the Great Financial Crisis of 2007 and 2008. Now the investment banks would be treated like commercial banks, regulated by the Federal Reserve (Fed officials have offices in each bank), a regulator with a record of achievement just good enough (the Fed didn't miss the Madoff scandal) to give investors some hope that at the very least the lunatics weren't in charge of the asylum.

So as officials from the Fed and the Treasury left for the holidays, a seeming break from the madness once again set in both on Wall Street and in Washington. The "crisis" seemed to be over. Depositors stopped yanking money out of Citigroup; confidence returned to the money market fund business after the government announced a voluntary program to insure money market funds. Perhaps it would be a quiet Christmas after all.

All that, however, changed with one telephone call.

Maybe he's looking for a little candy," Ben Bernanke told Kevin Warsh, a key Federal Reserve governor who had played a major role in the agency's response to the financial crisis, about the desperate call that Bank of America CEO Ken Lewis had just made to the Fed boss. It was December 15, and Lewis was desperate. "I need help," he pleaded to Bernanke. He relayed how surprised he had been—"stunned" was the word he used—to learn that Merrill Lynch's balance sheet problems were not over. Merrill was still losing money, lots of it. The way Lewis described it to officials, after shareholders had voted to approve the merger in early December, he had been informed by his finance people that the sludge on Merrill's books, undetected by his due diligence team, had eroded in value to such an extent that he could not go through with the merger on his own.

It wasn't necessarily subprime debt that had been hit, though there was still some of that on the firm's books. It was exposure to monolines, bonds that had been covered by the bond insurers Ambac and MBIA, which had lost their triple-A ratings during the crisis, plus other debt, which had combined to produce a toxic brew of losses that would eventually total $15 billion ($21.3 billion on a pretax basis).

Lewis was now threatening to walk away from the Merrill deal (or at the very least to renegotiate the terms in a major way) and government officials believed he would do it unless he could get federal assistance. It could still be done, of course. Both firms were technically free of each other, at least for two more weeks, when the merger would officially close and they would legally become one entity.

For Bernanke and Paulson, who also received a call from Lewis, it was an abrupt, strange about-face coming from a man once considered one of the best CEOs in banking to be making now, three months after announcing his purchase of Merrill with much fanfare. Paulson and Bernanke recalled the jubilation and confidence of senior executives at Bank of America after Lewis had finally gotten what he had coveted for so long: Merrill's mighty army of brokers, sixteen thousand strong, to add to his already formidable bank, now the largest in the country in market value and deposits.

They also recalled the memo written by Lisa A. White, a senior official in the Richmond Federal Reserve Bank, which regulates Bank of America, which suggested that Lewis had known what he was getting in Merrill, both the good and the bad. White had said that Lewis and his team felt "a much higher level of comfort with Merrill than it did with Lehman," which the bank had briefly considered buying in September, "specifically with the value of the franchise and the marks on the assets," such as the CDOs and the trading positions still on Merrill's books that Lewis now said were blowing up.

One of the concerns among Bank of America shareholders was the hefty price John Thain had extracted from Lewis—$29 a share in a deal worth around at the time $50 billion for a bank that would in all likelihood go bankrupt the following Monday. But Lewis, according to White's memo, said he hadn't overpaid. Quite the contrary, he thought he was getting the nation's largest brokerage firm at a steal, or a "30–50 percent discount," based on his valuation of all the company's assets.

Lewis now sounded a lot as Pandit had a few weeks earlier, and BofA was looking a lot like Citi in terms of its balance sheet woes. Regulators didn't take Lewis's threats at face value; Bernanke's "candy" remark underscored the initial consensus at the Fed that Lewis might have been bluffing, making the problem out to be worse than it was in order to receive some federal aid as Citi just had.

But the more the regulators dug into BofA's books, the deeper the problem facing the bank seemed to be. It wasn't just the losses from Merrill, which were staggering enough. The bank was struggling with its own, pre-Merrill balance sheet of soured loans and bonds tied to real estate and consumers, such as credit card receivables that were large enough, when combined with the Merrill mess, to imperil Bank of America's existence once they were disclosed to shareholders, depositors, and creditors.

More than that, Lewis appeared adamant about his intention to walk away unless the government came through with some money. "I'm not sure we can go through with it," Lewis reiterated. Though he didn't know the full extent of the losses at Merrill, his accountants were telling him they were huge, several times larger than the roughly $8 billion his analysts had been predicting, so that his own bank would be in danger of collapse.

What he did know was how to unwind the deal. Lewis said he was weighing whether to enforce a rarely used clause that can be found in most merger contracts, known as the material adverse change clause, or MAC, which allows a company to cancel a deal if something extraordinarily negative develops that evades normal due diligence procedures. One reason MACs are rarely enforced is because there aren't many events that can reasonably escape due diligence, except outright fraud or natural disasters.

Yet here was Ken Lewis telling the nation's top regulators that he had a legal opinion that the losses at Merrill were so high that they were MAC eligible.

Paulson and Bernanke, who thought they had dodged a bullet when Lewis agreed to buy Merrill, now found themselves staring down the barrel of a gun once again.

A meeting was set for December 21. While Lewis huddled with his advisers, Bernanke, Paulson, and Geithner huddled as well. Their conclusion: that either Lewis had lost his mind or he was bluffing and looking to take advantage of the government's new mandate to save "systemically important financial institutions," in Paulson's words, and get some free money to make the merger work.

Either way, they were irate. This was, after all, the same CEO who had purchased the poster child of bad subprime lending Countrywide Financial and then, with less than two days of meetings and due diligence, had wanted to buy Merrill Lynch in a deal worth $50 billion and was now looking for a handout to pay for it all.

For Paulson, Bernanke, and Geithner, the prospect of systemic risk was back. What would happen if BofA walked? The nightmare scenario, discussed during private conversations among the three men and their staffs, went something like this: Merrill Lynch would go into receivership without the federal government stepping in and buying the company out-right. Even if Merrill were saved by the government (and that would be the biggest bailout yet; the federal government would have likely had to take a majority interest in the firm, something neither Paulson or Bernanke wanted), all hell might still break loose, given the size of the firm's balance sheet and the uncertainty it might unleash in the already jittery global fi-nancial system. "We could see a scenario where asset values across the globe plummeted to zero as mass selling met with no buying," said one in-vestment banker who worked closely with government officials at the time.

It was, of course, a nightmare scenario, but as far as policy makers were concerned it wasn't just a bad dream but a possible reality if Lewis carried out his threat to walk away. That's why Paulson was about to carry out a threat of his own.

As the meeting at the Fed's Washington, D.C., headquarters got under way, Paulson and Bernanke were convinced that Lewis was no longer bluffing—their own analysis indicated that the losses at Merrill were stag-gering, so much so that, combined with BofA's own problems, the bank might not survive. But that didn't temper their response when Lewis, clearly tense and uncomfortable, sat across from the nation's top banking regulators and repeated in a slightly trembling voice that "We can't go through with it . . . not without some help" and that he was considering the use of the MAC as a way out.

Paulson didn't mince words with Lewis: invoke the MAC clause, he said, and you're finished. Not only would Lewis be sued by Merrill share-holders as their stock crashed to near zero because the clause didn't pro-vide an out for lack of due diligence (or, in this case, stupidity), but it would be reckless, bringing the financial crisis one step closer to the doomsday scenario they were fighting so hard to avoid.

Paulson told Lewis to consider something else: there would be a pen-alty for engaging in reckless behavior. Bernanke had the authority to remove not just him as CEO but also all his cronies on the Bank of America board of directors. One of the issues regulators were now grappling with as a major contributing factor to the financial crisis was just how incompetent the boards of the big financial firms had been, providing minimal oversight and almost never standing up to management. BofA's appeared to have been among the least engaged of them all.

Lewis didn't seem much more engaged, regulators concluded, which only added to their frustration. "No one at the Fed threatened Lewis with his job," said one senior Fed official, but the official said they didn't have to, because Paulson, after a long career at the snake pit known as Goldman Sachs, could threaten with the best of them, and while he did, Bernanke simply sat there and nodded in agreement, though he held out the notion that if it was money Lewis was looking for to make the deal work, the government was ready to act.

Lewis now had two choices: if he backed out of the deal, he was definitely toast—Paulson and Bernanke would do the deed—but if he made the deal, Bank of America would be in Citigroup territory, a virtual ward of the federal government. Given the size of the losses and the amount of money that the feds would have to inject into the bank, thus diluting shareholders, this must have made his stomach churn. BofA's stock price would be shattered, and the feds would be demanding ownership of a piece of the bank as they had with Citigroup, which was now trading at a paltry $3 a share and heading lower. But at least Lewis would have a job for now.

Over a Christmas break that resembled anything but a holiday, officials at the Fed, the Treasury, and Bank of America worked around the clock to hammer out a deal. Within the Fed, there was a debate over whether Bank of America would be better off trying to raise cash through some type of public offering, maybe with the government's assistance, to avoid the market fallout of becoming Citigroup, Part II.

But Paulson and Bernanke would not chance it. It was too big a gamble, they concluded; the markets weren't just skittish, they were terrified. What investor in this environment would buy a piece of a company whose biggest asset had just lost $15 billion, completely undetected during due diligence? "If they go it alone and they fail, the result would be catastrophic," Paulson said during one meeting.

On January 16, just days before the new administration was to take office, the old administration, in the person of Treasury Secretary Hank Paulson, announced a plan to "save" Bank of America by giving it an additional $20 billion from TARP and protecting it against losses on $118 billion in potentially troubled assets. To understand how far Bank of America had fallen, consider that when the deal for Merrill was struck in September, the combined bank would have been worth $179 billion. One month after the deal was consummated, the combined company had a market value of just $39 billion. That's $11 billion less than Lewis had agreed to pay for Merrill.

As this book goes to press, both Lewis and Pandit are still running their twin exemplars of corporate mismanagement, and even though their banks are beginning to show signs of life with meager profits, their future, as well as that of the banking system, is hardly assured.

One thing Wall Street firms can always do is find a way to pay their people, even if the only reason they have a job is the fact that the federal government assured their firms' continued existence. To be sure, Blankfein at Goldman and Mack at Morgan Stanley declined to take bonuses amid some of the worst losses in their company's history. But other big firms that were on life support from the federal government, such as AIG, Citigroup, and Merrill Lynch (even as part of Bank of America), and even those that weren't dependent on government handouts, still handed out huge bonuses to their stars.

John Thain, now the president of the Merrill Lynch unit of BofA, took this largesse to a new level. Ken Lewis, who had told Paulson and Bernanke that he still had confidence in Thain even after the massive losses were revealed, quickly began to have second thoughts.

Around the same time the losses were beginning to mount, Lewis heard that Thain had tried (unsuccessfully) to convince Merrill's board, in particular the board's compensation chief John Finnegan, that he should be rewarded with a $30 million to $40 million bonus for all his good work in 2008. Finnegan's response: no way. Thain then sent out feelers to Finnegan that a smaller bonus might be in order and got the same response. When it became clear that taking any money at all would jeopardize his relationship with Lewis, he took nothing.

Thain, through a spokesman, denied having asked for any specific bonus amount, but by now his relationship with Lewis was already damaged. A press report—a leak from a board member—in the *Wall Street Journal* detailed his bonus follies, embarrassing the BofA honcho, who by now was embarrassed enough by his purchase of Merrill.

Thain had for months told key Merrill executives that he was Lewis's likely successor—an odd prediction since Lewis was relatively young and the CEOs of acquired firms often leave to "pursue other interests" once the deal is done. Even odder was the exodus of management talent, not because these players despised the notion of working for a bank but because they couldn't stomach working any longer for Thain. Bob McCann, who was told by Lewis that he would have a home as head of the brokerage department, left after Thain embarrassed him with a quip to brokers alleging McCann to be a publicity hound. Bess Levin of the DealBreaker blog then reported that the person most responsible for the BofA-Merrill deal, Greg Fleming, was leaving as well. Finally Fleming and McCann found some common ground: they were both sick of working for Thain. If the departures, taken together with the bonus fiasco, hadn't put Thain on thin ice with his new boss, something else would have sealed his fate. Thain's high-end office was soon to be a secret no more after I received a leak about the makeover, all $1.2 million of it, complete with the $85,000 rugs, the $35,000 "commode on legs" and other big-ticket items. As CNBC was reporting the office spending, Thain was sitting in that very office watching the business news station only to learn one additional fact about his future: he was toast. Ken Lewis was on a plane to New York to fire him.

Since the firing, Thain has mounted a vigorous public relations campaign to restore his credibility. Friends of Thain told me that it wasn't really his idea to redo the office but that of his wife, Carmen, who saw the cold, sterile style of desk and chairs Stan O'Neal had left behind and prodded her CEO husband to make it look more livable.

But it didn't seem to help. Thain apologized for the office splurge in the oddest ways, saying in his final memo to Merrill employees that "the expenses were incurred over a year ago in a very different environment," as if 2007 had been such a great time to be spending shareholders' money on expensive rugs. The general consensus from people I speak to on Wall Street is that Thain will likely never run a major firm again.

Though no one will ever know whether Paulson's and Bernanke's predictions of global economic meltdown would have materialized if the various bailout plans hadn't taken place (we should all be thankful it didn't), the final tally of years of risk taking and leverage was indeed daunting for Wall Street and Main Street alike: two investment banks blown up, a third forced into a merger, billions of dollars in shareholder value destroyed, and a titanic destruction of the wealth of average Americans—trillions of dollars lost as housing prices fell and the Dow Jones Industrial Average fell from its October 2007 high of around 14,000 to close to 6,500 in March 2009.

Even without a total meltdown, Bernanke's prediction to House Republicans that the Wall Street recession would have broader ramifications was painfully accurate, as he was aware. Few people at the Fed or the Treasury believed Lehman deserved to be bailed out, for all the reasons enumerated in this book, but the decision to let the company die once Barclays backed out of a purchase was the one mistake both Paulson and Bernanke would now openly fess up to as they assessed the damage of those fateful months in 2008 and how economic despair had spread from the Street to the broader economy. Even with the billions in bailout money spent since Lehman's demise to prevent an even bigger panic, bank lending and the economy remained anemic through 2009 (as this book goes to press, banks are still saddled with as much as $7 trillion in possibly problematic loans, bonds, and trades on their books). So, too, did the market to package everything from home loans to credit card payments that had begun with Lew Ranieri and Larry Fink in the early 1980s and grown nearly every year ever since, until now. While some see that as proof that government bailouts don't work, consider what might have happened if the government hadn't acted.

In fact, even as the stock market rebounded off its March lows, the "Great Recession," as the economic collapse was now known, showed only a few signs of letting up as unemployment hovered around 9.5 percent. The financial crisis may have been abated, and the economy may have hit bottom in the summer of 2009, but most economists predict a feeble recovery because the extent of the economic damage was so extreme. In addition to the implosion of three major investment banks, a number of commercial banks, the automobile industry, and countless other businesses, more than $13 trillion in household wealth was destroyed by the end of 2008, according to the most recent statistics released by the Fed. Globally, some estimates place the decline at close to $50 trillion.

The billions of dollars spent on saving the banking system from three decades of reckless risk taking, through either direct cash payments or guarantees against losses, couldn't turn the economy around, but it did begin to strengthen the banking system. For all the second-guessing in congressional hearings where Bernanke, Paulson, and the heads of the remaining banks were grilled for hours about their motives (Paulson and Bernanke's alleged threat to Ken Lewis that he'd better buy Merrill Lynch or else made great theater in July 2009), at least the banking system began to stabilize.

The Fed's decision to cut its base rate to close to 0 percent, together with its decision to turn investment houses into banks and promise to bail out systemically important institutions meant that banks could now borrow for almost nothing, make increasingly risky bets, and invest in higher-yielding securities. The risk taking and carry trades that had doomed Wall Street were now its savior, as was something else: the banks were given a pass on their toxic assets. For months the accounting industry's standard-bearer, the Financial Accounting Standards Board, defended the mark-to-market accounting that had forced the Wall Street firms and banks to book all those losses on their bad debt. But in late 2008, it began to relax some of those standards, giving the banks a chance to stop taking the massive losses that had led the system into a crash and in fact to start marking up those bonds into gains.

With that even Citigroup and Bank of America began showing profits, while JPMorgan continued to show strong results. But no one did better than Goldman Sachs, whose bond trading produced more than $3 billion in profits in the second quarter of 2009. Goldman took full advantage of the government's subsidy of its borrowing and risk taking and partied as if it were 1999, increasing its leverage and risk taking more than any other firm on Wall Street. The difference was that Goldman isn't like any other firm on the Street. There are, of course, more than a few urban legends about the great Goldman profit machine: how it uses its influence in government to make all that money; how it corners the various markets with unfair trading advantages. I'm sure that there are some who believe Goldman created the swine flu to corner the market on drug stocks.

But what all these notions of conspiracy leave out is the fact that Goldman is just better than the competition. During the great risk-taking binge covered in this book, from roughly 1980 through 2008, just about every firm on the Street tried to emulate Goldman's business plan and each came up woefully short. It is true that Goldman has had its share of screw-ups, its big losses in 1998 around the LTCM fiasco and its exposure to mortgage debt tied to AIG being the most prominent. But Goldman prospered when other firms floundered primarily because (in the words of one *New York Times* reporter) it reduced risk taking to an art form. It's important to note that while Goldman cranked out $3 billion in profits in just three months in 2009, Morgan Stanley reported a loss over that same period because John Mack cut back on risk taking dramatically, finally realizing that Morgan Stanley is no Goldman Sachs.

For all the Obama administration's attacks on the corporate greed that had flourished during the free-market reign of the Republicans—firms that took TARP money faced executive pay restrictions, and as of this writing a handful, including Goldman, Morgan Stanley, and JPMorgan, have paid back the money they received—Obama's policies in dealing with the financial crisis appeared remarkably similar to those of his predecessor. Too Big to Fail remained the operating philosophy, albeit expanded to other industries such as automobile manufacturing. In the ultimate sign of consistency, Obama named former New York Fed president Tim Geithner to replace Hank Paulson as Treasury secretary.

As of this writing, the president's support of Ben Bernanke appears as strong as ever as the Fed chairman continues to keep interest rates at historically low levels in the hope that cheap borrowing rates will spur lending by banks (they really haven't, though, because the banks continue to hold some $7 trillion in bad debt and loans on their books), even though this raises the possibility of inflation at some later date.

The president and his supporters in Congress have also had very little to say about how taxpayers were subsidizing the increased risk taking at Goldman. For all Goldman's skill at making money, there is something unsettling about a major bank being subsidized by the federal government with barely any outrage from those now in power.

Some have argued that the president's silence is related to Goldman's power in the new administration or its role as an underwriter of the debt the new administration plans to issue to pay for its massive expansion of the federal government he is undertaking, from nationalized health care to other measures.

Maybe so, but the return of risk would indeed be much debated during the days that followed Goldman's massive profit, which will no doubt lead to massive year-end bonuses for all those involved. On the one hand were the firm's defenders, who made the valid point that Goldman was doing what the government wanted it to do: use the system and government help to return to health. On the other was the equally valid point that the federal government was subsidizing the same type of risk taking (albeit not to the same degree, given Goldman's lower leverage) that had led to the financial crisis in the first place. And Goldman's resurgence was lopsidedly self-serving; Goldman may be classified as a commercial bank, but in practice it still doesn't take in deposits and make loans. In essence, it's a big hedge fund subsidized by the U.S. government and taxpayers.

With Goldman's strong showing and the vast majority of Wall Street firms showing profits, the question turned to the future: could the great risk taking of the past three decades return, and could the country face another implosion of the financial system? The answer is undoubtedly yes. All the remaining investment banks are now commercial banks and thus have tougher capital requirements, regulated not by the dysfunctional SEC but by the more functional Federal Reserve, with offices now inside the banks, so they can have a bird's-eye view of the risk taking.

But regulation has proved to be fairly inept at rooting out excesses that lead to bubbles or just about any major fraud. The accounting scandal that led to the implosion of Enron and the government's indictment of its auditor, Arthur Andersen, led to a massive overhaul of how corporations account for various assets (remember that little law known as the Sarbanes-Oxley Act) and allegedly more disclosure of opaque investments; yet, about the same time those measures took effect, Wall Street was engaging in the biggest risk-taking binge in its history, largely undetected by regulators, watchdogs, and the press until it was too late. Maybe the biggest scandal in financial history, Bernie Madoff's $50 billion Ponzi scheme, went undetected for years, despite numerous complaints about Madoff's activities that were ignored by the SEC.

Yet regulation is here to stay, something economic libertarians (a category in which I count myself) should accept is necessary to right the system. As much as I believe in free markets, the problem is that you can't have a business as vital as banking (without banks making loans, there would be no economy) without oversight. You also can't have a business that is deemed too big to fail (meaning that taxpayers subsidize this business for the greater good of the nation) without someone watching the store.

But we must have effective regulation, and that should start with eliminating one of the most ineffective watchdogs of all time: the Securities and Exchange Commission. The SEC suffered a massive blow to its image when it bungled its "investigation" (I use that term loosely) into Bernie Madoff's Ponzi scheme, but in reality, it has been screwing up for years, examples of which could fill enough pages for a book on its own.

With that in mind, the SEC should be disbanded because of the false sense of security it provides and because of the conflicted nature of its own business model. SEC staffers can't wait to leave government to work on Wall Street or Wall Street law firms. This conflict of interest may make the SEC a good training ground for a career at a brokerage firm, but it makes for a pretty lousy investigator. Why would any SEC examiner crack down on Merrill's excessive risk taking or think twice about auditing Bernie Madoff's trading records (which would have uncovered his scam instantly, since he hadn't made a trade in years) if he or she might want to work for Merrill or Madoff down the road?

And maybe without a watchdog designed specifically to root out investment fraud and excess, average investors might become more suspicious of Wall Street and not so willing to believe sales pitches like Madoff's, which guaranteed 12 percent returns every year for close to 20 years. Too often the SEC has seen its job as encouraging investor confidence when a wiser regulator should have been encouraging investor caution.

Another watchdog that should be abolished is the rating agencies. If there were a group of people more universally incompetent than the SEC it would have to be the bond raters, which had missed just about every major market blowup over the past three decades, until the last one they missed nearly destroyed the world economy.

Much of the problem comes down to the raters' business model: they are paid by the municipalities, the corporations, and, in the case of mortgage bonds, the investment bankers doing the deals that they are rating.

This conflict must come to an end. Of course, the rating agencies will tell you that if corporations or bankers don't pay, no one will. Sophisticated investors would simply refuse to shell out enough money for their research (maybe they should get the hint). If that's the case, good riddance: Any business that rates a CDO as the equivalent of a U.S. Treasury bond (which the raters consistently did before the crisis erupted) shouldn't exist as a business in the first place.

With the bond raters and the SEC out of the way (the SEC, it should be noted, has authority to bring only civil cases, which doesn't exactly scare its targets), and with the Fed now focusing on risk taking, the burden of uncovering white-collar crime should fall to the criminal authorities, the FBI, and the state attorney generals. For all his faults (including pushing out Hank Greenberg of AIG, which set the stage for the company's demise), Eliot Spitzer for a time proved a pretty effective cop of Wall Street malfeasance, not only because he wasn't looking for a job on Wall Street down the road but also because he had the authority to prosecute people and possibly put them in jail and regularly threatened to do so.

Meanwhile, directing some of the budget that has gone to the SEC to the FBI would allow that agency to devote more resources to Wall Street abuse. The very fact that the FBI can put people in jail might be the best deterrent of them all.

While I have never been a proponent of class warfare, particularly when it comes to executive compensation, it's befuddling to hear my friends and colleagues on the right attack curbs on executive pay at banks that have been bailed out by taxpayers. One of the certainties of the past three decades of excessive risk taking on Wall Street is the role of compensation behind the excess. In 2006, when Wall Street top executives made more money than ever before, excessive risk taking reached its height and the unraveling of the nation's financial system began. If taxpayers are subsidizing that risk taking, whether implicitly or explicitly, because the banking system (rightly, I believe) is deemed too important too fail, the bankers need to accept government intrusion in return for that subsidy.

The CEOs of Merrill, Lehman, Bear Stearns, and nearly every firm created a system in which risk taking was rewarded to the extreme and losses rarely penalized, except if you consider being fired a stiff penalty, which it isn't as most traders are simply rehired by other firms. A real penalty might include some kind of mandatory clawback provision, wherein the decision makers—those ordering the risk taking at the CEO level and just below—are penalized for massive losses by having to return their huge salaries, egregious bonuses, and golden parachutes. Who will determine what is or isn't excessive risk taking? I don't know. I hope it's at the board level (although many boards have hardly shown themselves effective guardians of shareholders' interests) and not in Congress. But hitting the risk takers where they will feel it the most is the only way to make sure those who feel the urge to bet the ranch, whether on the CDO market or the next wonderful investment to come out of Wall Street, think twice.

The federal government should finally cut back on subsidizing risk, as it has done repeatedly and consistently through the past thirty years and appears to be doing again. Many libertarians have also debated the necessity of the bailout actions taken by Bernanke and Paulson during the heat of the crisis, arguing, at times persuasively, that doing away with the guarantees and bailouts and simply letting the natural forces of capitalism work would not just have taught Wall Street a lesson about risk (since nearly every firm would probably have imploded) but led to a faster recovery.

But what if asset values had fallen to rock-bottom levels after natural market forces caused Goldman, Morgan, and maybe Citigroup and BofA to follow Bear, Lehman, and Merrill (had it not been for BofA) and fall into liquidation, as Paulson and Bernanke feared? Wouldn't investors have started taking more reasonable risks, buying stocks and bonds and the assets of those companies, thus bidding up values and helping the economy and the markets recover on their own?

Yes, some say. *Maybe*, is my answer.

What if there hadn't been any buyers because so many people—from public and private pension funds to individuals such as Joe Lewis, the billionaire who took his lumps on Bear Stearns, to the Chinese, who hold not just U.S. Treasury bonds but also lots of mortgage debt—had loaded up on so many of those esoteric investments, from CDOs to interest rate swaps, that the toxicity was everywhere?

In other words, there wouldn't have been any buyers for the system to repair itself in a reasonable amount of time, or at least not enough of them.

A more measured approach would be for a moratorium on future bailouts. As I pointed out earlier in this book, the various bailouts from the losses in 1986, 1994, 1998, and the most recent mother of all crises only fed Wall Street's appetite for risk, to the point that up to almost the minute Lehman filed for bankruptcy, senior executives there, including CEO Dick Fuld, believed the government would come to its rescue.

Goldman's line of credit that allowed it to take enough risks earn $3 billion in the second quarter of 2009 should be cut off immediately, and the firm should be forced to turn in its credentials as a commercial bank under the regulation of the Fed and its too-big-to-fail agenda. If Lloyd Blankfein wants to take risk without government regulation, he and his traders should do so as a private partnership and play with their own money.

All of which brings me to an interview I had with the great financier Teddy Forstmann in the course of writing this book, in which he asked me about the title, *The Sellout*.

Anyone who knows Forstmann knows he's been holding the Street accountable for years, most famously during the risk binge of the 1980s, when he labeled junk bonds "funny money." I told Teddy that in my mind, *The Sellout* was all about Wall Street and the financial system having to "sell out" to survive after its three-decade binge on risk and leverage, by being bailed out first with capital coming from foreign sovereign wealth funds (thus diluting existing shareholders' investments and giving foreign governments a greater say in our financial system) and then ultimately by the federal government.

In other words, not only had Wall Street literally had to sell out because it had embraced excessive risk taking, it had also sold out its principles: greed had become its business model, not all the factors that make Wall Street important to society, such as raising money for businesses and providing access to the stock markets to the middle class. The mortgage bond market was the best example of my thesis because it was initially designed to allow banks to extend loans to people and families of limited means, but along the way it was perverted. Greed took something fundamentally good and made it into something that was fundamentally bad.

Forstmann got a good chuckle out of that. "So what you're saying is that somewhere along the line, Wall Street as an institution had some principles to sell out," he said with a laugh. "I am here to tell you Wall Street never had principles."

Teddy's critique of my title and of the Wall Street culture forced me to search long and hard for people on Wall Street who had principles and who refused to sell out. I found them, and their stories are here as well, but unfortunately for Wall Street, and for the rest of us, they were clearly outnumbered. And I fear that will remain the case for some time.

GLOSSARY

ABX Index: Launched in January 2006, this index serves as a benchmark for the subprime mortgage-backed securities market. ABX contracts are commonly used by investors to speculate on or to hedge against rising defaults in securities backed by subprime loans.

Alt-A: Alt-A mortgages are loans that are considered riskier than A-loans (aka prime loans) but less risky than subprime loans. Frequently, these types of loans do not require full documentation on the part of the borrower. Alt-A mortgages were created in the mid-1990s as a way to serve borrowers whose credit fell just short of prime-loan status but was above subprime standards.

Balance sheet: A financial statement that provides a snapshot of a company's financial position. It lists a company's assets, liabilities, and shareholders' equity at a specific point in time. In short, it shows what a company owns and what it owes to various creditors.

Bond: A debt investment in which investors lend money to either a corporation or a government for a set period of time at a fixed interest rate. A bond's **coupon** is the interest rate payment that is made over the life of the bond. The **principal** is the face value of the bond and is the amount on which the creditor receives interest. Bond investors are considered creditors, while equity (stock) investors are considered owners.

Carry trade: An investment strategy in which investors borrow at a low, short-term rate and then proceed to invest in higher-yielding longer-term debt. Investors using such a strategy make money by pocketing the difference between the short-term rate at which they borrow and the long-term rate at which they invest. The carry trade was at the heart of the investment banks' strategy to generate profits during the mortgage boom.

Collateralized debt obligation (CDO): A security backed by a pool of other debt securities, such as, but not limited to, mortgages. Car loans and credit card debt are two examples of other types of debt that might be packed into a CDO. A CDO is typically divided into various risk categories, or tranches, which are then sold to investors. The higher-rated, more senior, tranches are considered the less risky, while the lower-rated, or junior, ones pay higher interest. A **CDO squared** is a CDO that is backed by other CDOs.

Collateralized mortgage obligation (CMO): A type of mortgage-backed security that is backed by a wide pool of mortgage loans. A CMO is typically divided into several classes, or "tranches," that lay claim to various cash flows to the bondholders. By separating the flows into different sections, investors can better gauge their prepayment risk.

Commercial bank: A financial institution that accepts customers' deposits and makes commercial loans. Commercial banks are regulated by the Federal Reserve and in exchange for being able to access their customers' deposits are required to take less risk than investment banks. Since the repeal of the Glass-Steagall Act, however, the difference between commercial and investment banks has decreased.

Commercial paper: Short-term (270 days or less) debt that corporations issue to fund their ongoing operations, commercial paper is a cheap source of financing for highly rated corporations. Money market funds are typically big buyers of commercial paper.

Credit default swap: A derivative contract whose value is linked to an underlying debt instrument. CDSs gain or lose value depending on the likelihood of default of anything from a corporate loan to a complex mortgage-backed security. A CDS essentially acts as a form of insurance on a bond or structured loan. If a bond or structured loan defaults, the seller of a CDS pays the buyer for the losses.

Derivative: A financial instrument whose value is derived from the price of another financial asset. Derivatives are contracts that can refer to a vast array of financial products, from standard equity options such as puts or calls to much more complex instruments such as credit default swaps, which are linked to the value of an underlying debt instrument. Derivatives are typically used for hedging or speculative purposes. Derivatives are contracts, not securities, and are not regulated as such by the SEC.

Fannie Mae/Freddie Mac: The most prominent of the government-sponsored enterprises, Fannie and its smaller sibling, Freddie Mac, played a crucial role in the expansion of the housing bubble. Fannie Mae, originally known as the Federal National Mortgage Association, was created during the Great Depression to increase the availability of mortgage lending. It was chartered as a public company in 1968. Freddie Mac, the Federal Home Loan Mortgage Corporation, was created two years later. Fannie and Freddie buy mortgages from various lenders and then either hold those mortgages in their portfolios as investments or repackage them with other loans to create mortgage-backed securities, which are then sold to various investors. At one point these two companies owned or insured half of America's $12 trillion in mortgages. In September 2008 both companies were taken over by the government due to massive losses from non-performing mortgage loans.

Glass-Steagall Act: Enacted in response to the collapse of the banking system during the Great Depression, the Glass-Steagall Act, or Banking Act of 1933, prohibited commercial banks from engaging in investment banking businesses, such as the trading or underwriting of securities. It established the FDIC, which insured commercial bank deposits, but it also divided the banking industry into two distinct areas: the banks, which took in deposits and made loans, and the brokers, which engaged in the riskier parts of the capital markets. Glass-Steagall was changed in 1999 with the passage of the Gramm-Leach-Bliley Act, which relaxed many of the restrictions on commercial banking activities.

Government-sponsored enterprises (GSEs): A group of government-chartered companies whose purpose is to increase the availability of credit for everything from home borrowing to student loans. These corporations include Fannie Mae and Freddie Mac, Sallie Mae, and Ginnie Mae.

Hedge fund: A lightly regulated investment fund that is typically available only to wealthy investors. Unlike mutual funds, which must follow closely regulated investment mandates, hedge funds can use a wider range of investment and trading techniques to generate profits. Contrary to their name, they are often the most exposed to risk due to their heavy use of leverage.

Hedging: Any investment strategy designed to hedge, or offset, risk against an existing position or investment.

Interest-only (IO) bond: A type of mortgage-backed security that is referred to as a "strip" because the two basic elements of a bond, principal and interest, are stripped and diverted to two securities, the IO and the PO. IOs capture only the interest payment stream and therefore perform better when rates rise. That's because, as rates rise, people refinance their mortgage loans less frequently, and that extends the payments to the IO holder, making for steadier cash flow. That results in an increase in the value of the IO security in a rising-interest-rate environment.

Interest rate swap: A type of derivative contract in which two parties can exchange, or swap, payment streams from a fixed-income security. These contracts are used to hedge against interest rate risk. Typically, one party wants to exchange the payments from an adjustable-rate debt instrument for an instrument with a fixed-rate cash flow, and the other wants to do the opposite.

Investment bank: A financial institution that is primarily engaged in the riskier aspects of the capital markets, such as the underwriting and trading and selling of securities. Investment banks offer an array of financial services, from merger advice to back-office services for other financial institutions. The key distinction between commercial and investment banks is that an investment bank does not handle customer deposits. Investment banks are regulated by the Securities and Exchange Commission.

League tables: The standard by which various investment banks measure their respective performances, league tables show in dollar terms the number and amount of deals each bank has completed in a given time period. The rankings measure everything from stock and debt offerings to mergers and acquisitions advisory work.

Leverage: The use of borrowed money to enhance investment returns. Leverage was at the heart of the financial crisis as many firms borrowed at enormous levels, in some cases $30 or more for every dollar they owned, to make speculative bets on the markets. Leverage can work both ways. On the way up, it increases returns, but on the way down, it can magnify losses. For example, if a trader has $10 and leverages at 10 to 1, he now has $100 to invest. If his bet goes up by 50 percent, he's made $50, or 500 percent, of his initial stake of $10. But if his bet goes down by 50 percent, he's now in the hole $40—four times the amount of money he has on hand.

Liquidity: Typically, liquidity refers to an asset's ability to be quickly sold or bought, but, more broadly, it measures the ability to access cash. The more liquid an asset is, the more easily it can be sold. Similarly, the more liquid an institution is, the more easily it can tap various sources of funding.

Margin: The use of borrowed money to finance the purchase of securities. When investors borrow from their broker to buy stocks or bonds, they are borrowing on margin and using the purchased securities as collateral for the loan. If the value of those securities falls, the investor may be required to post more money against that loan. This is referred to as a **margin call** and can often trigger a wave of selling as investors look to fulfill their margin obligations.

Mark to market: A type of accounting treatment where securities are "marked," or valued, based on prevailing market prices. Their value is determined by the price a certain security could fetch in the open market if it were sold at a moment's notice.

Mark to model: A type of accounting treatment that "marks," or values, securities based on financial models and certain assumptions, not prevailing market prices.

Moral hazard: A concept that argues that when a business is bailed out and protected from taking excessive risk, other similar businesses act differently by continuing to take excessive risk because they believe they too will be bailed out.

Mortgage-backed security: A type of bond that is backed by the underlying cash flow of various mortgages.

Off-balance-sheet: Off-balance-sheet items are assets and liabilities that are not on a company's balance sheet. They are frequently kept in a separate subsidiary of the company. Companies attempt to move assets off balance sheets to disguise their true financial situation. One of the benefits of securitization was that it allowed banks to move mortgages off their balance sheets by selling them to Wall Street firms, thus freeing up the banks' capital for additional lending.

Principal-only (PO) bond: A type of mortgage-backed security that is referred to as a "strip" because the two basic elements of a bond, principal and interest, are stripped and diverted to two securities, the IO and the PO. It is the mirror image of an IO bond. POs capture only the principal payment stream and thus perform better when rates fall. That's because as rates fall, people look to refinance, and that accelerates principal payments on the mortgage loan, making for faster cash flow. That results in an increase in the value of the PO security in a declining-interest-rate environment.

Repurchase agreement (repo): The market for repos is a vital source of inexpensive funding in the securities industry. In a standard repo, a borrower pledges securities to a lender in exchange for cash for a fixed period of time. When that term expires, the borrower repurchases the inventory and the lender returns the pledged collateral to the borrower. Repo is at the heart of any investment bank's ability to fund itself.

Return on equity (ROE): This statistic provides a snapshot of a corporation's profitability. It specifically measures how efficient a company is in generating profits on its existing assets. It is a key measure of growth in the financial services industry,

Run on a bank: The unexpected and rapid exit of deposits or cash balances from a financial institution. Bank runs occur when individuals lose confidence in the solvency of a particular financial institution. The most historically famous instances of runs on the bank occurred during the Great Depression, when rumors of the insolvency of a bank would lead to long lines of customers waiting to withdraw their deposits—leading to just such an insolvency. Runs on investment banks are similar, only the creditors are hedge funds and other large investors as opposed to ordinary individuals.

Savings and loans (S&Ls): Saving and loans are commonly called thrifts and are federally insured financial institutions that take in deposits for the purpose of making real estate or other consumer loans. In the 1980s and early 1990s, lax lending standards and a slowdown in real estate sales caused the failure of a number of federally insured S&Ls.

Securitization: The process of structuring and redirecting the cash flow of an asset to create multiple securities. Securitization is at the heart of the credit industry, as bankers look to capture the cash flow of any type of receivable, be it mortgages, credit card loans, or something else, and structure it into a separate security that can be bought or sold by investors.

Short sale: A trading strategy that allows an investor to profit from the decline in value of a security. A short sale is completed when an investor sells a borrowed security in the hope of buying that security back later at a lower price.

Short squeeze: A trading phenomenon in which a stock price rises sharply as investors try to cover their bearish bets by buying back their "short," or contrary, bets. Short squeezes are typically triggered by a tight supply of stock.

Structured finance: A broad term that refers to the structuring or redirecting of cash flows of fixed-income securities to meet investors' needs. The goal is to tailor an existing cash flow from a bond to match the risk appetite of the investor.

Structured investment vehicles (SIVs): Large pools of investment assets that were kept off the balance sheets of many of the banks. SIVs were funded by selling short-term debt to investors to finance the purchase of higher-yielding long-term debt. The goal was to benefit from the difference between short- and long-term rates. When credit losses began to occur, many banks had to take those assets back onto their balance sheets, resulting in massive losses.

Super-senior tranche (also see **Tranche**): The section or slice of cash flow of a structured financial security that is deemed "senior," or less risky, to other sections of the security. In theory, "senior" tranches should provide more stable and secure streams of cash flows (although at lower coupons than the riskier junior tranches). In a typical mortgage-backed security, the junior tranches must absorb any losses due to defaults before the more senior tranche(s) start to get hit.

Systemic risk: The threat of a collapse of the entire financial system, as opposed to the failure of a single company or financial entity. It was the threat of systemic risk that drove the Federal Reserve and Treasury to take extraordinary steps in the midst of the financial crisis.

Tranche: French for "slice," a tranche is a section of a mortgage bond's cash flow. The dividing of cash flows is an integral component of mortgage-related securities, as various cash flows correspond to various risk profiles.

Value at risk (VAR): A risk management technique frequently used by banks that is designed to give a snapshot of risk at any given moment. VAR measures the worst expected losses under normal market conditions. Developed in the early 1990s by a group of mathematicians, VAR expresses the maximum dollar value a firm can lose on a given day based on historical trading patterns.

ACKNOWLEDGMENTS

Acknowledgments can be controversial. You can leave out people who have contributed significantly to your book. You can include people who don't deserve more than the time of day. With that in mind, I want to apologize to those I have offended either on these grounds or others, and acknowledge the following people:

My wife, Virginia Juliano, for putting up with me for so long (I try not to think about the years because it makes me feel old), and inspiring me to be more than where I came from, because she knows what it's like to come from that place.

My assistant, Max Meyers, for being a damn good journalist and instrumental in making this book a reality. Max is an amazing TV producer. He produces two shows for CNBC: *Options Action* and *Fast Money* (where I met him), and he knows how to create lively television as well as anyone I have ever met. What most people at CNBC don't know (now they will) is that Max can write well, report as if he had spent years at the *Wall Street Journal*, and is a damn good editor. This book is as much his as it is mine. Enough said.

My editors on the book were first rate. Ethan Friedman converted my prose (I use that word loosely) into more than readable English. Hollis Heimbouch at HarperCollins was a difficult taskmaster. But now that I'm finished, I feel comfortable thanking her for her hard work, insight, and for being great to work with. John Jusino and Matt Inman of HarperCollins deserve kudos as well for translating my garbled fixes into the English language.

My brother, Dr. James Gasparino, has a real job: He saves people's lives. Yet he's interested in mine, and for that I am honored. Kevin Goldman of NBC is a flack like no other, and a great friend and editor. I thank him for all the guidance he provided me over the past two years in producing this book. Michael Salomon of *The Daily Beast* and John Carney of *Clusterstock* are great friends, but, more than that, they are amazing journalists, and I couldn't have completed *The Sellout* without them and their editing guidance.

My mother in law, Angela Juliano, is still my biggest fan and I wouldn't want it any other way. My "uncle" Benny Giannone told me recently to "enjoy the ride," and I will follow his logic, which comes straight from Brooklyn. My brother-in-law, Joe DiSalvo, is still my best friend even though he has two kids and can't hang out any longer. Tina Brown is an amazing editor, and I thank her for giving me the opportunity to write for *The Daily Beast*. I didn't work with Ed Felsenthal during my years at the *Wall Street Journal*, but I'm making up for it now at the *Beast*, and I'm having a ball. *New York Post* editors Col Allan and Mark Cunningham, of the editorial page, are wonderful journalists, and I thank them for running my stuff.

I want to thank Susan Krakower for believing in me through all the ups and downs of this job. She is a great friend and mentor. I also want to thank Mark Hoffman, Jeremy Pink, and Tyler Mathisen of CNBC, who gave me the time to complete this work. Jonathan Wald, who was my boss at CNBC, is my friend, and I thank him for what he has done for my career. We made a great team.

I am also honored to know many others who have helped me through the ups and downs of writing a book (and life in general), starting with a great family: Tina DiSalvo, Peter Giannone, Aunt Phyllis Giannone, Michael Giannone, Matthew DiSalvo, Mia DiSalvo, Nicole Giannone, Aunt Roseann Bergman, and family members who span two continents but now, after drinking two martinis, I can't remember their names. And some great friends: Mark Schwartz; Eric Starkman; Eddie Grant (he owns a great piano bar, Marie's Crisis Café); Paul Carlucci (who is publisher of a great newspaper, the *New York Post*); Bess Levin (who is a joy to read on Dealbreaker); the Gang from Campagnola; the Bruno Family of San Pietro and Sistina; my agent, Todd Shuster; Bruno from "Club A" Steakhouse; Bill Summers; and Elaine Kaufman, who, when she's not yelling at me for dining elsewhere, runs a great spot known simply as *Elaine's*.

NOTES

One of the many challenges in writing *The Sellout* was the need to provide readers with a historical basis for Wall Street's unraveling, which began in early 2007, and continued through the year and into 2008 and early 2009, when the world as most people on the Street knew it collapsed.

To tell so sweeping a story, I relied on a number of sources in the form of people, newspaper articles, and previously written books to guide me. One of the best sources was the inimitable Teddy Forstmann. I conducted several interviews with Forstmann, who during the 1980s famously warned about Wall Street's excesses. Forstmann's insight was invaluable to me in understanding the Wall Street psyche that I believe led to the Great Panic of 2008, and for that I thank him.

I would also like to thank others (in no particular order of importance) who led me through the maze of the past twenty-five years, including Larry Fink, John Mack, Linda Robinson, Jane Marie McFadden, Doug Kass, Jim Chanos, Paul Critchlow, Mike Mayo, Meredith Whitney, Lucas Van Praag (who despite working for Goldman Sachs isn't such a bad guy), Richard Dickey, Pat Dunlavy, John Linder, Harvey Goldschmid, Lynn Turner, Dick Bove, Brad Hintz, Sean Egan, Bill Dallas, Denis Coleman, Phil Cohen, Josh Rosner, Alan Blinder, Neil Baron, Jesse Derris, Rich Peterson of Standard & Poor's, and many others who preferred to remain anonymous but were also instrumental in helping me understand *The Sellout* in its entirety.

Several other sources bear mentioning. I can't say enough great things about Roger Lowenstein's masterpiece *When Genius Failed: The Rise and Fall of Long-Term Capital Management*. I had the privilege to work with Roger early in my career at the *Wall Street Journal*, and I also had the privilege to read his book and use it as source material, particularly in the sections of *The Sellout* that describe the implosion of Long-Term Capital Management, its central characters, and what LTCM's demise meant in the context of Wall Street's current problems. I also relied on at least two other classic accounts of Wall Street as source material: *Liar's Poker: Rising Through the Wreckage on Wall Street*, by Michael Lewis, for descriptions of some of the major players at a firm that helped usher in the era of risk taking, Salomon Brothers; and to a lesser extent *Barbarians at the Gate: The Fall of RJR Nabisco* by Bryan Burrough and John Helyar, for their insight into the 1980s (when, in my opinion, the era of risk found its roots) as well as for their descriptions of some of the top deal makers and news makers of the time, including Teddy Forstmann.

Late last year, after a CNBC appearance, I ran into former Bear Stearns strategist Leo M. Tilman, and it was one of the most productive chance encounters I have ever had. Tilman had just started his own advisory firm but more importantly he told me about his new book, something he described as the statistical version of what I was attempting to do with *The Sellout*.

The book, *Financial Darwinism*, should be required reading (after reading *The Sellout*, of course) for anyone interested in the origins of the financial collapse. Throughout *The Sellout*, I relied on Tilman's data and at times his analysis as to why Wall Street embraced risk. I hope I've given him proper credit.

1: FUN AND GAMES

10–13 Description of the Great Race from interview with Pat Dunlavy.

10 Meriwether's winning bet on New York City municipal bonds from author's interviews and Roger Lowenstein, *When Genius Failed* (New York: Random House, 2000).

10 Background on Meriwether and arbitrage group from author's interviews and Lowenstein, *When Genius Failed*.

17 Ranieri's description of Strauss as company "stiff" from Michael Lewis, *Liar's Poker: Rising Through the Wreckage on Wall Street*, (New York: W. W. Norton & Co., 1989) and author interviews with representatives of Ranieri.

19 Description of the first mortgage bond deal from author's interview and Catherine Kittle, "Banking on Securitization; An Increasingly Popular Way to Dump Unwanted Assets," *Barron's*, June 1, 1987.

18 Description of Ranieri from Mike McNamee, "The Great Innovators: Louis S. Ranieri; Your Mortgage Was His Bonds," *BusinessWeek*, November 29, 2004.

20 Ranieri politely told Gutfreund: Conversation between Ranieri and Gutfreund over Ranieri's move to the mortgage bond desk from interviews with people familiar with the conversation.

20–22 Ranieri's efforts to advance the mortgage bond market from author interviews and from Michael Lewis, *Liar's Poker*.

22 Description of Robert Dall ceding more responsibility to Ranieri from William Glaberson, "Life After Salomon Brothers," *New York Times*, October 11, 1987.

22 Larry Fink's early career and the creation of the CMO from author's interviews and Katrina Brooker, "The Economy in Crisis; Can This Man Save Wall Street?" *Fortune*, November 10, 2008, and "The First CMO Was a Decisive Moment in Freddie Mac's History," *National Mortgage News*, July 24, 1995.

25 Continued fee compression and the fact that the big Wall Street firms were now increasingly public companies further drives Wall Street's business model toward risk from Leo M. Tilman, *Financial Darwinism: Create Value or Self Destruct in a World of Risk* (New York: John Wiley & Sons, 2008).

25 Background on changes in the financial markets in the early 1980s from technological advancement to declining fees on transactions and Wall Street's retreat from the agency model toward a risk-oriented business model from author interviews as well as from Tilman, *Financial Darwinism*, and author interviews.

27–28 Description of the competition between Fink and Ranieri over underwriting rankings from Ann Monroe, "Salomon Brothers Makes First Boston Sweat for Victory—Mortgage Bond Rankings Had Competitors Vying to the Wire for Crown," *Wall Street Journal*, July 2, 1985.

36 Description of turf battles inside Salomon Brothers from Lowenstein, *When Genius Failed*; Lewis, *Liar's Poker*; and author's interviews.

2: POWER AND PERKS

38–39 Conversation between Jimmy Cayne and Larry Friedlander at opening of this chapter from author interview with a person with knowledge of the conversation. Cayne had no comment on the conversation.

39 Description of early Bear Stearns, including its founders and later the description of Theodore Low, from Yael Bizouati, "A Dying Bear: JPMorgan's Takeover of Bear Stearns Offers as Many Questions as Answers," *Investment Dealers Digest*, March 24, 2008, and Michael A. Hiltzik, "Bear Stearns Is Bullish in Its New Role: Trader; Plunges in Investment Banking," *Los Angeles Times*, October 4, 1987.

42 Jimmy Cayne's early life, including his childhood, details about his father's real last name, his first marrage, illness of his child and pre-Bear days, including his second marrage, from author interviews with Cayne, various family members, and from Pierre Paulden, "Dealer's Choice," *Institutional Investor*, November 16, 2006.

43 Cayne's conversation with Ace Greenberg about Bridge and how Ace could never be as good as Cayne at it from author interview with Cayne and William Cohan, "Rise and Fall of Jimmy Cayne," *Fortune*, August 25, 2008.

44 Description of Jimmy Cayne's bet on New York City bonds from author's interviews and Paulden, "Dealer's Choice."

47 Cy Lewis's death from author's interviews with former Bear partner Phil Cohen and Nelson Schwartz, "What 'the Bear' Meant for the Street," *New York Times*, March 30, 2008.

49 Greenberg running risk meetings and Jimmy Cayne's absence from them while meeting with clients and building his power base from author's interviews with people who were present at the meetings.

49 Rudy Giuliani's investigation into Bear Stearns from James Barron, "Wall St. Firms Subpoenaed on Contributions to Regan," *New York Times*, September 7, 1998.

50 Background on Warren Spector, including how he met Cayne and his own love of Bridge and introduction to mortgage market, from author interviews and various news accounts, including one by Bradley Keoun and Jody Shenn, "Bear Stearns Removes Spector After Debt Market Losses," *Bloomberg*, August 6, 2007.

3: SEX, DRUGS, AND DEBT

55–56 Background on Michael Milken from his role in the advancement of the junk bond market to his growing clout inside Drexel and his personal net worth as the state of the junk bond market declined during the 1980s from author's interviews; James Stewart, *Den of Thieves* (New York: Simon & Schuster, 1992); and Kurt Eichenwald, "Wages Even Wall St. Can't Stomach," *New York Times*, April 3, 1989.

56–60 Description of Teddy Forstmann from author's interviews and Brian Burrough and John Helyar, *Barbarians at the Gate: The Fall of RJR Nabisco* (New York: HarperCollins, 1990).

59 Forstmann's effort to get Goldman, Morgan Stanley, and Salomon Brothers to no longer participate in the junk-inspired LBO business and his meetings with top executives at each firm from author interviews with Forstmann.

60 Problems at Citigroup at the end of the 1980s from Anthony Bianco, "What Wriston Wrought; His Innovations Increased Banking Profits—and Risk," *Business-Week*, February 7, 2005.

61–64 The more expansive description of IOs and POs in this section and the description of the IO and PO crisis on Wall Street, including the trading losses of Howie Rubin and Merrill's response, from author interviews with traders at the time, including Larry Fink, and James Sterngold, "Anatomy of a Staggering Loss," *New York Times*, May 11, 1987, and author's interviews.

4: AN EDUCATION IN RISK

65–67 Larry Fink's losses from author's interviews with Fink and others and Katrina Brooker, "The Economy in Crisis; Can This Man Save Wall Street?" *Fortune*, November 10, 2008.

68 Taleb quote from *The Black Swan* (New York: Random House, 2007).

69 Larry Fink being ostracized at First Boston following losses from author interviews, including Fink.

71 Mike Mortara's nickname of "Fat Ankles" from Lewis, *Liar's Poker*, and author interview with representatives of Ranieri and Pat Dunlavy.

71–72 Conversation between Mike Mortara and Pat Dunlavy over the possible mismarking of Salomon's bond portfolio from interviews with Dunlavy; conversation between Dunlavy and Ranieri over what Dunlavy believed was Mortara mismarking trading positions also from interview with Dunlavy. Mortara is dead and Ranieri through his press representative didn't deny the account.

74 Gutfreund's "terse" memo over Ranieri's firing from interview with Pat Dunlavy.

74 Description of Lew Ranieri's firing from author's interviews and Steve Swartz, "Firms Are Damaged by Expensive Mistakes; Salomon's Big Upheaval—A Surfeit of Prima Donnas?" *Wall Street Journal*, July 27, 1987.

75 Salomon's losses in mortgage and municipal bonds following Ranieri's and Mortara's firing from James Sterngold, "A Sweeping Shift at Salomon," *New York Times*, July 24, 1987.

75 Mortara's departure from Salomon being a "pretty mutual decision" from Ann Monroe and Michael Siconolfi, "Salomon Unit Merges Trading Finance Areas—Move Aimed at Tightening Control of Businesses; Parent's Profit Slid 66%," *Wall Street Journal*, July 24, 1987.

75 Conversation between Dunlavy and Strauss about Mortara and Ranieri's ouster from interview with Pat Dunlavy. Strauss through a spokesman declined to deny Dunlavy's account. A senior executive at a competing firm also says part of the reason why Mortara and Ranieri left Salomon Brothers was because of market losses.

77 Bear Stearns' hiring of Howie Rubin from author interviews and various news accounts, including Michael Siconolfi, "Talented Outcasts: Bear Stearns Prospers Hiring Daring Traders That Rival Firms Shun—It Lets Them Make Big Bets, but Sharp-Eyed 'Ferrets' Watch Their Every Move—Grilled at the 'Cold-Sweat,'" *Wall Street Journal*, November 11, 1993.

78 Wall Street losses and job losses from various news accounts, including Steve

Malanga, "N.Y. Economy Bottoming Out, but Very Weak," *Crain's New York Business*, June 24, 1991.

78 Drexel's illiquidity and then its demise from author interviews and Laurie P. Cohen, "The Final Days: Drexel Itself Made Firm's Sudden Demise All but Inevitable; Internal Dissension, Greed, Lax Scrutiny of New Deals Exacerbated Junk Woes—a Phone Call from the Fed," *Wall Street Journal*, February 26, 1990.

82 Larry Fink's departure from First Boston and creation of BlackRock from author interviews with Fink and other executives involved in the move.

83–84 Description of the Salomon Brothers Treasury bond scandal and the trader who earned $23 million in one year from author interviews, various news accounts and Laurence Zuckerman, "At Salomon, Trader Earns $30 million," *New York Times*, November 9, 1994.

5: BIGGER IS BETTER

88–91, 100 Mike Madden's remarks from author interview with Madden.

89 Richard Fuld's takeover at Lehman from author's interviews and Michael Quint, "Company News; Co-President at Shearson Named No. 1," *New York Times*, March 30, 1993.

93, 98 Description of growth of leverage from statistics compiled by Standard & Poor's.

93 Description of Lehman's leverage from statistics compiled by Standard & Poor's.

93 Background on Fuld from interviews with people who worked with him; the "pencil" story from author interview with former Lehman CFO Brad Hintz.

94 Fuld throwing the papers off the desk of a supervisor from author interviews and Charles Gasparino, "Beat of the Street," *Trader Monthly*, April–May 2007.

94 General description of E. Gerald Corrigan from Michael Quint, "The Fed's Plumber: E. Gerald Corrigan; A Crisis Manager Takes On the Mechanics of the Market," *New York Times*, October 2, 1988.

95 Corrigan's warning about the growth of derivatives from author's interviews and various newspaper accounts, including Saul Hansell, "The Market Place; Group Approves the Use of Derivatives," *New York Times*, July 22, 1993.

96 Description of the growth of the derivative market from Tillman, *Financial Darwinism* and data provided by ISDA.

98 Description of growth of mortgage-bond and asset-backed market from Standard & Poor's.

98 Corrigan's clash with Greenspan over derivatives from Lowenstein, *When Genius Failed*.

99 Description of trading profits and losses at the mortgage desks of the major firms from author interviews, various news accounts and Michael Siconolfi, "Talented Outcasts: Bear Stearns Prospers Hiring Daring Traders That Rival Firms Shun—It Lets Them Make Big Bets, but Sharp-Eyed 'Ferrets' Watch Their Every Move—Grilled at the 'Cold-Sweat.'"

99 Impact of interest rate increases on small town in Pennsylvania from Charles Gasparino and Michael Moss, "Failing Returns; What Happened When a Small Town Trusted Local Financial Wizard; His Risky Investments Bring Fiscal Ruin, Dash Hopes for Improved Schools; Mr. Black Brought to Tears," *Wall Street Journal*, December 26, 1997.

99 Derivative losses at Procter & Gamble and Greenspan's support of the market from Roger Fillion, "Greenspan, Others Downplay Derivatives Risk," Reuters, May 25, 1994; and Suzanne McGee and Laurie Hays, "Derivatives Are a Tempting Option Again—Trading Is Up and Regulatory Fears Are Subsiding," *Wall Street Journal*, March 8, 1996.

99 Salomon losses in 1994 from Peter Truell, "Salomon Says it Will Report Another Loss," *New York Times*, July 12, 1995.

99 Losses at Kidder from its mortgage operations run by Michael Vranos, and Lehman's problems from Laura Jereski and Michael Siconolfi, including "Kidder Facelift Will Slash Its Wall Street Role," *Wall Street Journal*, October 7, 1994, and "Lehman Says Earnings Plummet in Quarter, Mulls Cost-Cuts," Reuters, September 21, 1994.

101 Fuld holds talks to merge with DLJ from author interviews and Peter Truell "While It's Merger Season on Wall Street, Some Question the Firm's Intentions," *New York Times*, June 3, 1997.

101 Dick Fuld avoiding appointing a president for years from Greg Cresci, "Lehman Fills President's Post After 7 Years," Reuters, May 24, 2004.

102 Fuld's description of the firm's underperformance as if cut back on risk from Anita Raghavan, "Wall Street's Fear of Risk Crimps Profits," *Wall Street Journal*, July 18, 1995.

6: FINANCIAL RENAISSANCE

103 Cayne's description of Howie Rubin and background on Bear Stearns from author's interviews and Michael Siconolfi, "Talented Outcasts: Bear Stearns Prospers Hiring Daring Traders That Rival Firms Shun—It Lets Them Make Big Bets, But Sharp-Eyed 'Ferrets' Watch Their Every Move—Grilled at the 'Cold-Sweat.'" *Wall Street Journal*.

103, 117 Warren Spector's salary from Stephen Taub, Nanette Byrnes, and David Carey, "The $650 Million Man (George Soros, Highest Paid Money Manager; Top 100 Money Managers)," *Financial World*, July 6, 1993.

106 Bear Stearns' relationship with the GSEs and background about Fannie and Freddie from author interviews and Evelyn Wallace, "Spate of Ginnie Maes Barely Ruffles Market," *American Banker*, August 17, 1987.

107–8 Description of Henry Cisneros at HUD from author's interviews with Cisneros and David Streitfeld and Gretchen Morgenson, "The Reckoning; Building Flawed American Dreams," *New York Times*, October 19, 2008.

109 Discussion between Cisneros and Mario Cuomo about Andrew Cuomo from author interview with Cisneros.

109 Conservative support of the Cuomo Commission report on homelessness from Celia W. Dugger, "Giuliani Calls Dinkins Indecisive on Housing and Homeless," *New York Times*, August 5, 1993.

109 Description of Boston Fed study on racial disparity in home loans from John Cushman, "U.S. to Use Agents to Detect Mortgage Bias," *New York Times*, May 6, 1993.

109 Boston Fed study and criticism of it by conservatives from author's interviews, various news accounts, and Terence Corcoran, "Quantum of Failures; Forget the Markets: Massive Government Failure Is Behind World Financial Chaos," *Financial Post*, October 25, 2008.

111 Cisneros's and Andrew Cuomo's dealings with the GSEs from author's interviews with both men and Wayne Barrett, "Andrew Cuomo and Fannie and Freddie:

How the Youngest Housing and Urban Development Secretary in History Gave Birth to the Mortgage Crisis," *Village Voice*, August 5, 2008.

111   Boston Fed study and criticism of it by conservatives from author's interviews, various news accounts, and Terence Corcoran, "Quantum of Failures; Forget the Markets: Massive Government Failure Is Behind World Financial Chaos," *Financial Post*, October 25, 2008.

112   Cayne and Greenberg's salary of around $15 million a year in the early 1990s from author interviews and Stephen Taub and Nanette Byrnes and David Carey, "The $650 Million Man," *Financial World*, July 6 1993.

113–14   This section involving some of the incidents at Bear as Jimmy Cayne fights for control of the firm, including the firm's use of so-called ferrets, risk management, the compensation of Jimmy Cayne and Ace Greenberg, the so-called Geisha girls incident, Cayne's comment about women traders in *M Magazine* (which he says was taken out of context), Greenberg's comments to Spector about risk, through Greenberg's removal as CEO from author interviews with various Bear executives, including Cayne, and from Michael Siconolfi, "Talented Outcasts: Bear Stearns Prospers Hiring Daring Traders That Rival Firms Shun—It Lets Them Make Big Bets, but Sharp-Eyed 'Ferrets' Watch Their Every Move—Grilled at the 'Cold-Sweat,'" *Wall Street Journal* and Allen Myerson, "Careful Player Moves Closer to the Top at Bear Stearns," *New York Times*, July 14, 1993.

116   Comments from Lyndon LaRouche, Henry Gonzalez, and Edward Markey on regulating derivatives from Michael Peltz, "Congress's Lame Assault on Derivatives," *Institutional Investor*, December 1, 1994.

116   Corrigan's change of heart on derivative regulation once he gets to Goldman Sachs from Peltz, "Congress's Lame Assault on Derivatives," *Institutional Investor*.

117   Warren Spector's salary and its growth from various news accounts and Stephen Taub and Nanette Byrnes and David Carey, "The $650 Million Man," *Financial World*.

122   Changing asset mix at the big firms after the 1994 bond market rout from SNL Financial.

117   The equity capital of Bear and Lehman compared to the rest of the big firms from Michael Siconolfi, "Lehman Sought Merger Partners—Some Analysts Cool on Stock," *Wall Street Journal*, June 20, 1996.

121   Leverage stats after the 1994 bond market rout from Standard & Poor's.

118–19   Background described here concerning Larry Fink and BlackRock from author interviews with executives, including Fink himself, and various news accounts, including Andy Serwer and Julia Boorstin, "The Hidden Beauty of Bonds: While You Were Watching the Stock Market, a Little-known Bond House Called BlackRock Grew into One of the Biggest Players on Wall Street," *Fortune*, March 19, 2001.

119   BlackRock managing the Kidder mortgage bond portfolio from author's interviews with Fink and Laura Jereski, "BlackRock to Liquidate Kidder Mortgage Portfolio," *Wall Street Journal*, November 16, 2001.

7: PAYING THE PRICE

122–23   Market conditions before the implosion of Long-Term Capital Management from various news accounts and Lowenstein, *When Genius Failed*.

123   Description of Fuld and Lehman before the LTCM crisis from Robert Clow, "Fuld's Gold," *Institutional Investor*, July 1, 1998.

122   Wall Street's embrace of leverage and risk after 1994 through the beginning of the LTCM crisis from various news accounts; leverage ratios of Wall Street during this time from Standard & Poor's.

124   Fuld threatening rival CEOs and calling Dick Grasso of the NYSE to study Lehman's books from author's interviews and Ianthe Jeanne Dugan, "Battling Rumors on Wall St.; Lehman Brothers Chairman Launches Aggressive Defense," *Washington Post*, October 10, 1998.

126   LTCM's use of VAR from Lowenstein, *When Genius Failed*.

126   Description of value at risk from Joe Nocera, "RISK Mismanagement," *New York Times*, January 4, 2009.

126   Description of the odds of various activities from "Easy Money: The Odds of Gambling," *Frontline*, June 10, 1997, and "Extra!" *Rocky Mountain News*, April 29, 2006.

126–28   General description of Wall Street's trading ties to LTCM, the losses suffered by the firms, and other matters related to the LTCM implosion from author's interviews with Wall Street executives directly involved in the matter; various news accounts; and Lowenstein, *When Genius Failed*; James A. Bianco, "Trading, 1999-style," *Futures*, December 1, 1998; Apu Sikri, "LTCM Caused Sea Change in Risk Management," Reuters, March 17, 2000; and Stanley Reed, Ian Katz, Patrica Kranz, "The Great Shakeout of 1998," *BusinessWeek*, November 9, 1998.

128–29   Richard Dickey's experience at Lehman during the LTCM crisis from author interviews.

8: ONE BIG HAPPY FAMILY

130   The Merrill CFO Stan O'Neal's concerns about Merrill's reliance on short-term means of funding from author interviews and confirmed by O'Neal.

131   Description of Jimmy Cayne and his breakfast partners from author's interviews; information regarding broker Richard Sachs from author interview with Sachs and Michael Siconolfi, "Bear Stearns Beats Rivals to Bag Highly Productive Brokerage Team," *Wall Street Journal*, March 17 1995.

132–33   Jimmy Cayne's recounting of the meeting at the Fed to bail out LTCM from interview with Cayne.

133–34   Description of LTCM discussions, including statements by David Komansky and Peter Karches, from author interviews with people with direct knowledge of the conversations.

135   The continued use of value at risk on Wall Street after LTCM from Emily Thornton, "Inside Wall Street's Culture of Risk; Investment Banks Are Placing Bigger Bets than Ever and Beating the Odds—at Least for Now," *BusinessWeek*, June 12, 2006.

136   Maughan's announcement of Salomon's merger from author's interviews.

139–40   Dunlavy's meeting with Jamie Dimon and his concerns about risk taking from author interviews with Dunlavy and confirmed by representatives of Dimon.

140–41   Description of Weill and Dimon relationship and description of the creation of Citigroup from author interviews, news accounts, and Skip Wollenberg, "Citicorp, Travelers Agree on Mammoth Merger," Associated Press, April 6, 1998.

140–42   Jamie Dimon's relationship with Tom Maheras, including his concern about his personal trading and risk taking, from author interviews with people who worked for both men and confirmed by representatives of Dimon.

145–46 Description of Robert Rubin pushing for risk from author's interviews with Rubin, various traders at Citigroup, and Eric Dash and Julie Creswell, "The Reckoning: Citigroup Pays for a Rush to Risk," *New York Times*, November 23, 2008.
146 Billy Heinzerling's concerns about risk and issues surrounding Tom Maheras's personal trading and the change in the culture of Travelers and later Citigroup as Tom Maheras takes control from author interviews with people who worked at the firm at the time. Maheras would not comment. Officials at Citigroup, both past and present, mentioned in this section, would not deny the description as it is presented in this book.

9: OPENING THE FLOODGATES

153 "Money is the mother's milk of politics" is a famous quote by former California Assembly Speaker Jesse Unruh.
154–55 Overturning of the Glass-Steagall Act and its importance in the spread of risk and leverage through the financial system from author's interviews, various news accounts, and Peter Baker, "It's Not About Bill," *New York Times*, May 31, 2009.
156 Andrew Cuomo pushes Fannie Mae and Freddie Mac to guarantee increasingly more risky mortgages from James Peterson, "Fannie Mae's Franklin Raines: 'We Believe in Big Goals,'" *ABA Banking Journal*, January 1, 2000.
156 Fannie Mae's growth statistic from Fannie Mae official.
159 The lax lending standards described in this section from author interviews with people involved; the definition of "liar loan" from Anne Flaherty, "House Votes to Outlaw 'Liar Loans,' Other Practices that Lawmakers Say Preyed on Consumers," Associated Press, May 7, 2009.
161 Mozilo's speech to executives at headquarters from author's interviews with people who were present.
161–62 Background on Countrywide and Angelo Mozilo, including statistics on expansion into subprime and other loan areas, from author's interviews and Gretchen Morgenson and Geraldine Fabrikant, "Countrywide's Chief Salesman And Defender," *New York Times*, November 11, 2007.
162 Mozilo's comments made during a February 2003 speech at the Joint Center for Housing Studies at Harvard University.
162 Increase in subprime lending at Countrywide from Gretchen Morgenson and Geraldine Fabrikant, "Countrywide's Chief Salesman And Defender," *New York Times*.

10: NO MORE MOTHER MERRILL

164 Conversation between William Dallas and Vincent Mora from author interview with Dallas; Mora wouldn't deny the account.
166 Asset-backed securitization statistics from Standard & Poor's.
167 Wall Street's buying of subprime originators from Paul Muolo and Mathew Padilla, "Stan's Ill-Fated Dream," *Registered Rep*, December 1, 2008.
168–69 Michael Blum's presentation to Ownit's board from William Dallas and confirmed by Merrill representatives.
170–71 Background on Stan O'Neal from author's interview with representatives of O'Neal and O'Neal himself, and John Cassidy, "Subprime Suspect: Annals of Business," *New Yorker*, March 31, 2008.

171   Dave Komansky's ethnicity from Peter Truell, "An Old Broker Takes the World Stage," *New York Times,* December 20, 1996.

174–75   Stan O'Neal's ascension at Merrill and his actions during 9/11 from author's interviews and Charles Gasparino, "Bull by the Horns: As Merrill Digs Out, No. 2 O'Neal Moves Fast to Reshape Firm—Cost-Cutting President Pushes Overhaul amid 'Trauma'; Komansky's Role Shrinks—'I Don't Need Your Advice,'" *Wall Street Journal,* November 2, 2001.

178   Lloyd Blankfein being named CEO of Goldman from author's interviews and Susanne Craig and Randall Smith, "Bush Taps Paulson as Treasury Chief—Firm Expected to Elevate Blankfein, Underscoring a Shift Toward Trading—Mr. Inside vs. Mr. Outside," *Wall Street Journal,* May 31, 2006.

178–79, 180   O'Neal's push to change the culture of Merrill and his obsession with Goldman Sachs and his rationale for pushing Merrill to take more risk from interviews with executives directly involved in the matter, including O'Neal. Underwriting fees and commission decline described in this section from Tilman, *Financial Darwinism.*

11: THE MONEY MACHINE

182   Background on John Mack from author interviews and various news accounts.

181–82   Conversation between John Mack and Stan Druckenmiller from author interview and confirmed by Mack

183   Stan Druckenmiller's net worth from Laura Kroll, Matthew Miller, and Tatiana Serafin, "The World's Billionaires," *Forbes,* March, 11, 2009.

184–86   Jamie Dimon's mystification at Citigroup's balance sheet from interview with Dimon.

185   Chuck Prince's rise to become CEO after research scandal implicated Sandy Weill from interviews with former Citigroup executives and former investigators for then NY AG Eliot Spitzer. Both Prince and Weill declined to comment.

186–87   Description of risk taking at Citigroup from author's interviews and Eric Dash and Julie Creswell, "The Reckoning: Citigroup Pays for a Rush to Risk," *New York Times,* November 23, 2008.

187   Heinzerling's comments at risk management meeting about bonds remaining on firm's books for years from former Citi executives.

189   SEC chief Bill Donaldson meeting with Citigroup CEO Chuck Prince over scandals at the bank from author's interviews and Charles Gasparino, "Tough Beat; William Donaldson Is Wall Street's Top Cop, Facing the Nearly Impossible Task of Policing a Sprawling, Fast-moving Financial System. But Is He Fighting the Right Battles?" *Newsweek,* October 11, 2004.

188   Jamie Dimon's comment about Chuck Prince from author interview and confirmed by Dimon.

189–90   Dimon cutting back on risk from author interviews.

190–91   Rubin's role in Citigroup's embrace of increasing amount of risk and Rubin's quote, "the only undervalued asset is risk," from author's interview with Rubin, various Citigroup executives with direct knowledge of the matter, and Eric Dash and Julie Creswell, "The Reckoning: Citigroup Pays for a Rush to Risk," *New York Times.*

191   Hiring of consultant to study Citigroup's risk from author interview with Robert Rubin.

191–92 Description of "The Machine" at Citigroup and the contribution of Drexel bankers and David X. Li to the CDO from author interviews with Citigroup executives and Matthew Philips, "Revenge of The Nerd; Paul Wilmott Is Out To Save Wall Street's Soul One Dork At a Time," *Newsweek*, June 8, 2009.

192 Citigroup profits from trading, underwriting, and carrying mortgage bonds from author's interviews.

192 Background on Ralph Cioffi and the creation of the Bear Stearns hedge funds from various news accounts and author interviews with Bear executives and Cioffi himself.

193–94 Cioffi's success and his soaring personal wealth from author interviews and William D. Cohan, "Inside The Bear Stearns Boiler Room," *Fortune*, March 4, 2009.

194 Leverage and returns of Cioffi's hedge funds from author interviews with representatives of Cioffi's clients.

194 "Live for today" as the motto of the Citigroup trading desk from author's interview with an executive who was there at the time.

12: PERVERSE INCENTIVES

196 Goldschmid statement about the changing capital rules for risk taking and his general assessment of the SEC from author's interview with Goldschmid and other past SEC officials.

199–202 Description of rating agency methods, lack of standards, and how the profitability of mortgage bond ratings and the way raters were compensated led to inflated assessments of mortgage debt from author's interviews with former raters and various news accounts.

203 Quotes from rating agency officials from congressional hearings on the role of the rating agencies in the financial collapse before House Committee on Oversight and Government Reform, October 22, 2008.

203 Martin Sullivan's comment, "and don't forget to have fun," from author's interview and Theo Francis and Ian McDonald, "Hank-less AIG: More Deliberative (and Fun)," *Wall Street Journal*, May 20, 2005.

204–5 AIG's credit default swap business and its increase after Spitzer investigation from a *Washington Post* series on AIG, including Brady Dennis and Robert O'Harrow, "A Crack in the System; 1998, AIG Financial Products had made hundreds of millions of dollars and had captured Wall Street's attention with its precise, finely balanced system for managing risk. Then it subtly turned in a dangerous direction," *Washington Post*, December 30, 2008; and "Downgrades and Downfall; How could a single unit of AIG cause the giant company's near-ruin and become a fulcrum of the global financial crisis? Straying from its own rules for managing risk and then failing to anticipate the consequences," December 31, 2008.

205 Hank Greenberg's assessment of Martin Sullivan's management abilities and his assertion about his "Irish" descent from author interview with Greenberg.

205 AIG's losing triple-A rating and increasing its issuance of CDSs from author's interviews and O'Harrow and Dennis, "Downgrades and Downfall."

13: TOP OF THE WORLD

206 Fuld's "War Speech" from executive who was present at the event.

206   Mark Walsh's role at Lehman and his activities accounting for 20 percent of Lehman's profits from author interviews with former Lehman executives and Devin Leonard, "How Lehman Got Its Real Estate Fix," *New York Times*, May 3, 2009.

207   Conversation between former Lehman investment banker Andrew Malik and Fuld from conversation with Malik. Fuld declined numerous attempts to comment for this book.

208   Lehman's leverage numbers from Standard & Poor's.

208   Fuld becoming more isolated despite his reputation as a hands-on manager from author interviews with former Lehman executives.

209–10   Larry Lindsey's warnings from former Lehman executives and confirmed by Lindsey.

210   Dick Fuld's lifestyle from Susanne Craig and Kelly Crow, "Fallen Tycoon to Auction Prized Works," *Wall Street Journal*, September 26, 2008.

210   Lehman president Joe Gregory's lifestyle from author's interviews and from Steve Fishman, "Burning Down His House; Is Lehman CEO Dick Fuld the True Villain in the Collapse of Wall Street, or Is He Being Sacrificed for the Sins of His Peers?" *New York*, November 30, 2008.

210   Joe Gregory's sale of the Range Rover from a former Lehman executive with knowledge of the matter. Gregory declined to comment through Lehman's former general counsel, Tom Russo.

211   Mike Gelband comments to Fuld and firing from author's interviews with former Lehman executives and Fishman, "Burning Down His House."

212   Lehman's holdings of various bonds, including mortgage holdings, from statistics compiled by SNL Financial. Bear's holdings of various bonds from statistics compiled by SNL Financial.

212   Lehman T-Shirt from author interviews with former Lehman executives.

213   Fuld's wealth from "25 People to Blame for the Financial Crisis," *Time*, February 12, 2009.

213   Jimmy Cayne reaches billionaire status from Susanne Craig, "The Biggest Fish on Wall Street? Probably Not Who You Think," *Wall Street Journal*, May 9, 2006.

213–14   Cayne's efforts to prop up Bear stock price and admonition from the SEC about talking up shares from author interviews with SEC officials and former Bear officials.

216   Jamie Dimon's comments about Cayne from author interviews and confirmed by Dimon.

218   "I hate structured products," O'Neal said at the time, and Stan O'Neal's comments on risk during the LTCM crisis from author's interviews and confirmed by O'Neal.

218   O'Neal's comments about how to attain 30% ROE with risk from author interview and confirmed by O'Neal.

219   David Trone's comments about Merrill from Michael J. Martinez, "Merrill Lynch Beats Views for 4Q, Year," Associated Press, January 25, 2005.

220   Jeff Kronthal's risk aversion and his battles with Merrill Lynch management from author's interviews with former Merrill executives and Susan Pulliam, Serena Ng, and Randall Smith, "Merrill Upped Ante as Boom in Mortgage Bonds Fizzled—Fresh $6 Billion Hit Is Expected as Toll of CDO Push Rises," *Wall Street Journal*, April 16, 2008.

221–25   Ahmass Fakahany's clout at Merrill from numerous interviews with senior Merrill executives; his issue involving child support from "Wall St. Dad's Cheap Way Out," *New York Post*, November 30, 2005.

224   Cayne's lunching with O'Neal from author interviews with both men.

14: CASH STOPS FLOWING

226–27   AIG's decision to stop issuing credit default swaps based on author's interviews with former AIG officials and from Brady Dennis and Robert O'Harrow, "A Crack in the System; 1998, AIG Financial Products had made hundreds of millions of dollars and had captured Wall Street's attention with its precise, finely balanced system for managing risk. Then it subtly turned in a dangerous direction," *Washington Post*, December 30, 2008; and Dennis and O'Harrow, "Downgrades and Downfall; How could a single unit of AIG cause the giant company's near-ruin and become a fulcrum of the global financial crisis? Straying from its own rules for managing risk and then failing to anticipate the consequences," December 31, 2008.

227   Jeff Kronthal's continued concern about CDO exposure and Dow Kim's comments and his push into CDOs from author's interviews and Deborah Orr, "Trading Place; Dow Kim, a Quiet Man with a Taste for Risk, Is One of Wall Street's Best-Kept Secrets," *Forbes Global*, September 19, 2005.

228   Pete Kelly's encounter with O'Neal over Semerci's promotion and Kronthal's firing from author interviews with former senior Merrill executives who discussed the account with Kelly; O'Neal says he doesn't recall the incident.

232   The size of the buildup of mortgage debt on the books of the big banks from "The Future for Investing in Mortgage Backed Securities, 2008," by Datamonitor Premium Research Reports, March 19, 2008

232   Former HUD Secretary Andrew Cuomo's prodding of GSEs to expand guarantees to more risky loans, including subprime, from author's interviews and Wayne Barrett, "Andrew Cuomo and Fannie and Freddie: How the Youngest Housing and Urban Development Secretary in History Gave Birth to the Mortgage Crisis," *Village Voice*, August 5, 2008.

232–33   HUD puts pressure on Fannie and Freddie to expand loans to the poor and minorities from author's interviews and Kathleen Day, "HUD Says Mortgage Policies Hurt Blacks; Home Loan Giants Cited," *Washington Post*, March 2, 2000.

232   The growth of the GSEs balance sheet from author's interviews and Terence Corcoran, "Quantum of Failures; Forget the Markets: Massive Government Failure Is Behind World Financial Chaos," *Financial Post*, October 25, 2008.

233   Former Louisiana Representative Richard Baker's role in trying to rein in the GSEs from author's interview with Baker and William McGurn, "Main Street: Shouting 'Fannie!' in a Crowded Congress," *Wall Street Journal*, October 14, 2008.

233   Barney Frank's statement about problems at GSEs being "overblown" from Wells, "How Mortgage Crisis Happened."

233, 424   Senator Chris Dodd's involvement in the "Friends of Angelo" program from various news accounts, including Ben Conery and Sean Lengell, "Countrywide's Founder Accused of Insider Trade," *Washington Times*, June 5, 2009.

234   Congressional prodding to push the GSEs into subprime and HUD Secretary Andrew Cuomo's comments about the GSEs' increased involvement in the subprime mortgage market from author's interviews and M. Jay Wells, "How Mortgage Crisis Happened: Good Intentions Paved Dire Path," *Investor's Business Daily*, November 3, 2008.

234–35   Bush's speech about expanding home ownership from "Homeownership Gains Will Fuel Economy," *Mortgage Servicing News*, November 15, 2002.

235 Growth of Merrill's CDO portfolio from internal Merrill documents.

235 Dow Kim and Osman Semerci's claim that Merrill's risk models showed potential of only minor losses from author's interview with Merrill executives.

235 Dow Kim's salary from Louise Story, "Bonuses Soared on Wall Street Even as Earnings Were Starting to Crumble," *New York Times*, December 19, 2008.

236 Details of BlackRock-Merrill deal from author's interviews and Randall Smith and Tom Luricella, "Merrill Board Weighs the Fate of CEO O'Neal—Overture to Wachovia, $8.4 Billion Credit Hit Spur Directors to Move," *Wall Street Journal*, October 27, 2007.

239 Josh Rosner's conversation with Nouriel Roubini from author's interview with Rosner, and confirmed by Roubini.

15: NO CLUE

246 Stan Druckenmiller's speech about mortgage market and Einhorn's reaction from author interviews.

247 Background on Einhorn from author's interview and Hugo Lindgren, "The Confidence Man; Hedge-fund manager David Einhorn believes his public attack on Lehman Brothers wasn't just about making money. So what was it about?" *New York*, June 15, 2008.

247 Einhorn's remarks in speech from copy of his speech.

247 Einhorn's investment in New Century and MDC from author's interviews and Lingling Wei, "Tables Turn on New Century—Lender Could Be Big Casualty of Subprime Squeeze," *Wall Street Journal*, March 6, 2007, and Lingling Wei, "Einhorn Resigns from New Century Board," *Dow Jones Newswires*, March 8, 2007.

248 Automated mortgage processing from author's interview and from Lynnley Browning, "Subprime Loan Machine; Automated Underwriting Software Helped Fuel a Mortgage Boom," *New York Times*, March 23, 2007.

248 New Century's bankruptcy from author's interview and Wei, "Tables Turn on New Century" and "Einhorn Resigns from New Century Board."

249 2006 Wall Street bonus numbers from author interviews and Stephen Fleischman, "Beware the rich getting richer: Today's obscene corporate bonuses part of house-of-cards capitalism," *Chicago Sun-Times*, January 7, 2007.

251 Cioffi's comments from Walden Siew, "Subprime Turmoil May Take Toll on CDOs," Reuters, February 28, 2007.

251–52 Cioffi's background from various news accounts and author's interviews, including with Cioffi.

253 Cioffi's salary from author's interviews, including with Cioffi.

254 Cioffi's "panic" and description of his actions as the subprime market and his funds began to crater from author's interviews with executives at Bear Stearns, including Cioffi; investor lawsuit filed against Bear and Cioffi; and the indictment of Cioffi and his co-manager Matthew Tannin by the U.S. Attorney's Office for the Eastern District of New York, June 18, 2008.

255 Fuld's comment about "warehousing" bonds from author's interview.

255 Bear's balance sheet exposure to risky mortgage debt from JPMorgan Chase CEO Jamie Dimon's statement in the firm's 2008 shareholder report, June 8, 2009.

257 Jimmy Cayne's decision to give investors and creditors "nothing" from author's interview with Cayne.

258   Goldman's dealings with Cioffi from former Bear Stearns executives.
259   Dow Kim and Greg Fleming speech at UBS conference from internal Merrill Lynch documents.
261   Dale Lattanzio's presentation to the board from internal Merrill Lynch documents. Both he and Osman Semerci have declined comment through intermediaries.
262   Stan O'Neal's conversation when Dow Kim announced his departure from Merrill from author's interview with O'Neal, and spokesman for Kim.
264–65, 267   Merrill's dealings with Cioffi from author's interviews with various Bear Stearns and Merrill Lynch executives directly involved in the matter and various news accounts.
266   Cioffi's plan to save the Bear Stearns hedge funds through a line of credit from author's interviews with Cioffi and Kate Kelly, "Bear to Beef Up Its Oversight—Two Funds' Woes Stir Stronger Risk Control, New Reporting Chain," *Wall Street Journal*, July 4, 2007.
267   Spector and Fleming conversation from author's interviews.
269   Cayne's meeting with senior executives about the Cioffi hedge funds and Spector's allegedly unauthorized cash infusion from author's interview with Cayne and various news accounts.
272   Molinaro's statement and Cayne's action during August conference call from author's interviews and reporting on the call itself.
272–74   Bear's reliance on mortgage debt to finance itself in the repo market from author's interviews with Bear Stearns executives and from William Cohan, "The Rise and Fall of Jimmy Cayne," *Fortune*, August 18, 2008.
276   SEC's "proctology exam" from author interviews with Bear officials.
278   Jimmy Cayne and Bear's deliberations with Citigroup and JPMorgan concerning a line of credit from author's interviews with executives directly involved in the discussions.
279   The slow approach of regulators in dealing with the credit crisis described here and throughout the book from author interviews with senior officials at the Fed, the Treasury, and various Congressional leaders involved in the process. Bernanke's reaction to credit crunch at the Jackson Hole event obtained from interviews with Fed officials.
282–83   Maheras's standing inside Citigroup and trading from author's interviews with former Citigroup executives with firsthand knowledge of the events. Maheras declined repeated requests to comment.
285   Maheras's risk management, assurances that the firm's holdings were safe, and CEO Charles Prince's review from author's interviews and Eric Dash and Julie Creswell, "The Reckoning: Citigroup Pays for a Rush to Risk," *New York Times*, November 23, 2008.
286   Discussions between Gary Crittenden and Alan Schwartz over line of credit for Bear from author interviews with people directly involved in the discussions.
287   Jimmy Cayne's brush with death from author's interview with Jimmy Cayne and Pat Cayne.

16: THE DANCING STOPS

290 Layoffs at Countrywide and Mozilo's plan to exit subprime lending business as the meltdown spreads from various news accounts and "Countrywide Posts Loss," Associated Press, October 27, 2007.

295 Stan O'Neal's discussion with John Breit from author interviews and confirmed by representative of O'Neal. Breit declined comment.

296–97 Details of O'Neal's plan to invest in Dow Kim's hedge fund from author interviews and Merrill Lynch press release, "Dow Kim to Leave Merrill Lynch and Establish Multi-Strategy, Private Investment Firm by Year-End; Plans to Affiliate New Firm with Company," May 16, 2007.

297 Stan O'Neal's conversation with Dow Kim from author's interviews of people with direct knowledge of the conversation.

298–99 O'Neal's proposal to sell Merrill to Bank of America in September 2007 from author's interviews with people with direct knowledge of the talks.

300 The conversations between O'Neal and Bob McCann; McCann and Semerci, and Fakahany's decision to dismiss Semerci and Lattanzio confirmed through author interviews with former Merrill executives with direct knowledge of the incidents.

305–7 Citigroup's SIV holdings and efforts to create an SIV bailout from Eric Dash, "$75 Billion Fund Is Seen as Stopgap," *New York Times*, November 1, 2007.

307–8 Weill's actions during Chuck Prince's final days at Citigroup, including his meetings with Prince Alwaleed and desire to return to Citigroup, from author's interviews and Robin Sidel, Monica Langley, and David Enrich, "Two Weeks That Shook the Titans of Wall Street—As O'Neal Tottered, Sandy Weill Turned on Protégé Prince," *Wall Street Journal*, November 9, 2007.

308 Stan O'Neal's decision to approach Wachovia for a possible deal from author's interviews and from Jenny Anderson and Landon Thomas Jr., "Merrill's Chief Is Said to Float a Bid to Merge," *New York Times*, October 26, 2007.

309 Size of Merrill's CDO holdings and presentation to the Merrill board from internal Merrill Lynch documents and author interviews.

312 Merrill Lynch earning $700 million in fees from CDOs in 2006 from Shawn Tully, "Wall Street's Money Machine Breaks Down," *Fortune*, November 17, 2007.

312 Stan O'Neal's dinner with Larry Fink and the fallout from the Wachovia talks from author interviews with both men and their representatives.

315 Bob McCann's discussion with O'Neal regarding O'Neal's resignation from author interviews with representatives of both men.

320–21 The situation involving Morgan Stanley CEO John Mack's firing of Zoe Cruz from author's interviews.

321 Disclosure of "Level 1," "Level 2," and "Level 3" assets and huge debt levels from Jesse Eisinger, "Wall Street Requiem," *Portfolio*, November, 2007.

321–22 John Paulson and Stan Druckenmiller's shorting both the mortgage market from author's interviews, various news accounts, and Charles Gasparino, "Greenspan's Conflict," *New York Post*, February 20, 2009.

322 Goldman's reliance on AIG insurance on its CDO portfolio from author interviews with senior Goldman officials and Gretchen Morgenson "Behind Insurer's Crisis, Blind Eye to a Web of Risk," *New York Times*, September 27, 2008.

322 Cassano's assurances that AIG's exposure to risky debt was minimal from author's interviews, various news accounts, and Lynnley Browning, "AIG's House of Cards," *Portfolio*, September 28, 2008.

324–25   Details on Einhorn's bet on Lehman from author's interviews and Hugo Lindgren, "The Confidence Man; Hedge-Fund Manager David Einhorn Believes His Public Attack on Lehman Brothers Wasn't Just About Making Money. So What Was It About?" *New York*, June 15, 2008.

326   Bear's business of catering for short sellers from author's interviews and Joe Nocera, "Apple's Snag, and Other Thoughts," *New York Times*, July 12, 2008.

327   Sexual harassment claim by female Bear Stearns employee from author's interviews, including one with Greenberg.

17: THE END BEGINS

331   Investor Bruce Sherman ends support of Jimmy Cayne from author's interviews with Bear Stearns board members and Kate Kelly, "Cayne to Step Down as Bear Stearns CEO—Executive, Under Fire, to Remain Chairman; 'Time to Pass the Baton,'" *Wall Street Journal*, January 8, 2008.

335   McKinsey consulting study to break up Citigroup from author's interview. Citigroup wouldn't deny the account as portrayed in the chapter.

337–39   The competition between Larry Fink and John Thain to be CEO of Merrill and the account of John Thain's eventual hiring as Merrill's CEO from author's interviews with Merrill executives, Larry Fink, representatives of Thain, Merrill board members, and various news accounts.

338–42   John Thain's belief in this section and throughout book that he could quickly return Merrill back to profitability from interviews with senior Merrill executives, Merrill board members and representatives of Thain.

343   Details of John Thain's office from internal Merrill Lynch documents detailing the expenses.

346   Losses of the banks that are tied to mortgage market of $100 billion from Andrew Hurst, "Calling the Bottom for Banks Seen a Mug's Game," Reuters, January 13, 2008.

347   Fuld's 2006 compensation from Andrew Ross Sorkin, "Lehman Chief Rakes in $22.1 Million in 2007," *New York Times* Dealbook blog, March 6, 2008.

350   Eric Dinallo's comments to Wall Street bankers about bailing out the bond insurers from author's interviews with Dinallo and Susanne Craig and Liam Pleven, "Former Spitzer Aide Drives Bond-Insurer Rescue Talks," *Wall Street Journal*, February 6, 2008.

352   Battle between Dinallo and short seller William Ackman from author interviews with both men and their representatives.

353–56   Private detective Steve Davis's account of investigating mortgage fraud from author's interview with Davis.

358   Leverage stats from SNL Financial.

359–61   Alan Best's presentation about prime brokerage business and Bear Stearns to Jim Chanos from author's interview with people with firsthand knowledge of the matter.

367   Details of Alan Schwartz's whereabouts from Bryan Burrough, "Bringing Down Bear Stearns," *Vanity Fair*, August 2008.

371   Alan Schwartz calling in the lawyers and bankers to make contingency plans as Bear teetered from author's interviews and Kate Kelly, "In Final Week, Bear Executives Swung from Hope to Despair—Rumors Crushed Lenders' Confidence and Spurred Clients' Rush for the Exits," *Wall Street Journal*, May 29, 2008.

371 Alan Schwartz's call to Jim Chanos from author interviews.

371, 373–74 Actions by regulators during Bear Stearns meltdown from author's interviews of federal officials and various news accounts, including John Cassidy, "Anatomy of a Meltdown; a Reporter at Large," *New Yorker*, December 1, 2008.

379, 388 Bear's strategy in dealing with JPMorgan Chase's $2-per-share offer and deliberations among senior management from author's interviews with former board members, senior Bear executives, and other participants with direct knowledge of the matter as well as from various news accounts, including Kate Kelly, "The Fall of Bear Stearns: Bear Stearns Neared Collapse Twice in Frenzied Last Days—Paulson Pushed Low-Ball Bid, Relented; a Testy Time for Dimon," *Wall Street Journal*, May 29, 2008.

384 Paulson's comments about Bear Stearns bailout from various television interviews, including one on NBC's *Today* show, March 18, 2008.

385 President Bush's comments following Bear Stearns collapse from "Bush: In the Long Term, We'll Be Just Fine,'" Market News International, March 18, 2008.

386 Paulson's attitude toward House Republicans versus how he felt about Barney Frank and Nancy Pelosi from interviews with senior House Republicans, former aides to Frank, and government officials.

387 Paulson removing bailout language from Bush's speech from Kate Kelly, *Street Fighters: The Last 72 Hours of Bear Stearns, the Toughest Firm on Wall Street* (New York: Portfolio, 2009).

388–89 Deliberations that led to JPMorgan upping its bid for Bear Stearns from various news accounts and author interviews with people directly involved in the discussions.

390 Predictions that the credit crisis was nearing its end from Susanne Craig, Jeffrey McCracken, Aaron Lucchetti, and Kate Kelly, "The Weekend That Wall Street Died; Ties That Long United Strongest Firms Unraveled as Lehman Sank Toward Failure," *Wall Street Journal*, December 29, 2008.

18: "THERE ARE RUMORS THAT YOU GUYS ARE IN TROUBLE"

393 Details of conversation between John Mack and Dick Fuld from author's interview with Mack.

395 Dick Fuld's calls to Lloyd Blankfein from author's interview and from Kate Kelly and Susanne Craig, "Goldman Is Queried About Bear's Fall; Manipulation Talk Worried Schwartz; Lehman Also Calls," *Wall Street Journal*, July 16, 2008.

398–400 Einhorn's tactics with Lehman from author's interview with Einhorn and Hugo Lindgren, "The Confidence Man; Hedge-Fund Manager David Einhorn Believes His Public Attack on Lehman Brothers Wasn't Just About Making Money. So What Was It About?" *New York*, June 15, 2008.

401 Erin Callan's statements at the UBS conference from interview with attendee.

401 Fuld's negotiations with Warren Buffett from Yalman Onaran and John Helyar, "Lehman's Last Days; Richard Fuld Reached Out to Warren Buffett, Banks, and the Government for Help. His Failure to Acknowledge Lehman's Decline Made It Impossible to Save the Firm," *Bloomberg Markets*, January 2009.

401 Fuld's vow to attack the short sellers from author's interviews and various news accounts, including Mark DeCambre, "Lehman's $4B Puts Shorts in Their Place," *New York Post*, April 2, 2008.

402   Description of Einhorn's speech from text of speech, interview with Einhorn, and Lindgren, "The Confidence Man."

405   Lehman looking far and wide for a buyer, including having discussions with investor Carlos Slim, from author interviews and Onaran and Helyar, "Lehman's Last Days."

409   Paulson's statement to Fuld "you've got to find another partner" from author's interview with a senior Federal Reserve official.

409, 429   Lehman's negotiations with Korea Development Bank from author's inter-· views, various news accounts, and Heather Landy, "Lehman's Senior Statesman; In Credit Crisis, Veteran CEO Fuld Faces Another Big Test," *Washington Post*, September 3, 2008, as well as Onaran and Helyar, "Lehman's Last Days."

408   Lehman bankers pressuring Fuld for management changes from author's interviews and Steve Fishman, "Burning Down His House; Is Lehman CEO Dick Fuld the True Villain in the Collapse of Wall Street, or Is He Being Sacrificed for the Sins of His Peers?" *New York*, November 30, 2008.

408   Bart McDade's analysis of Lehman's books and his surprise at the condition of the firm's balance sheet from author's interviews and Fishman, "Burning Down His House."

19: FREE FALL

412   Freddie and Fannie loss estimates from Friedman, Billing, Ramsey & Co, research.

414   Warnings from David Andrukonis of problems at GSEs from Richard Gross, "Public Lashing; Fannie, Freddie Chiefs Excoriated at House Hearing," *Washington Times*, December 10, 2008.

415   Ben Bernanke's prodding of Treasury Secretary Hank Paulson for a broader bailout of the financial system from author's interviews with senior Fed officials.

415   Bernanke's appearance before the U.S. Senate and statements made during the hearing from various news accounts, including John Cassidy, "Anatomy of a Meltdown; A Reporter at Large," *New Yorker*, December 1, 2008.

416   Conversation between Paul Critchlow and Margaret Tutwiler from author's interview with Critchlow.

417   Thain's statements about Merrill's financial condition from various news sources, including "Merrill Lynch: Unafraid of Contradictions," *Seeking Alpha*, August 5, 2008.

418   Board members debate about Thain's statements from author's interviews with Merrill executives.

420   John Thain's deteriorating relationship with Larry Fink that involves Thain weighing a sale of Merrill's BlackRock stake from author's interview with various people, including Fink and press officials representing Thain.

422   Subprime-related losses in banking system from "After the Fall: Recovery Time," *Fund Strategy*, July 21, 2008.

422   Pandit's defense of universal banking model from Citigroup investor/analyst presentation, May 9, 2008.

422   Estimate of value of Smith Barney brokerage from senior banking executives at JPMorgan and former Citigroup brokerage officials.

423   Ben Bernanke's statement from a *BusinessWeek* transcript of his speech at the

Federal Reserve Bank of Kansas City's Annual Economic Symposium, Jackson Hole, Wyoming, Aug. 22, 2008.

424 Details of Angelo Mozilo's meeting with Barney Frank from a Frank staffer who was present at the meeting.

425 Kronthal's return from author's interviews and Daniel Fisher, "No Thain, No Gain; The Man Who Saved the NYSE Sets About Reviving a Badly Mauled Merrill Lynch," *Forbes*, April 7, 2008.

427–28 Details of the Cioffi/Tannin defense from author interviews with people with direct knowledge of the matter.

429 Waning interest of Korea Development Bank from Yalman Onaran and John Helyar, "Lehman's Last Days; Richard Fuld Reached out to Warren Buffett, Banks and the Government for Help. His Failure to Acknowledge Lehman's Decline Made It Impossible to Save the Firm," *Bloomberg Markets*, January 2009.

432 Government takeover of Fannie and Freddie from David Ellis, "U.S. Seizes Fannie and Freddie; Treasury Chief Paulson Unveils Historic Government Takeover of Twin Mortgage Buyers. Top Executives Are Out," CNNMoney, September 7, 2008.

433–34 Steve Black's conversation with Fuld and entire issue of collateral call from author's interviews and Steve Fishman, "Burning Down His House; Is Lehman CEO Dick Fuld the True Villain in the Collapse of Wall Street, or Is He Being Sacrificed for the Sins of His Peers?" *New York*, November 30, 2008.

435 Analysts' comments about Lehman, including Dick Bove's comments from Sam Mamudi, "Analysts Hit Lehman When It's Down," MarketWatch, September 11, 2008.

436 Fuld's increasingly desperate efforts to save Lehman, including the creation of a "war room" inside the firm, from author's interviews and Eric Dash, "Unraveling Wall Street Titans; Crisis on Wall Street; Days of Mounting Fear and Pressure Lead to the Demise of 2 Legendary Firms and Worries About the Fate of Others," *New York Times*, September 16, 2008.

436 Actions of Merrill Lynch banking chief Greg Fleming in contacting Bank of America from author's interviews with reprentatives of Fleming, representatives of John Thain, and others and Susanne Craig, Jeffrey McCracken, Aaron Lucchetti, and Kate Kelly, "The Weekend That Wall Street Died; Ties That Long United Strongest Firms Unraveled as Lehman Sank Toward Failure," *Wall Street Journal*, December 29, 2008.

438 Merrill Lynch board member John Finnegan's testy exchange with John Thain from author interviews with representatives of both men.

439 Paulson losing confidence in Fuld from author interviews with government officials and former Lehman executives.

20: THE RESCUE

441–45 Details of meeting between government officials and Wall Street executives to save Lehman as well as Paulson's statement about Wall Street exposure to Lehman from author's interviews with participants and Deborah Solomon, Dennis K. Berman, Susanne Craig, and Carrick Mollenkamp, "Ultimatum by Paulson Sparked Frantic End," *Wall Street Journal*, September 15, 2008.

445–46 John Thain's initial conversations with Greg Fleming about Merrill/BofA deal from author interviews with representatives of both men.

447   Paulson's prodding of Thain to find a merger partner for Merrill based on interviews with government officials and Wall Street executives with direct knowledge of the matter.

450, 453, 457   The consensus among the Wall Street firms looking into the "bad bank" that Lehman executives had overvalued the toxic real estate assets from author interviews with executives involved in the matter and Carrick Mollenkamp, Susanne Craig, Jeffrey McCracken, and Jon Hilsenrath, "The Two Faces of Lehman's Failure; Private Talks of Raising Capital Belied Firm's Public Optimism," *Wall Street Journal,* October 6, 2008.

451   Fink's conversations with Fleming about bailout of Lehman from author's interviews with both men and various news accounts.

452   John Mack's negotiations with Thain about potential Merrill Lynch/Morgan deal from author interviews with people that had direct knowledge of the matter.

453   The failure of Lehman Brothers, including the FSA's rationale to prevent Barclays from buying Lehman and the government efforts and efforts by Dick Fuld to find a buyer before the firm was forced into bankruptcy, from author interviews with government officials and executives involved in the discussions and various news accounts.

454–55   Fleming's discussions with Peter Kraus from author's interview and Randall Smith, Susanne Craig, and Dan Fitzpatrick: "Merrill Sought Goldman Aid Before Sealing Sale to BofA," *Wall Street Journal,* September 16, 2008.

456   Fleming's negotiations of BofA from author interviews with people directly involved in the matter.

456   The angry post-Lehman Fleming/Fink conversation from interviews with both men.

459   The decline in Fuld's net worth from "A Lehman Lesson," *New York Sun,* September 15, 2008.

459   Joe Gregory's money problems from Caroline Waxler, "Lehman No. 2 Joe Gregory Lays Off Household Staff of 29," *Business Insider,* December 8, 2008.

461   Fuld's note to Lehman employees from Susanne Craig, Carrick Mollenkamp, Deborah Solomon, and Dan Fitzpatrick, "Street Scenes: The Players Remaking Financial World," *Wall Street Journal,* September 19, 2008.

21: BAILOUT NATION

465   Details about the discussion among the regulators to bail out AIG obtained from interviews with government officials.

466   John Mack's demeanor and his attempts to save Morgan Stanley obtained from author interviews with Mack and government officials.

469–70   Discussions between Paulson and Bernanke to create what would later be known as TARP from author's interviews with senior Fed officials and Wall Street executives.

470–74   Paulson and Bernanke's negotiations with Congressional leaders obtained from government officials with direct knowledge of the talks.

473   Paulson's statement to Congress to pass the TARP legislation from congressmen who attended the event, including Spencer Bachus, Republican from Alabama.

475   Mike Pence's statement regarding TARP from "Bailout Rejected: Angry Voters Put Pressure on Pols," *Seattle Times,* October 1, 2008.

477–78 Details of meeting where Treasury Secretary Hank Paulson forced banks to take TARP money and the opposition from Wells Fargo CEO Richard Kovacevich from author's interviews with executives at the meeting, Fed officials, and Damian Paletta, Jon Hilsenrath, and Deborah Solomon, "At Moment of Truth, US Forced Big Bankers to Blink," *Wall Street Journal*, October 15, 2008.

INDEX

332
GAS

Gasparino, Charles.

The sellout.

$27.99   1|10

DATE			

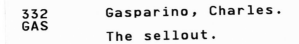